Any topic in the following lists may be tested using the properties of elementary functions. These functions include algebraic, trigonometric, exponential, logarithmic functions, and their inverses. The logarithmic and exponential functions may be to any arbitrary base a, where $a > 0$ and $a \neq 1$; i.e., $\log_a x$ and a^x.

Note: The examples that are given with some of the following topics are not intended to be exhaustive, but rather to illustrate the kinds of questions or functions that might appear on the test.

Column I. Topics covered in Calculus AB

Column II. Additional topics covered in Calculus BC

A. Functions

1. Properties of functions
   a. Definition, domain, and range
   b. Sum, product, quotient, and composition
   c. Absolute value, e.g., $|f(x)|$ and $f(|x|)$
   d. Inverse
   e. Odd and even
   f. Periodicity
   g. Graphs; symmetry and asymptotes
   h. Zeros of a function

   i. Vector functions and parametrically defined functions

   j. Graphs in polar coordinates

2. Limits
   a. Statement and applications of properties, e.g., limit of a constant, sum, product, and quotient
   b. The number e such that
   $$\lim_{n \to \infty} \left(1 + \frac{1}{n}\right)^n = e$$
   c. $\lim_{x \to 0} \dfrac{\sin x}{x} = 1$
   d. Nonexistent limits, e.g.,
   $$\lim_{x \to 0} \frac{1}{x^2}, \lim_{x \to 0} \sin \frac{1}{x}, \text{ and } \lim_{x \to 0} \frac{|x|}{x}$$
   are each, for different reasons, nonexistent.
   e. Continuity
   f. Statements and applications but not proofs of continuity theorems: if f is continuous on $[a, b]$, then f has a maximum and a minimum on $[a, b]$; the intermediate value theorem

   g. Epsilon-delta definition

B. Differential Calculus

1. The derivative
   a. Definitions of the derivative, e.g.,
   $$f'(a) = \lim_{x \to a} \frac{f(x) - f(a)}{x - a} \text{ and}$$
   $$f'(x) = \lim_{h \to 0} \frac{f(x + h) - f(x)}{h}$$

D1304795

Column I. Topics covered in Calculus AB

Column II. Additional topics covered in Calculus BC

    b. Derivatives of elementary functions

    c. Derivatives of sum, product, quotient

    d. Derivative of a composite function (chain rule)

    e. Derivative of an implicitly defined function

    f. Derivative of the inverse of a function (including Arcsin x and Arctan x)

    g. Logarithmic differentiation

    h. Derivatives of higher order

    i. Statement (without proof) of the Mean Value Theorem; applications and graphical illustrations

    j. Relation between differentiability and continuity

    k. Use of l'Hôpital's rule (quotient indeterminate forms)

                          l. Derivatives of vector functions and parametrically defined functions

                          m. l'Hôpital's rule (exponential and other indeterminate forms)

2. Applications of the derivative

    a. Slope of a curve; tangent and normal lines to a curve (including linear approximations)

                          h. Tangent lines to parametrically defined curves

    b. Curve sketching

        (1) Increasing and decreasing functions

        (2) Critical points; relative (local) and absolute maximum and minimum points

        (3) Concavity

        (4) Points of inflection

    c. Extreme value problems

    d. Velocity and acceleration of a particle moving along a line

                          i. Velocity and acceleration vectors for motion on a plane curve

    e. Average and instantaneous rates of change

    f. Related rates of change

    g. Newton's Method

C. Integral Calculus

1. Antiderivatives

2. Applications of antiderivatives

    a. Distance and velocity from acceleration with initial conditions

    b. Solutions of $y' = ky$ and applications to growth and decay

    c. Solutions of $f(y)dy = g(x)dx$ (variables separable)

Column I. Topics covered in Calculus AB

   3. Techniques of integration
- a. Basic integration formulas
- b. Integration by substitution (use of identities, change of variable)
- c. Simple integration by parts, e.g., $\int x \cos x \, dx$ and $\int \ln x \, dx$

   4. The definite integral
- a. Concept of the definite integral as an area
- b. Approximations to the definite integral by using rectangles or trapezoids
- c. Definition of the definite integral as the limit of a sum
- d. Properties of the definite integral
- e. Fundamental theorems:

  $\dfrac{d}{dx} \displaystyle\int_a^x f(t)dt = f(x)$ and

  $\displaystyle\int_a^b f(x)dx = F(b) - F(a)$,

  where $F'(x) = f(x)$

   5. Applications of the integral
- a. Average value of a function on an interval
- b. Area between curves
- c. Volume of a solid of revolution (disc, washer, and shell methods) about the x- and y-axes or lines parallel to the axes

Column II. Additional topics covered in Calculus BC

- d. Integration by trigonometric substitution; trigonometric integrals
- e. Repeated integration by parts, e.g., $\int e^x \cos x \, dx$
- f. Integration by partial fractions

- f. Simpson's Rule

- g. Recognition of limits of sums as definite integrals
- h. Functions defined by integrals, e.g.,

  $f(x) = \displaystyle\int_o^{x^2} e^{-t^2} dt$

- d. Area bounded by polar curves
- e. Volumes of solids with known cross sections
- f. Length of a path including parametric curves
- g. Improper integrals
- h. Work

D. Sequences and Series

   1. Sequences of real numbers and of functions; convergence

   2. Series of real numbers
- a. Geometric series
- b. Tests for convergence: comparison (including limit comparison), ratio, and integral tests
- c. Absolute and conditional convergence
- d. Alternating series and error approximation

Column I. Topics covered in Calculus AB

Column II. Additional topics covered in Calculus BC

3. Power series
   a. Radius of convergence; interval of convergence
   b. Representation of functions by power series; Taylor series
   c. Taylor polynomials with remainder and error approximation
   d. Manipulation of series, i.e., addition of series, substitution, term-by-term differentiation and integration

The Examinations

The Advanced Placement Calculus Examinations and the course descriptions are prepared by committees of teachers from colleges or universities and from secondary schools. The examinations "seek to determine how well a student has mastered the concepts and techniques of the subject matter of the corresponding course."

Each examination is three hours long and has two sections.

Section I consists of 45 multiple-choice questions "that test proficiency in a wide variety of topics." It is allotted one hour and thirty minutes.

Section II consists of six free-response questions that require "the student to demonstrate the ability to carry out proofs and solve problems involving a more extended chain of reasoning." It is also allotted one hour and thirty minutes. These problems call for detailed written solutions. The student is expected to show the methods used since the correctness of these methods will be considered in the grading along with the accuracy of the final answers obtained.

The Calculus AB and Calculus BC Examinations have no questions in common.

Scientific Calculators Required Beginning in May 1993

Beginning in May 1993, each student will be expected to bring a scientific calculator to the examination since there will be some multiple-choice questions on both the AB and the BC Examination for which "a scientific (non-graphing, non-programmable) calculator[†] is necessary or advantageous."

[†] The 1992 *Advanced Placement Course Description* notes that "it is anticipated that *graphing* calculators will soon become the norm in calculus classrooms" and that "the AP Calculus Development Committee has set May 1995 as a target date for introducing problems on the Calculus AB and Calculus BC Examinations for which a graphing calculator (without symbolic algebra capabilities) would be necessary or advantageous."

The following additional information is given about calculator-active questions (those requiring the use of a scientific calculator):

. . . involved and lengthy calculations usually will not be required. Students should be cautioned against the inappropriate use of a calculator. Most questions will not require the use of a calculator. For example, the answer choices may contain π, $\sqrt{2}$, Cos 7, or e. If a calculator-active question calls for an approximate answer, students should choose the best approximation to the exact answer from among the answer choices given.

Calculators may be used but will not be required on the free-response questions of the examinations. If students choose to give decimal approximations as answers to these questions, the answer should be rounded to three decimal places unless otherwise indicated in the question. Students should be cautioned against rounding values before calculations are concluded. Students will continue to be expected to show all work leading to the solutions of the problems posed. The current list of approved calculators is given below. Additions to this list may appear in the May 1993 Course Description that will be available in spring 1992.

Make	*Series*
Casio	FX82, FX115, FX250
Sharp	EL-506
Texas Instruments	TI30, TI34, TI35, TI36

Calculators will be checked prior to the test administration, so it is important that students have one of the calculators listed above. Students should be thoroughly familiar with the operation of the calculator they plan to use on the examinations. Calculators may not be shared.

Calculator-Active Questions

According to the 1992 *Advanced Placement Course Description:*

Approximately 10 percent of the multiple-choice questions will be calculator-active, and these questions will be scattered throughout the entire set of questions. The calculus examinations administered in May 1993 will be governed by the following principles and directions.

A SCIENTIFIC CALCULATOR WILL BE NEEDED FOR SOME QUESTIONS. CALCULATOR MEMORIES MUST BE CLEARED BEFORE BEGINNING WORK.

Directions: Solve each of the following problems, using the available space for scratchwork. Then decide which is the best of the choices given and fill in the corresponding oval on the answer sheet. No credit will be given for anything written in the test booklet. Do not spend too much time on any one problem.

Notes: (1) In this test, ln x denotes the natural logarithm of x (that is, logarithm to the base e). (2) Unless otherwise specified, the domain of a function f is assumed to be the set of all real numbers x for which $f(x)$ is a real number. (3) If the exact numerical answer does not appear among the choices, select from among the choices the number that best approximates the answer.

The following sample questions are intended to familiarize students with calculator-active questions. Answers and solutions are given on page xv, following the set of sample BC questions.

Sample Questions for Calculus AB

1. $\int_1^{1.5} \cos \frac{1}{2} x \, dx$ is equal approximately to

    (A) 0.009   **(B)** 0.101   **(C)** 0.202   **(D)** 0.404   **(E)** 0.959

2. The first-quadrant area lying under the curve $y = \dfrac{5}{3x + 2}$ and bounded at the right by $x = m$ is equal to 3 square units. Correct to three decimal places, m equals

    (A) 0.148   **(B)** 0.548   **(C)** 1.350   **(D)** 2.358   **(E)** 3.366

3. The value of c prescribed by the Mean Value Theorem for the function $f(x) = \sin x$ on the closed interval $[1, 3]$ is, correct to three decimal places,

    (A) 0.939   **(B)** 1.213   **(C)** 1.553   **(D)** 1.929   **(E)** 2.347

4. The inner diameter of a piece of copper tubing 3 m long is 1 cm, and the outer diameter is 1.5 cm. The approximate volume of copper in the tube, in cubic centimeters, is equal to

    (A) 3   **(B)** 59   **(C)** 118   **(D)** 295   **(E)** 1473

5. The half-life of uranium-232 is 74 years. In how many years will one third of the present quantity decay?

    (A) 43   **(B)** 49   **(C)** 81   **(D)** 97   **(E)** 117

Sample Questions for Calculus BC

1. Let $G(x) = \int_0^{x^2+1} \sqrt{1 + e^v} \, dv$. Then $G'(1)$, correct to three decimal places, equals

    (A) 0.000   **(B)** 2.896   **(C)** 3.857   **(D)** 4.458   **(E)** 5.793

2. If two equal subdivisions of the interval $[0, 2]$ are used, what is the trapezoidal approximation, correct to two decimal places, for the first-quadrant area bounded by the curve $y = \frac{\sin x}{x}$ and the line $x = 2$? (Assume for this question that $\frac{\sin x}{x}$ is continuous at $x = 0$; that is, that 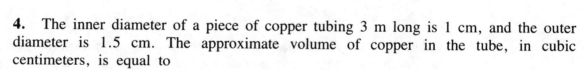 $\frac{\sin 0}{0} = 1 = \lim_{x \to 0} \frac{\sin x}{x}$.)

    (A) 0.53   **(B)** 0.95   **(C)** 1.05   **(D)** 1.57   **(E)** 3.14

3. The least number of terms needed to compute $\sum_{n=1}^{\infty} \dfrac{1}{n^4 + 1}$ with an error not exceeding 0.0001 is

    (A) 10   **(B)** 13   **(C)** 15   **(D)** 17   **(E)** 20

4. In 1940 the population of a city was 20,000; in 1990 it was 70,000. If the population continues to grow exponentially at a constant rate, what will the population be in the year 2000, rounded off to the nearest 1000?

    (A) 90,000   **(B)** 89,000   **(C)** 87,000   **(D)** 83,000   **(E)** 80,000

5. Correct to three decimal places, $\int_0^{0.4} \dfrac{dx}{1 + x^4}$ is equal, using series, to

    (A) 0.025   **(B)** 0.158   **(C)** 0.398   **(D)** 0.400   **(E)** 0.402

Answers and Solutions to Calculus AB Questions

(Questions are on page xiv.)

1. D. $\displaystyle\int_1^{1.5} \cos \tfrac{1}{2}x \; dx = 2 \sin \tfrac{1}{2}x \Big]_1^{1.5} = 2(\sin 0.75 - \sin 0.5)$

                                          $= 2(0.2022) = 0.404.$

    Be sure your calculator is in the radian mode.

2. E.   We seek m such that

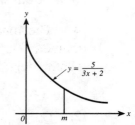

    $\displaystyle\int_0^m \dfrac{5}{3x + 2} \; dx = 3.$

    So $\dfrac{5}{3} \ln (3x + 2) \Big]_0^m$ must equal 3.

    $\dfrac{5}{3}[\ln (3m + 2) - \ln 2] = 3$ if

    $\ln (3m + 2) = \dfrac{9}{5} + \ln 2.$

    $3m + 2 = e^{\left(\frac{9}{5} + \ln 2\right)}$; $m = \dfrac{e^{\left(\frac{9}{5} + \ln 2\right)} - 2}{3} = 3.366.$ On your calculator,

    compute $\dfrac{9}{5}$ and add $\ln 2$; raise e to the preceding sum; subtract 2; and divide the result by 3.

3. D.   We want c for which $\cos c = \dfrac{\sin 3 - \sin 1}{3 - 1}$. Using the radian mode, compute the fraction at the right; then compute the inverse cosine ($\cos^{-1}$) of the result, to get 1.929 to three decimal places for c.

4. D.   The volume of copper equals $\pi(0.75^2 - 0.5^2) \cdot 300 \approx 295 \text{ cm}^3.$

5. A.   Use $Q = Q_0 e^{-kt}$, where Q_0 is the initial amount of uranium and k is the factor of proportionality (see pages 143 and 144). Since we are given that $Q = \dfrac{1}{2} Q_0$ when $t = 74$, we have

$$\tfrac{1}{2} Q_0 = Q_0 e^{-k \cdot 74};$$

$$\tfrac{1}{2} = e^{-74k}; \qquad \ln \tfrac{1}{2} = -74k;$$

$$k = 0.009366853.$$

    (*)

We must find t when $Q = \frac{2}{3} Q_0$:

$$\frac{2}{3} = e^{-kt};$$

$$\ln \frac{2}{3} = -kt; \qquad (**)$$

$$t = \frac{\ln \frac{2}{3}}{-k} \approx 43 \text{ years.} \qquad (***)$$

You need not copy down on paper the value of k found in (*); you can divide k, using (**), by $\ln \frac{2}{3}$, precede the result by a minus sign, and take the reciprocal to get 43 to the nearest year. Or you can store k in memory and find t from (***).

Answers and Solutions to Calculus BC Questions

(Questions are on page xiv.)

1. E.   If $G(x) = \int_0^{x^2+1} \sqrt{1 + e^v}\, dv$, then, applying Theorem (1) on page 172 and the chain rule, $G'(x) = 2x\sqrt{1 + e^{x^2+1}}$. So $G'(1) = 2\sqrt{1 + e^2} = 5.793$ to three decimal places.

2. D.   Recall the formula for the area of a trapezoid, $A = \frac{1}{2} h(b_1 + b_2)$, where h is the height and b_1 and b_2 are the parallel bases. (On page 179 we used Δx for h and the ordinates for the bases to derive the trapezoidal rule.)

$$\int_0^2 \frac{\sin x}{x}\, dx \approx \frac{1}{2} \cdot 1[y(0) + 2y(1) + y(2)]$$

$$\approx \frac{1}{2}\left(\frac{\sin 0}{0} + 2\frac{\sin 1}{1} + \frac{\sin 2}{2}\right)$$

$$\approx 1.57 \text{ to two decimal places.}$$

(Be sure to use the radian mode.)

3. C. $\sum_1^\infty \frac{1}{n^4 + 1}$ converges because $\frac{1}{n^4 + 1} \le \frac{1}{n^4}$ for all n, and $\frac{1}{n^4}$ converges by the integral test $\left(\int_1^\infty \frac{1}{x^4}\, dx = \frac{1}{3}\right)$. We seek n such that

$$R_n = \sum_{k=n}^\infty \frac{1}{k^4+1} < 0.0001.$$

Since $R_n < \int_n^\infty \frac{dx}{x^4} = \frac{1}{3n^3}$, n must satisfy the inequality $\frac{1}{3n^3} < 0.0001$. Thus $3n^3$ must exceed $\frac{1}{0.0001}$ or 10,000, and n must exceed $\sqrt[3]{\frac{10,000}{3}}$ or 14.9. Therefore at least 15 terms are needed for the specified accuracy.

4. A.   Use $N = N_0 e^{kt}$, where N_0 is the initial population (in 1940) and k is the factor of proportionality (see page 143). Thus $N = 20,000e^{kt}$ and k satisfies the equation $70,000 = 20,000e^{50k}$. With the calculator, we find k by computing

$$\ln \frac{7}{2} \div 50. \tag{*}$$

To find N when $t = 60$, we use $N = 20,000e^{60k}$. Without writing anything down, we multiply the value of k found in (*) by 60, raise e to this result, and then multiply by 20,000. The answer is $89,931^+$, which yields 90,000 to the nearest thousand.

5. C.   Long division yields a strictly alternating series:

$$\frac{1}{1+x^4} = 1 - x^4 + x^8 - \cdots,$$

$$\int_0^{0.4} \frac{dx}{1 + x^4} = x - \frac{x^5}{5} + \frac{x^9}{9} - \cdots \Bigg]_0^{0.4}$$

$$= 0.4 - \frac{(0.4)^5}{5} = 0.398,$$

to three decimal places. Theorem 5b, page 297, tells us that the error in using only the first two terms is less than $\frac{(0.4)^9}{9}$, which is less than 0.0001.

Grading the Examinations

Each completed AP examination paper receives a grade according to the following five-point scale:

5: Extremely well qualified
4: Well qualified
3: Qualified
2: Possibly qualified
1: No recommendation

Many colleges and universities accept a grade of 3 or better for credit or advanced placement or both; some also consider a grade of 2. Over 71% of the candidates who took the 1990 Calculus AB Examination earned grades of 3, 4, or 5. Over 81% of the 1990 BC candidates earned 3 or better.

The multiple-choice questions in Section I are scored by machine. To compensate for wild guessing on these questions, one fourth of the number of incorrect answers is subtracted from the number of correct ones. Blind or haphazard guessing is therefore likely to lower a student's grade. However, if one or more of the choices given for a question can be eliminated as clearly incorrect, then the chance of guessing the correct answer from among the remaining ones is increased.

The problems in Section II are graded by a group of carefully selected college professors and AP teachers called "readers." The answers in any one examination booklet are evaluated by different readers, and for each reader all scores given by preceding readers are concealed, as are the student's name and school. Readers are provided sample solutions for each problem, with detailed scoring scales and point distributions that allow partial credit for correct portions of a student's answer. Problems in Section II are all counted equally.

In the determination of the overall grade for each examination, the two sections are given equal weight. The total raw score is then converted into one of the five grades described above.

The following noteworthy comment about the AP Calculus examinations is in the *Advanced Placement Course Description, May 1992*. "Since the examinations are designed for full coverage of the subject matter, it is not expected that all students will be able to answer all the questions." It is also noted that the multiple-choice questions in Section I are "deliberately set at such a level of difficulty that students performing acceptably on the free-response section of an examination need to answer about 50 to 60 percent of the multiple-choice questions correctly to obtain a total grade of 3."

Great care is taken by all involved in the scoring and reading of papers to make certain that they are graded consistently and fairly so that a student's overall AP grade reflects as accurately as possible his or her achievement in the calculus.

The CLEP Examination in Calculus with Elementary Functions

Many colleges give credit to students who perform acceptably on tests offered by the College Level Examination Program (CLEP). The CLEP Examination in Calculus with Elementary Functions is one such test.

The College Board Guide to the CLEP Examinations (Rev. Ed., 1990) provides detailed descriptions of all examinations, test-taking tips, and lists of reference and supplementary materials. According to the *Guide,* the CLEP Examination on Calculus with Elementary Functions "includes approximately 45 multiple-choice questions to be answered in two separately timed 45-minute sections." Approximately 20% of this examination is on elementary functions, 40% on differential calculus, and about 40% on integral calculus. (A free-response test is available, but the student should *not* take it unless the college or university that he or she wishes to attend explicitly requires it.)

Since the content specifications for this CLEP examination are virtually the same as those for the Calculus AB Examination, the topics tested on the CLEP Calculus examination are those listed in the syllabus for the Advanced Placement AB course, appearing in Column I of pages ix through xi. The topics included in this list but *not* listed in the *Guide* as required for the CLEP examination are L'Hôpital's rule, linear approximation, Newton's method, and solutions of differential equations with variables separable (item C, 2, c on page x).

A candidate who plans to take the CLEP Examination in Calculus with Elementary Functions will benefit from a review of the Calculus AB topics covered in this book. The multiple-choice questions in Sections I of the four AB Practice Examinations provide excellent models of any CLEP Examination in Calculus with Elementary Functions for which the student may register in the near future.

According to the College Board's *Make Learning Pay with CLEP, 1991–92,* CLEP tests are administered during each month of the year at many colleges and universities. It is suggested that the student contact the college or university in which he or she is interested to obtain information on which CLEP tests it accepts for credit and on how to register for a test.

The publications named above and the College Board's *CLEP Colleges 1991–92* furnish further information about CLEP tests, test centers, fees, scores, and so on. The *Guide* can be ordered by writing to College Board Publications, Department N98, Box 886, New York, NY 10101-0886. *Make Learning Pay* and *CLEP Colleges* can be obtained by writing to CLEP, PO Box 6601, Princeton, NJ 08541-6601.

This Review Book

This book consists of the following two parts:

The "Topical Review" has ten chapters with notes on the main topics of the Calculus AB and BC syllabi and with numerous carefully worked-out illustrative examples. Each chapter concludes with (1) a set of multiple-choice questions, followed immediately by answers and solutions, and (2) a set of practice Section II problems, followed immediately by solutions.

The second part, "Further Practice," has two subparts: (1) a set of 100 multiple-choice questions on miscellaneous topics followed by the answer key; (2) eight complete Practice Examinations (four for Calculus AB, four for Calculus BC); each of these is also followed immediately by its solution key. Although the actual AB and BC examinations have no questions in common, we have chosen to include some identical questions on both types of practice examination because of the overlap in the content coverage of the two courses.

Although we were able to reprint actual AP Calculus Examinations in previous editions of this book, we have not done so here because the Educational Testing Service no longer permits this. AP Calculus teachers may obtain copies of the entire AP Calculus Examinations and Keys, AB and BC, from both 1985 and 1988, by ordering them from the Advanced Placement Program, CN 6670, Princeton, NJ 08541-6670.

In this book, review material on topics covered only in Calculus BC is preceded by an asterisk (*), as are both multiple-choice questions and free-response problems that are likely to occur only on a BC Examination. An optional topic or question not included in the BC Course Description is preceded by a dagger (†).

THE TEACHER WHO USES THIS BOOK WITH A CLASS may profitably do so in any of several ways. If the book is used throughout a year's course, the teacher can assign all or part of each set of multiple-choice questions and Section II problems after the topic has been covered. These sets can also be used for review purposes shortly before examination time. The Practice Examinations will also be very helpful in reviewing toward the end of the year. The teacher who wishes to give class "take-home" examinations throughout the year can assemble such examinations by choosing appropriate problems, on the material to be covered, from the sample free-response problems or Practice Examinations.

THE STUDENT WHO USES THIS BOOK INDEPENDENTLY will improve his or her performance by studying the illustrative examples carefully and trying to complete practice problems *before* referring to the solution keys.

Since many COLLEGES IN THEIR FIRST-YEAR MATHEMATICS COURSES follow syllabi very similar to the one proposed by CEEB for Advanced Placement high school courses, college students and teachers may also find the book useful.

TOPICAL
REVIEW

Chapter 1

Functions

Review of Definitions and Properties

A. Definitions

A1. A *function f* is a correspondence that associates with each element *a* of a set called the *domain* one and only one element *b* of a set called the *range*. We write $f(a) = b$; *b* is the *value* of *f* at *a*.

A vertical line cuts the graph of a function in at most one point.

Examples

1. The domain of $f(x) = x^2 - 2$ is the set of all real numbers; its range is the set of all reals greater than or equal to -2. Note that

$$f(0) = 0^2 - 2 = -2, \qquad f(-1) = (-1)^2 - 2 = -1,$$

$$f(\sqrt{3}) = (\sqrt{3})^2 - 2 = 1, \qquad f(c) = c^2 - 2,$$

$$f(x + h) - f(x) = [(x + h)^2 - 2] - [x^2 - 2]$$

$$= x^2 + 2hx + h^2 - 2 - x^2 + 2 = 2hx + h^2.$$

2. The domain of $f(x) = \dfrac{4}{x - 1}$ is the set of all reals except $x = 1$ (which we shorten to "$x \neq 1$").

The domain of $g(x) = \dfrac{x}{x^2 - 9}$ is $x \neq 3, -3$.

The domain of $h(x) = \dfrac{\sqrt{4 - x}}{x}$ is $x \leq 4$, $x \neq 0$ (which is a short way of writing $\{x \mid x \text{ is real}, x < 0 \text{ or } 0 < x \leq 4\}$).

A2. Two functions f and g with the same domain may be combined to yield

their *sum* and *difference:* $f(x) + g(x)$ and $f(x) - g(x)$;
their *product* and *quotient:* $f(x)g(x)$ and $f(x)/g(x)$.

The quotient is defined for all x in the shared domain except those for which $g(x)$, the denominator, equals zero.

Example 3.   If $f(x) = x^2 - 4x$ and $g(x) = x + 1$, then

$$\frac{f(x)}{g(x)} = \frac{x^2 - 4x}{x + 1} \text{ and has domain } x \neq -1;$$

$$\frac{g(x)}{f(x)} = \frac{x + 1}{x^2 - 4x} = \frac{x + 1}{x(x - 4)} \text{ and has domain } x \neq 0, 4.$$

A3. The *composition* (or *composite*) of f with g, written $f(g(x))$, is the function obtained by replacing x wherever it occurs in $f(x)$ by $g(x)$.

Example 4.   If $f(x) = 2x - 1$ and $g(x) = x^2$, then

$$f(g(x)) = 2(x^2) - 1 = 2x^2 - 1;$$

but

$$g(f(x)) = (2x - 1)^2 = 4x^2 - 4x + 1.$$

In general, $f(g(x)) \neq g(f(x))$.

A4. A function x is $\begin{smallmatrix}odd\\even\end{smallmatrix}$ if $\begin{smallmatrix}f(-x) = -f(x)\\f(-x) = f(x)\end{smallmatrix}$ The graph of an odd function is symmetric about the origin; the graph of an even function is symmetric about the y-axis.

Example 5.   The graphs of $f(x) = \frac{1}{2}x^3$ and $g(x) = 3x^2 - 1$ are shown in Figure N1–1; $f(x)$ is odd, $g(x)$ even.

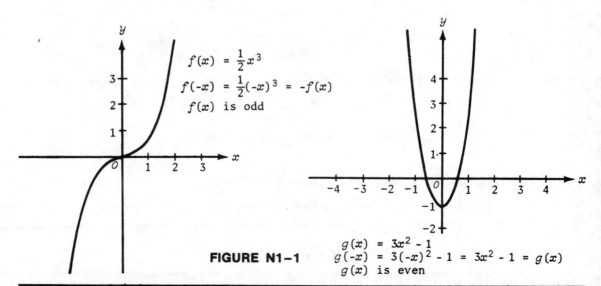

$$f(x) = \frac{1}{2}x^3$$
$$f(-x) = \frac{1}{2}(-x)^3 = -f(x)$$
$$f(x) \text{ is odd}$$

FIGURE N1–1

$$g(x) = 3x^2 - 1$$
$$g(-x) = 3(-x)^2 - 1 = 3x^2 - 1 = g(x)$$
$$g(x) \text{ is even}$$

A5. If, for any element b in the range of a function f, there is *only one* element in the domain of f (say a) such that $f(a) = b$, then f is *one-to-one*. Geometrically this means that any horizontal line $y = b$ cuts the graph of f in only one point. The function sketched at the left in Figure N1−1 is one-to-one; the function sketched at the right is not. A function that is increasing (or decreasing) on an interval I is one-to-one on that interval (see page 67 for definitions of increasing and decreasing functions).

A6. If f is one-to-one with domain X and range Y, then there is a function f^{-1}, with domain Y and range X, such that

$$f^{-1}(y_0) = x_0 \quad \text{if and only if} \quad f(x_0) = y_0.$$

The function f^{-1} is the *inverse* of f. It can be shown that f^{-1} is also one-to-one and that its inverse is f. The graphs of a function and its inverse are symmetric with respect to the line $y = x$.

> To find the inverse of $y = f(x)$
> solve for x in terms of y,
> then interchange x and y.

Example 6. Find the inverse of the one-to-one function $f(x) = x^3 - 1$.

Solve the equation for x: $\quad x = \sqrt[3]{y + 1}$.

Interchange x and y: $\quad y = \sqrt[3]{x + 1} = f^{-1}(x)$.

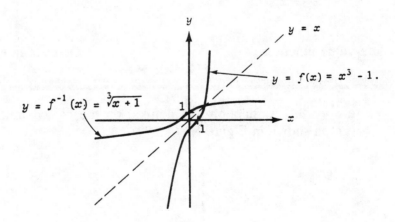

FIGURE N1−2

The graphs of $f(x)$ and $f^{-1}(x)$, shown in Figure N1−2, are mirror images, with the line $y = x$ as the mirror.

A7. The *zeros* of a function f are the values of x for which $f(x) = 0$; they are the x-intercepts of the graph of $y = f(x)$.

Example 7.   The zeros of $f(x) = x^4 - 2x^2$ are the x's for which $x^4 - 2x^2 = 0$. The function has three zeros, since $x^4 - 2x^2 = x^2(x^2 - 2)$ equals zero if $x = 0$, $+\sqrt{2}$, or $-\sqrt{2}$.

B.   Special Functions

The *absolute-value* function $f(x) = |x|$ and the *greatest-integer* function $g(x) = [x]$ are sketched in Figure N1–3.

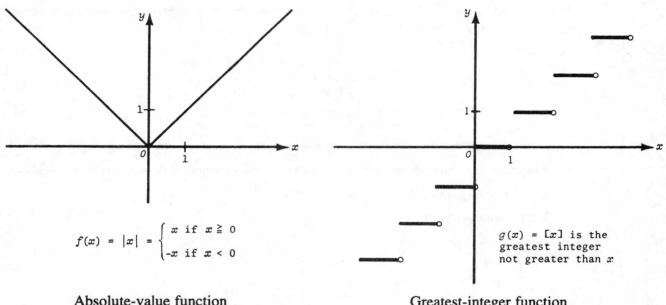

$$f(x) = |x| = \begin{cases} x & \text{if } x \geq 0 \\ -x & \text{if } x < 0 \end{cases}$$

$g(x) = [x]$ is the greatest integer not greater than x

Absolute-value function   Greatest-integer function

FIGURE N1–3

Example 8.   A function f is defined on the closed interval $[-2, 2]$ and has the graph shown in Figure N1–4.

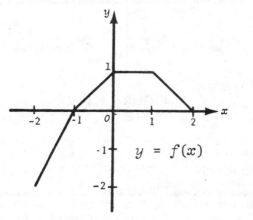

$y = f(x)$

FIGURE N1–4

(a) Sketch the graph of $y = |f(x)|$. (b) Sketch the graph of $y = f(|x|)$. (c) Sketch the graph of $y = -f(x)$. (d) Sketch the graph of $y = f(-x)$.

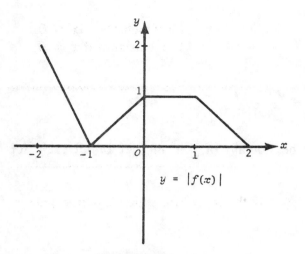

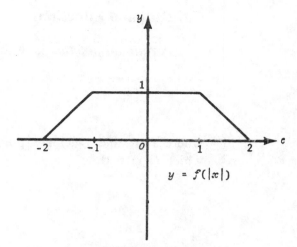

FIGURE N1−4a

FIGURE N1−4b

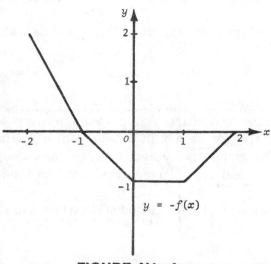

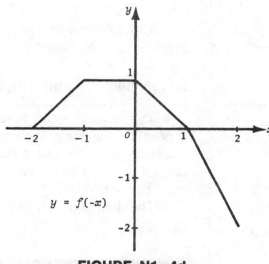

FIGURE N1−4c

FIGURE N1−4d

Note that the graph of $y = -f(x)$ is the reflection of the graph of $y = f(x)$ in the x-axis, whereas the graph of $y = f(-x)$ is the reflection of the graph of $y = f(x)$ in the y-axis.

C. Polynomial and Other Rational Functions___

C1. Polynomial Functions.

A *polynomial function* is of the form

$$f(x) = a_0x^n + a_1x^{n-1} + a_2x^{n-2} + \cdots + a_{n-1}x + a_n,$$

where n is a positive integer or zero, and the a's, the *coefficients*, are constants. If $a_0 \neq 0$, the degree of the polynomial is n.

$f(x) = mx + b$, a *linear* function, is of the first degree;

$f(x) = ax^2 + bx + c$, a *quadratic* function, has degree 2;

$f(x) = a_0x^3 + a_1x^2 + a_2x + a_3$, a *cubic*, has degree 3;

and so on. The domain of every polynomial function is the set of all reals.

C2. Rational Functions.

A *rational function* is of the form

$$f(x) = \frac{P(x)}{Q(x)},$$

where $P(x)$ and $Q(x)$ are polynomials. The domain of f is the set of all reals for which $Q(x) \neq 0$.

D. Trigonometric Functions

The fundamental trigonometric identities, graphs, and reduction formulas are given in the Appendix.

D1. Periodicity and Amplitude.

The trigonometric functions are periodic. A function f is *periodic* if there is a positive number p such that $f(x + p) = f(x)$ for each x in the domain of f. The smallest such p is called the *period* of f. The graph of f repeats every p units along the x-axis. The functions $\sin x$, $\cos x$, $\csc x$, and $\sec x$ have period 2π; $\tan x$ and $\cot x$ have period π.

The function $f(x) = A \sin bx$ has *amplitude* A and period $\frac{2\pi}{b}$; $g(x) = \tan cx$ has period $\frac{\pi}{c}$.

Examples

9. (a) For what value of k does $f(x) = \frac{1}{k} \cos kx$ have period 2?
(b) What is the amplitude of f for this k?

(a) Function f has period $\frac{2\pi}{k}$; since this must equal 2, we solve the equation $\frac{2\pi}{k} = 2$, getting $k = \pi$.

(b) It follows that the amplitude of f which equals $\frac{1}{k}$ has value $\frac{1}{\pi}$.

10. Find (a) the period and (b) the maximum value of $f(x) = 3 - \sin \frac{\pi x}{3}$. (c) What is the smallest positive x for which f is a maximum? (d) Sketch the graph.

(a) The period of f is $2\pi \div \frac{\pi}{3}$, or 6.

(b) Since the maximum value of $-\sin x$ is $-(-1)$ or $+1$, the maximum value of f is $3 + 1$ or 4.

(c) $-\left(\sin \frac{\pi x}{3}\right)$ equals $+1$ when $\sin \frac{\pi x}{3} = -1$, that is, when $\frac{\pi x}{3} = \frac{3\pi}{2}$. Solving yields $x = \frac{9}{2}$.

(d) See Figure N1–5.

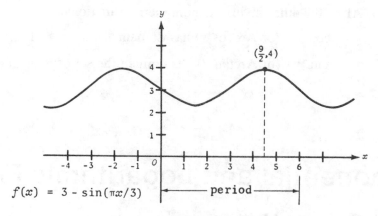

$$f(x) = 3 - \sin(\pi x/3)$$

Figure N1–5

D2. Inverses.

We obtain *inverses* of the trigonometric functions by limiting the domains of the latter so each trigonometric function is one-to-one over its restricted domain. For example, we restrict

$$\sin x \text{ to } \quad -\frac{\pi}{2} \leqq x \leqq \frac{\pi}{2},$$

$$\cos x \text{ to } \quad 0 \leqq x \leqq \pi,$$

$$\tan x \text{ to } \quad -\frac{\pi}{2} < x < \frac{\pi}{2}.$$

The graphs of $f(x) = \sin x$ on $\left[-\frac{\pi}{2}, \frac{\pi}{2} \right]$ and of its inverse $f^{-1}(x) = \sin^{-1} x$ are shown in Figure N1–6. The inverse trigonometric function $\sin^{-1} x$ is also commonly denoted by Arcsin x.*

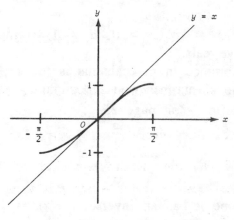

$y = \sin x$

domain: $-\frac{\pi}{2} \leqq x \leqq \frac{\pi}{2}$

range: $-1 \leqq y \leqq 1$

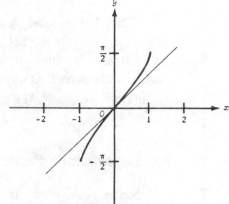

$y = \sin^{-1} x = \text{Arcsin } x$

domain: $-1 \leqq x \leqq 1$

range: $-\frac{\pi}{2} \leqq y \leqq \frac{\pi}{2}$

FIGURE N1–6

*Some authors reserve the notation Arcsin x (with a capital A) or $\text{Sin}^{-1} x$ (with a capital S) for the inverse *function* and use arcsin x for the relation obtained by interchanging x and y in the non-one-to-one sine function (with unrestricted domain).

Also, for other inverse trigonometric functions,

$y = \cos^{-1} x$ (or Arccos x) has domain $-1 \leq x \leq 1$ and range $0 \leq y \leq \pi$;

$y = \tan^{-1} x$ (or Arctan x) has domain the set of reals and range $-\dfrac{\pi}{2} < y < \dfrac{\pi}{2}$.

E. Exponential and Logarithmic Functions

E1. Exponential Functions.

The following laws of exponents hold for all rational m and n, provided that $a > 0$, $a \neq 1$:

$$a^0 = 1; \quad a^1 = a; \quad a^m \cdot a^n = a^{m+n}; \quad a^m \div a^n = a^{m-n};$$

$$(a^m)^n = a^{mn}; \quad a^{-m} = \frac{1}{a^m}.$$

These properties define a^x when x is rational. When x is irrational, we note that x is an infinite nonrepeating decimal; we then define a^x as the limit of an infinite sequence of rational numbers. For example,

$$3^{\sqrt{2}} = 3^{1.4142135\ldots}$$

is the limit of the unending sequence of rational numbers

$$3^1, \ 3^{1.4}, \ 3^{1.41}, \ 3^{1.414}, \ \ldots \ .$$

The *exponential function* $f(x) = a^x$ $(a > 0, \ a \neq 1)$ is thus defined for all real x; its range is the set of positive reals.

Of special interest and importance in the calculus is the exponential function $f(x) = e^x$, where e is an irrational number whose decimal approximation to five decimal places is 2.71828. We define e on page 25.

E2. Logarithmic Functions.

Since $f(x) = a^x$ is one-to-one it has an inverse, $f^{-1}(x) = \log_a x$, called the *logarithmic* function to the base a. We note that

$$y = \log_a x \quad \text{if and only if} \quad a^y = x.$$

The domain of $\log_a x$ is the set of positive reals; its range is the set of all reals. It follows that the graphs of the pair of mutually inverse functions $y = 2^x$ and $y = \log_2 x$ are symmetric to the line $y = x$; they are shown in Figure N1–7.

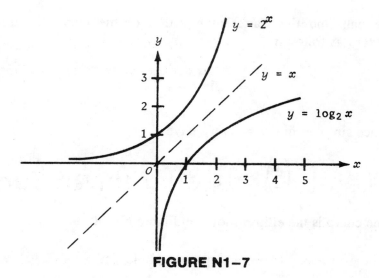

FIGURE N1–7

The logarithmic function $\log_a x$ $(a > 0, a \neq 1)$ has the following properties:

$$\log_a 1 = 0; \qquad \log_a a = 1; \qquad \log_a mn = \log_a m + \log_a n;$$

$$\log_a \frac{m}{n} = \log_a m - \log_a n; \qquad \log_a x^m = m \log_a x.$$

So important and convenient is the logarithmic base e in the calculus that we use a special symbol:

$$\log_e x = \ln x.$$

Logarithms to the base e are called *natural* logarithms. The graphs of the mutually inverse functions $\ln x$ and e^x are given in the Appendix, page 489.

*F. Parametrically Defined Functions_____

If the x- and y-coordinates of a point on a graph are given as functions f and g of a third variable, say t, then

$$x = f(t), \qquad y = g(t)$$

are called *parametric equations* and t is called the *parameter*.

Examples

11. From the parametric equations

$$x = 4 \sin t, \quad y = 5 \cos t \qquad (0 \leq t \leq 2\pi)$$

*An asterisk denotes a topic covered only in Calculus BC.

we can find the Cartesian equation of the curve by eliminating the parameter t as follows:

$$\sin t = \frac{x}{4}, \qquad \cos t = \frac{y}{5}.$$

Since $\sin^2 t + \cos^2 t = 1$, we have

$$\left(\frac{x}{4}\right)^2 + \left(\frac{y}{5}\right)^2 = 1 \qquad \text{or} \qquad \frac{x^2}{16} + \frac{y^2}{25} = 1.$$

The curve is the ellipse shown in Figure N1–8.

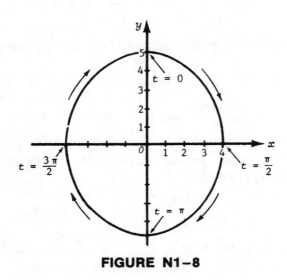

FIGURE N1–8

Note that, as t increases from 0 to 2π, a particle moving in accordance with the given parametric equations starts at point $(0, 5)$ (when $t = 0$) and travels in a clockwise direction along the ellipse, returning to $(0, 5)$ when $t = 2\pi$.

12. For the pair of parametric equations

$$x = 1 - t, \quad y = \sqrt{t} \qquad (t \geqq 0)$$

we can eliminate t by squaring the second equation and substituting for t in the first; then we have

$$y^2 = t \qquad \text{and} \qquad x = 1 - y^2.$$

At the left in Figure N1–9 we see the graph of the equation $x = 1 - y^2$. At the right we see only the upper part of this graph, the part defined by the parametric equations for which t and y are both restricted to nonnegative numbers.

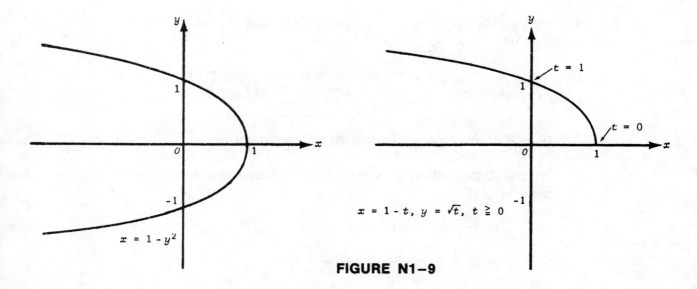

FIGURE N1–9

The function defined by the parametric equations here is $y = F(x) = \sqrt{1 - x}$, whose graph is at the right above; its domain is $x \leq 1$ and its range is the set of nonnegative reals.

Parametric equations give rise to vector functions, which will be discussed in connection with motion along a curve in Chapter 4.

Set 1: Multiple-Choice Questions on Functions

1. If $f(x) = x^3 - 2x - 1$, then $f(-2) =$ $(-2)^3 + 4 - 1 = -9 + 4$

   (A) -17 (B) -13 (C) -5 (D) -1 (E) 3

2. The domain of $f(x) = \frac{x - 1}{x^2 + 1}$ is

   (A) all $x \neq 1$ (B) all $x \neq 1, -1$ (C) all $x \neq -1$ (D) $x \geq 1$
   (E) all reals

3. The domain of $g(x) = \frac{\sqrt{x - 2}}{x^2 - x}$ is
   $x(x-1)$

   (A) all $x \neq 0, 1$ (B) $x \leq 2, x \neq 0, 1$ (C) $x \leq 2$ (D) $x \geq 2$
   (E) $x > 2$

4. If $f(x) = x^3 - 3x^2 - 2x + 5$ and $g(x) = 2$, then $g(f(x)) =$

   (A) $2x^3 - 6x^2 - 2x + 10$ (B) $2x^2 - 6x + 1$ (C) -6
   (D) -3 (E) 2

5. With the functions and choices as in Question 4, which choice is correct for $f(g(x))$?

6. If $f(x) = x^3 + Ax^2 + Bx - 3$ and if $f(1) = 4$ and $f(-1) = -6$, what is the value of $2A + B$?

   (A) 12   **(B)** 8   **(C)** 0   **(D)** -2
   (E) It cannot be determined from the given information.

7. Which of the following equations has a graph that is symmetric with respect to the origin?

   (A) $y = \dfrac{x - 1}{x}$ **(B)** $y = 2x^4 + 1$ **(C)** $y = x^3 + 2x$

   (D) $y = x^3 + 2$ **(E)** $y = \dfrac{x}{x^3 + 1}$

8. Let g be a function defined for all reals. Which of the following conditions is not sufficient to guarantee that g has an inverse function?

   (A) $g(x) = ax + b, a \neq 0$ **(B)** g is strictly decreasing

   (C) g is symmetric to the origin   **(D)** g is strictly increasing

   (E) g is one-to-one

9. Let $y = f(x) = \sin(\text{Arctan } x)$. Then the range of f is

   (A) $\{y \mid 0 < y \leq 1\}$ **(B)** $\{y \mid -1 < y < 1\}$ **(C)** $\{y \mid -1 \leq y \leq 1\}$

   (D) $\left\{y \mid -\dfrac{\pi}{2} < y < \dfrac{\pi}{2}\right\}$ **(E)** $\left\{y \mid -\dfrac{\pi}{2} \leq y \leq \dfrac{\pi}{2}\right\}$

10. Let $g(x) = |\cos x - 1|$. The maximum value attained by g on the closed interval $[0, 2\pi]$ is for x equal to

    (A) -1 **(B)** 0   **(C)** $\dfrac{\pi}{2}$ **(D)** 2   **(E)** π

11. Which of the following functions is not odd?

    (A) $f(x) = \sin x$ **(B)** $f(x) = \sin 2x$ **(C)** $f(x) = x^3 + 1$
    (D) $f(x) = \dfrac{x}{x^2 + 1}$ **(E)** $f(x) = \sqrt[3]{2x}$

12. The roots of the equation $f(x) = 0$ are 1 and -2. The roots of $f(2x) = 0$ are

    (A) 1 and -2 **(B)** $\dfrac{1}{2}$ and -1 **(C)** $-\dfrac{1}{2}$ and 1   **(D)** 2 and -4

    (E) -2 and 4

13. The set of zeros of $f(x) = x^3 + 4x^2 + 4x$ is

    (A) $\{-2\}$ **(B)** $\{0, -2\}$ **(C)** $\{0, 2\}$ **(D)** $\{2\}$ **(E)** $\{2, -2\}$

14. The values of x for which the graphs of $y = x + 2$ and $y^2 = 4x$ intersect are

    (A) -2 and 2   **(B)** -2 **(C)** 2   **(D)** 0   **(E)** none of those

15. The function whose graph is a reflection in the y-axis of the graph of $f(x) = 1 - 3^x$ is

(A) $g(x) = 1 - 3^{-x}$ (B) $g(x) = 1 + 3^x$ (C) $g(x) = 3^x - 1$
(D) $g(x) = \ln_3 (x - 1)$ (E) $g(x) = \ln_3 (1 - x)$

16. Let $f(x)$ have an inverse function $g(x)$. Then $f(g(x)) =$

(A) 1   (B) x (C) $\dfrac{1}{x}$ (D) $f(x) \cdot g(x)$ (E) none of those

17. The function $f(x) = 2x^3 + x - 5$ has exactly one real zero. It is between

(A) -2 and -1 (B) -1 and 0 (C) 0 and 1 (D) 1 and 2
(E) 2 and 3

18. The period of $f(x) = \sin \dfrac{2\pi}{3} x$ is

(A) $\dfrac{1}{3}$ (B) $\dfrac{2}{3}$ (C) $\dfrac{3}{2}$ (D) 3 (E) 6

19. The range of $y = f(x) = \ln(\cos x)$ is

(A) $\{y \mid -\infty < y \leqq 0\}$ (B) $\{y \mid 0 < y \leqq 1\}$ (C) $\{y \mid -1 < y < 1\}$
(D) $\left\{y \mid -\dfrac{\pi}{2} < y < \dfrac{\pi}{2}\right\}$ (E) $\{y \mid 0 \leqq y \leqq 1\}$

20. If $\log_b (3^b) = \dfrac{b}{2}$, then $b =$

(A) $\dfrac{1}{9}$ (B) $\dfrac{1}{3}$ (C) $\dfrac{1}{2}$ (D) 3 (E) 9

21. Let f^{-1} be the inverse function of $f(x) = x^3 + 2$. Then $f^{-1}(x) =$

(A) $\dfrac{1}{x^3 - 2}$ (B) $(x + 2)^3$ (C) $(x - 2)^3$ (D) $\sqrt[3]{x + 2}$
(E) $\sqrt[3]{x - 2}$

22. The set of x-intercepts of the graph of $f(x) = x^3 - 2x^2 - x + 2$ is

(A) $\{1\}$ (B) $\{-1, 1\}$ (C) $\{1, 2\}$ (D) $\{-1, 1, 2\}$
(E) $\{-1, -2, 2\}$

23. If the domain of f is restricted to the open interval $\left(-\dfrac{\pi}{2}, \dfrac{\pi}{2}\right)$, then the range of $f(x) = e^{\tan x}$ is

(A) the set of all reals   (B) the set of positive reals
(C) the set of nonnegative reals   (D) $\{y \mid 0 < y \leqq 1\}$
(E) none of the preceding

24. Which of the following is a reflection of the graph of $y = f(x)$ in the x-axis?

(A) $y = -f(x)$ (B) $y = f(-x)$ (C) $y = |f(x)|$ (D) $y = f(|x|)$
(E) $y = -f(-x)$

25. The smallest positive x for which the function $f(x) = \sin\left(\dfrac{x}{3}\right) - 1$ is a maximum is

(A) $\dfrac{\pi}{2}$ (B) π (C) $\dfrac{3\pi}{2}$ (D) 3π (E) 6π

26. $\tan\left(\text{Arccos}\left(-\dfrac{\sqrt{2}}{2}\right)\right) =$

(A) -1 (B) $-\dfrac{\sqrt{3}}{3}$ (C) $-\dfrac{1}{2}$ (D) $\dfrac{\sqrt{3}}{3}$ (E) 1

27. If $f^{-1}(x)$ is the inverse of $f(x) = 2e^{-x}$, then $f^{-1}(x) =$

(A) $\ln\left(\dfrac{2}{x}\right)$ (B) $\ln\left(\dfrac{x}{2}\right)$ (C) $\dfrac{1}{2}\ln x$ (D) $\sqrt{\ln x}$

(E) $\ln(2 - x)$

28. Which of the following functions does not have an inverse function?

(A) $y = \sin x \left(-\dfrac{\pi}{2} \leqq x \leqq \dfrac{\pi}{2}\right)$ (B) $y = x^3 + 2$ (C) $y = \dfrac{x}{x^2 + 1}$

(D) $y = \dfrac{1}{2}e^x$ (E) $y = \ln(x - 2)$ (where $x > 2$)

29. Suppose that $f(x) = \ln x$ for all positive x and $g(x) = 9 - x^2$ for all real x. The domain of $f(g(x))$ is

(A) $\{x \mid x \leqq 3\}$ (B) $\{x \mid |x| \leqq 3\}$ (C) $\{x \mid |x| > 3\}$
(D) $\{x \mid |x| < 3\}$ (E) $\{x \mid 0 < x < 3\}$

30. Suppose (as in Question 29) that $f(x) = \ln x$ for all positive x and $g(x) = 9 - x^2$ for all real x. The range of $y = f(g(x))$ is

(A) $\{y \mid y > 0\}$ (B) $\{y \mid 0 < y \leqq \ln 9\}$ (C) $\{y \mid y \leqq \ln 9\}$
(D) $\{y \mid y < 0\}$ (E) none of the preceding

ANSWERS AND SOLUTIONS TO MULTIPLE-CHOICE QUESTIONS

For each chapter set (Set 1 through Set 10), and for the set of miscellaneous multiple-choice questions, we give first a table showing the correct answers. Explanations for the individual questions follow.

Answers for Set 1: Functions

1.	C	**6.**	B	**11.**	C	**16.**	B	**21.**	E	**26.**	A
2.	E	**7.**	C	**12.**	B	**17.**	D	**22.**	D	**27.**	A
3.	D	**8.**	C	**13.**	B	**18.**	D	**23.**	B	**28.**	C
4.	E	**9.**	B	**14.**	E	**19.**	A	**24.**	A	**29.**	D
5.	D	**10.**	E	**15.**	A	**20.**	E	**25.**	C	**30.**	C

1. **C.** $f(-2) = (-2)^3 - 2(-2) - 1 = -5$.

2. **E.** The denominator, $x^2 + 1$, is never 0.

3. **D.** Since $x - 2$ may not be negative, $x \geqq 2$. The denominator equals 0 at $x = 0$ and $x = 1$, but these values are not in the interval $x \geqslant 2$.

4. **E.** Since $g(x) = 2$, g is a constant function. Thus, for *all* $f(x)$, $g(f(x)) = 2$.

5. **D.** $f(g(x)) = f(2) = -3$.

6. **B.** Solve the pair of equations
$$\left\{ \begin{array}{r} 4 = 1 + A + B - 3 \\ -6 = -1 + A - B - 3 \end{array} \right\}.$$
Add to get A; substitute in either equation to get B. $A = 2$ and $B = 4$.

7. **C.** The graph of $f(x)$ is symmetric to the origin if $f(-x) = -f(x)$. When we replace x by $-x$, we obtain $-y$ only in (C).

8. **C.** For g to have an inverse function it must be one-to-one. Note, on page 271, that although the graph of $y = xe^{-x^2}$ is symmetric to the origin it is not one-to-one.

9. **B.** Note that $-\frac{\pi}{2} < \text{Arctan } x < \frac{\pi}{2}$ and that the sine function varies from -1 to 1 as the argument varies from $-\frac{\pi}{2}$ to $\frac{\pi}{2}$.

10. **E.** The maximum value of g is 2, attained when $\cos x = -1$. On $[0, 2\pi]$, $\cos x = -1$ for $x = \pi$.

11. **C.** f is odd if $f(-x) = -f(x)$.

12. **B.** Since $f(q) = 0$ if $q = 1$ or $q = -2$, $f(2x) = 0$ if $2x$, a replacement for q, equals 1 or -2.

13. **B.** $f(x) = x(x^2 + 4x + 4) = x(x + 2)^2$.

14. **E.**   Solving simultaneously yields $(x + 2)^2 = 4x$; $x^2 + 4x + 4 = 4x$; $x^2 + 4 = 0$. There are no real solutions.

15. **A.**   **The reflection of $y = f(x)$ in the y-axis is $y = f(-x)$.**

16. **B.**   If g is the inverse of f, then f is the inverse of g. This implies that the function f assigns to each value $g(x)$ the number x.

17. **D.**   Since f is continuous (see page 26), then, if f is negative at a and positive at b, f must equal 0 at some intermediate point. Since $f(1) = -2$ and $f(2) = 13$, this point is between 1 and 2.

18. **D.**   The function $\sin bx$ has period $\frac{2\pi}{b}$. $2\pi \div \frac{2\pi}{3} = 3$.

19. **A.**   Since $\ln q$ is defined only if $q > 0$, the domain of $\ln \cos x$ is the set of x for which $\cos x > 0$, that is, when $0 < \cos x \le 1$. Thus $-\infty < \ln \cos x \le 0$.

20. **E.** $\log_b 3^b = \frac{b}{2}$ implies $b \log_b 3 = \frac{b}{2}$. Then $\log_b 3 = \frac{1}{2}$ and $3 = b^{1/2}$. So $3^2 = b$.

21. **E.**   Letting $y = f(x) = x^3 + 2$, we get
$$x^3 = y - 2, x = \sqrt[3]{y - 2} = f^{-1}(y).$$
So $f^{-1}(x) = \sqrt[3]{x - 2}$.

22. **D.**   Since $f(1) = 0$, $x - 1$ is a factor of f. $f(x)$ divided by $x - 1$ yields $x^2 - x - 2$. So
$$f(x) = (x - 1)(x + 1)(x - 2).$$

23. **B.**   If $-\frac{\pi}{2} < x < \frac{\pi}{2}$, then $-\infty < \tan x < \infty$ and $0 < e^{\tan x} < \infty$.

24. **A.**   The reflection of $f(x)$ in the x-axis is $-f(x)$.

25. **C.** $f(x)$ attains its maximum when $\sin\left(\frac{x}{3}\right)$ does. The maximum value of the sine function is 1; the smallest positive argument is $\frac{\pi}{2}$. Set $\frac{x}{3}$ equal to $\frac{\pi}{2}$.

26. **A.** $\text{Arccos}\left(\frac{-\sqrt{2}}{2}\right) = \frac{3\pi}{4}$; $\tan\left(\frac{3\pi}{4}\right) = -1$.

27. **A.**   Let $y = f(x) = 2e^{-x}$; then $\frac{y}{2} = e^{-x}$ and $\ln \frac{y}{2} = -x$. So
$$x = -\ln \frac{y}{2} = \ln \frac{2}{y} = f^{-1}(y).$$
Thus $f^{-1}(x) = \ln \frac{2}{x}$.

28. **C.**   The function in (C) is not one-to-one since for each y between $-\frac{1}{2}$ and $\frac{1}{2}$ there are two x's in the domain.

29. **D.**   The domain of the $\ln$ function is the set of positive reals. The function $g(x) > 0$ if $x^2 < 9$.

30. **C.**   Since the domain of $f(g)$ is $(-3, 3)$, $\ln(9 - x^2)$ takes on every real value less than or equal to $\ln 9$.

PRACTICE SECTION II PROBLEMS

Most of the chapters in this book end with a set of practice problems typical of those that have appeared in Section II of recent AB and BC Examinations. Questions preceded by an asterisk (*) are likely to occur only on the BC Examination.

In answering Section II questions of the AP Examinations the student is expected to give detailed written solutions. The methods used on these problems should be shown, since the work will be graded both on the correctness of the methods and on the accuracy of the final answers obtained.

The solutions for each set of practice Section II problems follow the problems immediately and are usually quite detailed and complete.

Practice Section II Problems on Functions

1. (a) Find an equation of the locus of a point $P(x, y)$ for which the product of the slopes of the lines joining it to $(2, -1)$ and $(4, 1)$ is a constant k.
   (b) What is the locus if $k = -1$?
   (c) What if $k < 0$ but $k \neq -1$?
   (d) What if $k = 0$?
   (e) What if $k > 0$ but $k \neq 1$?
   (f) What if $k = 1$?

2. A right triangle RST with hypotenuse ST is inscribed in the parabola $y = x^2$ so that R coincides with the vertex of the parabola. If ST intersects the axis of the parabola at Q, then show that Q is independent of the choice of right triangle.

*3. A rod RS of length c units slides with its ends R and S on the x- and y-axes respectively, as shown in the figure. A rod SP of length k units is rigidly attached to RS at S so that the points R, S, and P are collinear at all times.

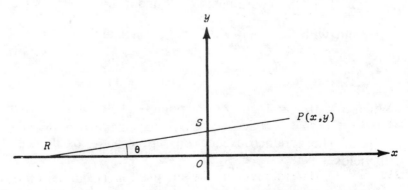

(a) Find the locus of P, using as parameter the angle θ shown.
(b) Find the Cartesian equation of the locus by eliminating θ.
(c) Name the curve.

Solutions for Section II Problems: Functions

1. (a) See Figure S1–1. An equation of the locus is

$$\frac{y-1}{x-4}\cdot\frac{y+1}{x-2} = k \quad \text{or} \quad y^2 - 1 = k(x^2 - 6x + 8).$$

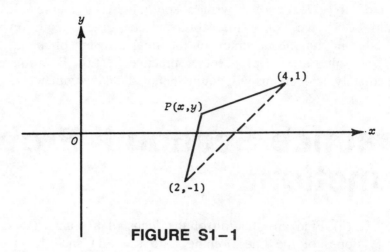

FIGURE S1–1

(b) If $k = -1$, we get

$$y^2 - 1 = -x^2 + 6x - 8 \quad \text{or} \quad x^2 - 6x + y^2 = -7.$$

Completing the square in x yields $(x - 3)^2 + y^2 = 2$. This curve is a circle with center at $(3, 0)$ and radius $\sqrt{2}$. Note that, if $k = -1$, the angle formed at P is a right angle and the given fixed points are the extremities of the diameter of the circle.

(c) If $k < 0$ but different from -1, the curve is an ellipse.

(d) If $k = 0$, the locus consists of the two lines $y = \pm 1$.

(e) If $k > 0$, $k \neq 1$, the curve is a hyperbola: $y^2 - kx^2 + 6kx = 8k + 1$.

(f) If $k = 1$, the locus is the pair of lines $y = \pm(x - 3)$.

2. Find the coordinates of S and T by solving simultaneously $y = x^2$ with $y = mx$ and with $y = -\frac{1}{m}x$. $S = (m, m^2)$ and $T = \left(-\frac{1}{m}, \frac{1}{m^2}\right)$. The equation of ST is

$$y - m^2 = \frac{m^4 - 1}{m(m^2 + 1)}(x - m).$$

When $x = 0$, $y = 1$. So Q is the point $(0, 1)$ for all nonzero m.

3. The parametric equations are $x = k \cos \theta$, $y = (c + k) \sin \theta$. Eliminating θ yields

$$\frac{x^2}{k^2} + \frac{y^2}{(c + k)^2} = 1;$$

the curve is an ellipse with major axis vertical.

Chapter 2

Limits and Continuity

Review of Definitions and Methods

*A. Definitions

1. $\lim_{x \to c} f(x) = L$ if, for any positive number ϵ, however small, there exists a positive number δ such that if $0 < |x - c| < \delta$ then $|f(x) - L| < \epsilon$.

Here c and L are finite numbers. Note that the function need not be defined at c and that x is *different* from c as x approaches c. This definition says precisely that, if $f(x)$ approaches the limit L as x approaches c, then the difference between $f(x)$ and L can be made arbitrarily small by taking x sufficiently close to c.

2. $\lim_{x \to c} f(x) = \infty$ if, for any positive number N, however large, there exists a positive number δ such that if $0 < |x - c| < \delta$ then $|f(x)| > N$ [if $f(x) > N$ then $\lim_{x \to c} f(x) = +\infty$, while if $f(x) < -N$ then $\lim_{x \to c} f(x) = -\infty$].

Here, again, $f(c)$ may or may not exist, and x does not equal c as $x \to c$. The function $f(x)$ is said to become infinite as x approaches c if $f(x)$ can be made numerically arbitrarily large by taking x sufficiently close to c. Note carefully that if $\lim_{x \to c} f(x) = \infty$ the limit *does not exist*.

3. (a) $\lim_{x \to +\infty} f(x) = L$ if, for any positive number ϵ, however small, there exists a positive number N such that if $x > N$ then $|f(x) - L| < \epsilon$.

(b) $\lim_{x \to -\infty} f(x) = L$ if, for any positive number ϵ, however small, there exists a positive number N such that if $x < -N$ then $|f(x) - L| < \epsilon$.

Here, L is again finite and $f(x)$ is said to approach the limit L as x becomes infinite. Under the notation "lim" the notation "$x \to \infty$" is sometimes used to mean indifferently "$x \to +\infty$" or "$x \to -\infty$."

4. $\lim_{x \to \infty} f(x) = \infty$ if, for any positive number N, however large, there exists a positive number M such that if $|x| > M$ then $|f(x)| > N$.

Here, $f(x)$ is said to become infinite as x becomes infinite, and, as in 2 above, *no limit exists*.

*An asterisk denotes a topic covered only in Calculus BC.

B.   Theorems on Limits

If c, k, R, S, U, and V are finite numbers and if

$$\lim_{x \to c} f(x) = R, \qquad \lim_{x \to c} g(x) = S, \qquad \lim_{x \to \infty} f(x) = U, \qquad \lim_{x \to \infty} g(x) = V,$$

then:

(1a) $\lim\limits_{x \to c} kf(x) = kR$ **(1b)** $\lim\limits_{x \to \infty} kf(x) = kU$

(2a) $\lim\limits_{x \to c} [f(x) + g(x)] = R + S$ **(2b)** $\lim\limits_{x \to \infty} [f(x) + g(x)] = U + V$

(3a) $\lim\limits_{x \to c} f(x)g(x) = RS$ **(3b)** $\lim\limits_{x \to \infty} f(x)g(x) = UV$

(4a) $\lim\limits_{x \to c} \dfrac{f(x)}{g(x)} = \dfrac{R}{S}$ (if $S \neq 0$)   **(4b)** $\lim\limits_{x \to \infty} \dfrac{f(x)}{g(x)} = \dfrac{U}{V}$ (if $V \neq 0$)

(5a) $\lim\limits_{x \to c} k = k$ **(5b)** $\lim\limits_{x \to \infty} k = k$

(6)   (The "Squeeze" Theorem.) If $f(x) \leqq g(x) \leqq h(x)$ and if $\lim\limits_{x \to c} f(x) = \lim\limits_{x \to c} h(x) = L$, then $\lim\limits_{x \to c} g(x) = L$.

Examples

1. $\lim\limits_{x \to 2} (5x^2 - 3x + 1) = 5 \lim\limits_{x \to 2} x^2 - 3 \lim\limits_{x \to 2} x + \lim\limits_{x \to 2} 1$

$\qquad\qquad\qquad\qquad = 5 \cdot 4 \quad\; - 3 \cdot 2 \quad\; + 1$

$\qquad\qquad\qquad\qquad = 15.$

2. $\lim\limits_{x \to 0} (x \cos 2x) = \lim\limits_{x \to 0} x \cdot \lim\limits_{x \to 0} (\cos 2x)$

$\qquad\qquad\qquad\qquad = 0 \qquad \cdot 1$

$\qquad\qquad\qquad\qquad = 0.$

3. $\lim\limits_{x \to -1} \dfrac{3x^2 - 2x - 1}{x^2 + 1} = \lim\limits_{x \to -1} (3x^2 - 2x - 1) \div \lim\limits_{x \to -1} (x^2 + 1)$

$\qquad\qquad\qquad\qquad = (3 + 2 - 1) \qquad\quad \div (1 + 1)$

$\qquad\qquad\qquad\qquad = 2.$

4. $\lim\limits_{x \to 3} \dfrac{x^2 - 9}{x - 3} = \lim\limits_{x \to 3} \dfrac{(x - 3)(x + 3)}{x - 3}.$ Since

$$\frac{(x - 3)(x + 3)}{x - 3} = x + 3 \qquad (x \neq 3)$$

and since by definition A1 above x must be different from 3 as $x \to 3$, it follows that

$$\lim_{x \to 3} \frac{(x - 3)(x + 3)}{x - 3} = \lim_{x \to 3} (x + 3) = 6,$$

where the factor $x - 3$ is removed *before* taking the limit.

5. $\lim\limits_{x \to -2} \dfrac{x^3 + 8}{x^2 - 4} = \lim\limits_{x \to -2} \dfrac{(x + 2)(x^2 - 2x + 4)}{(x + 2)(x - 2)} = \lim\limits_{x \to -2} \dfrac{x^2 - 2x + 4}{x - 2} =$

$\dfrac{4 + 4 + 4}{-4} = -3.$

6. $\lim\limits_{x \to 0} \dfrac{x}{x^3} = \lim\limits_{x \to 0} \dfrac{1}{x^2} = \infty$. As $x \to 0$, the numerator approaches 1 while the denominator approaches 0; the limit does *not* exist.

7. $\lim\limits_{x \to 1} \dfrac{x^2 - 1}{x^2 - 1} = \lim\limits_{x \to 1} 1 = 1.$

8. $\lim\limits_{\Delta x \to 0} \dfrac{(3 + \Delta x)^2 - 3^2}{\Delta x} = \lim\limits_{\Delta x \to 0} \dfrac{6\Delta x + \Delta x^2}{\Delta x} = \lim\limits_{\Delta x \to 0} 6 + \Delta x = 6.$

9. $\lim\limits_{h \to 0} \dfrac{1}{h}\left(\dfrac{1}{2 + h} - \dfrac{1}{2}\right) = \lim\limits_{h \to 0} \dfrac{2 - (2 + h)}{2h(2 + h)} = \lim\limits_{h \to 0} \dfrac{-h}{2h(2 + h)} =$
$\lim\limits_{h \to 0} -\dfrac{1}{2(2 + h)} = -\dfrac{1}{4}.$

C. Limit of a Quotient of Polynomials

To find $\lim\limits_{x \to \infty} \dfrac{P(x)}{Q(x)}$, where $P(x)$ and $Q(x)$ are polynomials in x, we can divide both numerator and denominator by the highest power of x that occurs and use the fact that $\lim\limits_{x \to \infty} \dfrac{1}{x} = 0$.

Examples

10. $\lim\limits_{x \to \infty} \dfrac{3 - x}{4 + x + x^2} = \lim\limits_{x \to \infty} \dfrac{\dfrac{3}{x^2} - \dfrac{1}{x}}{\dfrac{4}{x^2} + \dfrac{1}{x} + 1} = \dfrac{0 - 0}{0 + 0 + 1} = 0.$

11. $\lim\limits_{x \to \infty} \dfrac{4x^4 + 5x + 1}{37x^3 - 9} = \lim\limits_{x \to \infty} \dfrac{4 + \dfrac{5}{x^3} + \dfrac{1}{x^4}}{\dfrac{37}{x} - \dfrac{9}{x^4}} = \infty \text{ (no limit).}$

12. $\lim\limits_{x \to \infty} \dfrac{x^3 - 4x^2 + 7}{3 - 6x - 2x^3} = \lim\limits_{x \to \infty} \dfrac{1 - \dfrac{4}{x} + \dfrac{7}{x^3}}{\dfrac{3}{x^3} - \dfrac{6}{x^2} - 2} = \dfrac{1 - 0 + 0}{0 - 0 - 2} = -\dfrac{1}{2}.$

Rational Function Theorem.

We see from Examples 10, 11, and 12 that: if the degree of $Q(x)$ is higher than that of $P(x)$, then $\lim\limits_{x \to \infty} \dfrac{P(x)}{Q(x)} = 0$; if the degree of $P(x)$ is higher than that of $Q(x)$, then $\lim\limits_{x \to \infty} \dfrac{P(x)}{Q(x)} = \infty$ or $-\infty$ (i.e., does not exist); and if the degrees of $P(x)$ and $Q(x)$ are the same, then $\lim\limits_{x \to \infty} \dfrac{P(x)}{Q(x)} = \dfrac{a_n}{b_n}$, where a_n and b_n are the coefficients of the highest powers of x in $P(x)$ and $Q(x)$ respectively. These conclusions constitute the Rational Function Theorem.

Example 13. $\lim\limits_{x\to\infty} \dfrac{100x^2 - 19}{x^3 + 5x^2 + 2} = 0$; $\lim\limits_{x\to\infty} \dfrac{x^3 - 5}{1 + 6x + 81x^2} = \infty$ (no limit);

$\lim\limits_{x\to\infty} \dfrac{x - 4}{13 + 5x} = \dfrac{1}{5}$; $\lim\limits_{x\to\infty} \dfrac{4 + x^2 - 3x^3}{x + 7x^3} = -\dfrac{3}{7}$; $\lim\limits_{x\to\infty} \dfrac{x^3 + 1}{2 - x^2} = -\infty$ (no limit).

D. Absolute-Value and Greatest-Integer Functions

1. $\lim\limits_{x\to 0} |x| = 0$; $\lim\limits_{x\to c} |x| = c$ if $c > 0$; $\lim\limits_{x\to c} |x| = -c$ if $c < 0$. Note that $\lim\limits_{x\to c} |x|$ exists for every real number c (see the graph of the absolute-value function in Figure N1–3, page 6).

Example 14. $\lim\limits_{x\to 1} |x| = 1$; $\lim\limits_{x\to -3} |x| = 3$; $\lim\limits_{x\to \frac{1}{5}} |x| = \dfrac{1}{5}$; $\lim\limits_{x\to -0.01} |x| = 0.01$.

2. The greatest-integer function is graphed in Figure N1–3, page 6.

Example 15. $\lim\limits_{x\to 0.6} [x] = 0$; $\lim\limits_{x\to 0.9} [x] = 0$; $\lim\limits_{x\to 0.95} [x] = 0$;
$\lim\limits_{x\to \frac{3}{2}} [x] = 1$; $\lim\limits_{x\to 1.4} [x] = 1$; $\lim\limits_{x\to 1.1} [x] = 1$.

We note that $\lim\limits_{x\to 1} [x]$ does not exist, because (as the graph shows) $\lim\limits_{x\to 1^-} [x]$, the *left-hand* limit (as x approaches 1 through values *less* than 1), equals zero, whereas $\lim\limits_{x\to 1^+} [x]$, the *right-hand* limit (as x approaches 1 through values *greater* than 1), equals one.

If $\lim\limits_{x\to c^-} f(x) = L$ and $\lim\limits_{x\to c^+} f(x) = R$ but L and R are different, then $\lim\limits_{x\to c} f(x)$ does *not* exist.

If n is an integer, $\lim\limits_{x\to n} [x]$ does not exist; but the greatest-integer function does approach a limit at every nonintegral real number.

E. Other Basic Limits

E1. The basic trigonometric limit is the following:

$$\lim\limits_{\theta\to 0} \frac{\sin \theta}{\theta} = 1 \quad \text{if } \theta \text{ is measured in radians.}$$

Example 16. To find $\lim\limits_{x\to\infty} \dfrac{\sin x}{x}$ we let $x = \dfrac{1}{\theta}$. Then $\lim\limits_{x\to\infty} \dfrac{\sin x}{x} = \lim\limits_{\theta\to 0} \dfrac{\sin \frac{1}{\theta}}{\frac{1}{\theta}} =$

$\lim\limits_{\theta\to 0} \theta \sin \dfrac{1}{\theta}$.

Since

$$-1 \leqq \sin \frac{1}{\theta} \leqq 1 \quad \text{for all } \theta \neq 0,$$

$$-\theta \leqq \theta \sin \frac{1}{\theta} \leqq \theta \quad \text{if } \theta > 0$$

or

$$-\theta \geqq \theta \sin \frac{1}{\theta} \geqq \theta \quad \text{if } \theta < 0$$

In either case, as $\theta \to 0$, $\theta \sin \dfrac{1}{\theta} \to 0$ by the "Squeeze" Theorem (page 22). So

$$\lim\limits_{x\to\infty} \frac{\sin x}{x} = 0.$$

E2. The number e can be defined as follows:

$$e = \lim\limits_{n\to\infty} \left(1 + \frac{1}{n} \right)^n.$$

In Chapter 3, Example 36, page 44, and Example 45, page 45, show that

$$\lim\limits_{h\to 0} \frac{e^h - 1}{h} = 1.$$

F. Continuity

The function $f(x)$ is said to be *continuous* at $x = c$ if (1) $f(c)$ is a finite number, (2) $\lim\limits_{x\to c} f(x)$ exists, and (3) $\lim\limits_{x\to c} f(x) = f(c)$. A function is continuous over the closed interval $[a, b]$ if it is continuous at each x such that $a \leqq x \leqq b$. A function which is not continuous at $x = c$ is said to be discontinuous at that point.

Examples

17. $f(x) = \dfrac{x - 1}{x^2 + x} = \dfrac{x - 1}{x(x + 1)}$ is not continuous at $x = 0$ or $= -1$, since

the function is not defined for either of these numbers. Note also that neither $\lim\limits_{x\to 0} f(x)$ nor $\lim\limits_{x\to -1} f(x)$ exists.

18. $f(x) = \dfrac{x^2 - 4}{x - 2} = \dfrac{(x - 2)(x + 2)}{x - 2}$ is not continuous at $x = 2$ because $f(2)$ does not exist. Since, here, $\lim\limits_{x \to 2} f(x) = \lim\limits_{x \to 2} (x + 2) = 4$, we can define a new function

$$f(x) = \frac{x^2 - 4}{x - 2} \qquad (x \neq 2),$$

$$f(2) = 4,$$

which is continuous for all x, *including* 2.

19. If $f(x) = [x]$ (where $[x]$ is defined as the greatest integer not greater than x), then $f(x)$ is discontinuous at each integer (see Example 15, page 24).

Theorems on Continuous Functions.

If $f(x)$ is continuous on the closed interval $[a, b]$, then it follows that

(1) the function attains a minimum value and a maximum value somewhere on the interval;

(2) if M is a number such that $f(a) \leqq M \leqq f(b)$, then there is at least one number c between a and b such that $f(c) = M$ (the Intermediate Value Theorem);

(3) if x_1 and x_2 are numbers between a and b and if $f(x_1)$ and $f(x_2)$ have opposite signs, then there is at least one number c between x_1 and x_2 such that $f(c) = 0$ (this is a special case of the Intermediate Value Theorem).

If the functions f and g are continuous at $x = a$, so are their sum, their difference, their product, and (if the denominator is not zero at a) their quotient; also kf, where k is a constant, is continuous at a. Polynomials are everywhere continuous and so are rational functions $\dfrac{P(x)}{Q(x)}$, where P and Q are polynomials, except where Q equals zero.

Set 2: Multiple-Choice Questions on Limits and Continuity

1. $\lim\limits_{x \to 2} \dfrac{x^2 - 4}{x^2 + 4}$ is

    (A) 1   **(B)** 0   **(C)** $-\dfrac{1}{2}$ **(D)** -1 **(E)** ∞

2. $\lim\limits_{x \to \infty} \dfrac{4 - x^2}{x^2 - 1}$ is

   (A) 1   (B) 0   (C) -4 (D) -1 (E) ∞

3. $\lim\limits_{x \to 3} \dfrac{x - 3}{x^2 - 2x - 3}$ is

   (A) 0   (B) 1   (C) $\dfrac{1}{4}$ (D) ∞ (E) none of these

4. $\lim\limits_{x \to 0} \dfrac{x}{x}$ is

   (A) 1   (B) 0   (C) ∞ (D) -1 (E) nonexistent

5. $\lim\limits_{x \to 2} \dfrac{x^3 - 8}{x^2 - 4}$ is

   (A) 4   (B) 0   (C) 1   (D) 3   (E) ∞

6. $\lim\limits_{x \to \infty} \dfrac{4 - x^2}{4x^2 - x - 2}$ is

   (A) -2 (B) $-\dfrac{1}{4}$ (C) 1   (D) 2

   (E) nonexistent

7. $\lim\limits_{x \to \infty} \dfrac{5x^3 + 27}{20x^2 + 10x + 9}$ is

   (A) ∞ (B) $\dfrac{1}{4}$ (C) 3   (D) 0   (E) 1

8. $\lim\limits_{x \to \infty} \dfrac{3x^2 + 27}{x^3 - 27}$ is

   (A) 3   (B) ∞ (C) 1   (D) -1 (E) 0

9. $\lim\limits_{x \to \infty} \dfrac{2^{-x}}{2^x}$ is

   (A) -1 (B) 1   (C) 0   (D) ∞ (E) none of these

10. If $[x]$ is the greatest integer not greater than x, then $\lim\limits_{x \to 1/2} [x]$ is

   (A) $\dfrac{1}{2}$ (B) 1   (C) nonexistent   (D) 0   (E) none of these

11. (With the same notation) $\lim\limits_{x \to 2} [x]$ is

   (A) 0   (B) 1   (C) 2   (D) 3   (E) none of these

12. $\lim\limits_{x \to 0} \dfrac{\tan x}{x}$ is

   (A) 0   (B) 1   (C) π (D) ∞ (E) The limit does not exist.

13. $\lim\limits_{x \to 0} \dfrac{\sin 2x}{x}$ is

    (A) 1   (B) 2   (C) $\dfrac{1}{2}$ (D) 0   (E) ∞

14. $\lim\limits_{x \to \infty} \sin x$

    (A) is nonexistent   (B) is infinity   (C) oscillates between -1 and 1
    (D) is zero   (E) is 1 or -1

15. $\lim\limits_{x \to 0} \dfrac{\sin 3x}{\sin 4x}$ is

    (A) 1   (B) $\dfrac{4}{3}$ (C) $\dfrac{3}{4}$ (D) 0   (E) nonexistent

16. $\lim\limits_{x \to 0} \dfrac{1 - \cos x}{x}$ is

    (A) nonexistent   (B) 1   (C) 2   (D) ∞ (E) none of these

17. $\lim\limits_{x \to 0} \dfrac{\sin x}{x^2 + 3x}$ is

    (A) 1   (B) $\dfrac{1}{3}$ (C) 3   (D) ∞ (E) $\dfrac{1}{4}$

18. $\lim\limits_{x \to 0} \sin \dfrac{1}{x}$ is

    (A) ∞ (B) 1   (C) nonexistent   (D) -1 (E) none of these

19. $\lim\limits_{x \to 0} \dfrac{\tan \pi x}{x}$ is

    (A) $\dfrac{1}{\pi}$ (B) 0   (C) 1   (D) π (E) ∞

20. $\lim\limits_{x \to \infty} x^2 \sin \dfrac{1}{x}$

    (A) is 1   (B) is 0   (C) is ∞ (D) oscillates between -1 and 1
    (E) is none of these

21. $\lim\limits_{x \to 0} x \csc x$ is

    (A) $-\infty$ (B) -1 (C) 0   (D) 1   (E) ∞

22. $\lim\limits_{x \to \infty} \dfrac{2x^2 + 1}{(2 - x)(2 + x)}$ is

    (A) -4 (B) -2 (C) 1   (D) 2   (E) nonexistent

23. $\lim\limits_{x \to 0} \dfrac{|x|}{x}$ is

    (A) 0   (B) nonexistent   (C) 1   (D) -1 (E) none of these

24. $\lim\limits_{x \to \infty} x \sin \frac{1}{x}$ is

(A) 0   (B) ∞ (C) nonexistent   (D) -1 (E) 1

25. $\lim\limits_{x \to \pi} \dfrac{\sin(\pi - x)}{\pi - x}$ is

(A) 1   (B) 0   (C) ∞ (D) nonexistent   (E) none of these

26. If $[x]$ is the greatest integer in x, then what is $\lim\limits_{x \to -1} [x + 1]$?

(A) -1 (B) 0   (C) 1   (D) 2   (E) The limit does not exist.

27. Let $f(x) = \begin{cases} x^2 - 1 & \text{if } x \neq 1, \\ 4 & \text{if } x = 1. \end{cases}$

Which of the following statements, I, II, and III, are true?
I. $\lim\limits_{x \to 1} f(x)$ exists   II. $f(1)$ exists   III. f is continuous at $x = 1$

(A) only I   (B) only II   (C) I and II   (D) none of them
(E) all of them

*28. Suppose a function f is defined for all real numbers. If, for any positive number ϵ, there exists a positive number δ such that $|f(x) - 3| < \epsilon$ whenever $0 < |x - 1| < \delta$, then it follows that

(A) $\lim\limits_{x \to 1} f(x) = 3$ (B) $\lim\limits_{x \to 3} f(x) = 1$ (C) $\lim\limits_{x \to 0} f(x) = 3$
(D) $\lim\limits_{x \to \infty} f(x) = 3$ (E) $\lim\limits_{x \to 0} f(x) = 0$

29. If $\begin{cases} f(x) = \dfrac{x^2 - x}{2x} \text{ for } x \neq 0, \\ f(0) = k, \end{cases}$

and if f is continuous at $x = 0$, then $k =$

(A) -1 (B) $-\dfrac{1}{2}$ (C) 0   (D) $\dfrac{1}{2}$ (E) 1

30. Suppose $\begin{cases} f(x) = \dfrac{3x(x - 1)}{x^2 - 3x + 2} \text{ for } x \neq 1, 2, \\ f(1) = -3, \\ f(2) = 4. \end{cases}$

Then $f(x)$ is continuous

(A) except at $x = 1$ (B) except at $x = 2$ (C) except at $x = 1$ or 2
(D) except at $x = 0, 1,$ or 2   (E) at each real number

---

*Questions preceded by an asterisk are likely to appear only on the Calculus BC Examination.

Answers for Set 2: Limits and Continuity

1.	B	**6.**	B	**11.**	E	**16.**	E	**21.**	D	**26.**	E
2.	D	**7.**	A	**12.**	B	**17.**	B	**22.**	B	**27.**	C
3.	C	**8.**	E	**13.**	B	**18**	C	**23.**	B	**28.**	A
4.	A	**9.**	C	**14.**	A	**19.**	D	**24.**	E	**29.**	B
5.	D	**10.**	D	**15.**	C	**20.**	C	**25.**	A	**30.**	B

1. **B.** The limit as $x \to 2$ is $0 \div 8$.

2. **D.** Use the Rational Function Theorem (page 23). The degrees of $P(x)$ and $Q(x)$ are the same.

3. **C.** Remove the common factor $x - 3$ from numerator and denominator.

4. **A.** The fraction equals 1 for all nonzero x.

5. **D.** Note that $\dfrac{x^3 - 8}{x^2 - 4} = \dfrac{(x - 2)(x^2 + 2x + 4)}{(x - 2)(x + 2)}$.

6. **B.** Use the Rational Function Theorem, (page 23).

7. **A.** Use the Rational Function Theorem.

8. **E.** Use the Rational Function Theorem.

9. **C.** The fraction is equivalent to $\dfrac{1}{2^{2x}}$; the denominator approaches ∞.

10. **D.** See Figure N1–3 on page 6.

11. **E.** Note from Figure N1–3, page 6, that $\lim\limits_{x \to 2^-} [x] = 1$ but $\lim\limits_{x \to 2^+} [x] = 2$.

12. **B.** $\lim\limits_{x \to 0} \dfrac{\tan x}{x} = \lim\limits_{x \to 0} \dfrac{\sin x}{x} \cdot \dfrac{1}{\cos x} = \lim\limits_{x \to 0} \dfrac{\sin x}{x} \cdot \lim\limits_{x \to 0} \dfrac{1}{\cos x} = 1 \cdot 1 = 1.$

13. **B.** $\lim\limits_{x \to 0} \dfrac{\sin 2x}{x} = \lim\limits_{x \to 0} 2 \dfrac{\sin x}{x} \cos x = 2 \cdot 1 \cdot 1.$

14. **A.** As $x \to \infty$, the *function* $\sin x$ does, indeed, oscillate between -1 and 1.

15. **C.** $\dfrac{\sin 3x}{\sin 4x} = \dfrac{3}{4} \cdot \dfrac{\sin 3x}{3x} \cdot \dfrac{4x}{\sin 4x}$. As $x \to 0$, the latter product approaches $\dfrac{3}{4} \cdot 1 \cdot 1.$

16. **E.** The formula $1 - \cos x = 2 \sin^2 \dfrac{x}{2}$ can be used here to "resolve" the indeterminacy; then we get

$$\lim_{x\to 0} \frac{2\sin^2 \frac{x}{2}}{x} = \lim_{x\to 0} \frac{\sin^2 \frac{x}{2}}{\frac{x}{2}}$$

$$= \lim_{x\to 0} \left[\sin \frac{x}{2} \cdot \frac{\sin \frac{x}{2}}{\frac{x}{2}} \right]$$

$$= 0 \cdot 1 = 0.$$

17. B. Note that $\dfrac{\sin x}{x^2 + 3x} = \dfrac{\sin x}{x(x + 3)} = \dfrac{\sin x}{x} \cdot \dfrac{1}{x + 3} \to 1 \cdot \dfrac{1}{3}$.

18. C. As $x \to 0$, $\dfrac{1}{x}$ takes on varying finite values as it increases. Since the sine function repeats, $\sin \dfrac{1}{x}$ takes on, infinitely many times, each value between -1 and 1.

19. D. $\dfrac{\tan \pi x}{x} = \dfrac{\sin \pi x}{x \cos \pi x} = \dfrac{\pi \sin \pi x}{\pi x} \cdot \dfrac{1}{\cos \pi x}$, which approaches $\pi \cdot 1 \cdot 1$ as $x \to 0$.

20. C. $\lim_{x\to\infty} x^2 \sin \dfrac{1}{x} = \lim_{x\to\infty} x \dfrac{\sin (1/x)}{(1/x)} = \infty \cdot 1$.

21. D. $x \csc x = \dfrac{x}{\sin x}$.

22. B. $\dfrac{2x^2 + 1}{(2 - x)(2 + x),} = \dfrac{2x^2 + 1}{4 - x^2}$. Use the Rational Function Theorem, (page 23).

23. B. Since $|x| = x$ if $x > 0$ but equals $-x$ if $x < 0$,

$$\lim_{x\to 0^+} \frac{|x|}{x} = \lim_{x\to 0^+} \frac{|x|}{x} = 1 \quad \text{while} \quad \lim_{x\to 0^-} \frac{|x|}{x} = \lim_{x\to 0^-} \frac{-x}{x} = -1.$$

24. E. Note that $x \sin \dfrac{1}{x}$ can be rewritten as $\dfrac{\sin \frac{1}{x}}{\frac{1}{x}}$ and that, as $x \to \infty$, $\dfrac{1}{x} \to 0$.

25. A. As $x \to \pi$, $(\pi - x) \to 0$.

26. E. $\lim_{x\to -1^-} [x + 1] = -1$ but $\lim_{x\to -1^+} [x + 1] = 0$.

27. C. Since $f(x) = x + 1$ if $x \neq 1$, $\lim_{x\to 1} f(x)$ exists (and is equal to 2).

28. A. The question defines the limit in (A).

29. B. $f(x) = \dfrac{x(x - 1)}{2x} = \dfrac{x - 1}{2}$, for all $x \neq 0$. For x to be continuous at $x = 0$, $\lim_{x\to 0} f(x)$ must equal $f(0)$. $\lim_{x\to 0} f(x) = -\dfrac{1}{2}$.

30. B. Only $x = 1$ and $x = 2$ need be checked. Since $f(x) = \dfrac{3x}{x - 2}$ for $x \neq 1, 2$, and $\lim_{x\to 1} f(x) = -3 = f(1)$, f is continuous at $x = 1$. Since $\lim_{x\to 2} f(x)$ does not exist, f is not continuous at $x = 2$.

Differentiation

Review of Definitions and Methods

A. Definition of Derivative_____

At any x in the domain of the function $y = f(x)$, the derivative is defined to be

$$\lim_{\Delta x \to 0} \frac{f(x + \Delta x) - f(x)}{\Delta x} \quad \text{or} \quad \lim_{\Delta x \to 0} \frac{\Delta y}{\Delta x}.$$

The function is said to be *differentiable* at every x for which this limit exists, and its derivative may be denoted by $f'(x)$, y', $\frac{dy}{dx}$, or $D_x y$. Frequently Δx is replaced by h or some other symbol.

The derivative of $y = f(x)$ at $x = a$, denoted by $f'(a)$ or $y'(a)$, may be defined as follows:

$$f'(a) = \lim_{h \to 0} \frac{f(a + h) - f(a)}{h}.$$

If we replace $(a + h)$ by x, so that $h = x - a$, we get the equivalent expression

$$f'(a) = \lim_{x \to a} \frac{f(x) - f(a)}{x - a}.$$

The second derivative, denoted by $f''(x)$ or $\frac{d^2 y}{dx^2}$ or y'', is the (first) derivative of $f'(x)$. Also, $f''(a)$ is the second derivative of $f(x)$ at $x = a$.

B. Formulas

The formulas in this section for finding derivatives are so important that familiarity with them is essential. If a and n are constants and u and v are differentiable functions of x then:

$$\frac{da}{dx} = 0 \tag{1}$$

$$\frac{d}{dx} \, au = a \, \frac{du}{dx} \tag{2}$$

$$\frac{d}{dx} \, u^n = nu^{n-1} \, \frac{du}{dx} \quad \text{(the power rule)} \tag{3}$$

$$\frac{d}{dx} \, (u + v) = \frac{d}{dx} \, u + \frac{d}{dx} \, v \tag{4}$$

$$\frac{d}{dx} \, (uv) = u \, \frac{dv}{dx} + v \, \frac{du}{dx} \quad \text{(the product rule)} \tag{5}$$

$$\frac{d}{dx} \left(\frac{u}{v} \right) = \frac{v \, \dfrac{du}{dx} - u \, \dfrac{dv}{dx}}{v^2} \quad (v \neq 0) \quad \text{(the quotient rule)} \tag{6}$$

$$\frac{d}{dx} \, \sin u = \cos u \, \frac{du}{dx} \tag{7}$$

$$\frac{d}{dx} \, \cos u = -\sin u \, \frac{du}{dx} \tag{8}$$

$$\frac{d}{dx} \, \tan u = \sec^2 u \, \frac{du}{dx} \tag{9}$$

$$\frac{d}{dx} \, \cot u = -\csc^2 u \, \frac{du}{dx} \tag{10}$$

$$\frac{d}{dx} \, \sec u = \sec u \, \tan u \, \frac{du}{dx} \tag{11}$$

$$\frac{d}{dx} \, \csc u = -\csc u \, \cot u \, \frac{du}{dx} \tag{12}$$

$$\frac{d}{dx} \, \ln u = \frac{1}{u} \, \frac{du}{dx} \tag{13}$$

$$\frac{d}{dx} \, e^u = e^u \, \frac{du}{dx} \tag{14}$$

$$\frac{d}{dx} \, a^u = a^u \ln a \, \frac{du}{dx} \tag{15}$$

$$\frac{d}{dx} \, \sin^{-1} u = \frac{d}{dx} \, \text{Arcsin } u = \frac{1}{\sqrt{1 - u^2}} \, \frac{du}{dx} \tag{16}$$

$$\frac{d}{dx} \cos^{-1} u = \frac{d}{dx} \operatorname{Arccos} u = -\frac{1}{\sqrt{1 - u^2}} \frac{du}{dx} \qquad (17)$$

$$\frac{d}{dx} \tan^{-1} u = \frac{d}{dx} \operatorname{Arctan} u = \frac{1}{1 + u^2} \frac{du}{dx} \qquad (18)$$

$$\frac{d}{dx} \cot^{-1} u = \frac{d}{dx} \operatorname{Arccot} u = -\frac{1}{1 + u^2} \frac{du}{dx} \qquad (19)$$

$$\frac{d}{dx} \sec^{-1} u = \frac{d}{dx} \operatorname{Arcsec} u = \frac{1}{|u|\sqrt{u^2 - 1}} \frac{du}{dx} \qquad (20)$$

$$\frac{d}{dx} \csc^{-1} u = \frac{d}{dx} \operatorname{Arccsc} u = -\frac{1}{|u|\sqrt{u^2 - 1}} \frac{du}{dx} \qquad (21)$$

C.   The Derivative of a Composite Function; Chain Rule

If $y = f(u)$ and $u = g(x)$, then $y = f(g(x))$ is the composite function defined on page 4. If y and u are differentiable functions of u and x respectively, then

$$\frac{dy}{dx} = f'(u)g'(x) = f'(g(x))g'(x). \qquad (1)$$

This can also be written in the form

$$\frac{dy}{dx} = \frac{dy}{du} \cdot \frac{du}{dx}. \qquad (2)$$

In this form, it is called the *chain rule*. Note, in particular, that if we let $y = F(x)$ and if g is differentiable at x_0 and f is differentiable at u_0, **where $u_0 = g(x_0)$**, then we get, from equation (1),

$$F'(x_0) = f'(g(x_0))g'(x_0). \qquad (3)$$

Formulas (1)–(21) and many of the following illustrative examples use the chain rule stated above.

Examples

1. If $y = 4x^3 - 5x + 7$, then

$$\frac{dy}{dx} = 12x^2 - 5 \qquad \text{and} \qquad \frac{d^2y}{dx^2} = 24x.$$

2. If $u = \sqrt[3]{2x^2} - \frac{1}{\sqrt{3x}}$, then $u = (2x^2)^{1/3} - (3x)^{-1/2}$, and (for $x > 0$)

$$\frac{du}{dx} = \frac{1}{3}(2x^2)^{-2/3}(4x) + \frac{1}{2}(3x)^{-3/2} \cdot 3 = \frac{4}{3\sqrt[3]{4x}} + \frac{1}{2x\sqrt{3x}}.$$

3. If $y = \sqrt{3 - x - x^2}$, then $y = (3 - x - x^2)^{1/2}$ and

$$\frac{dy}{dx} = \frac{1}{2}(3 - x - x^2)^{-1/2}(-1 - 2x) = -\frac{1 + 2x}{2\sqrt{3 - x - x^2}}.$$

4. If $y = \dfrac{5}{\sqrt{(1 - x^2)^3}}$, then $y = 5(1 - x^2)^{-3/2}$ and

$$\frac{dy}{dx} = \frac{-15}{2}(1 - x^2)^{-5/2}(-2x) = \frac{15x}{(1 - x^2)^{5/2}}.$$

5. If $s = (t^2 + 1)^3(1 - t)^2$, then

$$\frac{ds}{dt} = (t^2 + 1)^3[2(1 - t)(-1)] + (1 - t)^2[3(t^2 + 1)^2(2t)]$$

$$= 2(t^2 + 1)^2(1 - t)(3t - 1 - 4t^2).$$

6. If $f(t) = e^{2t} \sin 3t$, then

$$f'(t) = e^{2t}(\cos 3t \cdot 3) + \sin 3t(e^{2t} \cdot 2)$$

$$= e^{2t}(3 \cos 3t + 2 \sin 3t).$$

7. If $z = \dfrac{v}{\sqrt{1 - 2v^2}}$, then

$$\frac{dz}{dv} = \frac{\sqrt{1 - 2v^2} - \dfrac{v(-4v)}{2\sqrt{1 - 2v^2}}}{1 - 2v^2} = \frac{\dfrac{(1 - 2v^2) + 2v^2}{\sqrt{1 - 2v^2}}}{1 - 2v^2} = \frac{1}{(1 - 2v^2)^{3/2}}.$$

Since $\dfrac{dz}{dv}$ equals $(1 - 2v^2)^{-3/2}$, note that

$$\frac{d^2z}{dv^2} = -\frac{3}{2}(1 - 2v^2)^{-5/2}(-4v) = \frac{6v}{(1 - 2v^2)^{5/2}}.$$

8. If $f(x) = \dfrac{\sin x}{x^2}$, $x \neq 0$, then

$$f'(x) = \frac{x^2 \cos x - \sin x \cdot 2x}{x^4} = \frac{x \cos x - 2 \sin x}{x^3}.$$

9. If $y = \tan (2x^2 + 1)$, then

$$y' = 4x \sec^2 (2x^2 + 1)$$

and

$$y'' = 4[x \cdot 2 \sec (2x^2 + 1) \cdot \sec (2x^2 + 1) \tan (2x^2 + 1) \cdot 4x + \sec^2 (2x^2 + 1)]$$

$$= 4 \sec^2 (2x^2 + 1)[8x^2 \tan (2x^2 + 1) + 1].$$

10. If $x = \cos^3 (1 - 3\theta)$, then

$$\frac{dx}{d\theta} = -3 \cos^2 (1 - 3\theta) \sin (1 - 3\theta)(-3)$$

$$= 9 \cos^2 (1 - 3\theta) \sin (1 - 3\theta).$$

11. If $y = e^{(\sin x) + 1}$, then

$$\frac{dy}{dx} = \cos x \cdot e^{(\sin x) + 1}.$$

12. If $y = (x + 1) \ln^2 (x + 1)$, then

$$\frac{dy}{dx} = (x + 1)\frac{2 \ln (x + 1)}{x + 1} + \ln^2 (x + 1) = 2 \ln (x + 1) + \ln^2 (x + 1).$$

13. If $f(\theta) = \tan^{-1} \frac{1 - \theta}{1 + \theta}$, then

$$f'(\theta) = \frac{\dfrac{(1 + \theta)(-1) - (1 - \theta)}{(1 + \theta)^2}}{1 + \left(\dfrac{1 - \theta}{1 + \theta}\right)^2}$$

$$= \frac{-1 - \theta - 1 + \theta}{1 + 2\theta + \theta^2 + 1 - 2\theta + \theta^2} = -\frac{1}{1 + \theta^2}.$$

14. If $y = \sin^{-1} x + x \sqrt{1 - x^2}$, then

$$y' = \frac{1}{\sqrt{1 - x^2}} + \frac{x(-2x)}{2\sqrt{1 - x^2}} + \sqrt{1 - x^2}$$

$$= \frac{1 - x^2 + 1 - x^2}{\sqrt{1 - x^2}} = 2\sqrt{1 - x^2}.$$

15. If $u = \ln \sqrt{v^2 + 2v - 1}$, then $u = \frac{1}{2} \ln (v^2 + 2v - 1)$ and

$$\frac{du}{dv} = \frac{1}{2} \frac{2v + 2}{v^2 + 2v - 1} = \frac{v + 1}{v^2 + 2v - 1}.$$

16. If $s = e^{-t}(\sin t - \cos t)$, then

$$s' = e^{-t}(\cos t + \sin t) + (\sin t - \cos t)(-e^{-t})$$

$$= e^{-t}(2 \cos t) = 2e^{-t} \cos t.$$

17. Let $y = 2u^3 - 4u^2 + 5u - 3$ and $u = x^2 - x$. Then

$$\frac{dy}{dx} = (6u^2 - 8u + 5)(2x - 1) = [6(x^2 - x)^2 - 8(x^2 - x) + 5](2x - 1).$$

This is an application of equation (2), the chain rule.

18. If $y = \sin(ax + b)$, then

$$\frac{dy}{dx} [\cos(ax + b)] \cdot a = a \cos(ax + b).$$

19. If $f(x) = ae^{kx}$ (with a and k constants), then

$$f'(x) = kae^{kx} \quad \text{and} \quad f'' = k^2 ae^{kx}.$$

20. Also, if $y = \ln(kx)$, where k is a constant, we can use both formula (13), page 33, and the chain rule to get

$$\frac{dy}{dx} = \frac{1}{kx} \cdot k = \frac{1}{x}.$$

Alternatively, we can rewrite the given function using a property of logarithms: $\ln(kx) = \ln k + \ln x$. So

$$\frac{dy}{dx} = 0 + \frac{1}{x} = \frac{1}{x},$$

as before.

21. If $f(u) = u^2 - u$ and $u = g(x) = x^3 - 5$, we can let $F(x) = f(g(x))$ and evaluate $F'(2)$ using equation (3) above. $F'(2) = f'(g(2))g'(2)$. Since $g(x) = x^3 - 5$, $g(2) = 3$. Since $f'(u) = 2u - 1$, $f'(g(2)) = f'(3) = 5$. Since $g'(x) = 3x^2$ and $g'(2) = 12$, we have

$$F'(2) = 5 \cdot 12 = 60.$$

---

*D. Derivatives of Parametrically Defined Functions

If $x = f(t)$ and $y = g(t)$ are differentiable functions of t, then

$$\frac{dy}{dx} = \frac{\dfrac{dy}{dt}}{\dfrac{dx}{dt}} \quad \text{and} \quad \frac{d^2y}{dx^2} = \frac{d}{dx}\left(\frac{dy}{dx}\right) = \frac{\dfrac{d}{dt}\left(\dfrac{dy}{dx}\right)}{\dfrac{dx}{dt}}$$

*An asterisk denotes a topic covered only in Calculus BC.

Example 22.   If $x = 2 \sin \theta$ and $y = \cos 2\theta$, then

$$\frac{dx}{d\theta} = 2 \cos \theta, \qquad \frac{dy}{d\theta} = -2 \sin 2\theta,$$

$$\frac{dy}{dx} = \frac{\dfrac{dy}{d\theta}}{\dfrac{dx}{d\theta}} = \frac{-2 \sin 2\theta}{2 \cos \theta} = -\frac{2 \sin \theta \cos \theta}{\cos \theta} = -2 \sin \theta.$$

Also,

$$\frac{d^2y}{dx^2} = \frac{\dfrac{d}{d\theta}\left(\dfrac{dy}{dx}\right)}{\dfrac{dx}{d\theta}} = \frac{-2 \cos \theta}{2 \cos \theta} = -1.$$

E.   Implicit Differentiation

This technique is used frequently when y is not defined explicitly in terms of x but is differentiable. In the examples below we differentiate both sides with respect to x, using appropriate formulas, and then solve for $\dfrac{dy}{dx}$.

Examples

23. If $x^2 - 2xy + 3y^2 = 2$, then

$$2x - 2\left(x\frac{dy}{dx} + y \cdot 1\right) + 6y\frac{dy}{dx} = 0 \qquad \text{and} \qquad \frac{dy}{dx}(6y - 2x) = 2y - 2x.$$

So

$$\frac{dy}{dx} = \frac{y - x}{3y - x}.$$

24. If $x \sin y = \cos (x + y)$, then

$$x \cos y \frac{dy}{dx} + \sin y = -\sin (x + y)\left(1 + \frac{dy}{dx}\right),$$

so that

$$\frac{dy}{dx} = -\frac{\sin y + \sin (x + y)}{x \cos y + \sin (x + y)}.$$

25. Find $\dfrac{dy}{dx}$ and $\dfrac{d^2y}{dx^2}$ using implicit differentiation on the equation $x^2 + y^2 = 1$.

$$2x + 2y\frac{dy}{dx} = 0 \qquad \rightarrow \qquad \frac{dy}{dx} = -\frac{x}{y}. \tag{1}$$

Then

$$\frac{d^2y}{dx^2} = -\frac{y \cdot 1 - x\left(\dfrac{dy}{dx}\right)}{y^2} = -\frac{y - x\left(-\dfrac{x}{y}\right)}{y^2} \tag{2}$$

$$= -\frac{y^2 + x^2}{y^3} = -\frac{1}{y^3}, \tag{3}$$

where we substituted for $\dfrac{dy}{dx}$ from (1) in (2), then used the given equation to simplify in (3).

26. To verify the formula for the derivative of the inverse sine function, $y = \sin^{-1} x = \text{Arcsin } x$, with domain $[-1, 1]$ and range $\left[-\dfrac{\pi}{2}, \dfrac{\pi}{2}\right]$, we proceed as follows:

$$y = \sin^{-1} x \qquad \leftrightarrow \qquad x = \sin y.$$

Now we differentiate with respect to x:

$$1 = \cos y \frac{dy}{dx},$$

$$\frac{dy}{dx} = \frac{1}{\cos y} = \frac{1}{+\sqrt{1 - \sin^2 y}} = \frac{1}{\sqrt{1 - x^2}},$$

where we chose the positive sign for $\cos y$ since $\cos y$ is nonnegative if $-\dfrac{\pi}{2} < y < \dfrac{\pi}{2}$. Note that this derivative exists only if $-1 < x < 1$.

Generalization of Formula (3), page 33.

If $y = u^{p/q}$ with u a differentiable function of x and p/q is a rational number, then implicit differentiation can be used to show that

$$\frac{dy}{dx} = \frac{p}{q}\left(u^{(p/q)-1}\right)\frac{du}{dx}.$$

This generalizes formula (3) (§B) so that, whereas as originally given n had to be an integer, now it can be any rational number.

F. Derivative of the Inverse of a Function _____

If $y = f(x)$ and $x = f^{-1}(y)$ are differentiable mutually inverse functions, then their derivatives are reciprocals:

$$\frac{dx}{dy} = \frac{1}{\dfrac{dy}{dx}}.$$

Examples

27. Let $y = f(x) = 2x^3 + 5x$. If $x = f^{-1}(y)$ is the inverse of f, then, since $\dfrac{dy}{dx} = 6x^2 + 5$, we have

$$\frac{dx}{dy} = \frac{1}{6x^2 + 5}$$

28. Let $y = f(x) = x^3 + x - 2$, and let g be the inverse function. Evaluate $g'(0)$.

Since $f'(x) = 3x^2 + 1$, $g'(y) = \dfrac{1}{3x^2 + 1}$. To find x when $y = 0$, we must solve the equation $x^3 + x - 2 = 0$. It is clear that $x = 1$. So

$$g'(0) = \frac{1}{3(1)^2 + 1} = \frac{1}{4}.$$

G. Logarithmic Differentiation _____

Many cumbersome exercises can be simplified by using the technique of logarithmic differentiation illustrated below.

Examples

29. To find $\dfrac{dy}{dx}$ when $y = \dfrac{\sqrt{x^2 + 1}}{(x^3 + 2)^{1/3}}$ we take the ln (natural logarithm) of both sides and use properties of the ln function:

$$\ln y = \frac{1}{2} \ln (x^2 + 1) - \frac{1}{3} \ln (x^3 + 2).$$

Now we differentiate implicitly with respect to x:

$$\frac{1}{y} \frac{dy}{dx} = \frac{1}{2} \frac{2x}{(x^2 + 1)} - \frac{1}{3} \frac{3x^2}{(x^3 + 2)},$$

$$\frac{dy}{dx} = y\left[\frac{x}{x^2 + 1} - \frac{x^2}{x^3 + 2}\right] = y \frac{x^4 + 2x - x^4 - x^2}{(x^2 + 1)(x^3 + 2)}.$$

Replacing y, we get

$$\frac{dy}{dx} = \frac{\sqrt{x^2 + 1}}{(x^3 + 2)^{1/3}} \frac{2x - x^2}{(x^2 + 1)(x^3 + 2)} = \frac{(2x - x^2)}{(x^3 + 2)^{4/3}(x^2 + 1)^{1/2}}.$$

30. To find $\frac{dy}{dx}$ if $y = x^x$ $(x > 0)$, we proceed similarly:

$$\ln y = x \ln x,$$

$$\frac{1}{y}\frac{dy}{dx} = x \cdot \frac{1}{x} + (1) \cdot \ln x = 1 + \ln x,$$

$$\frac{dy}{dx} = y(1 + \ln x) = x^x(1 + \ln x).$$

H. The Mean Value Theorem; Rolle's Theorem

If the function $f(x)$ is continuous at each point on the closed interval $a \leq x \leq b$ and has a derivative at each point on the open interval $a < x < b$, then there is a number c, $a < c < b$, such that $\frac{f(b) - f(a)}{b - a} = f'(c)$. This important theorem is illustrated in Figure N3–1. For the function sketched in the figure there are two numbers, c_1 and c_2, between a and b where the slope of the curve equals the slope of the chord PQ (i.e., where the tangent to the curve is parallel to the chord).

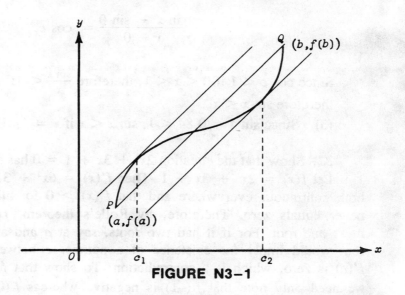

FIGURE N3–1

A special case of this theorem is embodied in *Rolle's theorem*. If in addition to the hypotheses of the Mean Value Theorem it is given that $f(a) = f(b) = 0$, then there is a number c between a and b such that $f'(c) = 0$. See Figure N3–2.

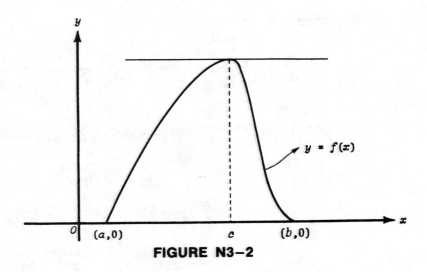

FIGURE N3–2

The Mean Value Theorem is one of the most useful laws when properly applied.

Examples

31. If $x > 0$, prove that $\sin x < x$.

Note that $\sin x \leqq 1$ for all x and consider the following cases if $x > 0$:

(1) $x > 1$. Here $\sin x < x$, since $\sin x \leqq 1$ and $1 < x$.

(2) $0 < x < 1$. Since $\sin x$ is continuous on $[0, 1]$ and since its derivative $\cos x$ exists at all x and therefore on the open interval $(0, 1)$, the mean-value theorem can be applied to the function, with $a = 0$ and $b = x$. Thus there is a number c, $0 < c < x < 1$, such that

$$\frac{\sin x - \sin 0}{x - 0} = \cos c.$$

Since $\cos c < 1$ if $0 < x < 1$, therefore $\frac{\sin x}{x} < 1$; and since $x > 0$, therefore $\sin x < x$.

(3) Since $\sin 1 \approx 0.84 < 1$, $\sin x < x$ if $x = 1$.

32. Show that the equation $2x^3 + 3x + 1 = 0$ has exactly one root.

Let $f(x) = 2x^3 + 3x + 1$. Then $f'(x) = 6x^2 + 3$. Note that f and f' are both continuous everywhere and that $f'(x) > 0$ for all x; in particular, $f'(x)$ never equals zero. Therefore, by Rolle's theorem, $f(x) = 0$ can have at most one root. For if it had two roots, say at a and at b, then Rolle's theorem would imply the existence of a number c between a and b such that $f'(c)$ is zero, which is a contradiction. To show that $f(x) = 0$ has one root, we need only note that $f(-1)$ is negative whereas $f(0)$ is positive. We use the Intermediate Value Theorem (page 26) to conclude that f is zero somewhere between $x = -1$ and $x = 0$. The equation therefore has exactly one root.

I. Differentiability and Continuity_____

It is important to remember that (1) if a function has a finite derivative at a point then it is also continuous at that point, but that (2) a function may be continuous at a point without having a derivative at that point.

Example 33. $f(x) = |x - 2|$ is everywhere (i.e., for all real x) continuous, but $f'(x)$ does *not* exist at $x = 2$. Note that $f'(x) = 1$ if $x > 2$, but $f'(x) = -1$ if $x < 2$.

> Differentiability implies continuity,
> but
> continuity does not imply differentiability.

J. L'Hôpital's Rule_____

To find the limit of an indeterminate form of the type $\frac{0}{0}$ or $\frac{\infty}{\infty}$, we apply L'Hôpital's rule, which involves taking derivatives. In the following, a is a finite number. The rule says:

(a) If $\lim_{x \to a} f(x) = \lim_{x \to a} g(x) = 0$ and if $\lim_{x \to a} \frac{f'(x)}{g'(x)}$ exists, then

$$\lim_{x \to a} \frac{f(x)}{g(x)} = \lim_{x \to a} \frac{f'(x)}{g'(x)};$$

if $\lim_{x \to a} \frac{f'(x)}{g'(x)}$ does not exist, then neither does $\lim_{x \to a} \frac{f(x)}{g(x)}$.

(b) If $\lim_{x \to a} f(x) = \lim_{x \to a} g(x) = \infty$, the same consequences follow as in case (a).

(c) If $\lim_{x \to \infty} f(x) = \lim_{x \to \infty} g(x) = 0$ and if $\lim_{x \to \infty} \frac{f'(x)}{g'(x)}$ exists, then

$$\lim_{x \to \infty} \frac{f(x)}{g(x)} = \lim_{x \to \infty} \frac{f'(x)}{g'(x)};$$

if $\lim_{x \to \infty} \frac{f'(x)}{g'(x)}$ does not exist, then neither does $\lim_{x \to \infty} \frac{f(x)}{g(x)}$. (Here the notation "$x \to \infty$" represents either "$x \to +\infty$" or "$x \to -\infty$.")

(d) If $\lim_{x \to \infty} f(x) = \lim_{x \to \infty} g(x) = \infty$, the same consequences follow as in case (c).

Examples

34. $\lim_{x \to 3} \frac{x^2 - 9}{x - 3}$ (Example 4, page 22) is of type $\frac{0}{0}$ and thus equals

$\lim_{x \to 3} \frac{2x}{1} = 6$, as before.

35. $\lim\limits_{x \to -2} \dfrac{x^3 + 8}{x^2 - 4}$ (Example 5, page 23) is of type $\dfrac{0}{0}$ and thus equals $\lim\limits_{x \to -2} \dfrac{3x^2}{2x} = -3$, as before.

36. $\lim\limits_{h \to 0} \dfrac{e^h - 1}{h}$ is of type $\dfrac{0}{0}$ and therefore equals $\lim\limits_{h \to 0} \dfrac{e^h}{1} = 1$.

37. $\lim\limits_{x \to \infty} \dfrac{x^3 - 4x^2 + 7}{3 - 6x - 2x^3}$ (Example 12, page 23) is of type $\dfrac{\infty}{\infty}$, so that it equals $\lim\limits_{x \to \infty} \dfrac{3x^2 - 8x}{-6 - 6x^2}$, which is again of type $\dfrac{\infty}{\infty}$. Apply L'Hôpital's rule twice more to get

$$\lim\limits_{x \to \infty} \frac{6x - 8}{-12x} = \lim\limits_{x \to \infty} \frac{6}{-12} = -\frac{1}{2},$$

as before.

38. $\lim\limits_{x \to \infty} \dfrac{\ln x}{x}$ is of type $\dfrac{\infty}{\infty}$ and equals $\lim\limits_{x \to \infty} \dfrac{1/x}{1} = 0$.

L'Hôpital's rule can also be applied to indeterminate forms of the types $0 \cdot \infty$ and $\infty - \infty$, if they can be transformed to either $\dfrac{0}{0}$ or $\dfrac{\infty}{\infty}$.

Example 39. $\lim\limits_{x \to \infty} x \sin \dfrac{1}{x}$ is of type $\infty \cdot 0$. Since $x \sin \dfrac{1}{x} = \dfrac{\sin 1/x}{1/x}$ and, as $x \to \infty$, the latter is $\dfrac{0}{0}$, we see that

$$\lim\limits_{x \to \infty} x \sin \frac{1}{x} = \lim\limits_{x \to \infty} \frac{-\dfrac{1}{x^2} \cos \dfrac{1}{x}}{-\dfrac{1}{x^2}} = \lim\limits_{x \to \infty} \cos \frac{1}{x} = 1.$$

---

*Other indeterminate forms, such as 0^0, 1^∞, and ∞^0, are handled by taking the natural logarithm and then applying L'Hôpital's rule.

Examples

*40. $\lim\limits_{x \to 0} (1 + x)^{1/x}$ is of type 1^∞. Let $y = (1 + x)^{1/x}$, so that $\ln y = \dfrac{1}{x} \ln (1 + x)$ and $\lim\limits_{x \to 0} \ln y$ is of type $\dfrac{0}{0}$. Thus

$$\lim\limits_{x \to 0} \ln y = \lim\limits_{x \to 0} \frac{\dfrac{1}{1 + x}}{1} = \frac{1}{1} = 1,$$

and since $\lim\limits_{x \to 0} \ln y = 1$, $\lim\limits_{x \to 0} y = e^1 = e$.

*An asterisk denotes a topic covered only in Calculus BC.

*41. $\lim\limits_{x \to \infty} x^{1/x}$ is of type ∞^0. Let $y = x^{1/x}$, so that $\ln y = \dfrac{1}{x} \ln x$ (which, as $x \to \infty$, is of type $\dfrac{\infty}{\infty}$). Then $\lim\limits_{x \to \infty} \ln y = \dfrac{1/x}{1} = 0$, and $\lim\limits_{x \to \infty} y = e^0 = 1$.

All of the multiple-choice questions in Set 2 on limits, beginning on page 26, can be done without recourse to L'Hôpital's rule. However, many of them can be handled more directly by applying it. Redo Set 2, applying L'Hôpital's rule wherever possible.

K.   Recognizing a Given Limit as a Derivative

It is often extremely useful to evaluate a limit by recognizing that it is merely an expression for the definition of the derivative of a specific function (often at a specific point). The relevant definition is

$$f'(c) = \lim_{h \to 0} \frac{f(c + h) - f(c)}{h}.$$

Examples

42. $\lim\limits_{h \to 0} \dfrac{(2 + h)^4 - 2^4}{h}$ is the derivative of $f(x) = x^4$ at the point where $x = 2$. Since $f'(x) = 4x^3$, the value of the given limit is $f'(2) = 4(2^3) = 32$.

43. $\lim\limits_{h \to 0} \dfrac{\sqrt{9 + h} - 3}{h} = f'(9)$, where $f(x) = \sqrt{x}$. The value of the limit is $\dfrac{1}{2} x^{-1/2}$ when $x = 9$, or $\dfrac{1}{6}$.

44. $\lim\limits_{h \to 0} \dfrac{1}{h}\left(\dfrac{1}{2 + h} - \dfrac{1}{2}\right) = f'(2)$, where $f(x) = \dfrac{1}{x}$. Verify that $f'(2) = -\dfrac{1}{4}$ and compare with Example 9, page 23.

45. $\lim\limits_{h \to 0} \dfrac{e^h - 1}{h} = f'(0)$, where $f(x) = e^x$. The limit has value e^0 or 1 (see also Example 36, on page 44).

46. $\lim\limits_{x \to 0} \dfrac{\sin x}{x}$ is $f'(0)$, where $f(v) = \sin v$. Here h in the definition of $f'(c)$ given above in §K has been replaced by x, and c is equal to zero. Note that, if $f(v) = \sin v$, we can write

$$f'(0) = \lim_{x \to 0} \frac{\sin (0 + x) - \sin 0}{x} = \lim_{x \to 0} \frac{\sin x}{x}.$$

The answer is 1, since $f'(v) = \cos v$ and $f'(0) = \cos 0 = 1$. Of course, we already know that the given limit is the basic trigonometric limit with value 1. Also, L'Hôpital's rule yields 1 as the answer immediately.

Set 3: Multiple-Choice Questions on Differentiation

In each of Questions 1–27 a function is given. Choose the alternative that is the derivative, $\dfrac{dy}{dx}$, of the function.

1. $y = (4x + 1)^2(1 - x)^3$

(A) $(4x + 1)^2(1 - x)^2(5 - 20x)$ (B) $(4x + 1)(1 - x)^2(4x + 11)$
(C) $5(4x + 1)(1 - x)^2(1 - 4x)$ (D) $(4x + 1)(1 - x)^2(11 - 20x)$
(E) $-24(4x + 1)(1 - x)^2$

2. $y = \dfrac{2 - x}{3x + 1}$

(A) $-\dfrac{7}{(3x + 1)^2}$ (B) $\dfrac{6x - 5}{(3x + 1)^2}$ (C) $-\dfrac{9}{(3x + 1)^2}$

(D) $\dfrac{7}{(3x + 1)^2}$ (E) $\dfrac{7 - 6x}{(3x + 1)^2}$

3. $y = \sqrt{3 - 2x}$

(A) $\dfrac{1}{2\sqrt{3 - 2x}}$ (B) $-\dfrac{1}{\sqrt{3 - 2x}}$ (C) $-\dfrac{(3 - 2x)^{3/2}}{3}$

(D) $-\dfrac{1}{3 - 2x}$ (E) $\dfrac{2}{3}(3 - 2x)^{3/2}$

4. $y = \dfrac{2}{(5x + 1)^3}$

(A) $-\dfrac{30}{(5x + 1)^2}$ (B) $-30(5x + 1)^{-4}$ (C) $\dfrac{-6}{(5x + 1)^4}$

(D) $-\dfrac{10}{3}(5x + 1)^{-4/3}$ (E) $\dfrac{30}{(5x + 1)^4}$

5. $y = 3x^{2/3} - 4x^{1/2} - 2$

(A) $2x^{1/3} - 2x^{-1/2}$ (B) $3x^{-1/3} - 2x^{-1/2}$ (C) $\dfrac{9}{5}x^{5/3} - 8x^{3/2}$

(D) $\dfrac{2}{x^{1/3}} - \dfrac{2}{x^{1/2}} - 2$ (E) $2x^{-1/3} - 2x^{-1/2}$

6. $y = 2\sqrt{x} - \dfrac{1}{2\sqrt{x}}$

   (A) $x + \dfrac{1}{x\sqrt{x}}$ (B) $x^{-1/2} + x^{-3/2}$ (C) $\dfrac{4x - 1}{4x\sqrt{x}}$

   (D) $\dfrac{1}{\sqrt{x}} + \dfrac{1}{4x\sqrt{x}}$ (E) $\dfrac{4}{\sqrt{x}} + \dfrac{1}{x\sqrt{x}}$

7. $y = \sqrt{x^2 + 2x - 1}$

   (A) $\dfrac{x + 1}{y}$ (B) $4y(x + 1)$ (C) $\dfrac{1}{2\sqrt{x^2 + 2x - 1}}$

   (D) $-\dfrac{x + 1}{(x^2 + 2x - 1)^{3/2}}$ (E) none of these

8. $y = \dfrac{x}{\sqrt{1 - x^2}}$

   (A) $\dfrac{1 - 2x^2}{(1 - x^2)^{3/2}}$ (B) $\dfrac{1}{1 - x^2}$ (C) $\dfrac{1}{\sqrt{1 - x^2}}$

   (D) $\dfrac{1 - 2x^2}{(1 - x^2)^{1/2}}$ (E) none of these

9. $y = \cos x^2$

   (A) $2x \sin x^2$ (B) $-\sin x^2$ (C) $-2 \sin x \cos x$
   (D) $-2x \sin x^2$ (E) $\sin 2x$

10. $y = \sin^2 3x + \cos^2 3x$

   (A) $-6 \sin 6x$ (B) 0 (C) $12 \sin 3x \cos 3x$
   (D) $6(\sin 3x + \cos 3x)$ (E) 1

11. $y = \ln \dfrac{e^x}{e^x - 1}$

   (A) $x - \dfrac{e^x}{e^x - 1}$ (B) $\dfrac{1}{e^x - 1}$ (C) $\dfrac{1}{1 - e^x}$ (D) 0

   (E) $\dfrac{e^x - 2}{e^x - 1}$

12. $y = \tan^{-1} \dfrac{x}{2}$

   (A) $\dfrac{4}{4 + x^2}$ (B) $\dfrac{1}{2\sqrt{4 - x^2}}$ (C) $\dfrac{2}{\sqrt{4 - x^2}}$ (D) $\dfrac{1}{2 + x^2}$

   (E) $\dfrac{2}{x^2 + 4}$

13. $y = \ln (\sec x + \tan x)$

(A) $\sec x$ (B) $\dfrac{1}{\sec x}$ (C) $\tan x + \dfrac{\sec^2 x}{\tan x}$ (D) $\dfrac{1}{\sec x + \tan x}$

(E) $-\dfrac{1}{\sec x + \tan x}$

14. $y = \cos^2 x$

(A) $-\sin^2 x$ (B) $2 \sin x \cos x$ (C) $-\sin 2x$ (D) $2 \cos x$
(E) $-2 \sin x$

15. $y = \dfrac{e^x - e^{-x}}{e^x + e^{-x}}$

(A) 0 (B) 1 (C) $\dfrac{2}{(e^x + e^{-x})^2}$ (D) $\dfrac{4}{(e^x + e^{-x})^2}$

(E) $\dfrac{1}{e^{2x} + e^{-2x}}$

16. $y = \ln (x\sqrt{x^2 + 1})$

(A) $1 + \dfrac{x}{x^2 + 1}$ (B) $\dfrac{1}{x\sqrt{x^2 + 1}}$ (C) $\dfrac{2x^2 + 1}{x\sqrt{x^2 + 1}}$ (D) $\dfrac{2x^2 + 1}{x(x^2 + 1)}$
(E)   none of these

17. $y = \ln (x + \sqrt{x^2 + 1})$

(A) $\dfrac{1}{x} + \dfrac{x}{x^2 + 1}$ (B) $\dfrac{1}{\sqrt{x^2 + 1}}$ (C) 1 (D) $\sqrt{x^2 + 1}$

(E) $\dfrac{1}{x} + \dfrac{1}{2\sqrt{x^2 + 1}}$

18. $y = x^2 \sin \dfrac{1}{x}$ $(x \neq 0)$

(A) $2x \sin \dfrac{1}{x} - x^2 \cos \dfrac{1}{x}$ (B) $-\dfrac{2}{x} \cos \dfrac{1}{x}$ (C) $2x \cos \dfrac{1}{x}$

(D) $2x \sin \dfrac{1}{x} - \cos \dfrac{1}{x}$ (E) $-\cos \dfrac{1}{x}$

19. $y = \dfrac{1}{2 \sin 2x}$

(A) $-\csc 2x \cot 2x$ (B) $\dfrac{1}{4 \cos 2x}$ (C) $-4 \csc 2x \cot 2x$

(D) $\dfrac{\cos 2x}{2\sqrt{\sin 2x}}$ (E) $-\csc^2 2x$

20. $y = x^{\ln x}$ $(x > 0)$

(A) $\dfrac{2}{x}$ (B) $2\dfrac{\ln x}{x}$ (C) $\dfrac{2(\ln x)y}{x}$ (D) $\dfrac{2y}{x}$ (E) $(\ln x)x^{\ln x - 1}$

21. $y = x\tan^{-1} x - \ln\sqrt{x^2 + 1}$

(A) 0 (B) $\dfrac{1}{\sqrt{1 - x^2}} - \dfrac{x}{x^2 + 1}$ (C) $\tan^{-1} x$

(D) $\dfrac{x}{1 + x^2} + \tan^{-1} x - x$ (E) $\dfrac{1 - x}{1 + x^2}$

22. $y = e^{-x}\cos 2x$

(A) $-e^{-x}(\cos 2x + 2\sin 2x)$ (B) $e^{-x}(\sin 2x - \cos 2x)$
(C) $2e^{-x}\sin 2x$ (D) $-e^{-x}(\cos 2x + \sin 2x)$ (E) $-e^{-x}\sin 2x$

23. $y = \sec^2\sqrt{x}$

(A) $\dfrac{\sec\sqrt{x}\tan\sqrt{x}}{\sqrt{x}}$ (B) $\dfrac{\tan\sqrt{x}}{\sqrt{x}}$ (C) $2\sec\sqrt{x}\tan^2\sqrt{x}$

(D) $\dfrac{\sec^2\sqrt{x}\tan\sqrt{x}}{\sqrt{x}}$ (E) $2\sec^2\sqrt{x}\tan\sqrt{x}$

24. $y = x\ln^3 x$

(A) $\dfrac{3\ln^2 x}{x}$ (B) $3\ln^2 x$ (C) $3x\ln^2 x + \ln^3 x$ (D) $3(\ln x + 1)$
(E) none of these

25. $y = \dfrac{1 + x^2}{1 - x^2}$

(A) $-\dfrac{4x}{(1 - x^2)^2}$ (B) $\dfrac{4x}{(1 - x^2)^2}$ (C) $\dfrac{-4x^3}{(1 - x^2)^2}$ (D) $\dfrac{2x}{1 - x^2}$

(E) $\dfrac{4}{1 - x^2}$

26. $y = \ln\left(\sqrt{2}\,x\right)$

(A) $\dfrac{\sqrt{2}}{x}$ (B) $\dfrac{1}{\sqrt{2}\,x}$ (C) $\dfrac{1}{2x}$ (D) $\dfrac{1}{x}$ (E) $\dfrac{1}{\sqrt{x}}$

27. $y = \sin^{-1} x - \sqrt{1 - x^2}$

(A) $\dfrac{1}{2\sqrt{1 - x^2}}$ (B) $\dfrac{2}{\sqrt{1 - x^2}}$ (C) $\dfrac{1 + x}{\sqrt{1 - x^2}}$ (D) $\dfrac{x^2}{\sqrt{1 - x^2}}$

(E) $\dfrac{1}{\sqrt{1 + x}}$

In each of Questions 28–31 a pair of equations is given which represents a curve parametrically. Choose the alternative that is the derivative $\dfrac{dy}{dx}$.

*28. $x = t - \sin t$ and $y = 1 - \cos t$

(A) $\dfrac{\sin t}{1 - \cos t}$ (B) $\dfrac{1 - \cos t}{\sin t}$ (C) $\dfrac{\sin t}{\cos t - 1}$ (D) $\dfrac{1 - x}{y}$

(E) $\dfrac{1 - \cos t}{t - \sin t}$

*29. $x = \cos^3 \theta$ and $y = \sin^3 \theta$

(A) $\tan^3 \theta$ (B) $-\cot \theta$ (C) $\cot \theta$ (D) $-\tan \theta$
(E) $-\tan^2 \theta$

*30. $x = 1 - e^{-t}$ and $y = t + e^{-t}$

(A) $\dfrac{e^{-t}}{1 - e^{-t}}$ (B) $e^{-t} - 1$ (C) $e^t + 1$ (D) $e^t - e^{-2t}$
(E) $e^t - 1$

*31. $x = \dfrac{1}{1 - t}$ and $y = 1 - \ln(1 - t)$ $(t < 1)$

(A) $\dfrac{1}{1 - t}$ (B) $t - 1$ (C) $\dfrac{1}{x}$ (D) $\dfrac{(1 - t)^2}{t}$ (E) $1 + \ln x$

In each of Questions 32–35, y is a differentiable function of x. Choose the alternative that is the derivative $\dfrac{dy}{dx}$.

32. $x^3 - xy + y^3 = 1$

(A) $\dfrac{3x^2}{x - 3y^2}$ (B) $\dfrac{3x^2 - 1}{1 - 3y^2}$ (C) $\dfrac{y - 3x^2}{3y^2 - x}$

(D) $\dfrac{3x^2 + 3y^2 - y}{x}$ (E) $\dfrac{3x^2 + 3y^2}{x}$

33. $x + \cos(x + y) = 0$

(A) $\csc(x + y) - 1$ (B) $\csc(x + y)$ (C) $\dfrac{x}{\sin(x + y)}$

(D) $\dfrac{1}{\sqrt{1 - x^2}}$ (E) $\dfrac{1 - \sin x}{\sin y}$

---

*Questions preceded by an asterisk are likely to appear only on the Calculus BC Examination.

34. $\sin x - \cos y - 2 = 0$

(A) $-\cot x$ (B) $-\cot y$ (C) $\dfrac{\cos x}{\sin y}$ (D) $-\csc y \cos x$

(E) $\dfrac{2 - \cos x}{\sin y}$

35. $3x^2 - 2xy + 5y^2 = 1$

(A) $\dfrac{3x + y}{x - 5y}$ (B) $\dfrac{y - 3x}{5y - x}$ (C) $3x + 5y$ (D) $\dfrac{3x + 4y}{x}$

(E) none of these

Individual instructions are given in full for each of the remaining questions of this set (Questions 36–65).

*36. If $x = t^2 - 1$ and $y = t^4 - 2t^3$, then when $t = 1$, $\dfrac{d^2y}{dx^2}$ is

(A) 1   (B) -1 (C) 0   (D) 3   (E) $\dfrac{1}{2}$

37. If $f(x) = x^4 - 4x^3 + 4x^2 - 1$, then the set of values of x for which the derivative equals zero is

(A) $\{1, 2\}$ (B) $\{0, -1, -2\}$ (C) $\{-1, +2\}$ (D) $\{0\}$
(E) $\{0, 1, 2\}$

38. If $f(x) = 16\sqrt{x}$, then $f'''(4)$ is equal to

(A) $\dfrac{3}{16}$ (B) -4 (C) $-\dfrac{1}{2}$ (D) 0   (E) 6

39. If $f(x) = \ln x$, then $f^{iv}(x)$ is

(A) $\dfrac{2}{x^3}$ (B) $\dfrac{24}{x^5}$ (C) $\dfrac{6}{x^4}$ (D) $-\dfrac{1}{x^4}$ (E) none of these

40. If a point moves on the curve $x^2 + y^2 = 25$, then, at $(0, 5)$, $\dfrac{d^2y}{dx^2}$ is

(A) 0   (B) $\dfrac{1}{5}$ (C) -5 (D) $-\dfrac{1}{5}$ (E) nonexistent

41. If $y = a \sin ct + b \cos ct$, where a, b, and c are constants, then $\dfrac{d^2y}{dt^2}$ is

(A) $ac^2(\sin t + \cos t)$ (B) $-c^2y$ (C) $-ay$ (D) $-y$
(E) $a^2c^2 \sin ct - b^2c^2 \cos ct$

42. If $f(x) = x^4 - 4x^2$, then $f^{iv}(2)$ equals

(A) 48   (B) 0   (C) 24   (D) 144   (E) 16

43. If $f(x) = \frac{x}{(x-1)^2}$, then the set of x's for which $f'(x)$ exists is

(A) all reals   (B) all reals except $x = 1$ and $x = -1$

(C) all reals except $x = -1$ (D) all reals except $x = \frac{1}{3}$ and $x = -1$

(E) all reals except $x = 1$

44. If $y = (x-1)^2 e^x$, then $\frac{d^2y}{dx^2}$ is equal to

(A) $e^x(x-1)^2$ (B) $e^x(x^2 - 2x - 1)$ (C) $e^x(x^2 + 2x - 1)$
(D) $2e^x(x-1)$ (E) none of these

45. If $f(x) = e^{-x} \ln x$, then, when $x = 1$, $\frac{df}{dx}$ is

(A) 0   (B) nonexistent   (C) $\frac{2}{e}$ (D) $\frac{1}{e}$ (E) e

46. If $y = \sqrt{x^2 + 1}$, then the derivative of y^2 with respect to x^2 is

(A) 1   (B) $\frac{x^2 + 1}{2x}$ (C) $\frac{x}{2(x^2 + 1)}$ (D) $\frac{2}{x}$ (E) $\frac{x^2}{x^2 + 1}$

47. If $f(x) = \frac{1}{x^2 + 1}$ and $g(x) = \sqrt{x}$, then the derivative of $f(g(x))$ is

(A) $\frac{-\sqrt{x}}{(x^2 + 1)^2}$ (B) $-(x + 1)^{-2}$ (C) $\frac{-2x}{(x^2 + 1)^2}$ (D) $\frac{1}{(x + 1)^2}$

(E) $\frac{1}{2\sqrt{x}(x + 1)}$

*48. If $x = e^\theta \cos \theta$ and $y = e^\theta \sin \theta$, then, when $\theta = \frac{\pi}{2}$, $\frac{dy}{dx}$ is

(A) 1   (B) 0   (C) $e^{\pi/2}$ (D) nonexistent   (E) -1

*49. If $x = \cos t$ and $y = \cos 2t$, then $\frac{d^2y}{dx^2}$ is

(A) $4 \cos t$ (B) 4   (C) $\frac{4y}{x}$ (D) -4 (E) $-4 \cot t$

50. If $y = x^2 + x$, then the derivative of y with respect to $\frac{1}{1 - x}$ is

(A) $(2x + 1)(x - 1)^2$ (B) $\frac{2x + 1}{(1 - x)^2}$ (C) $2x + 1$ (D) $\frac{3 - x}{(1 - x)^3}$
(E) none of these

51. $\displaystyle\lim_{h \to 0} \frac{(1 + h)^6 - 1}{h}$ is

(A) 0   (B) 1   (C) 6   (D) ∞ (E) nonexistent

52. $\displaystyle\lim_{h \to 0} \frac{\sqrt[3]{8 + h} - 2}{h}$ is

(A) 0   (B) $\dfrac{1}{12}$ (C) 1   (D) 192   (E) ∞

53. $\displaystyle\lim_{h \to 0} \frac{\ln (e + h) - 1}{h}$ is

(A) 0   (B) $\dfrac{1}{e}$ (C) 1   (D) e (E) nonexistent

54. $\displaystyle\lim_{x \to 0} \frac{\cos x - 1}{x}$ is

(A) -1 (B) 0   (C) 1   (D) ∞ (E) none of these

55. The function $f(x) = x^{2/3}$ on $[-8, 8]$ does not satisfy the conditions of the mean-value theorem because

(A) $f(0)$ is not defined   (B) $f(x)$ is not continuous on $[-8, 8]$
(C) $f'(-1)$ does not exist   (D) $f(x)$ is not defined for $x < 0$
(E) $f'(0)$ does not exist

56. If $f(a) = f(b) = 0$ and $f(x)$ is continuous on $[a, b]$, then

(A) $f(x)$ must be identically zero
(B) $f'(x)$ may be different from zero for all x on $[a, b]$
(C) there exists at least one number c, $a < c < b$, such that $f'(c) = 0$
(D) $f'(x)$ must exist for every x on (a, b)
(E) none of the preceding is true

57. If c is the number defined by Rolle's theorem, then, for $f(x) = 2x^3 - 6x$ on the interval $0 \leqq x \leqq \sqrt{3}$, c is

(A) 1   (B) -1 (C) $\sqrt{2}$ (D) 0   (E) $\sqrt{3}$

58. If h is the inverse function of f and if $f(x) = \dfrac{1}{x}$, then $h'(3) =$

(A) -9 (B) $-\dfrac{1}{9}$ (C) $\dfrac{1}{9}$ (D) 3   (E) 9

59. Suppose $y = f(x) = 2x^3 - 3x$. If h is the inverse function of f, then $h'(y) =$

(A) $\dfrac{1}{6y^2 - 3}$ (B) $\dfrac{1}{6x^2} - \dfrac{1}{3}$ (C) $\dfrac{1}{6x^2 - 3}$ (D) $-\dfrac{6x^2 - 3}{(2x^2 - 3)^2}$
(E) none of these

60. Suppose $y = f(x)$ and $x = f^{-1}(y)$ are mutually inverse functions. If $f(1) = 4$ and $\dfrac{dy}{dx} = -3$ at $x = 1$, then $\dfrac{dx}{dy}$ at $y = 4$ equals

(A) $-\dfrac{1}{3}$ (B) $-\dfrac{1}{4}$ (C) $\dfrac{1}{3}$ (D) 3 (E) 4

61. Let $y = f(x)$ and $x = h(y)$ be mutually inverse functions. If $f'(2) = 5$, then what is the value of $\dfrac{dx}{dy}$ at $y = 2$?

(A) -5 (B) $-\dfrac{1}{5}$ (C) $\dfrac{1}{5}$ (D) 5

(E) It cannot be determined from the information given.

62. If $f(x) = x^{\sin x}$ for $x > 0$, then $f'(x) =$

(A) $(\sin x)x^{\sin x - 1}$ (B) $x^{\sin x}(\cos x)(\ln x)$ (C) $\dfrac{\sin x}{x} + (\cos x)(\ln x)$

(D) $x^{\sin x}\left[\dfrac{\sin x}{x} + (\cos x)(\ln x)\right]$ (E) $x \cos x + \sin x$

63. Suppose $\lim\limits_{x \to 0} \dfrac{g(x) - g(0)}{x} = 1$. It follows necessarily that

(A) g is not defined at $x = 0$
(B) g is not continuous at $x = 0$
(C) The limit of $g(x)$ as x approaches 0 equals 1
(D) $g'(0) = 1$
(E) $g'(1) = 0$

64. If $\sin(xy) = x$, then $\dfrac{dy}{dx} =$

(A) $\sec(xy)$ (B) $\dfrac{\sec(xy)}{x}$ (C) $\dfrac{\sec(xy) - y}{x}$

(D) $-\dfrac{1 + \sec(xy)}{x}$ (E) $\sec(xy) - 1$

*65. $\lim\limits_{x \to 0^+} x^x =$

(A) 0 (B) $\dfrac{1}{e}$ (C) 1 (D) e (E) none of those

*Questions preceded by an asterisk are likely to appear only on the Calculus BC Examination.

Answers for Set 3: Differentiation

1.	C	**14.**	C	**27.**	C	**40.**	D	**53.**	B
2.	A	**15.**	D	**28.**	A	**41.**	B	**54.**	B
3.	B	**16.**	D	**29.**	D	**42.**	C	**55.**	E
4.	B	**17.**	B	**30.**	E	**43.**	E	**56.**	B
5.	E	**18.**	D	**31.**	C	**44.**	C	**57.**	A
6.	D	**19.**	A	**32.**	C	**45.**	D	**58.**	B
7.	A	**20.**	C	**33.**	A	**46.**	A	**59.**	C
8.	E	**21.**	C	**34.**	D	**47.**	B	**60.**	A
9.	D	**22.**	A	**35.**	B	**48.**	E	**61.**	E
10.	B	**23.**	D	**36.**	E	**49.**	B	**62.**	D
11.	C	**24.**	E	**37.**	E	**50.**	A	**63.**	D
12.	E	**25.**	B	**38.**	A	**51.**	C	**64.**	C
13.	A	**26.**	D	**39.**	E	**52.**	B	**65.**	C

NOTE: the formulas or rules in parentheses referred to in the following explanations are given on pages 33 and 34.

1. C. By the product rule, (5).

$$y' = (4x + 1)^2[3(1 - x)^2(-1)] + (1 - x)^3[2(4x + 1)(4)]$$
$$= -3(4x + 1)^2 (1 - x)^2 + 8(1 - x)^3(4x + 1)$$
$$= (4x + 1)(1 - x)^2(-12x - 3 + 8 - 8x)$$
$$= (4x + 1)(1 - x)^2(5 - 20x).$$

2. A. By the quotient rule, (6).

$$y' = \frac{(3x + 1)(-1) - (2 - x)(3)}{(3x + 1)^2} = \frac{-7}{(3x + 1)^2}.$$

3. B. Since $y = (3 - 2x)^{1/2}$, by the power rule, (3).

$$y' = \frac{1}{2}(3 - 2x)^{-1/2} \cdot (-2) = -\frac{1}{\sqrt{3 - 2x}}.$$

4. B. Since $y = 2(5x + 1)^{-3}$, $y' = -6(5x + 1)^{-4}(5)$.

5. E. $y' = 3\left(\frac{2}{3}\right)x^{-1/3} - 4\left(\frac{1}{2}\right)x^{-1/2}.$

6. D. Rewrite: $y = 2x^{1/2} - \frac{1}{2}x^{-1/2}$; so $y' = x^{-1/2} + \frac{1}{4}x^{-3/2}.$

7. A. Rewrite: $y = (x^2 + 2x - 1)^{1/2}$. (Use rule (3).)

8. **E.** Use the quotient rule:

$$y = \frac{\sqrt{1-x^2} \cdot 1 - x \cdot \dfrac{1 \cdot -2x}{2\sqrt{1-x^2}}}{1-x^2} = \frac{\dfrac{1-x^2+x^2}{\sqrt{1-x^2}}}{1-x^2}$$

$$= \frac{1}{(1-x^2)^{3/2}}.$$

9. **D.** Use formula (8): $y' = (-\sin x^2)(2x)$.

10. **B.** Note that $y = 1$.

11. **C.** Since

$$y = \ln e^x - \ln (e^x - 1),$$

$$y = x - \ln (e^x - 1),$$

$$y' = 1 - \frac{e^x}{e^x - 1} = \frac{e^x - 1 - e^x}{e^x - 1} = -\frac{1}{e^x - 1}.$$

12. **E.** Use formula (18): $y' = \dfrac{\dfrac{1}{2}}{1 + \dfrac{x^2}{4}}$.

13. **A.** Use formulas (13), (11), and (9):

$$y' = \frac{\sec x \tan x + \sec^2 x}{\sec x + \tan x} = \frac{\sec x \,(\tan x + \sec x)}{\sec x + \tan x}.$$

14. **C.** $y' = 2 \cos x \cdot (-\sin x)$.

15. **D.** By the quotient rule,

$$y' = \frac{(e^x + e^{-x})(e^x + e^{-x}) - (e^x - e^{-x})(e^x - e^{-x})}{(e^x + e^{-x})^2}$$

$$= \frac{(e^{2x} + 2 + e^{-2x}) - (e^{2x} - 2 + e^{-2x})}{(e^x + e^{-x})^2} = \frac{4}{(e^x + e^{-x})^2}.$$

16. **D.** Since $y = \ln x + \dfrac{1}{2} \ln (x^2 + 1)$,

$$y' = \frac{1}{x} + \frac{1}{2} \cdot \frac{2x}{x^2 + 1} = \frac{2x^2 + 1}{x(x^2 + 1)}.$$

17. **B.** $y' = \dfrac{1 + \dfrac{2x}{2\sqrt{x^2+1}}}{x + \sqrt{x^2+1}} = \dfrac{\dfrac{\sqrt{x^2+1}+x}{\sqrt{x^2+1}}}{x + \sqrt{x^2+1}} = \dfrac{1}{\sqrt{x^2+1}}$

18. **D.** $y' = x^2 \cos \dfrac{1}{x}\left(-\dfrac{1}{x^2}\right) + \sin \dfrac{1}{x} (2x)$.

19. **A.** Since $y = \dfrac{1}{2} \csc 2x$, $y' = \dfrac{1}{2}(-\csc 2x \cot 2x \cdot 2)$.

20. C.   Since $y = x^{\ln x}$, $\ln y = \ln x \ln x = (\ln x)^2$.

$$\frac{1}{y}\frac{dy}{dx} = 2\ln x\left(\frac{1}{x}\right).$$

21. C.   Since $\ln \sqrt{x^2 + 1} = \frac{1}{2}\ln(x^2 + 1)$, then

$$y' = x \cdot \frac{1}{1 + x^2} + \tan^{-1} x - \frac{1}{2}\cdot\frac{2x}{x^2 + 1}.$$

22. A. $y' = e^{-x}(-2\sin 2x) + \cos 2x(-e^{-x})$.

23. D.   Use formulas (3) and (11):

$$y' = 2\sec\sqrt{x}\cdot\sec\sqrt{x}\tan\sqrt{x}\cdot\left(\frac{1}{2\sqrt{x}}\right).$$

24. E. $y' = \frac{x(3\ln^2 x)}{x} + \ln^3 x$. The correct answer is $3\ln^2 x + \ln^3 x$.

25. B. $y' = \frac{(1 - x^2)(2x) - (1 + x^2)(-2x)}{(1 - x^2)^2}$.

26. D. $y = \ln\sqrt{2} + \ln x$; $y' = 0 + \frac{1}{x}$.

27. C. $y' = \frac{1}{\sqrt{1 - x^2}} - \frac{1\cdot(-2x)}{2\sqrt{1 - x^2}}$.

28. A. $\dfrac{dy}{dx} = \dfrac{\dfrac{dy}{dt}}{\dfrac{dx}{dt}} = \dfrac{\sin t}{1 - \cos t}$.

29. D. $\dfrac{dy}{dx} = \dfrac{\dfrac{dy}{d\theta}}{\dfrac{dx}{d\theta}} = \dfrac{3\sin^2\theta\cos\theta}{-3\cos^2\theta\sin\theta}$.

30. E. $\dfrac{dy}{dx} = \dfrac{\dfrac{dy}{dt}}{\dfrac{dx}{dt}} = \dfrac{1 - e^{-t}}{e^{-t}} = e^t - 1$.

31. C.   Since $\dfrac{dy}{dt} = \dfrac{1}{1 - t}$ and $\dfrac{dx}{dt} = \dfrac{1}{(1 - t)^2}$, then

$$\frac{dy}{dx} = 1 - t = \frac{1}{x}.$$

32. C.   Let y' be $\dfrac{dy}{dx}$; then $3x^2 - (xy' + y) + 3y^2 y' = 0$; $y'(3y^2 - x) = y - 3x^2$.

33. A. $1 - \sin(x + y)(1 + y') = 0$; $\dfrac{1 - \sin(x + y)}{\sin(x + y)} = y'$.

34. D. $\cos x + \sin y \cdot y' = 0$; $y' = -\dfrac{\cos x}{\sin y}$.

35. **B.** $6x - 2(xy' + y) + 10yy' = 0; y'(10y - 2x) = 2y - 6x.$

36. **E.** $\dfrac{dy}{dx} = \dfrac{4t^3 - 6t^2}{2t} = 2t^2 - 3t \ (t \neq 0); \dfrac{d^2y}{dx^2} = \dfrac{4t - 3}{2t}$. Replace t by 1.

37. **E.** $f'(x) = 4x^3 - 12x^2 + 8x = 4x(x - 1)(x - 2).$

38. **A.** $f'(x) = 8x^{-1/2}; f''(x) = -4x^{-3/2}; f'''(x) = 6x^{-5/2}$. Replace x by 4.

39. **E.** $f'(x) = \dfrac{1}{x}; f''(x) = -x^{-2}; f'''(x) = 2x^{-3}; f^{(4)}(x) = -6x^{-4}$. The correct answer is $-\dfrac{6}{x^4}$.

40. **D.** $2x + 2yy' = 0; y' = -\dfrac{x}{y}; y'' = -\dfrac{y - xy'}{y^2}$. At (0,5), $y'' = -\dfrac{5 - 0}{25}$.

41. **B.** $y' = ac \cos ct - bc \sin ct;$
$y'' = -ac^2 \sin ct - bc^2 \cos ct.$

42. **C.** $f'(x) = 4x^3 - 8x; f''(x) = 12x^2 - 8; f'''(x) = 24x; f^{(4)}(x) = 24.$

43. **E.**   Here $f(x)$ equals $\dfrac{-x - 1}{(x - 1)^3}$; note that $f'(-1) = 0.$

44. **C.**   When simplified, $y' = e^x(x^2 - 1)$. Use the product rule to obtain y''.

45. **D.** $f'(x) = \dfrac{e^{-x}}{x} - e^{-x} \ln x; f'(1) = e^{-1} - e^{-1} \cdot 0.$

46. **A.** $\dfrac{dy^2}{dx^2} = \dfrac{\dfrac{dy^2}{dx}}{\dfrac{dx^2}{dx}}$. Since $y^2 = x^2 + 1$, $\dfrac{dy^2}{dx^2} = \dfrac{2x}{2x}.$

47. **B.**   Note that $f(g(x)) = \dfrac{1}{x + 1}.$

48. **E.**   When simplified, $\dfrac{dy}{dx} = \dfrac{\cos \theta + \sin \theta}{\cos \theta - \sin \theta}.$

49. **B.**   Since

$$\dfrac{dy}{dt} = -2 \sin 2t = -4 \sin t \cos t \quad \text{and} \quad \dfrac{dx}{dt} = -\sin t,$$

then $\dfrac{dy}{dx} = 4 \cos t$. Then,

$$\dfrac{d^2y}{dx^2} = -\dfrac{4 \sin t}{-\sin t}.$$

50. **A.** $\dfrac{dy}{d\left(\dfrac{1}{1 - x}\right)} = \dfrac{\dfrac{dy}{dx}}{\dfrac{d\left(\dfrac{1}{1 - x}\right)}{dx}} = \dfrac{2x + 1}{\dfrac{1}{(1 - x)^2}}.$

51. **C.**   The limit given is the derivative of $f(x) = x^6$ at $x = 1$.

52. **B.**   The given limit is the definition for $f'(8)$, where $f(x) = \sqrt[3]{x}$;

$$f'(x) = \dfrac{1}{3x^{2/3}}.$$

53. B.   This is $f'(e)$, where $f(x) = \ln x$.

54. B.   The limit is the derivative of $f(x) = \cos x$ at $x = 0; f'(x) = -\sin x$.

55. E.   Since $f'(x) = \frac{2}{3x^{1/3}}, f'(0)$ is not defined.

56. B.   Sketch the graph of $f(x) = 1 - |x|$; note that $f(-1) = f(1) = 0$ and that f is continuous on $[-1, 1]$. Only (B) holds.

57. A.   Note that $f(0) = f(\sqrt{3}) = 0$ and that $f'(x)$ exists on the given interval. By Rolle's theorem, there is a number c in the interval such that $f'(c) = 0$. If $c = 1$, then $6c^2 - 6 = 0$. (-1 is not in the interval.)

58. B.   Since the inverse, h, of $f(x) = \frac{1}{x}$ is $h(x) = \frac{1}{x}$, $h'(x) = -\frac{1}{x^2}$. Replace x by 3.

59. C.   If $x = h(y)$ is the inverse function of $y = f(x)$, then

$$\frac{dx}{dy} = \frac{1}{\dfrac{dy}{dx}}.$$

Since $f'(x) = \dfrac{dy}{dx} = 6x^2 - 3, \dfrac{dx}{dy} = \dfrac{1}{6x^2 - 3}.$

60. A.   Use $\left(\dfrac{dx}{dy}\right)_{y = y_0} = \dfrac{1}{\left(\dfrac{dy}{dx}\right)_{x = x_0}}$, where $f(x_0) = y_0$.

$$\left(\frac{dx}{dy}\right)_{y = 4} = \frac{1}{\left(\dfrac{dy}{dx}\right)_{x = 1}} = \frac{1}{-3}.$$

61. E. $\dfrac{dx}{dy} = \dfrac{1}{\dfrac{dy}{dx}}$. However, to evaluate $\dfrac{dx}{dy}$ for a particular value of y, we need to evaluate $\dfrac{1}{\dfrac{dy}{dx}}$ at the corresponding value of x. This question does not say what value of x yields $y = 2$.

62. D.   If $y = x^{\sin x}$, $\ln y = (\sin x)(\ln x)$,

$$\frac{1}{y} \cdot y' = \frac{\sin x}{x} + (\cos x) \ln x.$$

Solve for y'.

63. D.   The given limit is the derivative of $g(x)$ at $x = 0$.

64. **C.** $\cos (xy)(xy' + y) = 1; x \cos (xy)y' = 1 - y \cos (xy);$

$$y' = \frac{1 - y \cos (xy)}{x \cos (xy)}.$$

65. **C.** Let $y = x^x$, take the ln of each side, and apply L'Hôpital's rule:

$$\ln y = x \ln x = \frac{\ln x}{\dfrac{1}{x}},$$

$$\lim_{x \to 0^+} \ln y = \lim_{x \to 0^+} \frac{\dfrac{1}{x}}{-\dfrac{1}{x^2}} = \lim_{x \to 0^+} (-x) = 0.$$

Since $\ln y \to 0$, $y \to e^0 = 1$.

Practice Section II Problems on Differentiation

1. Let the function f be defined by

$$\begin{cases} f(x) = x^2 \sin \dfrac{1}{x} & (x \neq 0) \\ f(0) = 0. \end{cases}$$

(a) Prove that f is continuous at $x = 0$.
(b) Prove that $f'(0)$ exists, and find it by using the definition of the derivative.
(c) Prove that f' is *not* continuous at $x = 0$.

2. (a) If $f'(x)$ is positive at each x in the interval $a \leq x \leq b$, prove that $f(x)$ is increasing over the interval [that is, that if $a \leq x_1 < x_2 \leq b$, then $f(x_1) < f(x_2)$].
(b) Prove that, if $0 < x < 2\pi$, then $\frac{\sin x}{x} < 1$.

3. Let the function f have a derivative for all real x. If $f(a) = 0$ and $f'(x) > 0$, prove that
(a) $f(x)$ is positive if $x > a$;
(b) $f(x)$ is negative if $x < a$.

4. Let the functions f and g be differentiable for all real x. If $f(a) = g(a)$ and $f'(x) > g'(x)$ for all x, prove that

$$\text{if } x > a, \quad f(x) > g(x);$$

$$\text{if } x < a, \quad f(x) < g(x).$$

Solutions for Section II Problems: Differentiation

1. (a) To prove that f is continuous at $x = 0$ we must show that $\lim_{x \to 0} f(x)$ exists and equals $f(0)$.

If $x \neq 0$, note that

$$-1 \leq \sin \frac{1}{x} \leq 1 \quad \text{and} \quad -x^2 \leq x^2 \sin \frac{1}{x} \leq x^2.$$

Since $\lim_{x \to 0} (-x^2) = \lim_{x \to 0} x^2 = 0$, it follows from the "Squeeze" Theorem (page 22) that $\lim_{x \to 0} x^2 \sin \frac{1}{x} = 0$; that is, $\lim_{x \to 0} f(x) = f(0)$.

(b) The definition of $f'(0)$ is $\lim_{h \to 0} \dfrac{f(0 + h) - f(0)}{h}$. Thus

$$f'(0) = \lim_{h \to 0} \frac{h^2 \sin \dfrac{1}{h}}{h} = \lim_{h \to 0} h \sin \frac{1}{h}.$$

Since $-1 \leq \sin \frac{1}{h} \leq 1$ $(h \neq 0)$, it follows that $h \sin \frac{1}{h}$ is squeezed between $-h$ and h as $h \to 0$, so that $\lim_{h \to 0} h \sin \frac{1}{h} = f'(0) = 0$.

(c) We can obtain $f'(x)$ for nonzero x by using the product rule. Thus

$$f'(x) = x^2 \cos \frac{1}{x} \left(-\frac{1}{x^2} \right) + 2x \sin \frac{1}{x} = -\cos \frac{1}{x} + 2x \sin \frac{1}{x}.$$

If f' is to be continuous at $x = 0$, then $\lim_{x \to 0} f'(x)$ must exist and be equal to $f'(0)$. Note, however, since $\lim_{x \to 0} \cos \frac{1}{x}$ fails to exist, that $f'(x)$ is not continuous at 0.

2. (a) To show that $f(x)$ is increasing on the closed interval $[a, b]$ we must show, if $a \leq x_1 < x_2 \leq b$, that $f(x_1) < f(x_2)$. Since $f'(x)$ is positive at each x in the given interval, the function is continuous at each x and we may apply the Mean Value Theorem. Thus, since $< x_2$, there exists a number c, $x_1 < c < x_2$, such that

$$\frac{f(x_2) - f(x_1)}{x_2 - x_1} = f'(c).$$

Since both $(x_2 - x_1)$ and $f'(c)$ are positive, so is $f(x_2) - f(x_1)$, and $f(x_1) < f(x_2)$.

(b) To prove $\frac{\sin x}{x} < 1$ if $0 < x < 2\pi$, note that $\sin x$ and its derivative, $\cos x$, are both continuous. By the Mean Value Theorem, then, there is a number c, $0 < c < x$, such that

$$\frac{\sin x - 0}{x - 0} = \cos c.$$

Since, if $0 < c < x < 2\pi$, then $\cos c < 1$, it follows that $\frac{\sin x}{x} < 1$. Actually this implies that $\frac{\sin x}{x} < 1$ for all positive x since $\lim_{x \to \infty} \frac{\sin x}{x} = 0$.

3.   Hint: Apply the Mean Value Theorem twice: first to $f(x)$ on $[a, x]$, where $x > a$, then to $f(x)$ on $[x, a]$, where $x < a$.

4. Let $H(x) = f(x) - g(x)$, and note that H satisfies the hypotheses of the Mean Value Theorem on $[a, x]$. Thus there is a number c, $a < c < x$, such that

$$\frac{H(x) - H(a)}{x - a} = H'(c).$$

Since $H'(c) = f'(c) - g'(c)$, and $f'(c) > g'(c)$ for all c, it follows that $H'(c) > 0$. Then, if $x > a$,

$$H(x) - H(a) = (x - a)H'(c) > 0.$$

Since

$$H(x) - H(a) = [f(x) - g(x)] - [f(a) - g(a)] \quad \text{and} \quad f(a) = g(a),$$

we see that if $x > a$ then $f(x) - g(x) > 0$, so that $f(x) > g(x)$. The proof that $f(x) < g(x)$ if $x < a$ is similar.

Chapter 4

Applications of Differential Calculus

Review of Definitions and Methods

A. Slope; Rate of Change_____

If the derivative of $y = f(x)$ exists at $P(x_1, y_1)$, then the *slope of the curve* at P (which is defined to be the slope of the tangent to the curve at P) is $f'(x_1)$, the derivative of $f(x)$ at $x = x_1$.

Any c such that $f'(c) = 0$ or such that $f'(c)$ does not exist is called a *critical point* or *critical value* of f. If f has a derivative everywhere, we find the critical points by solving the equation $f'(x) = 0$.

Examples

1. If $f(x) = 4x^3 - 6x^2 - 8$, then

$$f'(x) = 12x^2 - 12x = 12x(x - 1),$$

which equals zero if x is 0 or 1. So 0 and 1 are critical points.

2. If $f(x) = 3x^3 + 2x$, then

$$f'(x) = 9x^2 + 2.$$

Since $f'(x)$ never equals zero (indeed, it is always positive), f has no critical values.

3. If $f(x) = (x - 1)^{1/3}$, then

$$f'(x) = \frac{1}{3(x - 1)^{2/3}}.$$

Although f' is never zero, $x = 1$ is a critical value of f because f' does not exist at $x = 1$.

Average and Instantaneous Rate of Change.

If $y = f(x)$ is a function and if, as x varies from x_0 to $x_0 + \Delta x$, y varies from y_0 to $y_0 + \Delta y$, then

$$\frac{\Delta y}{\Delta x} = \frac{f(x_0 + \Delta x) - f(x_0)}{\Delta x}$$

is called the *average rate of change* of y (or of f) over the interval from x_0 to $x_0 + \Delta x$.

Thus, the *average velocity* of a moving object over some time interval is the change in distance divided by the change in time, the average rate of growth of a colony of fruit flies over some interval of time is the change in size of the colony divided by the time elapsed, the average rate of change in the profit of a company on some gadget with respect to production is the change in profit divided by the change in the number of gadgets produced.

The *instantaneous rate of change* (or just *rate of change*) of y (or f) with respect to x is the derivative y' or f'. The rate of change therefore is equal to

$$\lim_{\Delta x \to 0} \frac{\Delta y}{\Delta x} = \lim_{\Delta x \to 0} \frac{f(x + \Delta x) - f(x)}{\Delta x} = \lim_{\Delta x \to 0} (\text{average rate of change}).$$

On the graph of $y = f(x)$, the rate at which the y-coordinate changes with respect to the x-coordinate is $f'(x)$, the slope of the curve. The rate at which $s(t)$, the distance traveled by a particle in t seconds, changes with respect to time is $s'(t)$, the velocity of the particle; the rate at which a manufacturer's profit $P(x)$ changes relative to the production level x is $P'(x)$.

Example 4.   Let $G = 400(15 - t)^2$ be the number of gallons of water in a cistern t minutes after an outlet pipe is opened. Find the average rate of drainage during the first 5 minutes and the rate at which the water is running out at the end of 5 minutes.

The average rate of change during the first 5 minutes equals

$$\frac{G(5) - G(0)}{5} = \frac{400 \cdot 100 - 400 \cdot 225}{5} = -10,000 \text{ gal/min.}$$

The average rate of drainage during the first 5 minutes is 10,000 gallons per minute.

The instantaneous rate of change at $t = 5$ is $G'(5)$. Since

$$G'(t) = -800(15 - t),$$

$G'(5) = -800(10) = -8000$ gal/min. So the rate of drainage at the end of 5 minutes is 8000 gallons per minute.

B.   Tangents and Normals

The *equation of the tangent* to the curve $y = f(x)$ at point $P(x_1, y_1)$ is

$$y - y_1 = f'(x_1)(x - x_1).$$

Since the normal to the curve at P is the line through P which is perpendicular to the tangent, its slope is $-\dfrac{1}{f'(x_1)}$ and the *equation of the normal* is

$$y - y_1 = -\frac{1}{f'(x_1)}(x - x_1).$$

If the tangent to a curve is horizontal at a point, then the derivative at the point is 0. If it is vertical at a point, then the derivative does not exist at the point.

*Tangents to Parametrically Defined Curves.

If the curve is defined parametrically, say in terms of t (as in Chapter 3, §D, page 37), then we obtain the slope at any point from the parametric equations. We then evaluate the slope and the x- and y-coordinates by replacing t by the value specified in the question (see Example 9, page 66).

Examples

5. Find the equations of the tangent and normal to the curve of $f(x) = x^3 - 3x^2$ at the point $(1, -2)$.

Since $f'(x) = 3x^2 - 6x$ and $f'(1) = -3$, the equation of the tangent is

$$y + 2 = -3(x - 1) \quad \text{or} \quad y + 3x = 1,$$

and the equation of the normal is

$$y + 2 = \frac{1}{3}(x - 1) \quad \text{or} \quad 3y - x = -7.$$

*An asterisk denotes a topic covered only in Calculus BC.

6. Find the equation of the tangent to the curve of $x^2y - x = y^3 - 8$ at the point where $x = 0$.

Here we differentiate implicitly to get $\dfrac{dy}{dx} = \dfrac{1 - 2xy}{x^2 - 3y^2}$. Since $y = 2$ when $x = 0$ and the slope at this point is $\dfrac{1 - 0}{0 - 12} = -\dfrac{1}{12}$, the equation of the tangent is

$$y - 2 = -\frac{1}{12}x \qquad \text{or} \qquad 12y + x = 24.$$

7. Find the coordinates of any point on the curve of $y^2 - 4xy = x^2 + 5$ for which the tangent is horizontal.

Since $\dfrac{dy}{dx} = \dfrac{x + 2y}{y - 2x}$ and the tangent is horizontal when $\dfrac{dy}{dx} = 0$, then $x = -2y$. If we substitute this in the equation of the curve, we get

$$y^2 - 4y(-2y) = 4y^2 + 5.$$

Thus $y = \pm 1$ and $x = \mp 2$. The points, then, are $(2, -1)$ and $(-2, 1)$.

8. Find the abscissa of any point on the curve of $y = \sin^2 (x + 1)$ for which the tangent is parallel to the line $3x - 3y - 5 = 0$.

Since $\dfrac{dy}{dx} = 2 \sin (x + 1) \cos (x + 1) = \sin 2(x + 1)$ and since the given line has slope 1, we seek x such that $\sin 2(x + 1) = 1$. Then

$$2(x + 1) = \frac{\pi}{2} + 2n\pi \qquad (n \text{ an integer})$$

or

$$(x + 1) = \frac{\pi}{4} + n\pi \qquad \text{and} \qquad x = \frac{\pi}{4} + n\pi - 1.$$

9. Find the equation of the tangent to $x = \cos t$, $y = 2 \sin^2 t$ at the point where $t = \dfrac{\pi}{3}$.

Since $\dfrac{dx}{dt} = -\sin t$ and $\dfrac{dy}{dt} = 4 \sin t \cos t$, we see that

$$\frac{dy}{dx} = \frac{4 \sin t \cos t}{-\sin t} = -4 \cos t.$$

At $t = \dfrac{\pi}{3}$, $x = \dfrac{1}{2}$, $y = 2\left(\dfrac{\sqrt{3}}{2}\right)^2 = \dfrac{3}{2}$, and $\dfrac{dy}{dx} = -2$. The equation of the tangent is

$$y - \frac{3}{2} = -2\left(x - \frac{1}{2}\right) \qquad \text{or} \qquad 4x + 2y = 5.$$

C. Increasing and Decreasing Functions_____

Case I. Functions with Continuous Derivatives.

A function $y = f(x)$ is said to be $\frac{\text{increasing}}{\text{decreasing}}$ at $P(x_1, y_1)$ if its derivative $f'(x_1)$, the slope at P, is $\frac{\text{positive}}{\text{negative}}$. To find intervals over which $f(x)$ $\frac{\text{increases}}{\text{decreases}}$, that is, over which the curve $\frac{\text{rises}}{\text{falls}}$, compute $f'(x)$ and determine where it is $\frac{\text{positive}}{\text{negative}}$.

Example 10. If $f(x) = x^4 - 4x^3 + 4x^2$, then

$$f'(x) = 4x^3 - 12x^2 + 8x = 4x(x^2 - 3x + 2) = 4x(x - 1)(x - 2).$$

Since this derivative changes sign only at $x = 0$, 1, or 2, it has a constant sign in each of the intervals between these numbers. Thus:

if	$x < 0$,	then	$f'(x) < 0$ and f is decreasing;
	$0 < x < 1$,		$f'(x) > 0$ and f is increasing;
	$1 < x < 2$,		$f'(x) < 0$ and f is decreasing;
	$2 < x$,		$f'(x) > 0$ and f is increasing.

Case II. Functions Whose Derivatives Have Discontinuities.

Here we proceed as in Case I, but also consider intervals bounded by any points of discontinuity of f or f'.

Example 11. If $f(x) = \frac{1}{x + 1}$, then

$$f'(x) = -\frac{1}{(x + 1)^2}.$$

We note that neither f nor f' is defined at $x = -1$; furthermore, $f'(x)$ never equals zero. We need therefore examine only the signs of $f'(x)$ when $x < -1$ and when $x > -1$.

When $x < -1$, $f'(x) < 0$; when $x > -1$, $f'(x) < 0$. So f decreases on both intervals. The curve is a hyperbola whose center is at the point $(-1, 0)$.

D. Maximum, Minimum, and Inflection Points: Definitions

The curve of $y = f(x)$ has a *relative* or *local* $\begin{smallmatrix}maximum\\minimum\end{smallmatrix}$ at a point where $x = c$ if $\begin{smallmatrix}f(c) \geqq f(x)\\f(c) \leqq f(x)\end{smallmatrix}$ for all x in the immediate neighborhood of c. If a curve has a relative $\begin{smallmatrix}maximum\\minimum\end{smallmatrix}$ at $x = c$, then the curve changes from $\begin{smallmatrix}rising\ to\ falling\\falling\ to\ rising\end{smallmatrix}$ as x increases through c. If a function is differentiable on the closed interval $[a, b]$ and has a relative maximum or minimum at $x = c$ $(a < c < b)$, then $f'(c) = 0$. The converse of this statement is not true.

If $f(c)$ is either a relative or local maximum or a relative or local minimum, then $f(c)$ is called a *relative* or *local extreme value* or *extremum*. The plural of *extremum* is *extrema*.

The *absolute* $\begin{smallmatrix}maximum\\minimum\end{smallmatrix}$ of a function on $[a, b]$ occurs at $x = c$ if $\begin{smallmatrix}f(c) \geqq f(x)\\f(c) \leqq f(x)\end{smallmatrix}$ for all x on $[a, b]$.

A curve is said to be *concave* $\begin{smallmatrix}upward\\downward\end{smallmatrix}$ at a point $P(x_1, y_1)$ if the curve lies $\begin{smallmatrix}above\\below\end{smallmatrix}$ its tangent. If $\begin{smallmatrix}y'' > 0\\y'' < 0\end{smallmatrix}$ at P, the curve is concave $\begin{smallmatrix}up\\down\end{smallmatrix}$ In Figure N4–1, the curves sketched in (a) and (b) are concave downward at P while in (c) and (d) they are concave upward at P.

We call $f(c)$ an extreme value of f on $[a, b]$ if it is either the absolute maximum or the absolute minimum of f on $[a, b]$.

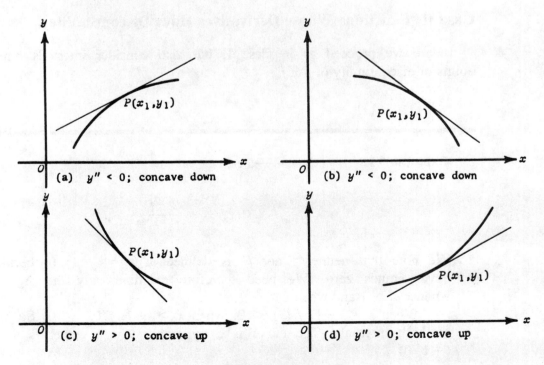

(a) $y'' < 0$; concave down

(b) $y'' < 0$; concave down

(c) $y'' > 0$; concave up

(d) $y'' > 0$; concave up

FIGURE N4–1

A *point of inflection* is a point where the curve changes its concavity from upward to downward or from downward to upward.

E.   Maximum, Minimum, and Inflection Points; Curve Sketching_____

Case I. Functions That Are Everywhere Differentiable.

The following procedure is suggested when seeking to determine any maximum, minimum, or inflection point of a curve and to sketch the curve.

(1)   Find y' and y''.

(2)   Find all x for which $y' = 0$; at each point corresponding to these abscissas the tangent to the curve is horizontal.

(3)   Let c be a number for which y' is 0; investigate the sign of y'' at c. If $y''(c) > 0$, the curve is concave up and c yields a relative minimum; if $y''(c) < 0$, the curve is concave down and c yields a relative maximum. See Figure N4–2. If $y''(c) = 0$, the second-derivative test fails and we must use the test in step (4) below.

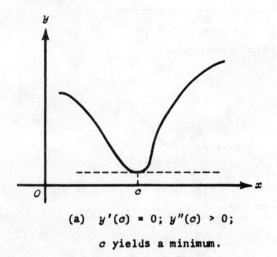

(a) $y'(c) = 0;\ y''(c) > 0;$

c yields a minimum.

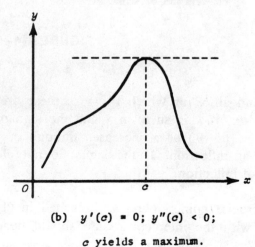

(b) $y'(c) = 0;\ y''(c) < 0;$

c yields a maximum.

FIGURE N4–2

(4)   If $y'(c) = 0$ and $y''(c) = 0$, investigate the signs of y' as x increases through c. If $y'(x) > 0$ for x's (just) less than c but $y'(x) < 0$ for x's (just) greater than c, then the situation is that indicated in Figure N4–3a, where the tangent lines have been sketched as x increases through c; here c yields a relative maximum. If the situation is reversed and the sign of y' changes from $-$ to $+$ as x increases through c then it yields a relative minimum. Figure N4–3b shows this case. The schematic sign pattern of y', $+ \, 0 \, -$ or $- \, 0 \, +$, describes each situation completely. If y' does not change sign as x increases through c, then c yields neither a relative maximum nor a relative minimum. Two examples of this appear in Figures N4–3c and N4–3d.

(a)

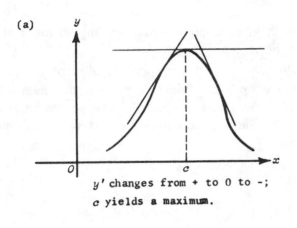

y' changes from + to 0 to -;
c yields a maximum.

(b)

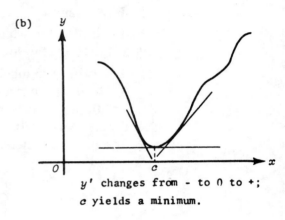

y' changes from - to 0 to +;
c yields a minimum.

(c)

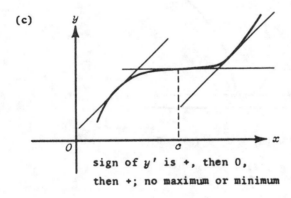

sign of y' is +, then 0,
then +; no maximum or minimum

(d)

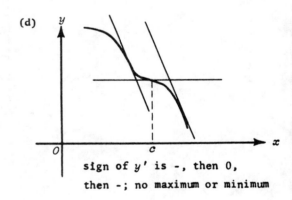

sign of y' is -, then 0,
then -; no maximum or minimum

FIGURE N4–3

(5) Find all x for which $y'' = 0$; these are abscissas of possible points of inflection. If c is such an x and the sign of y'' changes (from $+$ to $-$ or from $-$ to $+$) as x increases through c, then c is the x-coordinate of a point of inflection. If the signs do not change, then c does *not* yield a point of inflection.

The crucial points found as indicated in (1) through (5) above should be plotted along with the intercepts. Care should be exercised to ensure that the tangent to the curve is horizontal whenever $\dfrac{dy}{dx} = 0$ and that the curve has the proper concavity.

Examples

12. Find any maximum, minimum, or inflection points of $f(x) = x^3 - 5x^2 + 3x + 6$, and sketch the curve.

Steps:

(1) Here $f'(x) = 3x^2 - 10x + 3$ and $f''(x) = 6x - 10$.

(2) $f'(x) = (3x - 1)(x - 3)$, which is zero when $x = \frac{1}{3}$ or 3.

(3) Since $f''\left(\frac{1}{3}\right) < 0$, we know that the point $\left(\frac{1}{3}, f\left(\frac{1}{3}\right)\right)$ is a relative maximum; since $f''(3) > 0$, the point $(3, f(3))$ is a relative minimum. Thus, $\left(\frac{1}{3}, \frac{175}{27}\right)$ is a relative maximum and $(3, -3)$ a relative minimum.

(4) is unnecessary for this problem.

(5) $f''(x) = 0$ when $x = \frac{5}{3}$, and f'' does change sign as x increases through $\frac{5}{3}$, so that this x does yield an inflection point.

The curve is sketched in Figure N4-4.

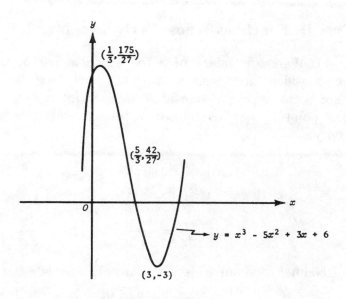

FIGURE N4-4

13. If we apply the procedure to $f(x) = x^4 - 4x^3$, we see that

(1) $f'(x) = 4x^3 - 12x^2$ and $f''(x) = 12x^2 - 24x$.

(2) $f'(x) = 4x^2(x - 3)$, which is zero when $x = 0$ or $x = 3$.

(3) Since $f''(x) = 12x(x - 2)$ and $f''(3) > 0$, the point $(3, -27)$ is a relative minimum. Since $f''(0) = 0$, the second-derivative test fails to tell us whether 0 yields a maximum or a minimum.

(4) Since $f'(x)$ does not change sign as x increases through 0, 0 yields neither a maximum nor a minimum.

(5) $f''(x) = 0$ when x is 0 or 2. It changes sign $(+\ 0\ -)$ as x increases through 0, and also $(-\ 0\ +)$ as x increases through 2. Thus both $(0, 0)$ and $(2, -16)$ are inflection points.

The curve is sketched in Figure N4−5.

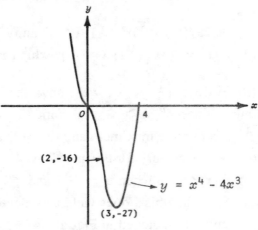

$(2, -16)$

$y = x^4 - 4x^3$

$(3, -27)$

FIGURE N4−5

Case II. Functions Whose Derivatives May Not Exist Everywhere.

If there are values of x for which a first or second derivative does not exist, we consider those values separately, recalling that a relative maximum or minimum point is one of transition between intervals of rise and fall and that an inflection point is one of transition between intervals of upward and downward concavity.

Examples

14. If $y = x^{2/3}$, then

$$\frac{dy}{dx} = \frac{2}{3x^{1/3}} \quad \text{and} \quad \frac{d^2y}{dx^2} = -\frac{2}{9x^{4/3}}.$$

Neither derivative is zero anywhere; both derivatives fail to exist when $x = 0$. As x increases through 0, $\frac{dy}{dx}$ changes from $-$ to $+$; $(0, 0)$ is therefore a minimum. Note that the tangent is vertical at the origin, and that since $\frac{d^2y}{dx^2}$ is negative everywhere except at 0, the curve is everywhere concave down. See Figure N4−6.

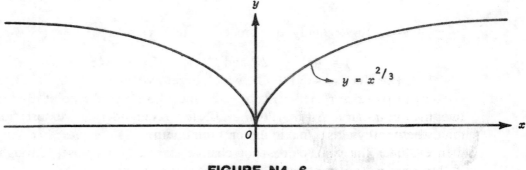

$y = x^{2/3}$

FIGURE N4−6

15. If $y = x^{1/3}$, then

$$\frac{dy}{dx} = \frac{1}{3x^{2/3}} \quad \text{and} \quad \frac{d^2y}{dx^2} = -\frac{2}{9x^{5/3}}.$$

As in Example 14, neither derivative ever equals zero and both fail to exist when $x = 0$. Here, however, as x increases through 0, $\frac{dy}{dx}$ does not change sign. Since $\frac{dy}{dx}$ is positive for all x except 0, the curve rises for all x different from zero and can have neither maximum nor minimum points. The tangent is again vertical at the origin. Note here that $\frac{d^2y}{dx^2}$ does change sign (from $+$ to $-$) as x increases through 0, so that $(0, 0)$ is a point of inflection. See Figure N4–7.

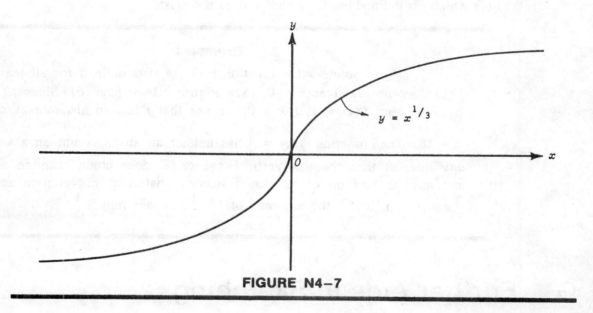

$y = x^{1/3}$

FIGURE N4–7

F. Absolute Maximum or Minimum

Case I. Differentiable Functions.

If a function f is differentiable on a closed interval $a \leq x \leq b$, then f is also continuous on the closed interval $[a, b]$ and we know that f attains both an absolute maximum (abs max) and an absolute minimum (abs min) on $[a, b]$. To determine these, we solve the equation $f'(x) = 0$ for critical points in the interval $[a, b]$, then evaluate f at each of those and also at $x = a$ and $x = b$. The largest value of f thus obtained is the abs max, the smallest the abs min.

Example 16.   Find the abs max and abs min of f on (a) $-2 \leqq x \leqq 3$, and (b) $0 \leqq x \leqq 3$, if $f(x) = 2x^3 - 3x^2 - 12x$.

(a) $f'(x) = 6x^2 - 6x - 12 = 6(x + 1)(x - 2)$, which equals zero if $x = -1$ or 2. Since $f(-2) = -4, f(-1) = 7, f(2) = -20$, and $f(3) = -9$, the abs max of f occurs at $x = -1$ and equals 7, and the abs min of f occurs at $x = 2$ and equals -20.

(b) Only the critical value 2 lies in $[0, 3]$. We now evaluate f at 0, 2, and 3. Since $f(0) = 0, f(2) = -20$, and $f(3) = -9$, the abs max of f equals 0 and the abs min equals -20.

Case II. Functions That Are Not Everywhere Differentiable.

We proceed as above but now evaluate f also at each point in a given interval for which f is defined but for which f' does not exist.

Examples

17. The absolute-value function $f(x) = |x|$ is defined for all real x, but $f'(x)$ does not exist at $x = 0$. (See Figure N1–3, page 6). Since $f'(x) = -1$ if $x < 0$, but $f'(x) = 1$ if $x > 0$, we see that f has an abs min at $x = 0$.

18. The function $f(x) = \frac{1}{x}$ has neither an abs max nor an abs min on *any* interval that contains zero. However, it does attain both an abs max and an abs min on every closed interval that does not contain zero. For instance, on $[2, 5]$, the abs max of f is $\frac{1}{2}$, the abs min is $\frac{1}{5}$.

G.   Further Aids in Sketching

It is often very helpful to investigate one or more of the following before sketching the graph of a function or of an equation:

(1) Intercepts. Set $x = 0$ and $y = 0$ to find any y- and x-intercepts respectively.

(2) Symmetry. Let the point (x, y) satisfy an equation. Then its graph is symmetric about

the x-axis if $(x, -y)$ also satisfies the equation;
the y-axis if $(-x, y)$ also satisfies the equation;
the origin if $(-x, -y)$ also satisfies the equation.

(3) Asymptotes. The line $y = b$ is a horizontal asymptote of the graph of a function f if either $\lim\limits_{x \to \infty} f(x) = b$ or $\lim\limits_{x \to -\infty} f(x) = b$. If $f(x) = \frac{P(x)}{Q(x)}$, inspect the degrees of $P(x)$ and $Q(x)$, then use the Rational Function Theorem, page 23. The line $x = c$ is a vertical asymptote of the rational function $\frac{P(x)}{Q(x)}$ if $Q(c) = 0$ but $P(c) \neq 0$.

(4) Points of discontinuity. Note points not in the domain of a function, particularly where a denominator equals zero.

Examples

19. To sketch the graph of $y = \dfrac{2x + 1}{x - 1}$, we note that, if $x = 0$, then $y = -1$, and that $y = 0$ when the numerator equals zero, which is when $x = -\frac{1}{2}$. A check shows that the graph does not possess any of the symmetries described above. Since $y \to 2$ as $x \to \pm\infty$, $y = 2$ is a horizontal asymptote; also, $x = 1$ is a vertical asymptote. We note immediately that the function is defined for all reals except $x = 1$; the latter is the only point of discontinuity.

If we rewrite the function as follows:

$$y = \frac{2x + 1}{x - 1} = \frac{2x - 2 + 3}{x - 1} = \frac{2(x - 1) + 3}{x - 1} = 2 + \frac{3}{x - 1},$$

it is easy to find derivatives:

$$y' = -\frac{3}{(x - 1)^2} \quad \text{and} \quad y'' = \frac{6}{(x - 1)^3}.$$

From y' we see that the function decreases (except at $x = 1$), and from y'' that the curve is concave down if $x < 1$, up if $x > 1$. See Figure N4−8.

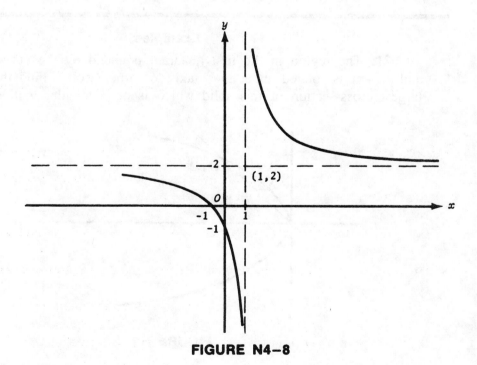

FIGURE N4−8

20. Describe any symmetries of the graphs of (a) $3y^2 + x = 2$; (b) $y = x + \frac{1}{x}$; (c) $x^2 - 3y^2 = 27$.

(a) Suppose point (x, y) is on this graph. Then so is point $(x, -y)$, since $3(-y)^2 + x = 2$ is equivalent to $3y^2 + x = 2$. So (a) is symmetric about the x-axis.

(b) Note that point $(-x, -y)$ satisfies the equation if point (x, y) does:

$$(-y) = (-x) + \frac{1}{(-x)} \quad \leftrightarrow \quad y = x + \frac{1}{x}.$$

Therefore the graph of this function is symmetric about the origin.

(c) This graph is symmetric about the x-axis, the y-axis, and the origin. It is easy to see that, if point (x, y) satisfies the equation, so do points $(x, -y)$, $(-x, y)$, and $(-x, -y)$.

H.   Problems Involving Maxima and Minima (Extreme-Value Problems)

The techniques described above can be applied to problems in which a function is to be maximized (or minimized). Often it helps to draw a figure. If y, the quantity to be maximized (or minimized), can be expressed explicitly in terms of x, then the procedure outlined above can be used. If the domain of y is restricted to some closed interval, one should always check the endpoints of this interval so as not to overlook possible extrema. Often, implicit differentiation, sometimes of two or more equations, is indicated.

Examples

21. The region in the first quadrant bounded by the curves of $y^2 = x$ and $y = x$ is rotated about the y-axis to form a solid. Find the area of the largest cross-section of this solid which is perpendicular to the y-axis.

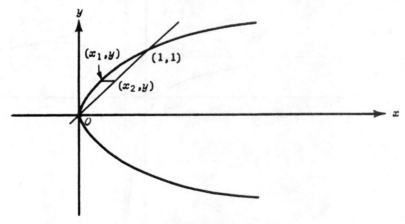

FIGURE N4–9

See Figure N4–9. The curves intersect at the origin and at $(1, 1)$, so $0 < y < 1$. A cross-section of the solid is a ring whose area A is the difference between the areas of two circles, one with radius x_2, the other with radius x_1. Thus

$$A = \pi x_2^2 - \pi x_1^2 = \pi(y^2 - y^4); \qquad \frac{dA}{dy} = \pi(2y - 4y^3) = 2\pi y(1 - 2y^2).$$

The only relevant zero of the first derivative is $y = \dfrac{1}{\sqrt{2}}$. Thus for the maximum area A we have

$$A = \pi\left(\frac{1}{2} - \frac{1}{4}\right) = \frac{\pi}{4}.$$

Note that $\dfrac{d^2A}{dy^2} = \pi(2 - 12y^2)$ and that this is negative when $y = \dfrac{1}{\sqrt{2}}$, assuring a maximum there. Note further that A equals zero at each endpoint of the interval $[0, 1]$ so that $\dfrac{\pi}{4}$ is the absolute maximum area.

22. The volume of a cylinder equals k in.^{3}, where k is a constant. Find the proportions of the cylinder that minimize the total surface area.

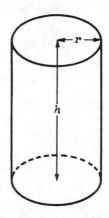

FIGURE N4–10

See Figure N4–10. We know that the volume is

$$V = \pi r^2 h = k, \tag{1}$$

where r is the radius and h the height. We seek to minimize S, the total surface area, where

$$S = 2\pi r^2 + 2\pi rh = 2\pi(r^2 + rh). \tag{2}$$

We can differentiate both (1) and (2) with respect to r, getting, from (1),

$$\pi\left(r^2 \cdot \frac{dh}{dr} + 2rh\right) = 0, \tag{3}$$

where we use the fact that $\dfrac{dk}{dr} = 0$, and, from (2),

$$2\pi\left(2r + r \cdot \frac{dh}{dr} + h\right) = 0, \tag{4}$$

where $\dfrac{dS}{dr}$ is set equal to zero because S is to be a minimum.

From (3) we see that $\dfrac{dh}{dr} = -\dfrac{2h}{r}$, and if we use this in (4) we get

$$2r + r\left(\frac{-2h}{r}\right) + h = 0 \quad \text{or} \quad h = 2r.$$

The total surface area of a cylinder of fixed volume is thus a minimum when its height equals its diameter.

(Note that we need not concern ourselves with the possibility that the value of r which renders $\frac{dS}{dr}$ equal to zero will produce a maximum surface area rather than a minimum one. With k fixed, we can choose r and h in such a way as to make S as large as we like.)

23. A charter bus company advertises a trip for a group as follows: At least 20 people must sign up. The cost when 20 participate is $80 per person. The price will drop by $2 per ticket for each member of the traveling group in excess of 20. If the bus can accommodate 28 people, how many participants will maximize the company's revenue?

Let x denote the number who sign up in excess of 20. Then $0 \leqq x \leqq 8$. The total number who agree to participate is $(20 + x)$, and the price per ticket is $(80 - 2x)$ dollars. So the revenue R, in dollars, is

$$R = (20 + x)(80 - 2x),$$

$$R'(x) = (20 + x)(-2) + (80 - 2x)\cdot 1$$

$$= 40 - 4x.$$

This is zero if $x = 10$. Although $x = 10$ yields maximum R—note that $R''(x) = -4$ and is always negative—this value of x is not within the restricted interval. We therefore evaluate R at the endpoints 0 and 8: $R(0) = 1600$ and $R(8) = 28\cdot 64 = 1792$. So 28 participants will maximize revenue.

I. Motion Along a Line

If a particle moves along a line according to the law $s = f(t)$, where s represents the position of the particle P on the line at time t, then the *velocity* v of P at time t is given by $\frac{ds}{dt}$ and its *acceleration* a by $\frac{dv}{dt}$ or by $\frac{d^2s}{dt^2}$. The *speed* of the particle is $|v|$, the magnitude of v. If the line of motion is directed positively to the right, then the motion of the particle P is subject to the following: At any instant,

(1)   if $v > 0$, then P is moving to the right and its distance s is increasing; if $v < 0$, then P is moving to the left and its distance s is decreasing;

(2)   if $a > 0$, then v is increasing; if $a < 0$, then v is decreasing;

(3)   if a and v are both positive or both negative, then (1) and (2) imply that the speed of P is increasing or that P is accelerating; if a and v have opposite signs, then the speed of P is decreasing or P is decelerating;

(4)   if s is a continuous function of t, then P reverses direction whenever v is zero and a is different from zero; note that zero velocity does not necessarily imply a reversal in direction.

Examples

24. A particle moves along a line according to the law $s = 2t^3 - 9t^2 + 12t - 4$, where $t \geqq 0$. (a) Find all t for which the distance s is increasing. (b) Find all t for which the velocity is increasing. (c) Find all t for which the speed of the particle is increasing. (d) Find the speed when $t = \frac{3}{2}$. (e) Find the total distance traveled between $t = 0$ and $t = 4$.

We have

$$v = \frac{ds}{dt} = 6t^2 - 18t + 12 = 6(t^2 - 3t + 2) = 6(t - 2)(t - 1)$$

and

$$a = \frac{dv}{dt} = \frac{d^2s}{dt^2} = 12t - 18 = 12\left(t - \frac{3}{2}\right)$$

The sign of v behaves as follows:

$$\begin{array}{llll}
\text{if} & t < 1 & \text{then} & v > 0, \\
& 1 < t < 2 & & v < 0, \\
& 2 < t & & v > 0.
\end{array}$$

For a, we have:

$$\begin{array}{llll}
\text{if} & t < \dfrac{3}{2} & \text{then} & a < 0, \\
& \dfrac{3}{2} < t & & a > 0.
\end{array}$$

These immediately yield the answers, as follows:

(a) s increases when $t < 1$ or $t > 2$.

(b) v increases when $t > \frac{3}{2}$.

(c) The speed $|v|$ is increasing when v and a are both positive, that is, for $t > 2$, and when v and a are both negative, that is, for $1 < t < \frac{3}{2}$.

(d) The speed when $t = \frac{3}{2}$ equals $|v| = \left| -\frac{3}{2} \right| = \frac{3}{2}$.

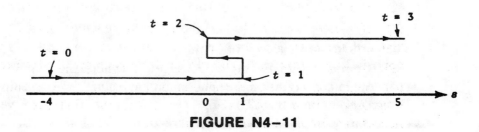

FIGURE N4–11

(e) P's motion can be indicated as shown in Figure N4–11. P moves to the right if $t < 1$, reverses its direction at $t = 1$, moves to the left when $1 < t < 2$, reverses again at $t = 2$, and continues to the right for all $t > 2$. The position of P at certain times t is shown in the following table:

t:	0	1	2	4
s:	-4	1	0	28

Thus P travels a total of 34 units between times $t = 0$ and $t = 4$.

25. Answer the questions of Example 24 if the law of motion is $s = t^4 - 4t^3$.

Since $v = 4t^3 - 12t^2 = 4t^2(t - 3)$ and $a = 12t^2 - 24t = 12t(t - 2)$, the signs of v and a are as follows:

if	$t < 3$,	then	$v < 0$
	$3 < t$		$v > 0$;
if	$t < 0$,	then	$a > 0$
	$0 < t < 2$		$a < 0$
	$2 < t$		$a > 0$.

Thus:

(a) s increases if $t > 3$.

(b) v increases if $t < 0$ or $t > 2$.

(c) Since v and a have the same sign if $0 < t < 2$ or if $t > 3$, the speed increases on these intervals.

(d) The speed when $t = \frac{3}{2}$ equals $|v| = \left| -\frac{27}{2} \right| = \frac{27}{2}$.

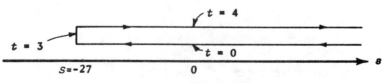

FIGURE N4–12

(e) The motion is shown in Figure N4–12. The particle moves to the left if $t < 3$ and to the right if $t > 3$, stopping instantaneously when $t = 0$ and $t = 3$, but reversing direction only when $t = 3$. Thus:

t:	0	3	4
s:	0	-27	0

The particle travels a total of 54 units between $t = 0$ and $t = 4$.

(Compare with Example 13, page 71, where the function $f(x) = x^4 - 4x^3$ is investigated for maximum and minimum values; also see the accompanying Figure N4–5 on page 72.)

*J.   Motion along a Curve: Velocity and Acceleration Vectors

J1. Derivative of Arc Length.

If the derivative of $y = f(x)$ is continuous, if Q is a fixed point on its curve and P any other point on it, and if s is the arc length from Q to P (see Figure N4-13), then the derivative of s is given by

$$\frac{ds}{dx} = \sqrt{1 + \left(\frac{dy}{dx}\right)^2} \qquad (1)$$

or

$$\frac{ds}{dy} = \sqrt{1 + \left(\frac{dx}{dy}\right)^2}, \qquad (2)$$

where it is assumed that s is increasing with x in (1), and with y in (2).

If the position of P is given parametrically by $x = f(t)$ and $y = g(t)$, then

$$\frac{ds}{dt} = \sqrt{\left(\frac{dx}{dt}\right)^2 + \left(\frac{dy}{dt}\right)^2}, \qquad (3)$$

where s increases as t does.

Formulas (1), (2), and (3) above can all be derived easily from the very simple formula

$$ds^2 = dx^2 + dy^2. \qquad (4)$$

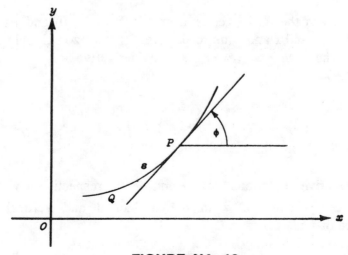

FIGURE N4-13

---

*An asterisk denotes a topic covered only in Calculus BC.

If, for instance, x is expressed in terms of y, making it convenient to use formula (2), note that it follows from (4) that

$$\frac{ds^2}{dy^2} = \frac{dx^2}{dy^2} + 1,$$

$$\left(\frac{ds}{dy}\right)^2 = \left(\frac{dx}{dy}\right)^2 + 1,$$

and

$$\frac{ds}{dy} = \sqrt{1 + \left(\frac{dx}{dy}\right)^2},$$

which is (2).

J2. Vector Functions: Velocity and Acceleration.

If a point P moves along a curve in accordance with the pair of parametric equations $x = f(t)$, $y = g(t)$, where t represents time, then the vector from the origin to P is called the *position vector*, often denoted by $[f(t), g(t)]$. Since a vector is defined for each t in the domain common to f and g, $[f(t), g(t)]$ is also called a *vector function*.

Vectors are symbolized either by boldface letters (thus: **R**, **i**, **j**) or by italic letters with an arrow written over them (thus: $\vec{R}$, $\vec{i}$, $\vec{j}$). When writing on a blackboard or with pencil and paper the arrow notation is simpler, but in print the boldface notation is clearer, and will be used here.

The position vector is denoted by **R**, and

$$\mathbf{R} = x\mathbf{i} + y\mathbf{j},$$

where **i** is the (unit) vector from $(0, 0)$ to $(1, 0)$ and **j** is the (unit) vector from $(0, 0)$ to $(0, 1)$, and x and y are respectively the horizontal and vertical components of **R**.

The *velocity vector* is the derivative of the vector function (the position vector):

$$\mathbf{v} = \frac{d\mathbf{R}}{dt} = \frac{dx}{dt}\mathbf{i} + \frac{dy}{dt}\mathbf{j}.$$

Alternative notations for $\frac{dx}{dt}$ and $\frac{dy}{dt}$ are respectively v_x and v_y, or $\dot{x}$ and $\dot{y}$; these are the components of **v** in the horizontal and vertical directions respectively. The slope of **v** is

$$\frac{\frac{dy}{dt}}{\frac{dx}{dt}} = \frac{dy}{dx},$$

which is the slope of the curve; the magnitude of **v**, denoted by $|\mathbf{v}|$, is

$$\sqrt{\left(\frac{dx}{dt}\right)^2 + \left(\frac{dy}{dt}\right)^2} = \sqrt{v_x^2 + v_y^2},$$

which equals $\frac{ds}{dt}$, the derivative of arc length. Thus, if the vector **v** is drawn initiating at P, it will be tangent to the curve at P and its magnitude will be the *speed of the particle at P*.

The *acceleration vector* **a** is $\frac{d\mathbf{v}}{dt}$ or $\frac{d^2\mathbf{R}}{dt^2}$, and can be obtained by a second differentiation of the components of **R**. Thus

$$\mathbf{a} = \frac{d^2x}{dt^2}\mathbf{i} + \frac{d^2y}{dt^2}\mathbf{j};$$

the direction of **a** is

$$\tan^{-1}\frac{\dfrac{d^2y}{dt^2}}{\dfrac{d^2x}{dt^2}};$$

and its magnitude is

$$|\mathbf{a}| = \sqrt{\left(\frac{d^2x}{dt^2}\right)^2 + \left(\frac{d^2y}{dt^2}\right)^2} = \sqrt{a_x^2 + a_y^2}.$$

For both **v** and **a** the quadrant is determined by the signs of their components. Vectors **i** and **j** are shown in Figure N4–14a; **R**, **v**, and **a**, and their components, are shown in Figure N4–14b. Note that v_x and v_y happen to be positive, so that ϕ is a first-quadrant angle, while $a_x < 0$ and $a_y > 0$ imply that θ is in the second quadrant.

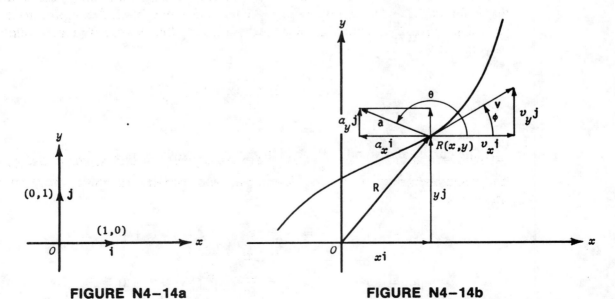

FIGURE N4–14a   **FIGURE N4–14b**

Examples

26. A particle moves according to the equations $x = 3\cos t$, $y = 2\sin t$. (a) Find a single equation in x and y for the path of the particle and sketch the curve. (b) Find the velocity and acceleration vectors at any time t, and show that $\mathbf{a} = -\mathbf{R}$ at all times. (c) Find $\mathbf{R}$, $\mathbf{v}$, and $\mathbf{a}$ when (1) $t_1 = \frac{\pi}{6}$, (2) $t_2 = \pi$, and draw them on the sketch. (d) Find the speed of the particle and the magnitude of its acceleration at each instant in (c). (e) When is the speed a maximum? a minimum?

(a) Since $\frac{x^2}{9} = \cos^2 t$ and $\frac{y^2}{4} = \sin^2 t$, therefore

$$\frac{x^2}{9} + \frac{y^2}{4} = 1$$

and the particle moves in a counterclockwise direction along an ellipse, starting, when $t = 0$, at $(3, 0)$ and returning to this point when $t = 2\pi$.

(b) We have

$$\mathbf{R} = 3\cos t\,\mathbf{i} + 2\sin t\,\mathbf{j},$$
$$\mathbf{v} = -3\sin t\,\mathbf{i} + 2\cos t\,\mathbf{j},$$
$$\mathbf{a} = -3\cos t\,\mathbf{i} - 2\sin t\,\mathbf{j} = -\mathbf{R}.$$

The acceleration, then, is always directed toward the center of the ellipse.

(c) At $t_1 = \frac{\pi}{6}$,

$$\mathbf{R}_1 = \frac{3\sqrt{3}}{2}\mathbf{i} + \mathbf{j},$$
$$\mathbf{v}_1 = -\frac{3}{2}\mathbf{i} + \sqrt{3}\mathbf{j},$$
$$\mathbf{a}_1 = -\frac{3\sqrt{3}}{2}\mathbf{i} - \mathbf{j}.$$

At $t_2 = \pi$,

$$\mathbf{R}_2 = -3\mathbf{i},$$
$$\mathbf{v}_2 = -2\mathbf{j},$$
$$\mathbf{a}_2 = 3\mathbf{i}.$$

The curve, and $\mathbf{v}$ and $\mathbf{a}$ at those instants, are sketched in Figure N4–15.

(d) At $t_1 = \frac{\pi}{6}$,   At $t_2 = \pi$,

$$|\mathbf{v}_1| = \sqrt{\frac{9}{4} + 3} = \frac{\sqrt{21}}{2}, \qquad |\mathbf{v}_2| = \sqrt{0 + 4} = 2,$$

$$|\mathbf{a}_1| = \sqrt{\frac{27}{4} + 1} = \frac{\sqrt{31}}{2}. \qquad |\mathbf{a}_2| = \sqrt{9 + 0} = 3.$$

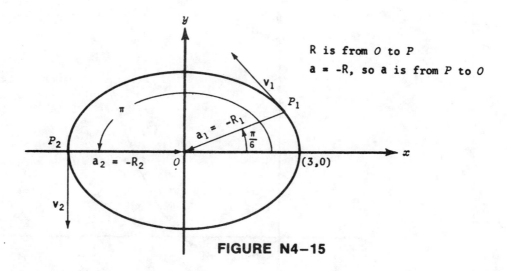

FIGURE N4–15

(e) For the speed $|\mathbf{v}|$ at any time t we have

$$\mathbf{v} = \sqrt{9 \sin^2 t + 4 \cos^2 t}$$
$$= \sqrt{4 \sin^2 t + 4 \cos^2 t + 5 \sin^2 t}$$
$$= \sqrt{4 + 5 \sin^2 t}.$$

We see immediately that the speed is a maximum when $t = \frac{\pi}{2}$ or $\frac{3\pi}{2}$, and a minimum when $t = 0$ or π. The particle goes fastest at the ends of the minor axis and most slowly at the ends of the major axis. Generally one can determine maximum or minimum speed by finding $\frac{d}{dt}|\mathbf{v}|$, setting it equal to zero, and applying the usual tests to sort out values of t that yield maximum or minimum speeds.

27. A particle moves along the parabola $y = x^2 - x$ with constant speed $\sqrt{10}$. Find $\mathbf{v}$ at $(2, 2)$.

Since

$$v_y = \frac{dy}{dt} = (2x - 1)\frac{dx}{dt} = (2x - 1)v_x \tag{1}$$

and

$$v_x^2 + v_y^2 = 10, \tag{2}$$

we have

$$v_x^2 + (2x - 1)^2 v_x^2 = 10. \tag{3}$$

Relation (3) holds at all times; specifically, at $(2, 2)$ $v_x^2 + 9v_x^2 = 10$ so that $v_x = \pm 1$. From (1), then, we see that $v_y = \pm 3$. Therefore $\mathbf{v}$ at $(2, 2)$ is either $\mathbf{i} + 3\mathbf{j}$ or $-\mathbf{i} - 3\mathbf{j}$. The former corresponds to counterclockwise motion along the parabola, as shown in Figure N4–16a; the latter to clockwise motion, indicated in Figure N4–16b.

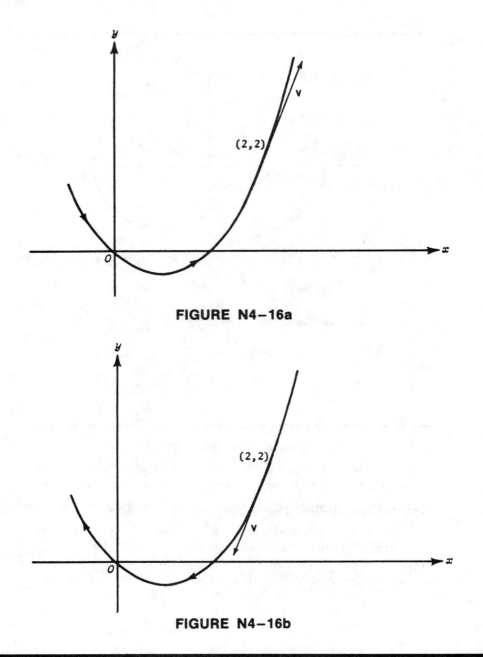

FIGURE N4–16a

FIGURE N4–16b

K.   Differentials; Linear Approximations_____

If $y = f(x)$ is differentiable, then the *differential dx* is defined to be a real variable and the *differential dy* is defined to be $f'(x)\ dx$. Every formula for derivatives in Chapter 3, page 33, gives rise to one for differentials by the mere replacement of the word "derivative" by the word "differential." Thus, for example we obtain from the rule "The derivative of a sum is the sum of the derivatives" the rule "The differential of a sum is the sum of the differentials"; in the same way, from "The derivative of a product is the first times the derivative of the second plus the second times the derivative of the first" we obtain "The differential of a product is the first times the differential of the second plus the second times the differential of the first."

Examples

28. If $y = xe^{-x}$, then

$$dy = x \, d(e^{-x}) + e^{-x} \, dx = -xe^{-x} \, dx + e^{-x} \, dx \quad \text{or} \quad e^{-x}(1 - x) \, dx.$$

29. If $\ln \sqrt{x^2 + y^2} = \tan^{-1} \dfrac{x}{y}$, then we can find $\dfrac{dy}{dx}$ by using differentials.

Since $\ln \sqrt{x^2 + y^2} = \dfrac{1}{2} \ln (x^2 + y^2)$, we get

$$\frac{1}{2} \frac{2x \, dx + 2y \, dy}{x^2 + y^2} = \frac{\dfrac{y \, dx - x \, dy}{y^2}}{1 + \dfrac{x^2}{y^2}}, \qquad \frac{x \, dx + y \, dy}{x^2 + y^2} = \frac{y \, dx - x \, dy}{y^2 + x^2},$$

$$x \, dx + y \, dy = y \, dx - x \, dy, \qquad \text{and} \qquad dy(y + x) = dx(y - x),$$

so that

$$\frac{dy}{dx} = \frac{y - x}{y + x}.$$

Approximations Using Differentials.

If dx is taken equal to Δx (and is sufficiently small), then dy is a reasonable approximation to Δy, as can be seen from Figure N4–17. Since dy is the change in y along the tangent line, it is called a *linear approximation*.

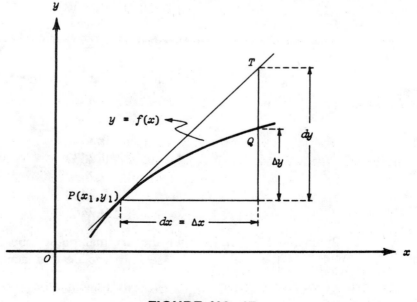

FIGURE N4–17

Examples

30. By approximately how much does the volume of a sphere of radius 4 in. change if the radius is decreased by 0.1 in.?

The volume V equals $\frac{4}{3}\pi r^3$ and $dV = 4\pi r^2 dr$. If we take $r = 4$ and $dr = -0.1$, then $dV = 4\pi(4)^2 \cdot (-0.1) = -6.4\pi$ in.^{3}

31. If the edge of a cube is increased by approximately 1%, find the approximate error in the surface area.

If we let S be the surface area and x the edge, then $S = 6x^2$ and $dS = 12x\,dx$. Since $dx = 0.01x$, $dS = 12x(0.01x) = 0.12x^2$. The relative error in S is thus $\frac{0.12x^2}{6x^2} = 0.02$, and the percentage error is 2%.

L.   Newton's Method for Approximating Roots

To approximate a root of the equation $f(x) = 0$, where $f'(x)$ exists, by Newton's method, we use the recursion formula

$$x_{n+1} = x_n - \frac{f(x_n)}{f'(x_n)}.$$

Examples

32. Find the real root of $f(x) = x^3 + 4x - 1 = 0$ to four decimal places.

Since $f(0) < 0$ while $f(1) > 0$, f has a root between 0 and 1. We choose $x_1 = 0.5$ as our first approximation. Since $f'(x) = 3x^2 + 4$, the formula above becomes

$$x_{n+1} = x_n - \frac{x_n^3 + 4x_n - 1}{3x_n^2 + 4}.$$

Retaining only four decimal places, we get the following table:

n	x_n
1	0.5
2	0.2632
3	0.2457
4	0.2463
5	0.2463

Since the four digits after the decimal point repeat, we conclude that $x = 0.2463$ is the real root we seek to the desired accuracy.

33. Find x in $[0, 2]$, correct to four decimal places, for which $x = \cos x$.

We let $f(x) = x - \cos x$ and choose x_1 equal to 1. Newton's method yields the formula

$$x_{n+1} = x_n - \frac{x_n - \cos x_n}{1 + \sin x_n}.$$

Our approximations, retaining four decimal places, are as follows:

n	x_n
1	1.
2	0.7066
3	0.7393
4	0.7391
5	0.7391

We see that the desired root is 0.7391, correct to four decimal places.

M. Related Rates

If several variables which are functions of time t are related by an equation, we can obtain a relation involving their (time) rates of change by differentiating with respect to t.

Example 34. If one leg AB of a right triangle increases at the rate of 2 in./sec, while the other leg AC decreases at 3 in./sec, find how fast the hypotenuse is changing when $AB = 6$ ft and $AC = 8$ ft.

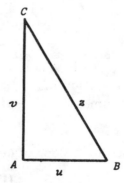

FIGURE N4–18

See Figure N4–18. Let u, v, and z denote the lengths respectively of AB, AC, and BC. We know that $\frac{du}{dt} = \frac{1}{6}$ (ft/sec) and $\frac{dv}{dt} = -\frac{1}{4}$. Since (at any time) $z^2 = u^2 + v^2$, then

$$2z \frac{dz}{dt} = 2u \frac{du}{dt} + 2v \frac{dv}{dt} \quad \text{and} \quad \frac{dz}{dt} = \frac{u \frac{du}{dt} + v \frac{dv}{dt}}{z}.$$

At the instant in question, $u = 6$, $v = 8$, and $z = 10$. So

$$\frac{dz}{dt} = \frac{6\left(\frac{1}{6}\right) + 8\left(-\frac{1}{4}\right)}{10} = -\frac{1}{10} \text{ ft/sec.}$$

Set 4: Multiple-Choice Questions on Applications of Differential Calculus

1. The slope of the curve $y^3 - xy^2 = 4$ at the point where $y = 2$ is

   (A) -2 (B) $\dfrac{1}{4}$ (C) $-\dfrac{1}{2}$ (D) $\dfrac{1}{2}$ (E) 2

2. The slope of the curve $y^2 - xy - 3x = 1$ at the point $(0, -1)$ is

   (A) -1 (B) -2 (C) $+1$ (D) 2   (E) -3

3. The equation of the tangent to the curve $y = x \sin x$ at the point $\left(\dfrac{\pi}{2}, \dfrac{\pi}{2}\right)$ is

   (A) $y = x - \pi$ (B) $y = \dfrac{\pi}{2}$ (C) $y = \pi - x$ (D) $y = x + \dfrac{\pi}{2}$

   (E) $y = x$

4. The tangent to the curve of $y = xe^{-x}$ is horizontal when x is equal to

   (A) 0   (B) 1   (C) -1 (D) $\dfrac{1}{e}$ (E) none of these

5. The point on the curve $y = \sqrt{2x + 1}$ at which the normal is parallel to the line $y = -3x + 6$ is

   (A) $(4, 3)$ (B) $(0, 1)$ (C) $(1, \sqrt{3})$ (D) $(4, -3)$
   (E) $(2, \sqrt{5})$

6. The minimum value of the slope of the curve $y = x^5 + x^3 - 2x$ is

   (A) 0   (B) 2   (C) 6   (D) -2 (E) none of these

7. The equation of the tangent to the curve $x^2 = 4y$ at the point on the curve where $x = -2$ is

   (A) $x + y - 3 = 0$ (B) $y - 1 = 2x(x + 2)$
   (C) $x - y + 3 = 0$ (D) $x + y - 1 = 0$
   (E) $x + y + 1 = 0$

8. The equation of the tangent to the hyperbola $x^2 - y^2 = 12$ at the point $(4, 2)$ on the curve is

   (A) $x - 2y + 6 = 0$ (B) $y = 2x$ (C) $y = 2x - 6$ (D) $y = \dfrac{x}{2}$
   (E) $x + 2y = 6$

9.   The tangent to the curve $y^2 - xy + 9 = 0$ is vertical when

   (A) $y = 0$ **(B)** $y = \pm\sqrt{3}$ **(C)** $y = \dfrac{1}{2}$ **(D)** $y = \pm 3$

   (E)   none of these

10. If differentials are used for computation, then the best approximation, in cubic inches, to the increase in volume of a sphere when the radius is increased from 3 to 3.1 in. is

   (A) 11.3   **(B)** 11.7   **(C)** 12.1   **(D)** 33.9   **(E)** 39.7

11. When $x = 3$, the equation $2x^2 - y^3 = 10$ has the solution $y = 2$. If differentials are used to compute, then, when $x = 3.04$, y equals approximately

   (A) 1.6   **(B)** 1.96   **(C)** 2.04   **(D)** 2.14   **(E)** 2.4

12. If the side e of a square is increased by 1%, then the area is increased approximately by

   (A) $0.02e$ **(B)** $0.02e^2$ **(C)** $0.01e^2$ **(D)** 1%   **(E)** $0.01e$

13. The edge of a cube has length 10 in., with a possible error of 1%. The possible error, in cubic inches, in the volume of the cube is

   (A) 3   **(B)** 1%   **(C)** 10   **(D)** 30   **(E)**   none of these

14. The function $f(x) = x^4 - 4x^2$ has

   (A)   one relative minimum and two relative maxima
   (B)   one relative minimum and one relative maximum
   (C)   two relative maxima and no relative minimum
   (D)   two relative minima and no relative maximum
   (E)   two relative minima and one relative maximum

15. The number of inflection points of the curve in Question 14 above is

   (A) 0   **(B)** 1   **(C)** 2   **(D)** 3   **(E)** 4

16. The maximum value of the function $y = -4\sqrt{2 - x}$ is

   (A) 0   **(B)** -4 **(C)** 2   **(D)** -2 **(E)**   none of these

17. The total number of relative maximum and minimum points of the function whose derivative, for all x, is given by $f'(x) = x(x - 3)^2(x + 1)^4$ is

   (A) 0   **(B)** 1   **(C)** 2   **(D)** 3   **(E)**   none of these

18. If $x \neq 0$, then the slope of $x \sin \dfrac{1}{x}$ equals zero whenever

   (A) $\tan \dfrac{1}{x} = x$ **(B)** $\tan \dfrac{1}{x} = -x$ **(C)** $\cos \dfrac{1}{x} = 0$

   (D) $\sin \dfrac{1}{x} = 0$ **(E)** $\tan \dfrac{1}{x} = \dfrac{1}{x}$

19. On the closed interval $[0, 2\pi]$, the maximum value of the function $f(x) = 4 \sin x - 3 \cos x$ is

(A) 3　　(B) 4　　(C) $\dfrac{24}{5}$　　(D) 5　　(E) none of these

20. If m_1 is the slope of the curve $xy = 2$ and m_2 is the slope of the curve $x^2 - y^2 = 3$, then at a point of intersection of the two curves

(A) $m_1 = -m_2$　　(B) $m_1 m_2 = -1$　　(C) $m_1 = m_2$　　(D) $m_1 m_2 = 1$
(E) $m_1 m_2 = -2$

21. The line $y = 3x + k$ is tangent to the curve $y = x^3$ when k is equal to

(A) 1 or -1　　(B) 0　　(C) 3 or -3　　(D) 4 or -4
(E) 2 or -2

22. The two tangents that can be drawn from the point $(3, 5)$ to the parabola $y = x^2$ have slopes

(A) 1 and 5　　(B) 0 and 4　　(C) 2 and 10　　(D) 2 and $-\dfrac{1}{2}$

(E) 2 and 4

In Questions 23–26, the motion of a particle on a straight line is given by $s = t^3 - 6t^2 + 12t - 8$.

23. The distance s is increasing for

(A) $t < 2$　　(B) all t except $t = 2$　　(C) $1 < t < 3$
(D) $t < 1$ or $t > 3$　　(E) $t > 2$

24. The minimum value of the speed is

(A) 1　　(B) 2　　(C) 3　　(D) 0　　(E) none of these

25. The acceleration is positive

(A) when $t > 2$　　(B) for all $t, t \neq 2$　　(C) when $t < 2$
(D) for $1 < t < 3$　　(E) for $1 < t < 2$

26. The speed of the particle is decreasing for

(A) $t > 2$　　(B) $t < 3$　　(C) all t　　(D) $t < 1$ or $t > 2$
(E) none of these

In Questions 27–29, a particle moves along a horizontal line according to the law $s = t^4 - 6t^3 + 12t^2 + 3$.

27. The particle is at rest when t is equal to

(A) 1 or 2　　(B) 0　　(C) $\dfrac{9}{4}$　　(D) 0, 2, or 3　　(E) none of these

28. The velocity, v, is increasing when

    (A) $t > 1$ **(B)** $1 < t < 2$ **(C)** $t < 2$ **(D)** $t < 1$ or $t > 2$
    (E) $t > 0$

29. The speed of the particle is increasing for

    (A) $0 < t < 1$ or $t > 2$ **(B)** $1 < t < 2$ **(C)** $t < 2$
    (D) $t < 0$ or $t > 2$ **(E)** $t < 0$

30. The displacement from the origin of a particle moving on a line is given by $s = t^4 - 4t^3$. The maximum displacement during the time interval $-2 \leq t \leq 4$ is

    (A) 27   **(B)** 3   **(C)** $12\sqrt{3} + 3$ **(D)** 48   **(E)** none of these

31. If a particle moves along a line according to the law $s = t^5 + 5t^4$, then the number of times it reverses direction is

    (A) 0   **(B)** 1   **(C)** 2   **(D)** 3   **(E)** 4

In Questions 32–35, the motion of a particle in a plane is given by the pair of equations $x = 2t$ and $y = 4t - t^2$.

*32. The particle moves along

    (A) an ellipse   **(B)** a circle   **(C)** a hyperbola   **(D)** a line
    (E) a parabola

*33. The speed of the particle at any time t is

    (A) $\sqrt{6 - 2t}$ **(B)** $2\sqrt{t^2 - 4t + 5}$ **(C)** $2\sqrt{t^2 - 2t + 5}$
    (D) $\sqrt{8}(|t - 2|)$ **(E)** $2(|3 - t|)$

*34. The minimum speed of the particle is

    (A) 2   **(B)** $2\sqrt{2}$ **(C)** 0   **(D)** 1   **(E)** 4

*35. The acceleration of the particle

    (A) depends on t **(B)** is always directed upward
    (C) is constant both in magnitude and in direction
    (D) never exceeds 1 in magnitude   **(E)** is none of these

In Questions 36–39, $\mathbf{R} = 3\cos\frac{\pi}{3}t\mathbf{i} + 2\sin\frac{\pi}{3}t\mathbf{j}$ is the (position) vector $x\mathbf{i} + y\mathbf{j}$ from the origin to a moving point $P(x, y)$ at time t.

*36. A single equation in x and y for the path of the point is

    (A) $x^2 + y^2 = 13$ **(B)** $9x^2 + 4y^2 = 36$ **(C)** $2x^2 + 3y^2 = 13$
    (D) $4x^2 + 9y^2 = 1$ **(E)** $4x^2 + 9y^2 = 36$

*Questions preceded by an asterisk are likely to appear only on the Calculus BC Examination.

*37.   When $t = 3$, the speed of the particle is

(A) $\dfrac{2\pi}{3}$ (B) 2   (C) 3   (D) π (E) $\dfrac{\sqrt{13}}{3}\pi$

*38.   The magnitude of the acceleration when $t = 3$ is

(A) 2   (B) $\dfrac{\pi^2}{3}$ (C) 3   (D) $\dfrac{2\pi^2}{9}$ (E) π

*39.   At the point where $t = \dfrac{1}{2}$, the slope of the curve along which the particle moves is

(A) $-\dfrac{2\sqrt{3}}{9}$ (B) $-\dfrac{\sqrt{3}}{2}$ (C) $\dfrac{2}{\sqrt{3}}$ (D) $-\dfrac{2\sqrt{3}}{3}$

(E)   none of these

*40.   If a particle moves along a curve with constant speed, then

(A)   the magnitude of its acceleration must equal zero
(B)   the direction of acceleration must be constant
(C)   the curve along which the particle moves must be a straight line
(D)   its velocity and acceleration vectors must be perpendicular
(E)   the curve along which the particle moves must be a circle

*41.   A particle is moving on the curve of $y = 2x - \ln x$ so that $\dfrac{dx}{dt} = -2$ at all time t. At the point $(1, 2)$, $\dfrac{dy}{dt}$ is

(A) 4   (B) 2   (C) -4 (D) 1   (E) -2

42.   A balloon is being filled with helium at the rate of 4 ft³/min. The rate, in square feet per minute, at which the surface area is increasing when the volume is $\dfrac{32\pi}{3}$ ft³ is

(A) 4π (B) 2   (C) 4   (D) 1   (E) 2π

43.   A circular conical reservoir, vertex down, has depth 20 ft and radius of the top 10 ft. Water is leaking out so that the surface is falling at the rate of $\dfrac{1}{2}$ ft/hr. The rate, in cubic feet per hour, at which the water is leaving the reservoir when the water is 8 ft deep is

(A) 4π (B) 8π (C) 16π (D) $\dfrac{1}{4\pi}$ (E) $\dfrac{1}{8\pi}$

44.   A vertical circular cylinder has radius r feet and height h feet. If the height and radius both increase at the constant rate of 2 ft/sec, then the rate, in square feet per second, at which the lateral surface area increases is

(A) $4\pi r$ (B) $2\pi(r + h)$ (C) $4\pi(r + h)$ (D) $4\pi rh$
(E) $4\pi h$

45. A relative minimum value of the function $y = \frac{e^x}{x}$ is

(A) $\frac{1}{e}$ (B) 1   (C) -1 (D) e (E) 0

46. The area of the largest rectangle that can be drawn with one side along the x-axis and two vertices on the curve of $y = e^{-x^2}$ is

(A) $\sqrt{\frac{2}{e}}$ (B) $\sqrt{2e}$ (C) $\frac{2}{e}$ (D) $\frac{1}{\sqrt{2e}}$ (E) $\frac{2}{e^2}$

47. A tangent drawn to the parabola $y = 4 - x^2$ at the point $(1, 3)$ forms a right triangle with the coordinate axes. The area of the triangle is

(A) $\frac{5}{4}$ (B) $\frac{5}{2}$ (C) $\frac{25}{2}$ (D) 1   (E) $\frac{25}{4}$

48. If the cylinder of largest possible volume is inscribed in a given sphere, the ratio of the volume of the sphere to that of the cylinder is

(A) $\sqrt{3} : 1$ (B) $\sqrt{3} : 3$ (C) $3 : 1$ (D) $2\sqrt{3} : 3$
(E) $3\sqrt{3} : 4$

49. A line is drawn through the point $(1, 2)$ forming a right triangle with the positive x- and y-axes. The slope of the line forming the triangle of least area is

(A) -1 (B) -2 (C) -4 (D) $-\frac{1}{2}$ (E) -3

50. The point(s) on the curve $x^2 - y^2 = 4$ closest to the point $(6, 0)$ is (are)

(A) $(2, 0)$ (B) $(\sqrt{5}, \pm 1)$ (C) $(3, \pm\sqrt{5})$ (D) $(\sqrt{13}, \pm\sqrt{3})$
(E) none of these

51. The sum of the squares of two positive numbers is 200; their minimum product is

(A) 100   (B) $25\sqrt{7}$ (C) 28   (D) $24\sqrt{14}$ (E) none of these

52. The first-quadrant point on the curve $y^2x = 18$ which is closest to the point $(2, 0)$ is

(A) $(2, 3)$ (B) $(6, \sqrt{3})$ (C) $(3, \sqrt{6})$ (D) $(1, 3\sqrt{2})$
(E) none of these

53. Two cars are traveling along perpendicular roads, car A at 40 mi/hr, car B at 60 mi/hr. At noon, when car A reaches the intersection, car B is 90 mi away, and moving toward it. At 1 P.M. the distance between the cars is changing, in miles per hour, at the rate of

(A) -40 (B) 68   (C) 4   (D) -4 (E) 40

54. For Problem 53, if t is the number of hours of travel after noon, then the cars are closest together when t is

(A) 0   (B) $\dfrac{27}{26}$ (C) $\dfrac{9}{5}$ (D) $\dfrac{3}{2}$ (E) $\dfrac{14}{13}$

55. If h is a small negative number, then the best approximation for $\sqrt[3]{27 + h}$ is

(A) $3 + \dfrac{h}{27}$ (B) $3 - \dfrac{h}{27}$ (C) $\dfrac{h}{27}$ (D) $-\dfrac{h}{27}$ (E) $3 - \dfrac{h}{9}$

56. If $f(x) = xe^{-x}$, then at $x = 0$

(A) f is increasing   (B) f is decreasing
(C) f has a relative maximum   (D) f has a relative minimum
(E) f' does not exist

57. A function f has a derivative for each x such that $|x| < 2$ and has a local minimum at $(2, -5)$. Which statement below must be true?

(A) $f'(2) = 0$.
(B) f' exists at $x = 2$.
(C) The graph is concave up at $x = 2$.
(D) $f'(x) < 0$ if $x < 2$, $f'(x) > 0$ if $x > 2$.
(E) $f'(x) > 0$ if $x < 2$, $f'(x) < 0$ if $x > 2$.

58. The height of a rectangular box is 10 in. Its length increases at the rate of 2 in./sec; its width decreases at the rate of 4 in./sec. When the length is 8 in. and the width is 6 in., the volume of the box is changing, in cubic inches per second, at the rate of

(A) 200   (B) 80   (C) -80 (D) -200 (E) -20

59. A cube whose edge is x is contracting. When its surface area is changing at a rate which is equal to 6 times the rate of change of its edge, then the length of the edge is

(A) 2   (B) $\dfrac{3}{4}$ (C) 1   (D) $\dfrac{4}{3}$ (E) $\dfrac{1}{2}$

60. If $f(x) = ax^4 + bx^2$ and $ab > 0$, then

(A) the curve has no horizontal tangents
(B) the curve is concave up for all x
(C) the curve is concave down for all x
(D) the curve has no inflection point
(E) none of the preceding is necessarily true

61. At which point on the following graph do both $\frac{dy}{dx}$ and $\frac{d^2y}{dx^2}$ equal zero?

(A) P (B) Q (C) R (D) S (E) T

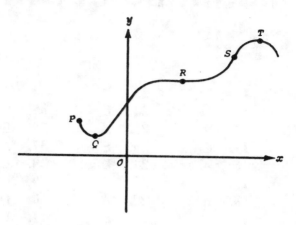

62. Which statement below is true about the curve $y = \dfrac{x^2 + 4}{2 + 7x - 4x^2}$?

(A)   The line $x = -\dfrac{1}{4}$ is a vertical asymptote.

(B)   The line $x = 1$ is a vertical asymptote.

(C)   The line $y = \dfrac{1}{4}$ is a horizontal asymptote.

(D)   The graph has no vertical or horizontal asymptotes.

(E)   The line $y = 2$ is a horizontal asymptote.

*63. The equation of the tangent to the curve with parametric equations $x = 2t + 1$, $y = 3 - t^3$ at the point where $t = 1$ is

(A) $2x + 3y = 12$ (B) $3x + 2y = 13$

(C) $6x + y = 20$ (D) $3x - 2y = 5$

(E)   none of these

64. The positive root of $x \ln x = 1$, to two decimal places, using Newton's method, is

(A) 1.7013   (B) 1.7462   (C) 1.7577   (D) 1.7632   (E) 1.7730

Answers for Set 4: Applications of Differential Calculus

1. D	12. B	23. B	34. A	45. D	55. A
2. A	13. D	24. D	35. C	46. A	56. A
3. E	14. E	25. A	36. E	47. E	57. D
4. B	15. C	26. E	37. A	48. A	58. D
5. A	16. A	27. B	38. B	49. B	59. E
6. D	17. B	28. D	39. D	50. C	60. D
7. E	18. E	29. A	40. D	51. E	61. C
8. C	19. D	30. D	41. E	52. C	62. A
9. D	20. B	31. C	42. C	53. D	63. B
10. A	21. E	32. E	43. B	54. B	64. D
11. C	22. C	33. B	44. C		

1. D.   Substituting $y = 2$ yields $x = 1$. We find y' implicitly.

$$3y^2y' - (2xyy' + y^2) = 0; \qquad (3y^2 - 2xy)y' - y^2 = 0.$$

Replace x by 1 and y by 2; solve for y'.

2. A. $2yy' - (xy' + y) - 3 = 0$. Replace x by 0 and y by -1; solve for y'.

3. E.   Find the slope of the curve at $x = \frac{\pi}{2}$: $y' = x \cos x + \sin x$. At $x = \frac{\pi}{2}$, $y' = \frac{\pi}{2} \cdot 0 + 1$. The equation is $y - \frac{\pi}{2} = 1\left(x - \frac{\pi}{2}\right)$.

4. B.   Since $y' = e^{-x}(1 - x)$ and $e^{-x} > 0$ for all x, $y' = 0$ when $x = 1$.

5. A.   Since the slope of the tangent to the curve is $y' = \dfrac{1}{\sqrt{2x + 1}}$, the slope of the normal is $-\sqrt{2x + 1}$. So $-\sqrt{2x + 1} = -3$ and $2x + 1 = 9$.

6. D.   The slope $y' = 5x^4 + 3x^2 - 2$. Let $g = y'$. Since $g'(x) = 20x^3 + 6x = 2x(10x^2 + 3)$, $g'(x) = 0$ only if $x = 0$. Since $g''(x) = 60x^2 + 6$, g'' is always positive, assuring that $x = 0$ yields the minimum slope. Find y' when $x = 0$.

7. E.   The slope $y' = \frac{2x}{4}$; at the given point $y' = -\frac{4}{4} = -1$ and $y = 1$. The equation is therefore

$$y - 1 = -1(x + 2) \qquad \text{or} \qquad x + y + 1 = 0.$$

8. C.   Since $2x - 2yy' = 0$, $y' = \frac{x}{y}$. At $(4, 2)$, $y' = 2$. The equation of the tangent at $(4, 2)$ is $y - 2 = 2(x - 4)$.

9. D. Since $y' = \dfrac{y}{2y - x}$, the tangent is vertical for $x = 2y$. Substitute in the given equation.

10. A. Since $V = \frac{4}{3}\pi r^3$, $dV = 4\pi r^2\,dr$. Replace r by 3 and dr by 0.1. To the nearest tenth, a calculator gives 11.3 for the increase.

11. C. We get, taking differentials, $4x\,dx - 3y^2\,dy = 0$, and use $x = 3$, $y = 2$, and $dx = 0.04$ to find dy. The answer is $y + dy$.

12. B. Since $A = e^2$, $dA = 2e\,de = 2e(0.01e)$.

13. D. Since $e = 10$ with a possible error of 1%, $de = \pm 0.1$. From $V = e^3$ and $dV = 3e^2\,de$, we get $dV = \pm 30$ or a possible error of 30 in.^{3} in the volume.

14. E. $f'(x) = 4x^3 - 8x = 4x(x^2 - 2)$. $f' = 0$ if $x = 0$ or $\pm\sqrt{2}$.
$f''(x) = 12x^2 - 8$; f'' is positive if $x = \pm\sqrt{2}$, negative if $x = 0$.

15. C. Since $f''(x) = 4(3x^2 - 2)$, it equals 0 if $x = \pm\sqrt{\dfrac{2}{3}}$. Since f'' changes sign as x increases through each of these, both are inflection points.

16. A. The domain of y is $\{x \mid x \le 2\}$. Note that y is negative for each x in the domain except 2, where $y = 0$.

17. B. $f'(x)$ changes sign only as x passes through zero.

18. E. The derivative equals $x\cos\dfrac{1}{x}\left(-\dfrac{1}{x^2}\right) + \sin\dfrac{1}{x}$.

19. D. Since $f'(x) = 4\cos x + 3\sin x$, the critical values of x are those for which $\tan x = -\dfrac{4}{3}$. For these values,

$$\text{when } \frac{\pi}{2} < x < \pi, \text{ then } \sin x = \frac{4}{5} \text{ and } \cos x = -\frac{3}{5};$$

but

$$\text{when } \frac{3\pi}{2} < x < 2\pi, \text{ then } \sin x = -\frac{4}{5} \text{ and } \cos x = \frac{3}{5}.$$

If these are used to determine the sign of $f''(x)$, we see that the second-quadrant x yield a negative second derivative, but the fourth-quadrant x give a positive second derivative. It follows that the function is a maximum for the former set, for which f has the value 5. Note that $f(0) = f(2\pi) = -3$.

20. B. We find the points of intersection. From the first curve we get $y = \dfrac{2}{x}$, which yields, in the second curve,

$$x^2 - \frac{4}{x^2} = 3 \qquad \text{or} \qquad x^4 - 3x^2 - 4 = 0.$$

$(x^2 - 4)(x^2 + 1) = 0$ if $x = \pm 2$. So the points of intersection are $(2, 1)$ and $(-2, -1)$. Since $m_1 = -\dfrac{2}{x^2}$ and $m_2 = \dfrac{x}{y}$, at each point $m_1 = -\dfrac{1}{2}$ and $m_2 = 2$.

21. E. The slope of $y = x^3$ is $3x^2$. It is equal to 3 when $x = \pm 1$. At $x = 1$, the equation of the tangent is

$$y - 1 = 3(x - 1) \qquad \text{or} \qquad y = 3x - 2.$$

At $x = -1$, the equation is

$$y + 1 = 3(x + 1) \qquad \text{or} \qquad y = 3x + 2.$$

22. C. Let the tangent to the parabola from $(3, 5)$ meet the curve at (x_1, y_1). Its equation is $y - 5 = 2x_1(x - 3)$. Since the point (x_1, y_1) is on both the tangent and the parabola, we solve simultaneously:

$$y_1 - 5 = 2x_1(x_1 - 3) \qquad \text{and} \qquad y_1 = x_1^2$$

The points of tangency are $(5, 25)$ and $(1, 1)$. The slopes, which equal $2x_1$, are 10 and 2.

23. B. The distance is increasing when v is positive. Since $v = \dfrac{ds}{dt} = 3(t - 2)^2$, $v > 0$ for all $t \neq 2$.

24. D. The speed $= |v|$. From Question 23, $|v| = v$. The least value of v is 0.

25. A. The acceleration $a = \dfrac{dv}{dt}$. From Question 23, $a = 6(t - 2)$.

26. E. The speed is decreasing when v and a have opposite signs. The answer is $t < 2$, since for all such t the velocity is positive while the acceleration is negative. For $t > 2$, both v and a are positive.

27. B. The particle is at rest when $v = 0$; $v = 2t(2t^2 - 9t + 12) = 0$ only if $t = 0$. Note that the discriminant of the quadratic factor $(b^2 - 4ac)$ is negative.

28. D. Since $a = 12(t - 1)(t - 2)$, we check the signs of a in the intervals $t < 1, 1 < t < 2$, and $t > 2$. We choose those where $a > 0$.

29. A. From Questions 27 and 28 we see that $v > 0$ it $t > 0$ and that $a > 0$ if $t < 1$ or $t > 2$. So both v and a are positive if $0 < t < 1$ or $t > 2$. There are no values of t for which both v and a are negative.

30. D. See Figure 4-1, which shows the motion of the particle during the

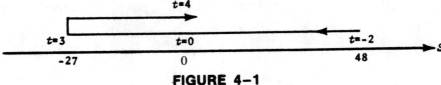

FIGURE 4–1

time interval $-2 \leq t \leq 4$. The particle is at rest when $t = 0$ or 3, but reverses direction only at 3. The endpoints need to be checked here, of course. Indeed, the maximum displacement occurs at one of those, namely, when $t = -2$

31. C.   Since $v = 5t^3(t + 4)$, $v = 0$ when $t = -4$ or 0. Note that v does change sign at each of these times.

32. E.   Eliminating t yields the equation $y = -\frac{1}{4}x^2 + 2x$.

33. B. $|\mathbf{v}| = \sqrt{\left(\dfrac{dx}{dt}\right)^2 + \left(\dfrac{dy}{dt}\right)^2} = \sqrt{2^2 + (4 - 2t)^2}.$

34. A.   Since $|\mathbf{v}| = 2\sqrt{t^2 - 4t + 5}$, $\dfrac{d|\mathbf{v}|}{dt} = \dfrac{2t - 4}{\sqrt{t^2 - 4t + 5}} = 0$ if $t = 2$. We note that, as t increases through 2, the signs of $|\mathbf{v}|'$ are $-$, 0, $+$, assuring a minimum of $|\mathbf{v}|$ at $t = 2$. Evaluate $|\mathbf{v}|$ at $t = 2$.

35. C.   The direction of $\mathbf{a}$ is $\tan^{-1}\dfrac{\dfrac{d^2y}{dt^2}}{\dfrac{d^2x}{dt^2}}$. Since $\dfrac{d^2x}{dt^2} = 0$ and $\dfrac{d^2y}{dt^2} = -2$, the acceleration is always directed downward. Its magnitude, $\sqrt{0^2 + (-2)^2}$, is 2 for all t.

36. E.   Since $x = 3 \cos\dfrac{\pi}{3}t$ and $y = 2 \sin\dfrac{\pi}{3}t$, we note that $\left(\dfrac{x}{3}\right)^2 + \left(\dfrac{y}{2}\right)^2 = 1.$

37. A.   Note that $\mathbf{v} = -\pi \sin\dfrac{\pi}{3}t\mathbf{i} + \dfrac{2\pi}{3} \cos\dfrac{\pi}{3}t\mathbf{j}.$ At $t = 3$,

$$|\mathbf{v}| = \sqrt{(-\pi \cdot 0)^2 + \left(\dfrac{2\pi}{3} \cdot -1\right)^2}.$$

38. B. $\mathbf{a} = -\dfrac{\pi^2}{3} \cos\dfrac{\pi}{3}t\mathbf{i} - \dfrac{2\pi^2}{9} \sin\dfrac{\pi}{3}t\mathbf{j}.$ At $t = 3$,

$$|\mathbf{a}| = \sqrt{\left(\dfrac{-\pi^2}{3} \cdot -1\right)^2 + \left(\dfrac{-2\pi^2}{9} \cdot 0\right)^2}.$$

39. D.   The slope of the curve is the slope of $\mathbf{v}$, namely, $\dfrac{\dfrac{dy}{dt}}{\dfrac{dx}{dt}}$. At $t = \dfrac{1}{2}$, the slope is equal to

$$\dfrac{\dfrac{2\pi}{3} \cdot \cos\dfrac{\pi}{6}}{-\pi \cdot \sin\dfrac{\pi}{6}} = -\dfrac{2}{3} \cot\dfrac{\pi}{6}.$$

40. **D.**   Using the notations v_x, v_y, a_x, and a_y, we are given that $|\mathbf{v}| = \sqrt{v_x^2 + v_y^2} = k$, where k is a constant. Then

$$\frac{d|\mathbf{v}|}{dt} = \frac{v_x a_x + v_y a_y}{|\mathbf{v}|} = 0 \qquad \text{or} \qquad \frac{v_x}{v_y} = -\frac{a_y}{a_x}.$$

41. **E.** $\dfrac{dy}{dt} = \left(2 - \dfrac{1}{x}\right)\dfrac{dx}{dt} = \left(2 - \dfrac{1}{x}\right)(-2).$

42. **C.**   Since $V = \dfrac{4}{3}\pi r^3$, $\dfrac{dV}{dt} = 4\pi r^2 \dfrac{dr}{dt}$. Since $\dfrac{dV}{dt} = 4$, $\dfrac{dr}{dt} = \dfrac{1}{\pi r^2}$. When $V = \dfrac{32\pi}{3}$, $r = 2$ and $\dfrac{dr}{dt} = \dfrac{1}{4\pi}$.

$$S = 4\pi r^2; \qquad \frac{dS}{dt} = 8\pi r \frac{dr}{dt};$$

when $r = 2$, $\dfrac{dS}{dt} = 8\pi(2)\left(\dfrac{1}{4\pi}\right) = 4$.

43. **B.**   See Figure S4–5 on page 112. Replace the printed measurements of the radius and height by 10 and 20, respectively. We are given here that $r = \dfrac{h}{2}$ and that $\dfrac{dh}{dt} = -\dfrac{1}{2}$. Since $V = \dfrac{1}{3}\pi r^2 h$, we have $V = \dfrac{\pi}{3}\dfrac{h^3}{4}$. So

$$\frac{dV}{dt} = \frac{\pi h^2}{4}\frac{dh}{dt} = \frac{-\pi h^2}{8}.$$

Replace h by 8.

44. **C.**   See Example 22 on pages 77 and 78. We know that $\dfrac{dh}{dt} = \dfrac{dr}{dt} = 2$. Since $S = 2\pi rh$,

$$\frac{dS}{dt} = 2\pi\left(r\frac{dh}{dt} + h\frac{dr}{dt}\right).$$

45. **D.** $y' = \dfrac{e^x(x - 1)}{x^2}$ $(x \neq 0)$. Since $y' = 0$ if $x = 1$ and changes from negative to positive as x increases through 1, $x = 1$ yields a minimum. Evaluate y at $x = 1$.

46. **A.**   The domain of y is $-\infty < x < \infty$. The graph of y, which is nonnegative, is symmetric to the y-axis. The inscribed rectangle has area $A = 2xe^{-x^2}$. Thus $A' = \dfrac{2(1 - 2x^2)}{e^{x^2}}$, which is 0 when the positive value of x is $\dfrac{\sqrt{2}}{2}$. This value of x yields maximum area. Evaluate A.

47. **E.**   The equation of the tangent is $y = -2x + 5$. Its intercepts are $\dfrac{5}{2}$ and 5.

48.   A.   See Figure 4–2, where V, the volume of the cylinder, equals $2\pi r^2 h$. Then $V = 2\pi(R^2h - h^3)$ is a maximum for $h^2 = \frac{R^2}{3}$, and the ratio of the volumes of sphere to cylinder is

$$\frac{4}{3}\pi R^3 \;:\; 2\pi \cdot \frac{2}{3}R^2 \cdot \frac{R}{\sqrt{3}} \qquad \text{or} \qquad \sqrt{3} \;:\; 1.$$

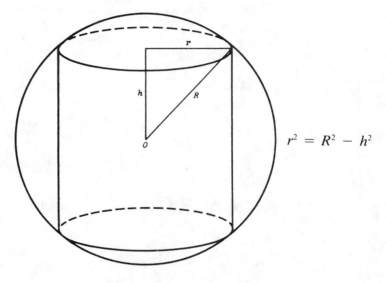

$$r^2 = R^2 - h^2$$

FIGURE 4–2

49.   B.   See Figure 4–3. If we let m be the slope of the line, then its equation is $y - 2 = m(x - 1)$ with intercepts as indicated in the figure.

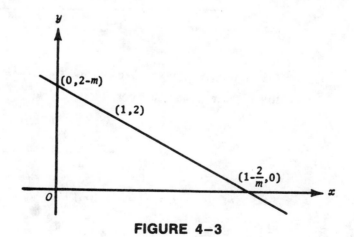

FIGURE 4–3

The area A of the triangle is given by

$$A = \frac{1}{2}(2 - m)\left(1 - \frac{2}{m}\right)$$

$$= \frac{1}{2}\left(4 - \frac{4}{m} - m\right).$$

Then $\frac{dA}{dm} = \frac{1}{2}\left(\frac{4}{m^2} - 1\right)$ and equals 0 when $m = \pm 2$; m must be negative.

50. C.   Let $q = (x - 6)^2 + y^2$ be the quantity to be minimized. Then

$$q = (x - 6)^2 + (x^2 - 4);$$

$q' = 0$ when $x = 3$. Note that it suffices to minimize the square of the distance.

51. E.   Minimize, if possible, xy, where $x^2 + y^2 = 200$ $(x, y > 0)$. The derivative of the product is $\dfrac{2(100 - x^2)}{\sqrt{200 - x^2}}$, which equals 0 for $x = 10$. But the signs of the derivative as x increases through 10 show that 10 yields a maximum product. No minimum exists.

52. C.   Minimize $q = (x - 2)^2 + \dfrac{18}{x}$. Since

$$q' = 2(x - 2) - \frac{18}{x^2} = \frac{2(x^3 - 2x^2 - 9)}{x^2},$$

$q' = 0$ if $x = 3$. The signs of q' about $x = 3$ assure a minimum.

53. D.   See Figure 4–4. At noon, car A is at O, car B at N; the cars are shown

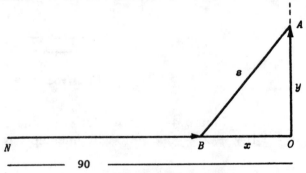

FIGURE 4–4

t hours after noon. We know that $\dfrac{dx}{dt} = -60$ and that $\dfrac{dy}{dt} = 40$. Using $s^2 = x^2 + y^2$, we get

$$\frac{ds}{dt} = \frac{x\dfrac{dx}{dt} + y\dfrac{dy}{dt}}{s} = \frac{-60x + 40y}{s}.$$

At 1 P.M., $x = 30$, $y = 40$, and $s = 50$.

54. B. $\dfrac{ds}{dt}$ (from Question 53) is zero when $y = \dfrac{3}{2}x$. Note that $x = 90 - 60t$ and $y = 40t$.

55. A.   Let $y = \sqrt[3]{x} = x^{1/3}$, and evaluate $y + dy$ for $x = 27$ and $dx = h$. Then

$$dy = \frac{1}{3x^{2/3}}\,dx = \frac{h}{3(27)^{2/3}} = \frac{h}{27}.$$

So the answer is $3 + \dfrac{h}{27}$.

56.   A.   Since $f'(x) = e^{-x}(1 - x), f'(0) > 0$.

57.   D.   Note that (E) is false and that no one of (A), (B), (C) is necessarily true.

58.   D.   Since $V = 10lw$, $V' = 10\left(l\frac{dw}{dt} + w\frac{dl}{dt}\right) = 10(8 \cdot -4 + 6 \cdot 2)$.

59.   E.   We must find x at the particular instant when $\frac{dS}{dt} = 6\frac{dx}{dt}$. Since $S = 6x^2$, $\frac{dS}{dt} = 12x\frac{dx}{dt}$ at any instant.

60.   D.   Since $ab > 0$, a and b have the same sign; therefore $f''(x) = 12ax^2 + 2b$ never equals 0. The curve has one horizontal tangent at $x = 0$.

61.   C.   Note that $\frac{dy}{dx} = 0$ at Q, R, and T. At Q, $\frac{d^2y}{dx^2} > 0$; at T, $\frac{d^2y}{dx^2} < 0$.

62.   A.   Note that, since $y = \frac{x^2 + 4}{(2 - x)(1 + 4x)}$, both $x = 2$ and $x = -\frac{1}{4}$ are vertical asymptotes. Also, $y = -\frac{1}{4}$ is a horizontal asymptote. (See page 74.)

63.   B.   Since $\frac{dy}{dx} = -\frac{3t^2}{2}$, therefore, at $t = 1$, $\frac{dy}{dx} = -\frac{3}{2}$. Also, $x = 3$ and $y = 2$.

64.   D.   With $x_1 = 1.8$, we get the successive approximations 1.7635, 1.7632, 1.7632.

Practice Section II Problems on Applications of Differentiation

1. Sketch the graph of $y = \frac{x^2}{x-2}$ after finding.

(a) intercepts;
(b) coordinates of any relative maximum or minimum points;
(c) values of x for which the graph is concave upward;
(d) any vertical or horizontal asymptotes.

2. Let the length of one side of a triangle be the constant $a,$ and let the measure of the opposite angle be the constant A. Find the proportions of the remaining sides of the triangle of maximum area.

3. Prove that the two tangents that can be drawn from any point on the line $y = -1$ to the parabola $x^2 = 4y$ are perpendicular.

4. An eaves trough is made by bending a flat piece of tin, which is 6 ft by 16 in., along two lines parallel to its length to form a rectangular cross-section. Find the largest cross-sectional area that can be thus obtained. Justify your answer.

5. The diameter and height of a paper cup in the shape of a cone are both 4 in. and water is leaking out at the rate of $\frac{1}{2}$ in.^{3}/sec. Find the rate at which the water level is dropping when the diameter of the surface is 2 in.

6. Find the point(s) on the parabola $2x = y^2$ which is (are) closest to the point $\left(0, \frac{3}{2}\right)$.

7. Sketch the graph of the function $f(x) = \frac{\sin x}{x}$ after considering

    (a) the domain;
    (b) the intercepts;
    (c) any relative maximum or minimum points;
    (d) symmetry;
    (e) $\lim f$ as $x \to 0$;
    (f) continuity of f at $x = 0$;
    (g) behavior for large x.

*8. The current i (in amp) in a certain circuit at time t (in sec) is given by

$$i = \frac{E}{R}(1 - e^{-(Rt/L)}),$$

where E, R, and L are positive constants.

    (a) Show that $i < \frac{E}{R}$ for all positive t.

    (b) Show that $\lim_{t \to \infty} i = \frac{E}{R}$.

    (c) If $E = 6$, $R = 5$, and $L = 0.1$, find the approximate change in current when t increases from $t = 0$ to $t = 0.01$.

    (d) If t, E, and L are held fixed while R varies, find, if it exists, $\lim_{R \to 0} i$.

9. (a) Prove that the curves of $xy = k$ and $y^2 - x^2 = h$ intersect in two points for every pair of reals k and h, provided that $hk \neq 0$.

    (b) Prove that, if $hk \neq 0$, every curve of the first family intersects every curve of the second family at right angles.

10. Find the equations of the lines that are tangent to the curve of $y = \frac{x}{1-x}$ and perpendicular to the line $4x + y + 4 = 0$.

11. Let P be a fixed point with coordinates (p, q). Prove analytically that the number of tangents that can be drawn from P to the parabola $y^2 = x$ depends on whether q^2 is greater than, less than, or equal to p, and determine the number in each of these cases.

12. A tangent drawn to the parabola $y = x^2$ at any point $P(x_1, y_1)$ $(x_1 > 0)$ intersects the axis of the parabola at T. If Q is the foot of the perpendicular from P to the axis of the parabola, prove that the area of the triangle QPT equals $x_1 y_1$.

---

*Questions preceded by an asterisk are likely to appear only on the Calculus BC Examination.

13. Find the constants a, b, c, d such that the curve of

$$y = ax^3 + bx^2 + cx + d$$

has a relative maximum at the point $(0, 1)$ and a point of inflection at the point $\left(1, \frac{1}{3}\right)$. Sketch the curve.

*14. A cycloid is given parametrically by $x = \theta - \sin \theta$, $y = 1 - \cos \theta$.
(a) Find the slope of the curve at the point where $\theta = \frac{2\pi}{3}$.
(b) Find the equation of the tangent to the cycloid at the point where $\theta = \frac{\pi}{3}$.

15. Find the area of the largest rectangle (with sides parallel to the coordinate axes) that can be inscribed in the region bounded by the graphs of $f(x) = 8 - 2x^2$ and $g(x) = x^2 - 4$.

16. Sketch the graph of $y = \ln(4 + x^2)$ after considering

(a) intercepts;
(b) symmetry;
(c) relative maximum or minimum points;
(d) possible inflection points.

17. The equation of a curve is

$$y^2 + ay = \frac{x + b}{cx^2 + dx + e},$$

where a, b, c, d, e are integers. Determine these integers if the curve exhibits all of the following characteristics:
(a) the curve goes through the origin but has no other intercepts;
(b) the curve is symmetric to the x-axis;
(c) the graph has two horizontal asymptotes, $y = \pm 1$, and one vertical asymptote, $x = 4$;
(d) the graph exists only if $x \leq 0$ or $x > 4$.

*18. A particle moves on the curve of $y^3 = 2x + 1$ so that its distance from the x-axis is increasing at the constant rate of 2 units per second. When $t = 0$, the particle is at $(0, 1)$.
(a) Find a pair of parametric equations $x = x(t)$ and $y = y(t)$ which describe the motion of the particle for nonnegative t.
(b) Find $|a|$, the magnitude of its acceleration, when $t = 1$.

19. Sketch the curve of $y^2 = x - x^3$ after finding
(a) intercepts;
(b) domain;
(c) symmetry;
(d) asymptotes;
(e) relative maxima and minima.

20. Let b be a constant. Find m in terms of b if the line $y = mx + b$ is to be tangent to the curve of $xy = 1$.

21. Sketch the graph of $y = x + \sin x$ after having found
(a) intercepts;
(b) relative maximum or minimum points;
(c) inflection points.

22. A cube is contracting so that its surface area decreases at the constant rate of 72 in.^{2}/sec. Determine how fast the volume is changing at the instant when the surface area is 54 ft^2.

*23. A particle starts when $t = 0$ at the point $(4, 0)$ and moves counterclockwise on the circle $x^2 + y^2 = 16$. Its speed at time t equals e^t.
(a)   Find when it first reaches the point $(-4, 0)$.
(b)   Write its acceleration **a** as a vector in terms of the unit vectors **i** and **j** at the time found in part (a).

24. Investigate the curve of $y = \ln \sin x$ on the interval $-2\pi < x < 2\pi$ for

(a) extent;
(b) intercepts;
(c) relative maximum or minimum points;
(d) concavity;
(e) asymptotes.
Sketch the curve.

Solutions for Section II Problems: Applications of Differentiation

1. (a) The origin is the only intercept.
(b) Note that y' and y'' can be found easily if we rewrite

$$y = \frac{(x^2 - 4) + 4}{x - 2} = \frac{x^2 - 4}{x - 2} + \frac{4}{x - 2}.$$

Then

$$y = (x + 2) + 4(x - 2)^{-1}, \qquad y' = 1 - 4(x - 2)^{-2}, \qquad \text{and} \qquad y'' = 8(x - 2)^{-3}.$$

To find any maxima or minima note that $y' = \dfrac{x^2 - 4x}{(x - 2)^2}$ and that $x = 0$ and $x = 4$ are both critical values. Since $y'' = \dfrac{8}{(x - 2)^3}$ and is negative for $x = 0$ but positive for $x = 4$, the points $(0, 0)$ and $(4, 8)$ are respectively a maximum and a minimum.

(c) Since $y'' > 0$ when $x > 2$, the curve is concave up for $x > 2$.

(d) Here, $x = 2$ is a vertical asymptote. The graph has no horizontal asymptote, but does approach the line $y = x + 2$ as $|x|$ becomes large. Note that the domain is all real numbers except 2.

The curve is sketched in Figure S4–1.

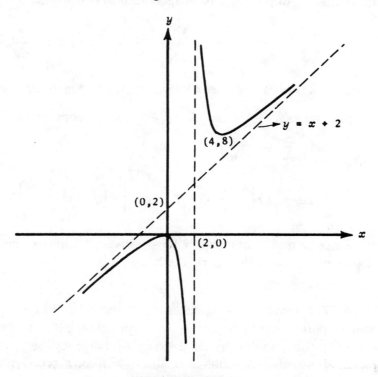

FIGURE S4–1

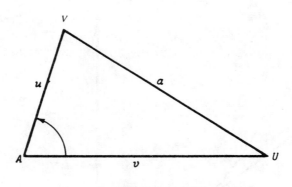

FIGURE S4–2

2. The triangle is sketched in Figure S4–2, where u and v represent the lengths of the remaining sides. The area

$$S = \frac{1}{2} uv \sin A \tag{1}$$

and, by the law of cosines,

$$a^2 = u^2 + v^2 - 2uv \cos A. \tag{2}$$

Differentiating S in (1) with respect to u yields

$$\frac{dS}{du} = \frac{1}{2} \sin A \left(u\frac{dv}{du} + v \right);$$

differentiating implicitly with respect to u in (2), we get

$$0 = 2u + 2v\frac{dv}{du} - 2\cos A \left(u\frac{dv}{du} + v \right), \tag{3}$$

where $\frac{d}{du} a^2 = 0$ because a^2 is a constant. Since we seek a maximum area for the triangle, we set $\frac{dS}{du} = 0$. Since $\frac{1}{2}\sin A$ is constant and is different from zero, it follows that $\frac{dv}{du} = -\frac{v}{u}$. Using this in (3), we get

$$0 = 2u + 2v\left(-\frac{v}{u}\right) - 2\cos A \left[u\left(-\frac{v}{u}\right) + v \right] \quad \text{or} \quad u^2 - v^2 = 0.$$

Since u and v must both be positive, it follows that the triangle has maximum area when it is isosceles, that is, when $u = v$.

3. The parabola is sketched in Figure S4–3. Let p be the x-coordinate of the random point on the line $y = -1$ from which the tangents are drawn to the parabola. Let the coordinates of a point of tangency be (x_1, y_1). We wish to find the slopes of the two tangents and show that their product equals -1.

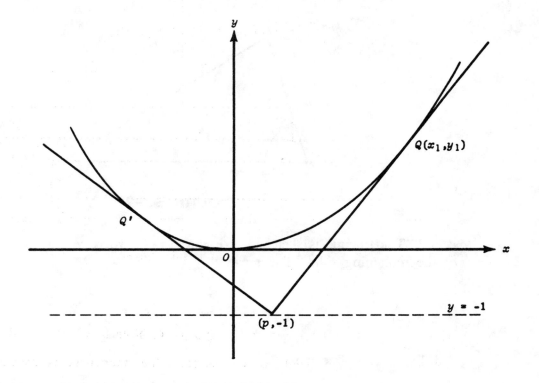

FIGURE S4–3

The equation of a tangent is $y - y_1 = \frac{x_1}{2}(x - x_1)$, where $\frac{x_1}{2}$ is the slope of the tangent and is obtained from $\frac{dy}{dx}$ evaluated at $x = x_1$. Since $(p, -1)$ is on the tangent, it follows that

$$-1 - y_1 = \frac{x_1}{2}(p - x_1).$$

Letting $y_1 = \frac{x_1^2}{4}$, we have

$$-1 - \frac{x_1^2}{4} = \frac{x_1}{2}(p - x_1),$$

or, simplifying, $x_1^2 - 2px_1 - 4 = 0$. By the quadratic formula we get two solutions for x_1: $x_1 = p \pm \sqrt{p^2 + 4}$.

We thus have the abscissas of the two points of tangency, Q and Q', and the slopes $\frac{p + \sqrt{p^2 + 4}}{2}$ and $\frac{p - \sqrt{p^2 + 4}}{2}$. Since their product $\frac{p^2 - (p^2 + 4)}{4} = -1$, the two tangents are perpendicular and this fact is independent of our choice of point on the line $y = -1$.

4. See Figure S4–4. The area A of a cross section is

$$A(x) = x(16 - 2x) = 2(8x - x^2).$$

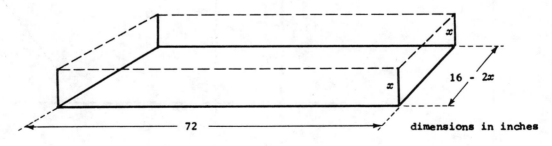

FIGURE S4–4

Then $\frac{dA}{dx} = 2(8 - 2x)$, which is 0 if $x = 4$. Note that $0 < x < 8$, that $A(x)$ is continuous, that $\frac{d^2A}{dx^2} = -4$, which indicates that the curve is concave down. It follows that the area is a maximum when $x = 4$ and that it equals 32 in². Note that the length of the trough is irrelevant for this problem.

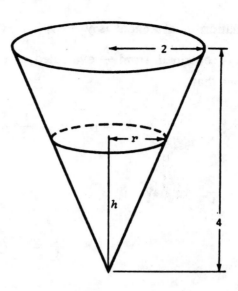

FIGURE S4–5

5. See Figure S4–5. We know that $\frac{dV}{dt} = -\frac{1}{2}$ and that $h = 2r$. Here, $V = \frac{1}{3}\pi^2 h = \frac{\pi h^3}{12}$. So

$$\frac{dV}{dt} = \frac{\pi h^2}{4}\frac{dh}{dt} \qquad \text{and} \qquad \frac{dh}{dt} = -\frac{1}{2}\frac{4}{\pi h^2} = -\frac{2}{\pi h^2} \qquad \text{at any time.}$$

When the diameter is 2 in., so is the height, and $\frac{dh}{dt} = -\frac{1}{2\pi}$. The water level is thus dropping at the rate of $\frac{1}{2\pi}$ in./sec.

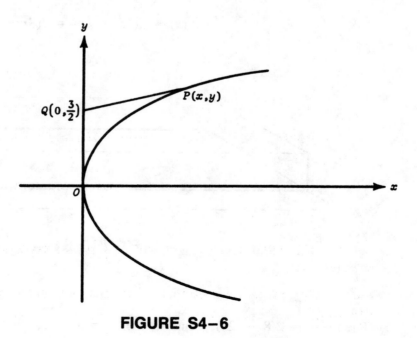

FIGURE S4–6

6. The parabola $2x = y^2$ is shown in Figure S4–6. If we let $s = \overline{PQ}^2 = x^2 + \left(y - \frac{3}{2}\right)^2$, then

$$\frac{ds}{dx} = 2x + 2\left(y - \frac{3}{2}\right)\frac{dy}{dx}.$$

Since $2x = y^2$,

$$2 = 2y\frac{dy}{dx} \qquad \text{and} \qquad \frac{dy}{dx} = \frac{1}{y}.$$

Setting $\frac{ds}{dx}$ equal to 0 yields

$$x + \left(y - \frac{3}{2}\right)\frac{1}{y} = 0,$$

and (since P is on the curve)

$$\frac{y^2}{2} + \frac{y - \frac{3}{2}}{y} = 0.$$

Then $y^3 + 2y - 3 = 0$, and $y = 1$ is the only critical value.

Note that (1) the distance $\overline{PQ}$ cannot be maximized; (2) if P is the origin, $\overline{PQ} = \frac{3}{2}$; (3) if P is $\left(\frac{1}{2}, 1\right)$, $\overline{PQ} = \frac{\sqrt{2}}{2}$. We see that $\left(\frac{1}{2}, 1\right)$ is the point on the curve of $2x = y^2$ closest to the point $\left(0, \frac{3}{2}\right)$.

7. (a) The domain of the function is all nonzero reals.

(b) $\frac{\sin x}{x} = 0$, where $\sin x = 0$; that is, for $x = n\pi$, n a nonzero integer.

(c) If $x \neq 0$, $f'(x) = \frac{x \cos x - \sin x}{x^2}$ and $f'(x) = 0$ when $x \cos x - \sin x = 0$, that is, when $\tan x = x$.

(d) Since $f(-x) = f(x)$, the graph is symmetric to the y-axis.

(e) $\lim\limits_{x \to 0} \frac{\sin x}{x} = 1$.

(f) Although $\lim\limits_{x \to 0} f(x)$ exists, f is not continuous at $x = 0$, because $f(0)$ does not exist.

(g) Since $\sin x$ oscillates between -1 and 1, remaining finite, as $|x| \to \infty$, $\frac{\sin x}{x} \to 0$.

The graphs of $y = x$ and $y = \tan x$ are sketched in Figure S4–7. R, O, and S are three points of the infinite set of intersections of these graphs. The slope of the curve equals zero at O, and at R, S, and all other points for which $\tan x = x$.

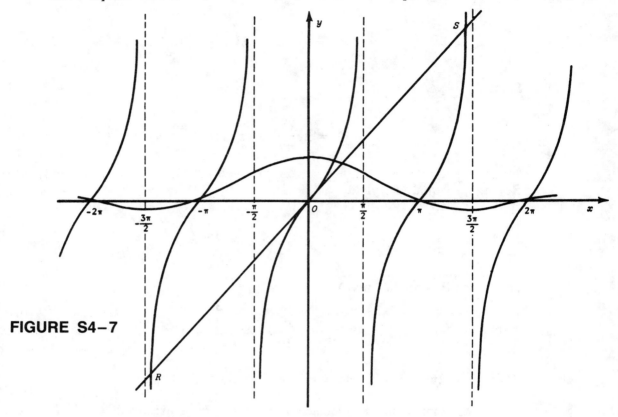

FIGURE S4–7

8. (a) Since

$$i = \frac{E}{R}(1 - e^{-(Rt/L)}) \quad \text{and} \quad e^{-(Rt/L)} < 1 \quad \text{if } t > 0,$$

then $i < \dfrac{E}{R}$.

(Note that $\dfrac{d}{dt}e^{-(Rt/L)} = -\dfrac{R}{L}e^{-(Rt/L)}$, indicating that this function decreases for all t; when $t = 0$, $e^{-(Rt/L)} = 1$.)

(b) As $t \to \infty$, $e^{-(Rt/L)} \to 0$; so $\displaystyle\lim_{t \to \infty} i = \frac{E}{R}$.

(c) For small changes in t, the change in i is approximately equal to di. Thus $\Delta i \simeq di = \dfrac{d}{dt}i\,\Delta t$, where $\dfrac{di}{dt}$ is to be evaluated for $t = 0$ and $\Delta t = 0.01$. So

$$\Delta i \simeq \frac{R}{L}\cdot\frac{E}{R}\,e^{-(Rt/L)}\Delta t = \frac{E}{L}e^0(0.01) = 0.01\,\frac{E}{L} = 0.6.$$

(d) If E, L, and t are all held fixed, while R varies, we can write

$$\lim_{R \to 0} i = \lim_{R \to 0} (-E)\frac{(e^{-(Rt/L)} - 1)}{R} = -E \lim_{R \to 0} \frac{e^{-(Rt/L)} - 1}{R}.$$

Recalling that $f'(0)$ is defined as

$$\lim_{h \to 0} \frac{f(h) - f(0)}{h} \quad \text{or} \quad \lim_{R \to 0} \frac{f(R) - f(0)}{R},$$

we see that $\lim\limits_{R\to 0} i$ equals $-Ef'(0)$, where $f(R) = e^{-(Rt/L)}$. Remember that t and L are constants here. Thus

$$\lim_{R\to 0} i = -E\left(\frac{-t}{L}\right)e^{-(Rt/L)},$$

and when $R = 0$ this is $\frac{Et}{L}$.

Alternatively, we may note that

$$\lim_{R\to 0} i = E \lim_{R\to 0} \frac{1 - e^{-(Rt/L)}}{R} = E \lim_{R\to 0} \frac{u(R)}{v(R)},$$

where $u(R)$ and $v(R)$ both approach zero as R does. We can then apply L'Hôpital's rule:

$$E \lim_{R\to 0} \frac{1 - e^{-(Rt/L)}}{R} = E \lim_{R\to 0} \frac{\dfrac{t}{L}e^{-(Rt/L)}}{1} = \frac{Et}{L}.$$

9. (a) In Figure S4–9 a curve of each family is sketched, with h and k both taken as positive. To find possible points of intersection we solve simultaneously the equations $xy = k$ and $y^2 - x^2 = h$, getting $\frac{k^2}{x^2} - x^2 = h$ or

$$x^4 + hx^2 - k^2 = 0. \tag{1}$$

The quadratic formula here yields

$$x^2 = \frac{-h \pm \sqrt{h^2 + 4k^2}}{2}. \tag{2}$$

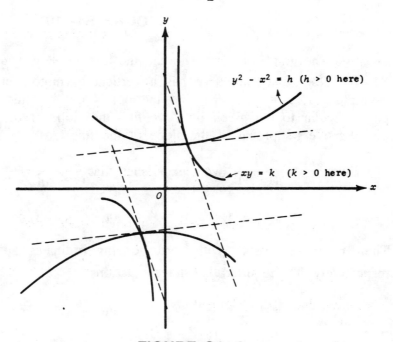

FIGURE S4–9

If $hk \neq 0$, $\sqrt{h^2 + 4k^2} > |h|$ and $-h + \sqrt{h^2 + 4k^2}$ is positive. There are, thus, if $hk \neq 0$, two distinct reals which satisfy (1) and therefore two distinct points of intersection of every pair of the given families of curves.

(b) Note that for $xy = k$ the slope $\dfrac{dy}{dx} = -\dfrac{y}{x}$, while for $y^2 - x^2 = h$ the slope $\dfrac{dy}{dx} = \dfrac{x}{y}$. At a point of intersection (x_1, y_1), the existence of which is guaranteed by (2) above, the product of these slopes is $-\dfrac{y_1}{x_1} \cdot \dfrac{x_1}{y_1} = -1$, indicating that the curves intersect at right angles.

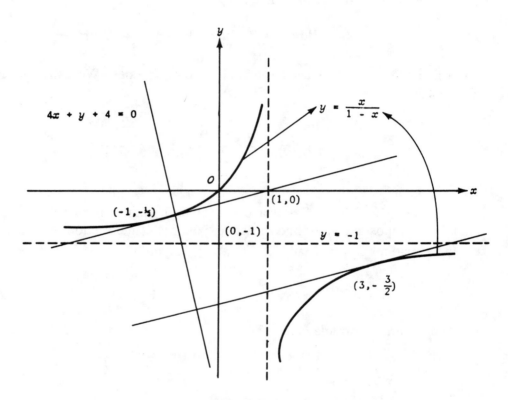

FIGURE S4–10

10. The graph of $y = \dfrac{x}{1-x}$ and the line $4x + y + 4 = 0$ are shown in Figure S4–10. Note that the curve has a vertical asymptote at $x = 1$ and a horizontal asymptote at $y = -1$. To find the points on the curve at which the tangent is perpendicular to the given line, we find the slope of the curve and set it equal to the negative reciprocal of the slope of the line.

Since $\dfrac{dy}{dx} = \dfrac{1}{(1-x)^2}$, we seek x_1 such that $\dfrac{1}{(1-x_1)^2} = \dfrac{1}{4}$. We get

$$(1 - x_1)^2 = 4 \quad \text{or} \quad 1 - x_1 = \pm 2.$$

Then $x_1 = -1$ or 3, with corresponding ordinates on the curve $-\dfrac{1}{2}$ and $-\dfrac{3}{2}$, respectively. The required lines have equations

$$y + \frac{1}{2} = \frac{1}{4}(x + 1) \quad \text{and} \quad y + \frac{3}{2} = \frac{1}{4}(x - 3).$$

A reasonably careful sketch will expose unreasonable answers.

11. The parabola is sketched in Figure S4–11a. If we let $Q(x_1, y_1)$ be a point of tangency, then we can find Q by using the fact that it is both on the tangent PQ and

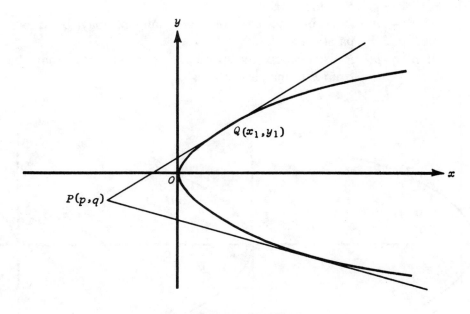

FIGURE S4–11a

on the curve $y^2 = x$. The equation of the tangent from P to the curve is

$$y - q = \frac{1}{2y_1}(x - p), \tag{1}$$

where the slope is the derivative of $y^2 = x$ when $y = y_1$. Since $Q(x_1, y_1)$ is on this tangent, (1) becomes

$$y_1 - q = \frac{1}{2y_1}(x_1 - p)$$

or

$$2y_1^2 - 2qy_1 = x_1 - p. \tag{2}$$

Since Q is on the parabola, we can use $y_1^2 = x_1$ in (2) to get $2y_1^2 - 2qy_1 = y_1^2 - p$, or

$$y_1^2 - 2qy_1 + p = 0. \tag{3}$$

Applying the quadratic formula to solve (3), we see that

$$y_1 = \frac{2q \pm \sqrt{4q^2 - 4p}}{2} = q \pm \sqrt{q^2 - p}. \tag{4}$$

The number of possible solutions for y_1 thus depends on q and p. There are none, one, or two according to whether q^2 is less than, equal to, or greater than p.

In Figure S4–11b the three cases are shown, and demonstrate the following:

If $q^2 < p$: P is within the parabola; no tangents may be drawn from P to the parabola.

If $q^2 = p$: P is on the parabola; one tangent may be drawn to the parabola at P.

If $q^2 > p$: P is outside the parabola; two tangents may be drawn from P to the parabola.

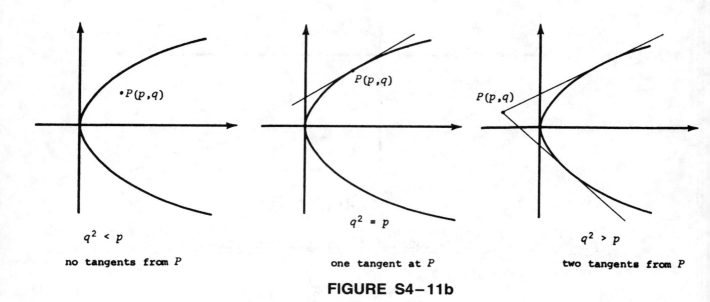

$q^2 < p$

no tangents from P

$q^2 = p$

one tangent at P

$q^2 > p$

two tangents from P

FIGURE S4–11b

12. The parabola and a tangent to it at P are sketched in Figure S4–12. The equation of the tangent at P is $y - y_1 = 2x_1(x - x_1)$, where the slope $2x_1$ is the

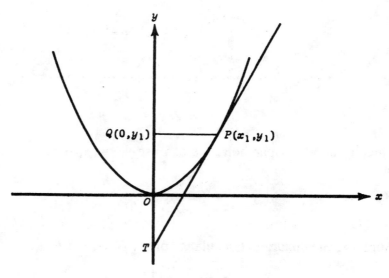

FIGURE S4–12

derivative of $y = x^2$ at $x = x_1$. To find the ordinate of T, we let $x = 0$, getting

$$y - y_1 = -2x_1^2 \quad \text{or} \quad y = y_1 - 2x_1^2.$$

Since (x_1, y_1) is on the parabola, $y_1 = x_1^2$ and the ordinate of T is $-y_1$. The area of triangle QPT equals

$$\tfrac{1}{2}\overline{QP} \cdot \overline{TQ} = \tfrac{1}{2}\, x_1 \cdot (2y_1) = x_1 y_1.$$

13. Since $y = ax^3 + bx^2 + cx + d$,

$$y' = 3ax^2 + 2bx + c \qquad \text{and} \qquad y'' = 6ax + 2b.$$

If the curve is to have a relative maximum at $(0, 1)$ and a point of inflection at $\left(1, \tfrac{1}{3}\right)$, then it must follow that

$$
\begin{aligned}
y'(0) &= 0, & c &= 0; \\
y''(1) &= 0, & 6a + 2b &= 0; \\
y(0) &= 1, & d &= 1; \\
y(1) &= \tfrac{1}{3}, & a + b + c + d &= \tfrac{1}{3}.
\end{aligned}
$$

The values for c and d given by the first and third of the equations on the right enable us to solve the second and fourth of the equations simultaneously, yielding $a = \tfrac{1}{3}$ and $b = -1$.

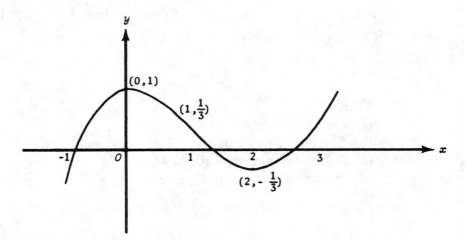

FIGURE S4–13

The graph of $y = \tfrac{1}{3}x^3 - x^2 + 1$ is sketched in Figure S4–13. Note that the curve has a relative minimum at $\left(2, -\tfrac{1}{3}\right)$.

14. Since $x = \theta - \sin \theta$ and $y = 1 - \cos \theta$,

$$dx = (1 - \cos \theta)\, d\theta \qquad \text{and} \qquad dy = \sin \theta\, d\theta.$$

(a) The slope at any point is given by $\frac{dy}{dx}$, which here equals $\frac{\sin \theta}{1 - \cos \theta}$. When $\theta = \frac{2\pi}{3}$,

$$\frac{dy}{dx} = \frac{\sqrt{3}/2}{1 - \left(-\frac{1}{2}\right)} = \frac{\sqrt{3}}{3}.$$

(b) When $\theta = \frac{2\pi}{3}$, $x = \frac{2\pi}{3} - \frac{\sqrt{3}}{2}$ and $y = 1 - \left(-\frac{1}{2}\right) = \frac{3}{2}$. The equation of the tangent is

$$9y - 3\sqrt{3}x = 18 - 2\pi\sqrt{3}.$$

15. Draw a sketch of the region bounded above by $y_1 = 8 - 2x^2$ and below by $y_2 = x^2 - 4$, and inscribe a rectangle in this region as described in the problem. If (x, y_1) and (x, y_2) are the vertices of the rectangle in quadrants I and II, respectively, then the area

$$A = 2x(y_1 - y_2) = 2x(12 - 3x^2), \qquad \text{or} \qquad A(x) = 24x - 6x^3.$$

Then $A'(x) = 24 - 18x^2 = 6(4 - 3x^2)$, which equals 0 when $x = \frac{2}{\sqrt{3}} = \frac{2\sqrt{3}}{3}$. Check to verify that $A''(x) < 0$ at this point. This assures that this value of x yields maximum area, which is given by $\frac{4\sqrt{3}}{3} \times 8$.

16. See Figure S4–16.
  (a) $x = 0$, $y = \ln 4$.
  (b) Symmetric to y-axis.
  (c) Relative minimum at $(0, \ln 4)$.
  (d) Inflection points at $(\pm 2, \ln 8)$.

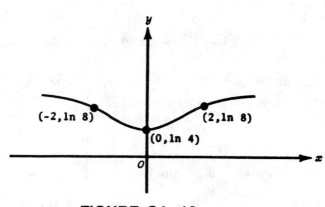

FIGURE S4–16

17. We know from (a) that, when $x = 0$, $y = 0$. So $b = 0$. From (b) it follows that y cannot occur to an odd power; that is, a must be 0. This yields

$$y^2 = \frac{x}{cx^2 + dx + e}.$$

Since there are two horizontal asymptotes $y = \pm 1$, c must equal 0 and d must be 1. Now we have

$$y^2 = \frac{x}{x + e}.$$

Since $x = 4$ is a vertical asymptote, e must be -4. Note that the right-hand side must be nonnegative. The equation of the curve is

$$y^2 = \frac{x}{x - 4}.$$

18. (a) Since $\frac{dy}{dt} = 2$, $y = 2t + 1$ and $x = 4t^3 + 6t^2 + 3t$.

(b) Since $\frac{d^2y}{dt^2} = 0$ and $\frac{d^2x}{dt^2} = 24t + 12$, then, when $t = 1$, $|a| = 6$.

19. (a) The intercepts are $(0, 0)$, $(\pm 1, 0)$.

(b) The domain is $\{x \mid x \leq -1 \text{ or } 0 \leq x \leq 1\}$.

(c) The curve is symmetric to the x-axis.

(d) There are no vertical or horizontal asymptotes.

(e) The curve has a relative maximum at $\left(\frac{\sqrt{3}}{3}, \frac{\sqrt{2\sqrt{3}}}{3}\right)$ and a relative minimum at $\left(\frac{\sqrt{3}}{3}, -\frac{\sqrt{2\sqrt{3}}}{3}\right)$.

The curve is sketched in Figure S4–19.

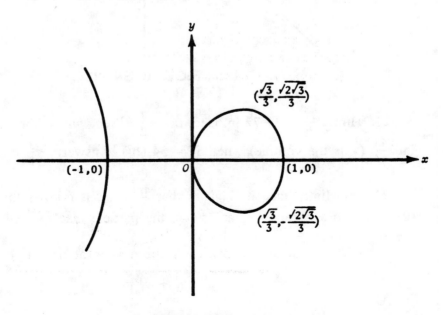

FIGURE S4–19

20. Hint: Let (x_1, y_1) be a common point of tangency. Then the following three equations hold:

$$y_1 = mx_1 + b; \qquad x_1y_1 = 1; \qquad -\frac{1}{x_1^2} = m,$$

the last of these deriving from the fact that the slope of the curve at (x_1, y_1) equals the slope of the line. Simultaneous solution of the equations yields $m = -\frac{b^2}{4}$.

21. (a) The origin is the only intercept.

(b) There are no relative maxima or minima. Although $y' = 1 + \cos x$ is zero for $x = (2n + 1)\pi$, n an integer, note that y' does not change sign ($y' \geq 0$ for all x).

(c) The curve has inflection points when $x = n\pi$, n an integer. See Figure S4–21.

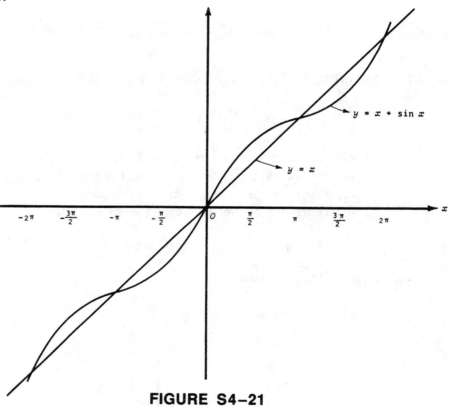

FIGURE S4–21

22. Hint: $\frac{dS}{dt} = -72$ in.²/sec, or $-\frac{1}{2}$ ft²/sec, where the surface area $S = 6x^2$. Find $\frac{dV}{dt}$ (V is the volume) when $S = 54$ (ft²). Answer: $-\frac{3}{8}$ ft³/sec.

23. (a) Hint: Since it is given that $\frac{ds}{dt} = e^t$, it follows that $s = e^t + C$, where $s(0) = 0$. So $s = e^t - 1$, and when the particle reaches $(-4, 0)$ first we note that $s = 4\pi$. Thus $t = \ln(4\pi + 1)$.

(b) Note since the particle's speed is e^t at time t, that,

$$v_x^2 + v_y^2 = e^{2t}. \tag{1}$$

Also,

$$xv_x + yv_y = 0, \tag{2}$$

where the latter is obtained by differentiating the equation of the circle with respect to t. At $(-4, 0)$ we get from (2) that $v_x = 0$ and thus, from (1), that $v_y = -e^t$. We differentiate (1) with respect to t, getting $v_x a_x + v_y a_y = e^{2t}$, so that a_y at $(-4, 0)$ equals $-e^t$ or $-(4\pi + 1)$.

If we differentiate (2) with respect to t, we get

$$xa_x + ya_y + v_x^2 + v_y^2 = 0 \qquad \text{and} \qquad xa_x + ya_y = -e^{2t}.$$

At $(-4, 0)$, then, $-4a_x = -e^{2t}$, so

$$a_x = \frac{e^{2t}}{4} = \frac{(4\pi + 1)^2}{4}.$$

Thus

$$\mathbf{a} = \frac{(4\pi + 1)^2}{4}\mathbf{i} - (4\pi + 1)\mathbf{j}.$$

24. (a) Since $\ln u$ is defined only if $u > 0$, the domain of $\ln \sin x$ is $-2\pi < x < -\pi$ or $0 < x < \pi$; since $0 < \sin x \leq 1$ on this domain, $\ln \sin x \leq 0$.

(b) $\ln \sin x = 0$ if $x = -\frac{3\pi}{2}$ or $\frac{\pi}{2}$.

(c) Note, since $y' = \cot x$, that it is zero when $x = -\frac{3\pi}{2}$ or $\frac{\pi}{2}$. Since $y'' = -\csc^2 x$, we see that y'' is always negative, so that $\left(-\frac{3\pi}{2}, 0\right)$ and $\left(\frac{\pi}{2}, 0\right)$ are relative maxima.

(d) The curve is everywhere concave down.

(e) Since $\sin x$ is zero at -2π, $-\pi$, 0, and π, we see that $\ln \sin x$ becomes negatively infinite as x approaches any of these numbers. The curve has vertical asymptotes corresponding to each of these x's.

See Figure S4–24 for a sketch of the function.

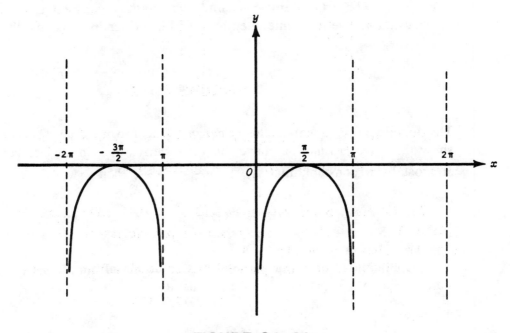

FIGURE S4–24

Chapter 5

Integration

Review of Definitions and Methods

A. Antiderivatives_____

The *antiderivative* or *indefinite integral* of a function $f(x)$ is a function $F(x)$ whose derivative is $f(x)$. Since the derivative of a constant equals zero, the antiderivative of $f(x)$ is not unique; that is, if $F(x)$ is an integral of $f(x)$, then so is $F(x) + C$, where C is any constant. The arbitrary constant C is called the *constant of integration*. The indefinite integral of $f(x)$ is written as $\int f(x)\, dx$; thus

$$\int f(x)\, dx = F(x) + C \quad \text{if } \frac{dF(x)}{dx} = f(x).$$

The function $f(x)$ is called the *integrand*. The Mean Value Theorem can be used to show that, if two functions have the same derivative, on an interval, then they differ at most by a constant; that is, if $\frac{dF(x)}{dx} = \frac{dG(x)}{dx}$, then

$$F(x) - G(x) = C \qquad (C \text{ a constant}).$$

Applications of antiderivatives and solutions of simple differential equations are considered in §F, page 141, of this chapter.

B.   Basic Formulas

Familiarity with the following fundamental integration formulas is essential.

$$\int k f(x)\, dx = k \int f(x)\, dx \qquad (k \neq 0) \tag{1}$$

$$\int [f(x) + g(x)]\, dx = \int f(x)\, dx + \int g(x)\, dx \tag{2}$$

$$\int u^n\, du = \frac{u^{n+1}}{n+1} + C \qquad (n \neq -1) \tag{3}$$

$$\int \frac{du}{u} = \ln |u| + C \tag{4}$$

$$\int \cos u\, du = \sin u + C \tag{5}$$

$$\int \sin u\, du = -\cos u + C \tag{6}$$

$$\int \tan u\, du = \ln |\sec u| + C$$
$$\text{or} -\ln |\cos u| + C \tag{7}$$

$$\int \cot u\, du = \ln |\sin u| + C$$
$$\text{or} -\ln |\csc u| + C \tag{8}$$

$$\int \sec^2 u\, du = \tan u + C \tag{9}$$

$$\int \csc^2 u\, du = -\cot u + C \tag{10}$$

$$\int \sec u \tan u\, du = \sec u + C \tag{11}$$

$$\int \csc u \cot u\, du = -\csc u + C \tag{12}$$

$$\int \sec u\, du = \ln |\sec u + \tan u| + C \tag{13}$$

$$\int \csc u\, du = \ln |\csc u - \cot u| + C \tag{14}$$

$$\int e^u\, du = e^u + C \tag{15}$$

$$\int a^u\, du = \frac{a^u}{\ln a} + C \qquad (a > 0,\ a \neq 1) \tag{16}$$

$$\int \frac{du}{\sqrt{a^2 - u^2}} = \sin^{-1} \frac{u}{a} + C$$
$$\text{or Arcsin} \frac{u}{a} + C \tag{17}$$

$$\int \frac{du}{a^2 + u^2} = \frac{1}{a} \tan^{-1} \frac{u}{a} + C$$

$$\text{or } \frac{1}{a} \text{ Arctan } \frac{u}{a} + C \qquad (18)$$

$$\int \frac{du}{u\sqrt{u^2 - a^2}} = \frac{1}{a} \sec^{-1} \left| \frac{u}{a} \right| + C$$

$$\text{or } \frac{1}{a} \text{ Arcsec } \left| \frac{u}{a} \right| + C \qquad (19)$$

All the references in the following set of examples are to the preceding basic formulas.

Examples

1. $\int 5x \, dx = 5 \int x \, dx$ by (1), $= 5 \cdot \frac{x^2}{2} + C$ by (3).

2. $\int \left(x^4 + \sqrt[3]{x^2} - \frac{2}{x^2} - \frac{1}{3\sqrt[3]{x}} \right) dx = \int \left(x^4 + x^{2/3} - 2x^{-2} - \frac{1}{3} x^{-1/3} \right) dx$

$= \int x^4 \, dx + \int x^{2/3} \, dx - 2 \int x^{-2} \, dx - \frac{1}{3} \int x^{-1/3} \, dx$ by (1) and (2),

$= \frac{x^5}{5} + \frac{x^{5/3}}{\frac{5}{3}} - \frac{2x^{-1}}{-1} - \frac{1}{3} \frac{x^{2/3}}{\frac{2}{3}} + C$ by (3), $= \frac{x^5}{5} + \frac{3}{5}x^{5/3} + \frac{2}{x} - \frac{1}{2}x^{2/3} + C.$

3. Similarly, $\int (3 - 4x + 2x^3) \, dx = \int 3 \, dx - 4 \int x \, dx + 2 \int x^3 \, dx =$

$3x - \frac{4x^2}{2} + \frac{2x^4}{4} + C = 3x - 2x^2 + \frac{x^4}{2} + C.$

4. $\int 2(1 - 3x)^2 \, dx = 2 \int (1 - 3x)^2 \, dx = 2 \int (1 - 6x + 9x^2) \, dx$

$= 2 \left(x - \frac{6x^2}{2} + \frac{9x^3}{3} \right) + C = 2x - 6x^2 + 6x^3 + C.$ We have expanded here, and used the fact that the integral of a sum is the sum of the integrals ((2) above). Alternatively, we can let $u = 1 - 3x$; then $du = -3 \, dx$. Since

$$2 \int (1 - 3x)^2 \, dx = \frac{2}{-3} \int (1 - 3x)^2 (-3) \, dx,$$

we can now apply formula (3), getting

$$-\frac{2}{3} \int (1 - 3x)^2 (-3) \, dx = -\frac{2}{3} \int u^2 \, du = -\frac{2}{9} u^3 + C \text{ by (3),}$$

$$= -\frac{2}{9} (1 - 3x)^3 + C.$$

If this last function is expanded, note that the two different answers to the given integral differ only by a constant; in fact, the second function is precisely the first function plus the constant $-\frac{2}{9}$.

5. $\int (2x^3 - 1)^5 \cdot x^2 \, dx = \frac{1}{6}\int (2x^3 - 1)^5 \cdot 6x^2 \, dx = \frac{1}{6}\int u^5 \, du$ (where

$u = 2x^3 - 1$), $= \frac{1}{6}\frac{u^6}{6} + C$ by (3), $= \frac{1}{36}(2x^3 - 1)^6 + C$.

6. $\int \sqrt[3]{1 - x} \, dx = \int (1 - x)^{1/3} \, dx = -\int (1 - x)^{1/3}(-1) \, dx = -\int u^{1/3} \, du$

(where $u = 1 - x$), $= -\frac{u^{4/3}}{\frac{4}{3}} + C$ by (3), $= -\frac{3}{4}(1 - x)^{4/3} + C$.

7. $\int \frac{x}{\sqrt{3 - 4x^2}} \, dx = \int (3 - 4x^2)^{-1/2} \cdot x \, dx = -\frac{1}{8}\int (3 - 4x^2)^{-1/2}(-8x) \, dx$

$= -\frac{1}{8}\int u^{-1/2} \, du$ (where $u = 3 - 4x^2$), $= -\frac{1}{8}\frac{u^{1/2}}{\frac{1}{2}} + C$ by (3), $=$

$-\frac{1}{4}\sqrt{3 - 4x^2} + C$.

8. $\int \frac{4x^2}{(x^3 - 1)^3} \, dx = 4\int (x^3 - 1)^{-3} \cdot x^2 \, dx = \frac{4}{3}\int (x^3 - 1)^{-3}(3x^2) \, dx =$

$\frac{4}{3}\frac{(x^3 - 1)^{-2}}{-2} + C = -\frac{2}{3}\frac{1}{(x^3 - 1)^2} + C$.

9. $\int \frac{(1 + \sqrt{x})^4}{\sqrt{x}} \, dx = \int (1 + x^{1/2})^4 \cdot \frac{1}{x^{1/2}} \, dx$. Now let $u = 1 + x^{1/2}$, and

note that $du = \frac{1}{2}x^{-1/2} \, dx$; this gives $2\int (1 + x^{1/2})^4 \frac{1}{2x^{1/2}} \, dx = \frac{2}{5}(1 + \sqrt{x})^5$

$+ C$.

10. $\int (2 - y)^2 \cdot \sqrt{y} \, dy = \int (4 - 4y + y^2) \cdot y^{1/2} \, dy = \int (4y^{1/2} - 4y^{3/2} + y^{5/2}) \, dy$

$= 4 \cdot \frac{2}{3}y^{3/2} - 4 \cdot \frac{2}{5}y^{5/2} + \frac{2}{7} \cdot y^{7/2} + C$ by (2), $= \frac{8}{3}y^{3/2} - \frac{8}{5}y^{5/2} + \frac{2}{7}y^{7/2} + C$.

11. $\int \frac{x^3 - x - 4}{2x^2} \, dx = \frac{1}{2}\int \left(x - \frac{1}{x} - \frac{4}{x^2}\right) dx = \frac{1}{2}\left(\frac{x^2}{2} - \ln |x| + \frac{4}{x}\right) + C$.

Formulas (1), (2), (3), and (4) have all been used.

12. $\int \frac{3x - 1}{\sqrt[3]{1 - 2x + 3x^2}} \, dx = \int (1 - 2x + 3x^2)^{-1/3}(3x - 1) \, dx$

$= \frac{1}{2}\int (1 - 2x + 3x^2)^{-1/3}(6x - 2) \, dx$ (where we have let $u = 1 - 2x + 3x^2$

and noted that $du = (-2 + 6x) \, dx$), $= \frac{1}{2} \cdot \frac{3}{2}(1 - 2x + 3x^2)^{2/3} + C$ by (3),

$= \frac{3}{4}(1 - 2x + 3x^2)^{2/3} + C$.

13. $\int \frac{2x^2 - 4x + 3}{(x - 1)^2} \, dx = \int \frac{2x^2 - 4x + 3}{x^2 - 2x + 1} \, dx = \int \left(2 + \frac{1}{(x - 1)^2}\right) dx =$

$\int 2 \, dx + \int \frac{dx}{(x - 1)^2} = 2x - \frac{1}{x - 1} + C$. This example illustrates the follow-

ing principle: *If the degree of the numerator of a rational function is not less than that of the denominator, divide until a remainder of lower degree is obtained.*

14. $\int \frac{du}{u - 3} = \ln |u - 3| + C$ by (4).

15. $\int \frac{z\,dz}{1-4z^2} = -\frac{1}{8}\int \frac{-8z\,dz}{1-4z^2} = -\frac{1}{8}\ln|1-4z^2| + C$ by (4) with $u = 1 - 4z^2$.

16. $\int \frac{\cos x}{5+2\sin x}\,dx = \frac{1}{2}\int \frac{2\cos x}{5+2\sin x}\,dx = \frac{1}{2}\ln(5+2\sin x) + C$ by (4) with $u = 5 + 2\sin x$. The absolute-value sign is not necessary here since $(5 + 2\sin x) > 0$ for all x.

17. $\int \frac{e^x}{1-2e^x}\,dx = -\frac{1}{2}\int \frac{-2e^x}{1-2e^x}\,dx = -\frac{1}{2}\ln|1-2e^x| + C$ by (4) with $u = 1 - 2e^x$.

18. $\int \frac{x}{1-x}\,dx = \int \left(-1 + \frac{1}{1-x}\right)\,dx$ by long division $= -x - \ln|1-x| + C$.

19. $\int \sin(1-2y)\,dy = -\frac{1}{2}\int \sin(1-2y)(-2\,dy) = -\frac{1}{2}[-\cos(1-2y)] + C$ by (6), $= \frac{1}{2}\cos(1-2y) + C$.

20. $\int \sin^2\frac{x}{2}\cos\frac{x}{2}\,dx = 2\int \sin^2\frac{x}{2}\cos\frac{x}{2}\frac{dx}{2} = \frac{2}{3}\sin^3\frac{x}{2} + C$ by (3) with $u = \sin\frac{x}{2}$ so that $du = \cos\frac{x}{2}\left(\frac{1}{2}\,dx\right)$.

21. $\int \frac{\sin x}{1+3\cos x}\,dx = -\frac{1}{3}\int \frac{-3\sin x}{1+3\cos x}\,dx = -\frac{1}{3}\ln|1+3\cos x| + C$ by (4) with $u = 1 + 3\cos x$.

22. $\int e^{\tan y}\sec^2 y\,dy = e^{\tan y} + C$ by (15) with $u = \tan y$.

23. $\int e^x \tan e^x\,dx = -\ln|\cos e^x| + C$ by (7) with $u = e^x$.

24. $\int \frac{\cos z}{\sin^2 z}\,dz = \int \csc z \cot z\,dz = -\csc z + C$ by (12).

25. $\int \tan t \sec^2 t\,dt = \frac{\tan^2 t}{2} + C$ by (3) with $u = \tan t$ and $du = \sec^2 t\,dt$.

26. $\int \sec^4 x\,dx = \int \sec^2 x \sec^2 x\,dx = \int(\tan^2 x + 1)\sec^2 x\,dx = \int \tan^2 x \sec^2 x\,dx + \int \sec^2 x\,dx = \frac{\tan^3 x}{3} + \tan x + C$ by (3) and (9).

27. (a) $\int \frac{dz}{\sqrt{9-z^2}} = \sin^{-1}\frac{z}{3} + C$ by (17) with $u = z$ and $a = 3$.

(b) $\int \frac{z\,dz}{\sqrt{9-z^2}} = -\frac{1}{2}\int (9-z^2)^{-1/2}(-2z\,dz) = -\frac{1}{2}\frac{(9-z^2)^{1/2}}{1/2} + C$ by (3) $\left(\text{with } u = 9 - z^2,\ n = -\frac{1}{2}\right) = -\sqrt{9-z^2} + C$.

(c) $\int \frac{z\,dz}{9-z^2} = -\frac{1}{2}\ln|9-z^2| + C$ by (4) with $u = 9 - z^2$.

(d) $\int \frac{z\,dz}{(9-z^2)^2} = \frac{1}{2(9-z^2)} + C$ by (3).

(e) $\int \frac{dz}{9+z^2} = \frac{1}{3}\tan^{-1}\frac{z}{3} + C$ by (18) with $u = z$ and $a = 3$.

28. $\int \dfrac{dx}{\sqrt{x}(1 + 2\sqrt{x})} = \ln(1 + 2\sqrt{x}) + C$ by (4) with $u = 1 + 2\sqrt{x}$ and $du = \dfrac{dx}{\sqrt{x}}$.

29. $\int \sin x \, \cos x \, dx = \dfrac{1}{2} \sin^2 x + C$ by (3) with $u = \sin x$; OR $= -\dfrac{1}{2} \cos^2 x + C$ by (3) with $u = \cos x$; OR $= -\dfrac{1}{4} \cos 2x + C$ by (6), where we use the trigonometric identity $\sin 2x = 2 \sin x \cos x$.

30. $\int \dfrac{\cos \sqrt{x}}{\sqrt{x}} \, dx = 2 \sin \sqrt{x} + C$ by (5) with $u = \sqrt{x}$.

31. $\int \sin^2 y \, dy = \int \left(\dfrac{1}{2} - \dfrac{\cos 2y}{2}\right) dy = \dfrac{y}{2} - \dfrac{\sin 2y}{4} + C$.

32. $\int \cos^2 z \sin^3 z \, dz = \int \cos^2 z \sin^2 z \sin z \, dz =$
$\int \cos^2 z \, (1 - \cos^2 z) \sin z \, dz = \int \cos^2 z \sin z \, dz - \int \cos^4 z \sin z \, dz =$
$-\dfrac{\cos^3 z}{3} + \dfrac{\cos^5 z}{5} + C$, by repeated application of (3) with $u = \cos z$ and $du = -\sin z \, dz$.

33. $\int \dfrac{x \, dx}{x^4 + 1} = \dfrac{1}{2} \tan^{-1} x^2 + C$ by (18) with $u = x^2$ and $a = 1$.

34. $\int \dfrac{dy}{\sqrt{6y - y^2}} = \int \dfrac{dy}{\sqrt{9 - (y^2 - 6y + 9)}} = \sin^{-1} \dfrac{y - 3}{3} + C$ by (17) with $u = y - 3$ and $a = 3$.

35. $\int \dfrac{e^x}{3 + e^{2x}} \, dx = \dfrac{1}{\sqrt{3}} \tan^{-1} \dfrac{e^x}{\sqrt{3}} + C$ by (18) with $u = e^x$ and $a = \sqrt{3}$.

36. $\int \dfrac{e^x - e^{-x}}{e^x + e^{-x}} \, dx = \ln(e^x + e^{-x}) + C$ by (4) with $u = e^x + e^{-x}$.

37. To evaluate $\int \dfrac{x + 1}{x^2 + 4x + 13} \, dx$, we let $u = x^2 + 4x + 13$, note that du is $(2x + 4) \, dx$, and rewrite the integral as

$$\dfrac{1}{2} \int \dfrac{2x + 2}{x^2 + 4x + 13} \, dx = \dfrac{1}{2} \int \dfrac{2x + 2 + 2 - 2}{x^2 + 4x + 13} \, dx$$

$$= \dfrac{1}{2} \int \dfrac{2x + 4}{x^2 + 4x + 13} \, dx - \int \dfrac{dx}{x^2 + 4x + 13}$$

$$= \dfrac{1}{2} \ln(x^2 + 4x + 13) - \int \dfrac{dx}{(x + 2)^2 + 3^2}$$

$$= \dfrac{1}{2} \ln(x^2 + 4x + 13) - \dfrac{1}{3} \tan^{-1} \dfrac{x + 2}{3} + C,$$

where we have used (4) for the first integral and (18) for the second.

38. $\int \dfrac{2x - 1}{\sqrt{8 - 2x - x^2}}\, dx = -\int \dfrac{-2x + 1}{\sqrt{8 - 2x - x^2}}\, dx$

$= -\int \dfrac{-2x - 2 + 1 + 2}{\sqrt{8 - 2x - x^2}}\, dx = -\int \dfrac{-2x - 2}{\sqrt{8 - 2x - x^2}}\, dx - 3\int \dfrac{dx}{\sqrt{8 - 2x - x^2}}$

$= -\int (8 - 2x - x^2)^{-1/2}(-2x - 2)\, dx - 3\int \dfrac{dx}{\sqrt{9 - (x^2 + 2x + 1)}}$

$= -2\sqrt{8 - 2x - x^2} \text{ (by (3))} - 3 \sin^{-1} \dfrac{x + 1}{3} \text{ (by (17))} + C.$

39. $\int \dfrac{dt}{\sin^2 2t} = \int \csc^2 2t\, dt = \dfrac{1}{2}\int \csc^2 2t\,(2\,dt) = -\dfrac{1}{2}\cot 2t + C \text{ by (10).}$

40. $\int \cos^2 4z\, dz = \int \left(\dfrac{1}{2} + \dfrac{\cos 8z}{2}\right) dz$ by a trigonometric identity

$= \dfrac{z}{2} + \dfrac{\sin 8z}{16} + C.$

41. $\int \dfrac{\sin 2x}{1 + \sin^2 x}\, dx = \ln (1 + \sin^2 x) + C$ by (4) with $u = 1 + \sin^2 x$
and $du = 2 \sin x \cos x\, dx = \sin 2x\, dx.$

42. $\int x^2 e^{-x^3}\, dx = -\dfrac{1}{3} e^{-x^3} + C$ by (15) with $u = -x^3.$

43. $\int \dfrac{dy}{y\sqrt{1 + \ln y}} = 2\sqrt{1 + \ln y} + C$ by (3) with $u = 1 + \ln y,$

$du = \dfrac{dy}{y}, \text{ and } n = -\dfrac{1}{2}.$

44. $\int \dfrac{\sqrt{x - 1}}{x}\, dx.$ We let $u = \sqrt{x - 1},\ u^2 = x - 1.$ Then $2u\, du = dx.$
Then we substitute, as follows:

$\int \dfrac{\sqrt{x - 1}}{x}\, dx = 2\int \dfrac{u^2\, du}{u^2 + 1} = 2\int \dfrac{u^2 + 1 - 1}{u^2 + 1}\, du = 2\int \left(1 - \dfrac{1}{u^2 + 1}\right) du$

$= 2(u - \tan^{-1} u) + C = 2(\sqrt{x - 1} - \tan^{-1} \sqrt{x - 1}) + C.$

*C. Trigonometric Substitutions

Examples 1–44 illustrate the technique of integration by substitution. Trigonometric substitutions are also effective for certain integrals, as we now show.

(1) Integrals involving $\sqrt{a^2 - u^2}$. Let $u = a \sin \theta$ and replace $a^2 - u^2$ by $a^2 \cos^2 \theta.$

(2) Integrals involving $\sqrt{a^2 + u^2}$. Let $u = a \tan \theta$ and replace $a^2 + u^2$ by $a^2 \sec^2 \theta.$

(3) Integrals involving $\sqrt{u^2 - a^2}$. Let $u = a \sec \theta$ and replace $u^2 - a^2$ by $a^2 \tan^2 \theta.$

We thus obtain a new integrand, in each case involving trigonometric functions of θ. This technique is illustrated in the following examples.

*An asterisk denotes a topic covered only in Calculus BC.

Examples

45. $\int \dfrac{x^2}{\sqrt{4-x^2}}\, dx$. Let $x = 2 \sin \theta$ $\left(-\dfrac{\pi}{2} < \theta < \dfrac{\pi}{2}\right)$. Then

$$dx = 2 \cos \theta\, d\theta \qquad \text{and} \qquad \sqrt{4-x^2} = 2 \cos \theta,$$

where we have noted that $\cos \theta > 0$ on the prescribed interval. Then

$$\int \frac{x^2\, dx}{\sqrt{4-x^2}} = \int \frac{4 \sin^2 \theta}{2 \cos \theta}\, 2 \cos \theta\, d\theta = 4 \int \sin^2 \theta\, d\theta$$

$$= 4 \int \left(\frac{1}{2} - \frac{\cos 2\theta}{2}\right) d\theta = 4\left(\frac{\theta}{2} - \frac{\sin 2\theta}{4}\right) + C$$

$$= 2\theta - \sin 2\theta + C \qquad \text{or} \qquad 2\theta - 2 \sin \theta \cos \theta + C.$$

If we draw an appropriate right triangle involving θ and x (see Figure N5–1), we can easily obtain the answer in terms of x.

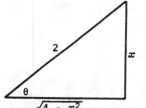

FIGURE N5–1

Thus

$$\int \frac{x^2\, dx}{\sqrt{4-x^2}} = 2 \sin^{-1} \frac{x}{2} - 2\, \frac{x}{2}\, \frac{\sqrt{4-x^2}}{2} + C$$

$$= 2 \sin^{-1} \frac{x}{2} - \frac{x\sqrt{4-x^2}}{2} + C.$$

46. $\int \dfrac{dx}{x\sqrt{4x^2+9}}$. Let $2x = 3 \tan \theta$ $\left(-\dfrac{\pi}{2} < \theta < \dfrac{\pi}{2}\right)$. Then

$$2\, dx = 3 \sec^2 \theta\, d\theta \qquad \text{and} \qquad \sqrt{4x^2+9} = 3 \sec \theta,$$

where $\sec \theta > 0$ on the above interval. So

$$\int \frac{dx}{x\sqrt{4x^2+9}} = \int \frac{3 \sec^2 \theta\, d\theta}{2 \cdot \dfrac{3 \tan \theta}{2} \cdot 3 \sec \theta}$$

$$= \frac{1}{3} \int \frac{\sec \theta}{\tan \theta}\, d\theta = \frac{1}{3} \int \csc \theta\, d\theta$$

$$= \frac{1}{3} \ln |\csc \theta - \cot \theta| + C$$

by (14), page 125. We use Figure N5–2 to express the answer in terms of x:

$$\int \frac{dx}{x\sqrt{4x^2 + 9}} = \frac{1}{3} \ln \left| \frac{\sqrt{4x^2 + 9}}{2x} - \frac{3}{2x} \right| + C$$

$$= \frac{1}{3} \ln \left| \frac{\sqrt{4x^2 + 9} - 3}{x} \right| + C',$$

where we have replaced $C - \frac{1}{3} \ln 2$ by C' in the final answer.

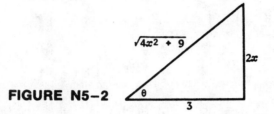

FIGURE N5–2

47. $\int \frac{du}{\sqrt{u^2 - 1}}$. Let $u = \sec \theta$; here $|u| > 1$. Then

$$du = \sec \theta \tan \theta \, d\theta \quad \text{and} \quad \sqrt{u^2 - 1} = \quad \tan \theta \quad \text{if } u > 1,$$
$$= -\tan \theta \quad \text{if } u < -1.$$

Then

$$\int \frac{du}{\sqrt{u^2 - 1}} = \pm \int \frac{\sec \theta \tan \theta \, d\theta}{\tan \theta} = \quad \int \sec \theta \, d\theta \quad \text{if } u > 1,$$
$$= -\int \sec \theta \, d\theta \quad \text{if } u < -1;$$

this yields $\ln |\sec \theta + \tan \theta| + C$ for $u > 1$, $-\ln |\sec \theta + \tan \theta| + C$ for $u < -1$, by (13). Note that $\sec \theta = u$, $\tan \theta = \pm\sqrt{u^2 - 1}$. Figure N5–3 enables us to write our answer as $\ln |u + \sqrt{u^2 - 1}| + C$ for $u > 1$. For $u < -1$ we get $-\ln |u - \sqrt{u^2 - 1}| + C$.

Since the latter can be rewritten as

$$\ln \left| \frac{1}{u - \sqrt{u^2 - 1}} \cdot \frac{u + \sqrt{u^2 - 1}}{u + \sqrt{u^2 - 1}} \right| + C = \ln |u + \sqrt{u^2 - 1}| + C,$$

we see that, for all u, $|u| > 1$,

$$\int \frac{du}{\sqrt{u^2 - 1}} = \ln |u + \sqrt{u^2 - 1}| + C.$$

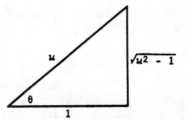

FIGURE N5–3

*D. Integration by Partial Fractions⎯⎯⎯⎯⎯⎯

The method of partial fractions makes it possible to express a rational function $\frac{f(x)}{g(x)}$ as a sum of simpler fractions. Here $f(x)$ and $g(x)$ are real polynomials in x and it is assumed that $\frac{f(x)}{g(x)}$ is a proper fraction; that is, that $f(x)$ is of lower degree than $g(x)$. If not, we divide $f(x)$ by $g(x)$ to express the given rational function as the sum of a polynomial and a proper rational function. Thus

$$\frac{x^4 + x^2 - 4}{x(x^2 + 1)} = x - \frac{4}{x(x^2 + 1)},$$

where the fraction on the right is proper.

Theoretically, every real polynomial can be expressed as a product of (powers of) real linear factors and (powers of) real quadratic factors. The particular form of the partial fractions depends both on the nature of these factors (i.e., linear versus quadratic) and on the power (i.e., the degree) to which each factor occurs. A real quadratic factor is irreducible if it cannot be decomposed into real linear factors.

In the following cases, all the capital letters denote constants to be determined.

(a) For each *distinct linear factor* $(x - a)$ of $g(x)$ we set up one partial fraction of the type $\frac{A}{x - a}$.

(b) For each *distinct irreducible quadratic factor* $(x^2 + bx + c)$ of $g(x)$ we set up one partial fraction of the form $\frac{Bx + C}{x^2 + bx + c}$. Note here that $b^2 - 4c$ is negative if $(x^2 + bx + c)$ is irreducible.

(c) If a linear factor $(x - a)$ of $g(x)$ occurs to the nth degree (i.e., is repeated n times), we set up n fractions for it as follows:

$$\frac{A_1}{x - a} + \frac{A_2}{(x - a)^2} + \frac{A_3}{(x - a)^3} + \cdots + \frac{A_n}{(x - a)^n}.$$

Note that (a) above is a special case of this where $n = 1$.

(d) If a quadratic factor $(x^2 + bx + c)$ is repeated m times, we set up m fractions for it as follows:

$$\frac{B_1 x + C_1}{x^2 + bx + c} + \frac{B_2 x + C_2}{(x^2 + bx + c)^2} + \frac{B_3 x + C_3}{(x^2 + bx + c)^3} + \cdots + \frac{B_m x + C_m}{(x^2 + bx + c)^m}.$$

The techniques for determining the unknown constants above are illustrated in the following examples.

Examples

48. Find $\int \dfrac{x^2 - x + 4}{x^3 - 3x^2 + 2x}\, dx$.

We factor the denominator and then set

$$\frac{x^2 - x + 4}{x(x - 1)(x - 2)} = \frac{A}{x} + \frac{B}{x - 1} + \frac{C}{x - 2}, \tag{1}$$

where the constants A, B, and C are to be determined. It follows that

$$x^2 - x + 4 = A(x - 1)(x - 2) + Bx(x - 2) + Cx(x - 1). \tag{2}$$

Since the polynomial on the right in (2) is to be identical to the one on the left, we can find the constants by either of the following methods:

METHOD ONE.   Expand and combine on the right in (2), getting

$$x^2 - x + 4 = (A + B + C)x^2 - (3A + 2B + C)x + 2A.$$

We then *equate coefficients of like powers in x* and solve simultaneously. Thus:

using the coefficients of x^2, we get $1 = A + B + C$;

using the coefficients of x, we get $-1 = -(3A + 2B + C)$;

using the constant coefficient, $4 = 2A$.

These equations yield $A = 2$, $B = -4$, $C = 3$.

METHOD TWO.   Although equation (1) above is meaningless for $x = 0$, $x = 1$, or $x = 2$, it is still true that equation (2) must hold even for these special values. We see, in (2), that:

if $x = 0$,   then $4 = 2A$ and $A = 2$;

if $x = 1$,   then $4 = -B$ and $B = -4$;

if $x = 2$,   then $6 = 2C$ and $C = 3$.

The second method is shorter than the first and most convenient when the denominator of the given fraction can be decomposed into distinct linear factors. Sometimes a combination of these methods is effective.

Finally, then, the original integral equals

$$\int \left(\frac{2}{x} - \frac{4}{x - 1} + \frac{3}{x - 2} \right) dx = 2 \ln |x| - 4 \ln |x - 1| + 3 \ln |x - 2| + C'$$

$$= \ln \frac{x^2 |x - 2|^3}{(x - 1)^4} + C'.$$

[The symbol "C'" appears here for the constant of integration because "C" was used in simplifying the original rational function.]

49. Integrate $\int \dfrac{12 - 8x}{x^2(x^2 + 4)}\, dx$.

Since the linear factor x occurs twice in the denominator and since $x^2 + 4$ is an irreducible quadratic factor, we write

$$\frac{12 - 8x}{x^2(x^2 + 4)} = \frac{A}{x} + \frac{B}{x^2} + \frac{Cx + D}{x^2 + 4}.$$

Then

$$12 - 8x = Ax(x^2 + 4) + B(x^2 + 4) + (Cx + D)x^2.$$

We can let $x = 0$ and get $12 = 4B$, so that $B = 3$. We can then equate coefficients of like powers to determine the remaining constants A, C, and D. The following equations are obtained by using the coefficients of x^3, x^2, and x respectively:

$$0 = A + C,$$
$$0 = B + D,$$
$$-8 = 4A.$$

Thus, $A = -2$, $B = 3$, $C = 2$, and $D = -3$, and

$$\frac{12 - 8x}{x^2(x^2 + 4)} = -\frac{2}{x} + \frac{3}{x^2} + \frac{2x - 3}{x^2 + 4}.$$

Therefore the original integral equals

$$\int \left(-\frac{2}{x} + \frac{3}{x^2} + \frac{2x - 3}{x^2 + 4} \right) dx$$

$$= -2 \ln |x| - \frac{3}{x} + \ln (x^2 + 4) - \frac{3}{2} \tan^{-1} \frac{x}{2} + C'.$$

50. Integrate $\int \dfrac{2x^4 - 1}{(x^2 + 1)^2}\, dx$.

Since the numerator is not of lower degree than the denominator, we divide as indicated:

$$
\begin{array}{r}
2 \\
x^4 + 2x^2 + 1\overline{\smash{\big)}\ 2x^4 - 1} \\
\underline{2x^4 + 4x^2 + 2} \\
- 4x^2 - 3 \ .
\end{array}
$$

So

$$\frac{2x^4 - 1}{(x^2 + 1)^2} = 2 - \frac{4x^2 + 3}{(x^2 + 1)^2}.$$

Since the (irreducible) quadratic factor $x^2 + 1$ occurs twice in the denominator of the proper rational function on the right, we set

$$\frac{4x^2 + 3}{(x^2 + 1)^2} = \frac{Ax + B}{x^2 + 1} + \frac{Cx + D}{(x^2 + 1)^2},$$

whence

$$4x^2 + 3 = (Ax + B)(x^2 + 1) + Cx + D.$$

We now equate coefficients of like powers, respectively those of x^3, x^2, x, and x^0, to get

$$0 = A,$$

$$4 = B,$$

$$0 = A + C,$$

$$3 = B + D.$$

Thus $A = C = 0$, $B = 4$, and $D = -1$. So

$$\int \frac{2x^4 - 1}{(x^2 + 1)^2} \, dx = \int 2 \, dx - \left[\int \frac{4 \, dx}{x^2 + 1} - \int \frac{dx}{(x^2 + 1)^2} \right]$$

$$= 2x - 4 \tan^{-1} x + \int \frac{dx}{(x^2 + 1)^2}.$$

To evaluate the last integral, we let $x = \tan \theta$, so that $dx = \sec^2 \theta \, d\theta$ and

$$\int \frac{dx}{(x^2 + 1)^2} = \int \frac{\sec^2 \theta \, d\theta}{\sec^4 \theta} = \int \cos^2 \theta \, d\theta$$

$$= \frac{1}{2} \theta + \frac{\sin 2\theta}{4} + C'.$$

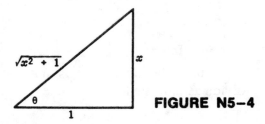

FIGURE N5–4

Figure N5–4 enables us to write this part of the answer as $\frac{1}{2} \tan^{-1} x + \frac{1}{4} \frac{2x}{(1 + x^2)} + C'$. Finally, then,

$$\int \frac{2x^4 - 1}{(x^2 + 1)^2} \, dx = 2x - \frac{7}{2} \tan^{-1} x + \frac{x}{2(1 + x^2)} + C'.$$

E.   Integration by Parts

The Parts Formula stems from that for the differential of a product: $d(uv) = u\,dv + v\,du$, or $u\,dv = d(uv) - v\,du$. Integrating, we get

$$\int u\,dv = uv - \int v\,du,$$

the Parts Formula. Success in using this important technique depends on being able to separate a given integral into parts u and dv so that

(a) dv can be integrated, and

(b) $\int v\,du$ is no more difficult to calculate than the original integral (and hopefully simpler).

Examples

51. To integrate $\int x \cos x\,dx$, we let $u = x$ and $dv = \cos x\,dx$. Then $du = dx$ and $v = \sin x$. Thus the Parts Formula yields

$$\int x \cos x\,dx = x \sin x - \int \sin x\,dx = x \sin x + \cos x + C.$$

52. To integrate $\int x^4 \ln x\,dx$, we let $u = \ln x$ and $dv = x^4\,dx$. Then $du = \frac{dx}{x}$ and $v = \frac{x^5}{5}$. Thus

$$\int x^4 \ln x\,dx = \frac{x^5}{5} \ln x - \frac{1}{5}\int x^4\,dx = \frac{x^5}{5} \ln x - \frac{x^5}{25} + C.$$

*53. To integrate $\int x^2 e^x\,dx$ we let $u = x^2$ and $dv = e^x\,dx$. Then $du = 2x\,dx$ and $v = e^x$. So $\int x^2 e^x\,dx = x^2 e^x - 2 \int x e^x\,dx$. We use the Parts Formula again, letting $u = x$ and $dv = e^x\,dx$ so that $du = dx$ and $v = e^x$. Thus

$$\int x^2 e^x\,dx = x^2 e^x - 2(x e^x - \int e^x\,dx) = x^2 e^x - 2x e^x + 2e^x + C.$$

*54. Let $I = \int e^x \cos x\,dx$. To integrate, we can let $u = e^x$ and $dv = \cos x\,dx$; then $du = e^x\,dx$, $v = \sin x$. So

$$I = e^x \sin x - \int e^x \sin x\,dx.$$

To evaluate the integral on the right we again let $u = e^x$, $dv = \sin x\,dx$, so that $du = e^x\,dx$ and $v = -\cos x$. So

$$I = e^x \sin x - (-e^x \cos x + \int e^x \cos x\,dx)$$

$$= e^x \sin x + e^x \cos x - I,$$

$$2I = e^x(\sin x + \cos x),$$

$$I = \frac{1}{2}e^x(\sin x + \cos x) + C.$$

55. Let $I = \int x \tan^{-1} x \, dx$. To find I let $u = \tan^{-1} x$ and $dv = x \, dx$; then $du = \dfrac{dx}{1 + x^2}$ and $v = \dfrac{x^2}{2}$. We see that

$$I = \frac{1}{2} x^2 \tan^{-1} x - \frac{1}{2} \int \frac{x^2}{1 + x^2} \, dx$$

$$= \frac{1}{2} x^2 \tan^{-1} x - \frac{1}{2} \int \frac{1 + x^2 - 1}{1 + x^2} \, dx$$

$$= \frac{1}{2} x^2 \tan^{-1} x - \frac{1}{2} \int \left(1 - \frac{1}{1 + x^2} \right) dx$$

$$= \frac{1}{2} x^2 \tan^{-1} x - \frac{x}{2} + \frac{1}{2} \tan^{-1} x + C$$

$$= \frac{x^2 + 1}{2} \tan^{-1} x - \frac{x}{2} + C.$$

*56. Integration by parts is frequently useful in obtaining reduction formulas, which express an integral in terms of simpler ones.

Let $I_m = \int \sin^m x \, dx$; find a reduction formula for I_m. We let $u = \sin^{m-1} x$ and $dv = \sin x \, dx$. Then $du = (m - 1)\sin^{m-2} x \cos x \, dx$ and $v = -\cos x$. So

$$I_m = -\sin^{m-1} x \cos x + (m - 1) \int \sin^{m-2} x \cos^2 x \, dx$$

$$= -\sin^{m-1} x \cos x + (m - 1) \int \sin^{m-2} x \, (1 - \sin^2 x) \, dx$$

$$= -\sin^{m-1} x \cos x + (m - 1) \int \sin^{m-2} x \, dx - (m - 1) \int \sin^m x \, dx$$

$$= -\sin^{m-1} x \cos x + (m - 1)I_{m-2} - (m - 1)I_m.$$

From this we get

$$mI_m = -\sin^{m-1} x \cos x + (m - 1)I_{m-2},$$

$$I_m = -\frac{\sin^{m-1} x \cos x}{m} + \frac{m - 1}{m} \int \sin^{m-2} x \, dx.$$

The given integral has thus been expressed in terms of a simpler integral (i.e., with exponent reduced). The reduction formula can then be applied to the integral on the right.

*The Tic-Tac-Toe Method.[†]

This method of integrating by parts is extremely useful when repeated integration is necessary. To integrate $\int u(x)v(x) \, dx$, we construct a table as follows:

[†] This method is described by K. W. Folley in Vol. 54 (1947) of the *American Mathematical Monthly* and was referred to in the movie *Stand and Deliver*.

$$
\begin{array}{ccc}
u(x) & \xrightarrow{\;+\;} & v(x) \\
u'(x) & \xrightarrow{\;-\;} & v_1(x) \\
u''(x) & \xrightarrow{\;+\;} & v_2(x) \\
u'''(x) & \xrightarrow{\;-\;} & v_3(x) \\
u'^{\,v}(x) & \xrightarrow{\;+\;} & v_4(x)
\end{array}
$$

Here the column at the left contains the successive derivatives of $u(x)$, and the column at the right contains the successive antiderivatives of $v(x)$ (always with $C = 0$): that is, $v_1(x)$ is the antiderivative of $v(x)$, $v_2(x)$ of $v_1(x)$, and so on. The diagonal arrows join the pairs of terms whose products form the successive terms of the formula; above each arrow is given the sign of that product. By the tic-tac-toe method we have

$$
\int u(x)v(x)\,dx \;=\; u(x)v_1(x) - u'(x)v_2(x) + u''(x)v_3(x) - u'''(x)v_4(x) + \cdots .
$$

We will first show how to use this formula, then demonstrate where it comes from.

Examples

*57. To integrate $\int x^4 \cos x\,dx$ by the tic-tac-toe method, we let $u(x) = x^4$ and $v(x) = \cos x$, and get the following table:

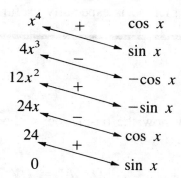

$$
\begin{array}{ccc}
x^4 & \xrightarrow{\;+\;} & \cos x \\
4x^3 & \xrightarrow{\;-\;} & \sin x \\
12x^2 & \xrightarrow{\;+\;} & -\cos x \\
24x & \xrightarrow{\;-\;} & -\sin x \\
24 & \xrightarrow{\;+\;} & \cos x \\
0 & & \sin x
\end{array}
$$

The method yields

$$
\int x^4 \cos x = x^4 \sin x - (-4x^3 \cos x) + (-12x^2 \sin x) - 24x \cos x + 24 \sin x + C
$$

$$
= x^4 \sin x + 4x^3 \cos x - 12x^2 \sin x - 24x \cos x + 24 \sin x + C.
$$

With the ordinary method we would have had to apply the Parts Formula four times to perform this integration.

Note that, after gathering like terms in the answer and differentiating, we obtain the given integrand, thus verifying the correctness:

$$
\frac{d}{dx}\,[\sin x(x^4 - 12x^2 + 24) + \cos x\,(4x^3 - 24x) + C]
$$

$$
= \sin x(4x^3 - 24x) + \cos x(x^4 - 12x^2 + 24)
$$
$$
- \sin x(4x^3 - 24x) + \cos x(12x^2 - 24)
$$

$$
= x^4 \cos x.
$$

*58. To find $\int x^3 e^x \, dx$, we let $u(x) = x^3$ and $v(x) = e^x$. We construct the table:

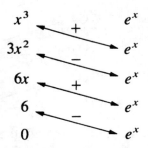

The method then yields

$$\int x^3 e^x \, dx = x^3 e^x - 3x^2 e^x + 6x e^x - 6e^x + C.$$

Check:

$$\frac{d}{dx} [e^x(x^3 - 3x^2 + 6x - 6) + C] = e^x(3x^2 - 6x + 6) + e^x(x^3 - 3x^2 + 6x - 6)$$

$$= x^3 e^x.$$

By the ordinary method we would have had to use the Parts Formula three times.

Note that this method is especially useful when some derivative of $u(x)$ equals 0.

We now show how the tic-tac-toe method results from successive applications of the Parts Formula,

$$\int U \, dV = UV - \int V \, dU.$$

To integrate $\int u(x)v(x) \, dx$, we let $U = u(x)$ and $dV = v(x) \, dx$. Then

$$dU = u'(x) \, dx;$$

$$V = \int v(x) \, dx = v_1(x).$$

So

$$\int u(x)v(x) \, dx = u(x)v_1(x) - \int u'(x)v_1(x) \, dx.$$

For $\int u'(x)v_1(x) \, dx$ we redefine U and V: $U = u'(x)$, $dV = v_1(x) \, dx$. Then

$$dU = u''(x) \, dx;$$

$$V = \int v_1(x) \, dx = v_2(x).$$

So

$$\int u(x)v(x) \ dx \ = \ u(x)v_1(x) \ - \ [u'(x)v_2(x) \ - \ \int u''(x)v_2(x) \ dx].$$

For the latter integral we again redefine U and V: $U = u''(x)$ and $dV = v_2(x) \ dx$. Then $dU = u'''(x) \ dx$ and $V = \int v_2(x) \ dx = v_3(x)$; and

$$\int u(x)v(x) \ dx \ = \ u(x)v_1(x) \ - \ u'(x)v_2(x) \ + \ [u''(x)v_3(x) \ - \ \int u''' \ (x)v_3(x) \ dx].$$

We continue integrating by parts to obtain

$$\int u(x)v(x) \ dx \ = \ u(x)v_1(x) \ - \ u'(x)v_2(x) \ + \ u''(x)v_3(x) \ - \ u'''(x)v_4(x) \ + \ \cdots.$$

F. Applications of Antiderivatives; Differential Equations

The following examples show how we use given conditions to determine constants of integration.

Examples

59. If $f'(x) = 3x^2$ and $f(1) = 6$, then

$$f(x) = \int 3x^2 \ dx = x^3 + C.$$

Since $f(1) = 6$, $1^3 + C$ must equal 6; so C must equal $6 - 1$ or 5, and $f(x) = x^3 + 5$.

60. If the slope of a curve at each point (x, y) equals the reciprocal of the abscissa and if the curve contains the point $(e, -3)$, we are given that $\frac{dy}{dx} = \frac{1}{x}$ and that $y = -3$ when $x = e$. This equation is also solved by integration. Since $\frac{dy}{dx} = \frac{1}{x}$,

$$dy \ = \ \frac{1}{x} \ dx, \qquad \text{and} \qquad y \ = \ \int \frac{1}{x} dx.$$

Thus $y = \ln x + C$. We now use the given condition, by substituting the point $(e, -3)$, to determine C. Since $-3 = \ln e + C$, we have $-3 = 1 + C$, and $C = -4$. So the solution of the differential equation subject to the given condition is

$$y = \ln x - 4.$$

Differential Equations: Motion Problems.

An equation involving a derivative or differential is called a *differential equation*. In Examples 59 and 60 we solved two simple differential equations. In Examples 61 and 62 we use the velocity (or acceleration) of a particle moving on a line to find the position of the particle. Note especially how the initial (or other given) conditions are used to evaluate constants of integration.[†]

Examples

61. The velocity of a particle moving along a line is given by $v(t) = 4t^3 - 3t^2$ at time t. If initially the particle is at $x = 3$ on the line, find its position when $t = 2$.

Since

$$v(t) = \frac{dx}{dt} = 4t^3 - 3t^2,$$

we have $dx = (4t^3 - 3t^2)\, dt$ **and** $x = \int(4t^3 - 3t^2)\, dt = t^4 - t^3 + C$. **Since** $x(0) = 0^4 - 0^3 + C = 3$, **we see that** $C = 3$, **and that the position function is** $x(t) = t^4 - t^3 + 3$. **When** $t = 2$, **we see that**

$$x(2) = 2^4 - 2^3 + 3 = 16 - 8 + 3 = 11.$$

62. Suppose that $a(t)$, the acceleration of a particle at time t, is given by $a(t) = 4t - 3$, that $v(1) = 6$, and that $f(2) = 5$, where $f(t)$ is the position function. (a) Find $v(t)$ and $f(t)$. (b) Find the position of the particle when $t = 1$.

(a)
$$a(t) = v'(t) = \frac{dv}{dt} = 4t - 3,$$

$$dv = (4t - 3)\, dt,$$

$$v = \int(4t - 3)\, dt = 2t^2 - 3t + C_1.$$

Using $v(1) = 6$, we get $6 = 2(1)^2 - 3(1) + C_1$, and $C_1 = 7$, from which it follows that $v(t) = 2t^2 - 3t + 7$. Since

$$v(t) = f'(t) = \frac{df}{dt},$$

$$f(t) = \int(2t^2 - 3t + 7)\, dt = \frac{2t^3}{3} - \frac{3t^2}{2} + 7t + C_2.$$

[†] See also Chapter 9, §C, page 246, for further uses of integration in finding the distance traveled by a particle moving on a line.

Using $f(2) = 5$, we get $5 = \frac{2}{3}(2)^3 - \frac{3}{2}(2)^2 + 7(2) + C_2, 5 = \frac{16}{3} - 6 + 14 + C_2,$ so $C_2 = -\frac{25}{3}$. Thus

$$f(t) = \frac{2}{3}t^3 - \frac{3}{2}t^2 + 7t - \frac{25}{3}.$$

(b) When $t = 1$,

$$f(1) = \frac{2}{3} - \frac{3}{2} + 7 - \frac{25}{3} = -\frac{13}{6}.$$

Differential Equations: Exponential Growth and Decay.

An interesting special differential equation with wide applications is illustrated in Example 63.

Example 63. The bacteria in a certain culture increase continuously at a rate proportional to the number present. (a) If the number triples in 6 hr, how many are there in 12 hr? (b) In how many hours will the original number quadruple?

We let N be the number at time t and N_0 the number initially. Then

$$\frac{dN}{dt} = kN, \qquad \frac{dN}{N} = k\,dt, \qquad \ln N = kt + C, \qquad \text{and} \qquad \ln N_0 = 0 + C,$$

so that $C = \ln N_0$. The general solution is then $N = N_0 e^{kt}$, with k still to be determined.

Since $N = 3N_0$ when $t = 6$, we see that $3N_0 = N_0 e^{6k}$ and that $k = \frac{1}{6}\ln 3$. Thus

$$N = N_0 e^{(t \ln 3)/6}.$$

(a) When $t = 12$, $N = N_0 e^{2\ln 3} = N_0 e^{\ln 3^2} = N_0 e^{\ln 9} = 9N_0$.

(b) We let $N = 4N_0$ in the centered equation above, and get

$$4 = e^{(t \ln 3)/6}, \qquad \ln 4 = \frac{t}{6}\ln 3, \qquad \text{and} \qquad t = \frac{6\ln 4}{\ln 3} = 7.6 \text{ hr.}$$

When a quantity y increases or decreases at time t at a rate proportional to the amount y present at that time, the quantity satisfies the differential equation

$$\frac{dy}{dt} = kt,$$

where k is the factor of proportionality. We see from Example 63 that the solution of this equation is

$$y = ce^{kt}, \tag{1}$$

where c is the initial amount (at time $t = 0$) of the quantity y. If the quantity grows with time, then $k > 0$; if it decays (diminishes or decomposes), then $k < 0$. Equation (1) is often referred to as the law of natural growth or decay.

Example 64.   Radium-226 decays at a rate proportional to the quantity present. Its half-life is 1612 years. How long will it take for one quarter of a given quantity of radium-226 to decay?

If $Q(t)$ is the amount present at time t, then it satisfies the equation

$$Q(t) = Q_0 e^{kt}, \tag{2}$$

where Q_0 is the initial amount and k is the (negative) factor of proportionality. Since it is given that $Q = \frac{1}{2} Q_0$ when $t = 1612$, equation (2) yields

$$\frac{1}{2} Q_0 = Q_0 e^{k(1612)};$$

$$\frac{1}{2} = e^{1612k};$$

$$k = \frac{\ln \frac{1}{2}}{1612} = -0.00043$$

$$= -0.043\% \text{ to the nearest thousandth of a percent.}$$

We now have

$$Q = Q_0 e^{-0.00043t}. \tag{3}$$

When one quarter of Q_0 has decayed, three quarters of the initial amount remains. We use this fact in equation (3) to find t:

$$\frac{3}{4} Q_0 = Q_0 e^{-0.00043t};$$

$$\frac{3}{4} = -0.00043t;$$

$$t = \frac{\ln \frac{3}{4}}{-0.00043} \approx 669 \text{ years.}$$

Differential Equations with Separable Variables.

A differential equation has variables separable if it is of the form

$$\frac{dy}{dx} = \frac{f(x)}{g(y)} \qquad \text{or} \qquad g(y)\,dy - f(x)\,dx = 0. \qquad (4)$$

The general solution is

$$\int g(y)\,dy - \int f(x)\,dx = C \qquad (C \text{ arbitrary}). \qquad (5)$$

Examples

65. Solve the differential equation $\frac{dy}{dx} = \frac{x}{y}$ if $y = 3$ when $x = 2$.
We separate the variables to get $y\,dy = x\,dx$; then

$$y\,dy = \int x\,dx \qquad \text{or} \qquad \frac{1}{2}y^2 = \frac{1}{2}x^2 + C.$$

To evaluate C, we replace y by 3 and x by 2:

$$\frac{1}{2}(3^2) = \frac{1}{2}(2^2) + C; \quad C = \frac{5}{2}.$$

The solution is $y^2 = x^2 + 5$.

66. If $\frac{ds}{dt} = \sqrt{st}$ and $t = 0$ when $s = 1$, find s when $t = 9$.

We separate variables:

$$\frac{ds}{\sqrt{s}} = \sqrt{t}\,dt;$$

then, integration yields

$$2s^{1/2} = \frac{2}{3}t^{3/2} + C.$$

Using $s = 1$ and $t = 0$, we get $2 = \frac{2}{3} \cdot 0 + C$, so $C = +2$. Then

$$2s^{1/2} = \frac{2}{3}t^{3/2} + 2 \qquad \text{or} \qquad s^{1/2} = \frac{1}{3}t^{3/2} + 1.$$

When $t = 9$, we find that $s^{1/2} = 9 + 1$. So $s = 100$.

67. If $(\ln y)\,\dfrac{dy}{dx} = \dfrac{y}{x}$, and $y = e$ when $x = 1$, find the value of y greater than 1 that corresponds to $x = e^4$.

Separating, we get $\dfrac{\ln y}{y}\,dy = \dfrac{dx}{x}$. We integrate:

$$\frac{1}{2}\ln^2 y = \ln |x| + C.$$

Using $y = e$ when $x = 1$ yields $C = \dfrac{1}{2}$. So

$$\frac{1}{2}\ln^2 y = \ln |x| + \frac{1}{2}.$$

When $x = e^4$, we have $\dfrac{1}{2}\ln^2 y = 4 + \dfrac{1}{2}$, $\ln^2 y = 9$, $\ln y = 3$ (where we chose $\ln y > 0$ because $y > 1$). So $y = e^3$.

68. Find the general solution of the differential equation $\dfrac{du}{dv} = e^{v-u}$.
We rewrite:

$$\frac{du}{dv} = \frac{e^v}{e^u};\ e^u\,du = e^v\,dv.$$

Taking antiderivatives yields $e^u = e^v + C$.

Set 5: Multiple-Choice Questions on Integration

1. $\int (3x^2 - 2x + 3)\,dx =$

(A) $x^3 - x^2 + C$ (B) $3x^3 - x^2 + 3x + C$ (C) $x^3 - x^2 + 3x + C$

(D) $\dfrac{1}{2}(3x^2 - 2x + 3)^2 + C$ (E) none of these

2. $\int \left(x - \dfrac{1}{2x}\right)^2 dx =$

(A) $\dfrac{1}{3}\left(x - \dfrac{1}{2x}\right)^3 + C$ (B) $x^2 - 1 + \dfrac{1}{4x^2} + C$

(C) $\dfrac{x^3}{3} - 2x - \dfrac{1}{4x} + C$ (D) $\dfrac{x^3}{3} - x - \dfrac{4}{x} + C$ (E) none of these

3. $\int \sqrt{4 - 2t}\, dt =$

   (A) $-\dfrac{1}{3}(4 - 2t)^{3/2} + C$ (B) $\dfrac{2}{3}(4 - 2t)^{3/2} + C$

   (C) $-\dfrac{1}{6}(4 - 2t)^3 + C$ (D) $+\dfrac{1}{2}(4 - 2t)^2 + C$

   (E) $\dfrac{4}{3}(4 - 2t)^{3/2} + C$

4. $\int (2 - 3x)^5\, dx =$

   (A) $\dfrac{1}{6}(2 - 3x)^6 + C$ (B) $-\dfrac{1}{2}(2 - 3x)^6 + C$

   (C) $\dfrac{1}{2}(2 - 3x)^6 + C$ (D) $-\dfrac{1}{18}(2 - 3x)^6 + C$ (E)   none of these

5. $\int \dfrac{1 - 3y}{\sqrt{2y - 3y^2}}\, dy =$

   (A) $4\sqrt{2y - 3y^2} + C$ (B) $\dfrac{1}{4}(2y - 3y^2)^2 + C$

   (C) $\dfrac{1}{2}\ln \sqrt{2y - 3y^2} + C$ (D) $\dfrac{1}{4}(2y - 3y^2)^{1/2} + C$

   (E) $\sqrt{2y - 3y^2} + C$

6. $\int \dfrac{dx}{3(2x - 1)^2} =$

   (A) $\dfrac{-3}{2x - 1} + C$ (B) $\dfrac{1}{6 - 12x} + C$ (C) $+\dfrac{6}{2x - 1} + C$

   (D) $\dfrac{2}{3\sqrt{2x - 1}} + C$ (E) $\dfrac{1}{3}\ln |2x - 1| + C$

7. $\int \dfrac{2\, du}{1 + 3u} =$

   (A) $\dfrac{2}{3}\ln |1 + 3u| + C$ (B) $-\dfrac{1}{3(1 + 3u)^2} + C$

   (C) $2 \ln |1 + 3u| + C$ (D) $\dfrac{3}{(1 + 3u)^2} + C$ (E)   none of these

8. $\int \dfrac{t}{\sqrt{2t^2 - 1}}\, dt =$

   (A) $\dfrac{1}{2}\ln \sqrt{2t^2 - 1} + C$ (B) $4 \ln \sqrt{2t^2 - 1} + C$

   (C) $8\sqrt{2t^2 - 1} + C$ (D) $-\dfrac{1}{4(2t^2 - 1)} + C$ (E) $\dfrac{1}{2}\sqrt{2t^2 - 1} + C$

9. $\int \cos 3x \, dx =$

    (A) $3 \sin 3x + C$ **(B)** $-\sin 3x + C$ **(C)** $-\dfrac{1}{3} \sin 3x + C$

    (D) $\dfrac{1}{3} \sin 3x + C$ **(E)** $\dfrac{1}{2} \cos^2 3x + C$

10. $\int \dfrac{x \, dx}{1 + 4x^2} =$

    (A) $\dfrac{1}{8} \ln (1 + 4x^2) + C$ **(B)** $\dfrac{1}{8(1 + 4x^2)^2} + C$

    (C) $\dfrac{1}{4} \sqrt{1 + 4x^2} + C$ **(D)** $\dfrac{1}{2} \ln |1 + 4x^2| + C$ **(E)** $\dfrac{1}{2} \tan^{-1} 2x + C$

11. $\int \dfrac{dx}{1 + 4x^2} =$

    (A) $\tan^{-1} (2x) + C$ **(B)** $\dfrac{1}{8} \ln (1 + 4x^2) + C$ **(C)** $\dfrac{1}{8(1 + 4x^2)^2} + C$

    (D) $\dfrac{1}{2} \tan^{-1} (2x) + C$ **(E)** $\dfrac{1}{8x} \ln |1 + 4x^2| + C$

12. $\int \dfrac{x}{(1 + 4x^2)^2} \, dx =$

    (A) $\dfrac{1}{8} \ln (1 + 4x^2)^2 + C$ **(B)** $\dfrac{1}{4} \sqrt{1 + 4x^2} + C$

    (C) $-\dfrac{1}{8(1 + 4x^2)} + C$ **(D)** $-\dfrac{1}{3(1 + 4x^2)^3} + C$

    (E) $-\dfrac{1}{(1 + 4x^2)} + C$

13. $\int \dfrac{x \, dx}{\sqrt{1 + 4x^2}} =$

    (A) $\dfrac{1}{8} \sqrt{1 + 4x^2} + C$ **(B)** $\dfrac{\sqrt{1 + 4x^2}}{4} + C$ **(C)** $\dfrac{1}{2} \sin^{-1} 2x + C$

    (D) $\dfrac{1}{2} \tan^{-1} 2x + C$ **(E)** $\dfrac{1}{8} \ln \sqrt{1 + 4x^2} + C$

14. $\int \dfrac{dy}{\sqrt{4 - y^2}} =$

    (A) $\dfrac{1}{2} \sin^{-1} \dfrac{y}{2} + C$ **(B)** $-\sqrt{4 - y^2} + C$ **(C)** $\sin^{-1} \dfrac{y}{2} + C$

    (D) $-\dfrac{1}{2} \ln \sqrt{4 - y^2} + C$ **(E)** $-\dfrac{1}{3(4 - y^2)^{3/2}} + C$

15. $\int \dfrac{y\,dy}{\sqrt{4-y^2}} =$

(A) $\dfrac{1}{2}\sin^{-1}\dfrac{y}{2} + C$ (B) $-\sqrt{4-y^2} + C$ (C) $\sin^{-1}\dfrac{y}{2} + C$

(D) $-\dfrac{1}{2}\ln\sqrt{4-y^2} + C$ (E) $2\sqrt{4-y^2} + C$

16. $\int \dfrac{2x+1}{2x}\,dx =$

(A) $x + \dfrac{1}{2}\ln|x| + C$ (B) $1 + \dfrac{1}{2}x^{-1} + C$ (C) $x + 2\ln|x| + C$

(D) $x + \ln|2x| + C$ (E) $\dfrac{1}{2}\left(2x - \dfrac{1}{x^2}\right) + C$

17. $\int \dfrac{(x-2)^3}{x^2}\,dx =$

(A) $\dfrac{(x-2)^4}{4x^2} + C$ (B) $\dfrac{x^2}{2} - 6x + 6\ln|x| - \dfrac{8}{x} + C$

(C) $\dfrac{x^2}{2} - 3x + 6\ln|x| + \dfrac{4}{x} + C$ (D) $-\dfrac{(x-2)^4}{4x} + C$

(E) none of these

18. $\int \left(\sqrt{t} - \dfrac{1}{\sqrt{t}}\right)^2 dt =$

(A) $t - 2 + \dfrac{1}{t} + C$ (B) $\dfrac{t^3}{3} - 2t - \dfrac{1}{t} + C$ (C) $\dfrac{t^2}{2} + \ln|t| + C$

(D) $\dfrac{t^2}{2} - 2t + \ln|t| + C$ (E) $\dfrac{t^2}{2} - t - \dfrac{1}{t^2} + C$

19. $\int (4x^{1/3} - 5x^{3/2} - x^{-1/2})\,dx =$

(A) $3x^{4/3} - 2x^{5/2} - 2x^{1/2} + C$ (B) $3x^{4/3} - 2x^{5/2} + 2x^{1/2} + C$

(C) $6x^{2/3} - 2x^{5/2} - \dfrac{1}{2}x^2 + C$ (D) $\dfrac{4}{3}x^{-2/3} - \dfrac{15}{2}x^{1/2} + \dfrac{1}{2}x^{-3/2} + C$

(E) none of these

20. $\int \dfrac{x^3 - x - 1}{(x+1)^2}\,dx =$

(A) $(x-2) + \dfrac{2x+1}{(x+1)^2} + C$ (B) $x^2 - 2x + \dfrac{1}{2}\ln(x^2 + 2x + 1) + C$

(C) $\dfrac{1}{2}x^2 - 2x + \ln|x+1|^2 - \dfrac{1}{x+1} + C$

(D) $\dfrac{1}{2}(x-2)^2 + 2\ln|x+1| + \dfrac{1}{x+1} + C$ (E) none of these

21. $\int \dfrac{dy}{\sqrt{y}\,(1 - \sqrt{y})} =$

(A) $4\sqrt{1 - \sqrt{y}} + C$

(B) $\dfrac{1}{2} \ln |1 - \sqrt{y}| + C$

(C) $2 \ln (1 - \sqrt{y}) + C$

(D) $2\sqrt{y} - \ln |y| + C$

(E) $-2 \ln |1 - \sqrt{y}| + C$

22. $\int \dfrac{u\,du}{\sqrt{4 - 9u^2}} =$

(A) $\dfrac{1}{3} \sin^{-1} \dfrac{3u}{2} + C$

(B) $-\dfrac{1}{18} \ln \sqrt{4 - 9u^2} + C$

(C) $2\sqrt{4 - 9u^2} + C$

(D) $\dfrac{1}{6} \sin^{-1} \dfrac{3}{2} u + C$

(E) $-\dfrac{1}{9}\sqrt{4 - 9u^2} + C$

23. $\int \sin \theta \cos \theta \, d\theta =$

(A) $-\dfrac{\sin^2 \theta}{2} + C$

(B) $-\dfrac{1}{4} \cos 2\theta + C$

(C) $\dfrac{\cos^2 \theta}{2} + C$

(D) $\dfrac{1}{2} \sin 2\theta + C$

(E) $\cos 2\theta + C$

24. $\int \dfrac{\sin \sqrt{x}}{\sqrt{x}}\,dx =$

(A) $-2 \cos^{1/2} x + C$

(B) $-\cos \sqrt{x} + C$

(C) $-2 \cos \sqrt{x} + C$

(D) $\dfrac{3}{2} \sin^{3/2} x + C$

(E) $\dfrac{1}{2} \cos \sqrt{x} + C$

25. $\int t \cos (2t)^2 \, dt =$

(A) $\dfrac{1}{8} \sin (4t^2) + C$

(B) $\dfrac{1}{2} \cos^2 (2t) + C$

(C) $-\dfrac{1}{8} \sin (4t^2) + C$

(D) $\dfrac{1}{4} \sin (2t)^2 + C$

(E) none of these

*26. $\int \cos^2 2x \, dx =$

(A) $\dfrac{x}{2} + \dfrac{\sin 4x}{8} + C$

(B) $\dfrac{x}{2} - \dfrac{\sin 4x}{8} + C$

(C) $\dfrac{x}{4} + \dfrac{\sin 4x}{4} + C$

(D) $\dfrac{x}{4} + \dfrac{\sin 4x}{16} + C$

(E) $\dfrac{1}{4}(x + \sin 4x) + C$

---

*Questions preceded by an asterisk are likely to appear only on the Calculus BC Examination.

27. $\int \sin 2\theta \, d\theta =$

(A) $\dfrac{1}{2} \cos 2\theta + C$ (B) $-2 \cos 2\theta + C$ (C) $-\sin^2 \theta + C$

(D) $\cos^2 \theta + C$ (E) $-\dfrac{1}{2} \cos 2\theta + C$

28. $\int x \cos x \, dx =$

(A) $x \sin x + C$ (B) $x \sin x + \cos x + C$ (C) $x \sin x - \cos x + C$

(D) $\cos x - x \sin x + C$ (E) $\dfrac{x^2}{2} \sin x + C$

29. $\displaystyle\int \frac{du}{\cos^2 3u} =$

(A) $-\dfrac{\sec 3u}{3} + C$ (B) $\tan 3u + C$ (C) $u + \dfrac{\sec 3u}{3} + C$

(D) $\dfrac{1}{3} \tan 3u + C$ (E) $\dfrac{1}{3 \cos 3u} + C$

30. $\displaystyle\int \frac{\cos x \, dx}{\sqrt{1 + \sin x}} =$

(A) $-\dfrac{1}{2}(1 + \sin x)^{1/2} + C$ (B) $\ln \sqrt{1 + \sin x} + C$

(C) $2\sqrt{1 + \sin x} + C$ (D) $\ln |1 + \sin x| + C$

(E) $\dfrac{2}{3(1 + \sin x)^{3/2}} + C$

31. $\displaystyle\int \frac{\cos (\theta - 1) d\theta}{\sin^2 (\theta - 1)}$

(A) $2 \ln \sin |\theta - 1| + C$ (B) $-\csc (\theta - 1) + C$

(C) $-\dfrac{1}{3} \sin^{-3} (\theta - 1) + C$ (D) $-\cot (\theta - 1) + C$

(E) $\csc (\theta - 1) + C$

32. $\displaystyle\int \sec \frac{t}{2} \, dt =$

(A) $\ln \left| \sec \dfrac{t}{2} + \tan \dfrac{t}{2} \right| + C$ (B) $2 \tan^2 \dfrac{t}{2} + C$ (C) $2 \ln \cos \dfrac{t}{2} + C$

(D) $\ln |\sec t + \tan t| + C$ (E) $2 \ln \left| \sec \dfrac{t}{2} + \tan \dfrac{t}{2} \right| + C$

33. $\displaystyle\int \frac{\sin 2x \, dx}{\sqrt{1 + \cos^2 x}} =$

(A) $-2\sqrt{1 + \cos^2 x} + C$ (B) $\dfrac{1}{2} \ln (1 + \cos^2 x) + C$

(C) $\sqrt{1 + \cos^2 x} + C$ (D) $-\ln \sqrt{1 + \cos^2 x} + C$

(E) $2 \ln |\sin x| + C$

*34. $\int \sec^{3/2} x \tan x \, dx =$

(A) $\dfrac{2}{5} \sec^{5/2} x + C$ (B) $-\dfrac{2}{3} \cos^{-3/2} x + C$ (C) $\sec^{3/2} x + C$

(D) $\dfrac{2}{3} \sec^{3/2} x + C$ (E)   none of these

35. $\int \tan \theta \, d\theta =$

(A) $-\ln |\sec \theta| + C$ (B) $\sec^2 \theta + C$ (C) $\ln |\sin \theta| + C$
(D) $\sec \theta + C$ (E) $-\ln |\cos \theta| + C$

36. $\displaystyle\int \dfrac{dx}{\sin^2 2x} =$

(A) $\dfrac{1}{2} \csc 2x \cot 2x + C$ (B) $-\dfrac{2}{\sin 2x} + C$ (C) $-\dfrac{1}{2} \cot 2x + C$

(D) $-\cot x + C$ (E) $-\csc 2x + C$

*37. $\displaystyle\int \dfrac{\tan^{-1} y}{1 + y^2} \, dy =$

(A) $\sec^{-1} y + C$ (B) $(\tan^{-1} y)^2 + C$ (C) $\ln (1 + y^2) + C$
(D) $\ln (\tan^{-1} y) + C$ (E)   none of these

*38. $\int \sin^3 \theta \cos^3 \theta \, d\theta =$

(A) $\dfrac{\sin^4 \theta}{4} - \dfrac{\sin^6 \theta}{6} + C$ (B) $\dfrac{\cos^4 \theta}{4} - \dfrac{\cos^6 \theta}{6} + C$ (C) $\dfrac{\sin^4 \theta}{4} + C$

(D) $-\dfrac{\cos^4 \theta}{4} + C$ (E) $\dfrac{\sin^4 \theta \cos^4 \theta}{16} + C$

39. $\displaystyle\int \dfrac{\sin 2t}{1 - \cos 2t} \, dt =$

(A) $\dfrac{2}{(1 - \cos 2t)^2} + C$ (B) $-\ln |1 - \cos 2t| + C$
(C) $\ln \sqrt{|1 - \cos 2t|} + C$ (D) $\sqrt{1 - \cos 2t} + C$
(E) $2 \ln |1 - \cos 2t| + C$

40. $\int \cot 2u \, du =$

(A) $\ln |\sin u| + C$ (B) $\dfrac{1}{2} \ln |\sin 2u| + C$ (C) $-\dfrac{1}{2} \csc^2 2u + C$

(D) $-\sec 2u + C$ (E) $2 \ln |\sin 2u| + C$

41. $\displaystyle\int \dfrac{e^x}{e^x - 1} \, dx =$

(A) $x + \ln |e^x - 1| + C$ (B) $x - e^x + C$ (C) $x - \dfrac{1}{(e^x - 1)^2} + C$

(D) $1 + \dfrac{1}{e^x - 1} + C$ (E) $\ln |e^x - 1| + C$

*42. $\int \frac{x-1}{x(x-2)} dx =$

(A) $\frac{1}{2} \ln |x| + \ln |x - 2| + C$ (B) $\frac{1}{2} \ln \left| \frac{x-2}{x} \right| + C$

(C) $\ln |x - 2| + \ln |x| + C$ (D) $\ln \sqrt{|x^2 - 2x|} + C$

(E) none of these

43. $\int xe^{x^2} dx =$

(A) $\frac{1}{2} e^{x^2} + C$ (B) $e^{x^2}(2x^2 + 1) + C$ (C) $2e^{x^2} + C$

(D) $e^{x^2} + C$ (E) $\frac{1}{2} e^{x^2+1} + C$

44. $\int \cos \theta \, e^{\sin \theta} d\theta =$

(A) $e^{\sin \theta + 1} + C$ (B) $e^{\sin \theta} + C$ (C) $-e^{\sin \theta} + C$ (D) $e^{\cos \theta} + C$

(E) $e^{\sin \theta}(\cos \theta - \sin \theta) + C$

45. $\int e^{2\theta} \sin e^{2\theta} d\theta =$

(A) $\cos e^{2\theta} + C$ (B) $2e^{4\theta}(\cos e^{2\theta} + \sin e^{2\theta}) + C$

(C) $-\frac{1}{2} \cos e^{2\theta} + C$ (D) $-2 \cos e^{2\theta} + C$ (E) none of these

46. $\int \frac{e^{\sqrt{x}}}{\sqrt{x}} dx =$

(A) $2\sqrt{x}(e^{\sqrt{x}} - 1) + C$ (B) $2e^{\sqrt{x}} + C$ (C) $\frac{e^{\sqrt{x}}}{2}\left(\frac{1}{x} + \frac{1}{x\sqrt{x}}\right) + C$

(D) $\frac{1}{2} e^{\sqrt{x}} + C$ (E) none of these

47. $\int xe^{-x} dx =$

(A) $e^{-x}(1 - x) + C$ (B) $\frac{e^{1-x}}{1-x} + C$ (C) $-e^{-x}(x + 1) + C$

(D) $-\frac{x^2}{2} e^{-x} + C$ (E) $e^{-x}(x + 1) + C$

*48. $\int x^2 e^x dx =$

(A) $e^x(x^2 + 2x) + C$ (B) $e^x(x^2 - 2x - 2) + C$

(C) $e^x(x^2 - 2x + 2) + C$ (D) $e^x(x - 1)^2 + C$

(E) $e^x(x + 1)^2 + C$

49. $\int \frac{e^x + e^{-x}}{e^x - e^{-x}} dx =$

(A) $x - \ln |e^x - e^{-x}| + C$ (B) $x + 2 \ln |e^x - e^{-x}| + C$

(C) $-\frac{1}{2}(e^x - e^{-x})^{-2} + C$ (D) $\ln |e^x - e^{-x}| + C$

(E) $\ln (e^x + e^{-x}) + C$

50. $\int \frac{e^x}{1 + e^{2x}} dx =$

(A) $\tan^{-1} e^x + C$ (B) $\frac{1}{2} \ln (1 + e^{2x}) + C$ (C) $\ln (1 + e^{2x}) + C$

(D) $\frac{1}{2} \tan^{-1} e^x + C$ (E) $2 \tan^{-1} e^x + C$

51. $\int \frac{\ln v \, dv}{v} =$

(A) $\ln |\ln v| + C$ (B) $\ln \frac{v^2}{2} + C$ (C) $\frac{1}{2} (\ln v)^2 + C$

(D) $2 \ln v + C$ (E) $\frac{1}{2} \ln v^2 + C$

52. $\int \frac{\ln \sqrt{x}}{x} dx =$

(A) $\frac{\ln^2 \sqrt{x}}{\sqrt{x}} + C$ (B) $\ln^2 x + C$ (C) $\frac{1}{2} \ln |\ln x| + C$

(D) $\frac{(\ln \sqrt{x})^2}{2} + C$ (E) $\frac{1}{4} \ln^2 x + C$

53. $\int \xi^3 \ln \xi \, d\xi =$

(A) $\xi^2(3 \ln \xi + 1) + C$ (B) $\frac{\xi^4}{16}(4 \ln \xi - 1) + C$

(C) $\frac{\xi^4}{4}(\ln \xi - 1) + C$ (D) $3\xi^2\left(\ln \xi - \frac{1}{2}\right) + C$ (E) none of these

54. $\int \ln \eta \, d\eta =$

(A) $\frac{1}{2} \ln^2 \eta + C$ (B) $\eta(\ln \eta - 1) + C$ (C) $\frac{1}{2} \ln \eta^2 + C$

(D) $\ln \eta(\eta - 1) + C$ (E) $\eta \ln \eta + \eta + C$

55. $\int \ln x^3 \, dx =$

(A) $\frac{3}{2} \ln^2 x + C$ (B) $3x(\ln x - 1) + C$ (C) $3 \ln x (x - 1) + C$

(D) $\frac{3x \ln^2 x}{2} + C$ (E) none of these

56. $\int \dfrac{\ln y}{y^2} dy =$

(A) $\dfrac{1}{y}(1 - \ln y) + C$ (B) $\dfrac{1}{2y}\ln^2 y + C$ (C) $-\dfrac{1}{3y^3}(4 \ln y + 1) + C$

(D) $-\dfrac{1}{y}(\ln y + 1) + C$ (E) $\dfrac{\ln y}{y} - \dfrac{1}{y} + C$

57. $\int \dfrac{dv}{v \ln v} =$

(A) $\dfrac{1}{\ln v^2} + C$ (B) $-\dfrac{1}{\ln^2 v} + C$ (C) $-\ln |\ln v| + C$

(D) $\ln \dfrac{1}{v} + C$ (E) $\ln |\ln v| + C$

58. $\int \dfrac{y - 1}{y + 1} dy =$

(A) $y - 2 \ln |y + 1| + C$ (B) $1 - \dfrac{2}{y + 1} + C$

(C) $\ln \dfrac{|y|}{(y + 1)^2} + C$ (D) $1 - 2 \ln |y + 1| + C$

(E) $\ln \left| \dfrac{e^y}{y + 1} \right| + C$

*59. $\int t\sqrt{t + 1}\, dt =$

(A) $\dfrac{2}{3}(t + 1)^{3/2} + C$ (B) $\dfrac{2}{15}(3t - 2)(t + 1)^{3/2} + C$

(C) $2\left[\dfrac{(t + 1)^{5/2}}{5} + \dfrac{(t + 1)^{3/2}}{5} \right] + C$ (D) $2t(t + 1) + C$

(E) none of these

60. $\int \sqrt{x}(\sqrt{x} - 1)\, dx =$

(A) $2(x^{3/2} - x) + C$ (B) $\dfrac{x^2}{2} - x + C$ (C) $\dfrac{1}{2}(\sqrt{x} - 1)^2 + C$

(D) $\dfrac{1}{2}x^2 - \dfrac{2}{3}x^{3/2} + C$ (E) $x - 2\sqrt{x} + C$

*61. $\int e^{\theta} \cos \theta\, d\theta =$

(A) $e^{\theta}(\cos \theta - \sin \theta) + C$ (B) $e^{\theta} \sin \theta + C$

(C) $\dfrac{1}{2}e^{\theta}(\sin \theta + \cos \theta) + C$ (D) $2 e^{\theta}(\sin \theta + \cos \theta) + C$

(E) $\dfrac{1}{2}e^{\theta}(\sin \theta - \cos \theta) + C$

62. $\int \frac{(1 - \ln t)^2}{t} dt =$

(A) $\frac{1}{3}(1 - \ln t)^3 + C$ (B) $\ln t - 2 \ln^2 t + \ln^3 t + C$

(C) $-2(1 - \ln t) + C$ (D) $\ln t - \ln^2 t + \frac{\ln t^3}{3} + C$

(E) $-\frac{(1 - \ln t)^3}{3} + C$

*63. $\int u \sec^2 u \, du =$

(A) $u \tan u + \ln |\cos u| + C$ (B) $\frac{u^2}{2} \tan u + C$

(C) $\frac{1}{2} \sec u \tan u + C$ (D) $u \tan u - \ln |\sin u| + C$

(E) $u \sec u - \ln |\sec u + \tan u| + C$

*64. $\int \frac{2x + 1}{4 + x^2} dx =$

(A) $\ln (x^2 + 4) + C$ (B) $\ln (x^2 + 4) + \tan^{-1} \frac{x}{2} + C$

(C) $\frac{1}{2} \tan^{-1} \frac{x}{2} + C$ (D) $\ln (x^2 + 4) + \frac{1}{2} \tan^{-1} \frac{x}{2} + C$

(E) none of these

*65. $\int \frac{x + 2}{x^2 + 2x + 10} dx =$

(A) $\frac{1}{2} \ln (x^2 + 2x + 10) + C$ (B) $\frac{1}{3} \tan^{-1} \frac{x + 1}{3} + C$

(C) $\frac{1}{2} \ln (x^2 + 2x + 10) + \frac{1}{3} \tan^{-1} \frac{x + 1}{3} + C$ (D) $\tan^{-1} \frac{x + 1}{3} + C$

(E) $\frac{1}{2} \ln |x^2 + 2x + 10| - 2 \tan^{-1} \frac{x + 1}{3} + C$

*66. $\int \frac{2x - 1}{\sqrt{4x - 4x^2}} dx =$

(A) $4 \ln \sqrt{4x - 4x^2} + C$ (B) $\sin^{-1} (1 - 2x) + C$

(C) $\frac{1}{2}\sqrt{4x - 4x^2} + C$ (D) $-\frac{1}{4} \ln (4x - 4x^2) + C$

(E) $-\frac{1}{2}\sqrt{4x - 4x^2} + C$

67. $\int \frac{e^{2x}}{1 + e^x} dx =$

    (A) $\tan^{-1} e^x + C$ (B) $e^x - \ln(1 + e^x) + C$

    (C) $e^x - x + \ln|1 + e^x| + C$ (D) $e^x + \frac{1}{(e^x + 1)^2} + C$

    (E) none of these

68. $\int \frac{\cos \theta}{1 + \sin^2 \theta} d\theta =$

    (A) $\sec \theta \tan \theta + C$ (B) $\sin \theta - \csc \theta + C$

    (C) $\ln(1 + \sin^2 \theta) + C$ (D) $\tan^{-1}(\sin \theta) + C$

    (E) $-\frac{1}{(1 + \sin^2 \theta)^2} + C$

*69. $\int x \tan^{-1} x \, dx =$

    (A) $\frac{x^2}{2} \tan^{-1} x - \frac{x}{2} + C$ (B) $\frac{1}{2}[x^2 \tan^{-1} x + \ln(1 + x^2)] + C$

    (C) $\left(\frac{x^2 + 1}{2}\right) \tan^{-1} x + C$ (D) $\frac{1}{2x}(\tan^{-1} x) + C$

    (E) $\frac{1}{2}[(x^2 + 1)\tan^{-1} x - x] + C$

*70. $\int \frac{dx}{1 - e^x} =$

    (A) $-\ln|1 - e^x| + C$ (B) $x - \ln|1 - e^x| + C$ (C) $\frac{1}{(1 - e^x)^2} + C$

    (D) $e^{-x} \ln|1 - e^x| + C$ (E) none of these

71. $\int \frac{(2 - y)^2}{4\sqrt{y}} dy =$

    (A) $\frac{1}{6}(2 - y)^3 \sqrt{y} + C$ (B) $2\sqrt{y} - \frac{2}{3}y^{3/2} + \frac{8}{5}y^{5/2} + C$

    (C) $\ln|y| - y + 2y^2 + C$ (D) $2y^{1/2} - \frac{2}{3}y^{3/2} + \frac{1}{10}y^{5/2} + C$

    (E) none of these

72. $\int e^{2 \ln u} \, du =$

    (A) $\frac{1}{3}e^{u^3} + C$ (B) $e^{(1/3 u^3)} + C$ (C) $\frac{1}{3}u^3 + C$ (D) $\frac{2}{u}e^{2 \ln u} + C$

    (E) $e^{1 + 2 \ln u} + C$

73. $\int \dfrac{dy}{y(1 + \ln y^2)} =$

    (A) $\dfrac{1}{2} \ln |1 + \ln y^2| + C$ **(B)** $-\dfrac{1}{(1 + \ln y^2)^2} + C$

    (C) $\ln |y| + \dfrac{1}{2} \ln |\ln y| + C$ **(D)** $\tan^{-1} (\ln |y|) + C$

    (E) none of these

74. $\int (\tan \theta - 1)^2 \, d\theta =$

    (A) $\sec \theta + \theta + 2 \ln |\cos \theta| + C$ **(B)** $\tan \theta + 2 \ln |\cos \theta| + C$
    (C) $\tan \theta - 2 \sec^2 \theta + C$ **(D)** $\sec \theta + \theta - \tan^2 \theta + C$
    (E) $\tan \theta - 2 \ln |\cos \theta| + C$

*75. $\int \dfrac{d\theta}{1 + \sin \theta} =$

    (A) $\sec \theta - \tan \theta + C$ **(B)** $\ln (1 + \sin \theta) + C$
    (C) $\ln |\sec \theta + \tan \theta| + C$ **(D)** $\theta + \ln |\csc \theta - \cot \theta| + C$
    (E) none of these

76. A particle starting at rest at $t = 0$ moves along a line so that its acceleration at time t is $12t$ ft/sec/sec. How much distance does the particle cover during the first 3 sec?

    (A) 16 ft   **(B)** 32 ft   **(C)** 48 ft   **(D)** 54 ft   **(E)** 108 ft

77. The equation of the curve whose slope at point (x, y) is $x^2 - 2$ and which contains the point $(1, -3)$ is

    (A) $y = \dfrac{1}{3}x^3 - 2x$ **(B)** $y = 2x - 1$ **(C)** $y = \dfrac{1}{3}x^3 - \dfrac{10}{3}$

    (D) $y = \dfrac{1}{3}x^3 - 2x - \dfrac{4}{3}$ **(E)** $3y = x^3 - 10$

78. A particle moves along a line with acceleration $2 + 6t$ at time t. When $t = 0$, its velocity equals 3 and it is at position $s = 2$. When $t = 1$, it is at position $s =$

    (A) 2   **(B)** 5   **(C)** 6   **(D)** 7   **(E)** 8

79. If $f'(x) = 2f(x)$ and $f(2) = 1$, then $f(x) =$

    (A) e^{2x-4} **(B)** $e^{2x} + 1 - e^4$ **(C)** e^{4-2x} **(D)** e^{2x+1} **(E)** e^{x-2}

80. The population of a county increases at a rate proportional to the existing population. If the population doubles in 20 years, then the factor of proportionality is

    (A) $\ln 2$ **(B)** $\ln 20$ **(C)** $\dfrac{1}{20} \ln 2$ **(D)** $\dfrac{1}{2} \ln 20$ **(E)** $2e^{-20}$

81. A solution of the differential equation $y \, dy = x \, dx$ is

(A) $x^2 - y^2 = 4$ (B) $x^2 + y^2 = 4$ (C) $y^2 = 4x^2$
(D) $x^2 - 4y^2 = 0$ (E) $x^2 = 9 - y^2$

82. If $\dfrac{dy}{dx} = \dfrac{y}{2\sqrt{x}}$ and $y = 1$ when $x = 4$, then

(A) $y^2 = 4\sqrt{x} - 7$ (B) $\ln y = 4\sqrt{x} - 8$ (C) $\ln y = \sqrt{x - 2}$
(D) $y = e^{\sqrt{x}}$ (E) $y = e^{\sqrt{x-2}}$

83. If $\dfrac{dy}{dx} = e^y$ and $y = 0$ when $x = 1$, then

(A) $y = \ln |x|$ (B) $y = \ln |2 - x|$ (C) $e^{-y} = 2 - x$
(D) $y = -\ln |x|$ (E) $e^{-y} = x - 2$

84. If $\dfrac{dy}{dx} = \dfrac{x}{\sqrt{9 + x^2}}$ and $y = 5$ when $x = 4$, then

(A) $y = \sqrt{9 + x^2} - 5$ (B) $y = \sqrt{9 + x^2}$

(C) $y = 2\sqrt{9 + x^2} - 5$ (D) $y = \dfrac{\sqrt{9 + x^2} + 5}{2}$

(E) none of these

85. If $\dfrac{ds}{dt} = \sin^2 \dfrac{\pi}{2} s$ and if, when $t = 0$, $s = 1$, then, when $s = \dfrac{3}{2}$, t is equal to

(A) $\dfrac{1}{2}$ (B) $\dfrac{\pi}{2}$ (C) 1 (D) $\dfrac{2}{\pi}$ (E) $-\dfrac{2}{\pi}$

86. The general solution of the differential equation $x \, dy = y \, dx$ is a family of

(A) circles (B) hyperbolas (C) parallel lines (D) parabolas
(E) lines passing through the origin

87. The general solution of the differential equation $\dfrac{dy}{dx} = y$ is a family of

(A) parabolas (B) straight lines (C) hyperbolas (D) ellipses
(E) none of these

88. A function $f(x)$ which satisfies the equations $f(x)f'(x) = x$ and $f(0) = 1$ is

(A) $f(x) = \sqrt{x^2 + 1}$ (B) $f(x) = \sqrt{1 - x^2}$ (C) $f(x) = x$
(D) $f(x) = e^x$ (E) none of these

89. The curve that passes through the point $(1, 1)$ and whose slope at any point (x, y) is equal to $\dfrac{3y}{x}$ has the equation

(A) $3x - 2 = y$ (B) $y^3 = x$ (C) $y = |x^3|$ (D) $3y^2 = x^2 + 2$
(E) $3y^2 - 2x = 1$

90. If radium decomposes at a rate proportional to the amount present, then the amount R left after t years, if R_0 is present initially and c is the negative constant of proportionality, is given by

   (A) $R = R_0 ct$ (B) $R = R_0 e^{ct}$ (C) $R = R_0 + \frac{1}{2} ct^2$

   (D) $R = e^{R_0 ct}$ (E) $R = e^{R_0 + ct}$

91. The population of a city increases continuously at a rate proportional, at any time, to the population at that time. The population doubles in 50 years. After 75 years the ratio of the population P to the initial population P_0 is

   (A) $\dfrac{9}{4}$ (B) $\dfrac{5}{2}$ (C) $\dfrac{4}{1}$ (D) $\dfrac{2\sqrt{2}}{1}$ (E) none of these

92. If a substance decomposes at a rate proportional to the amount of the substance present, and if the amount decreases from 40 gm to 10 gm in 2 hr, then the constant of proportionality is

   (A) $-\ln 2$ (B) $-\dfrac{1}{2}$ (C) $-\dfrac{1}{4}$ (D) $\ln \dfrac{1}{4}$ (E) $\ln \dfrac{1}{8}$

93. If $\frac{dy}{dx} = \frac{k}{x}$, k a constant, and if $y = 2$ when $x = 1$ and $y = 4$ when $x = e$, then, when $x = 2$, $y =$

   (A) 2   (B) 4   (C) $\ln 8$ (D) $\ln 2 + 2$ (E) $\ln 4 + 2$

94. If $(g'(x))^2 = g(x)$ for all real x and $g(0) = 0$, $g(4) = 4$, then $g(1)$ equals

   (A) $\dfrac{1}{4}$ (B) $\dfrac{1}{2}$ (C) 1   (D) 2   (E) 4

Answers for Set 5: Integration

1. C	20. D	39. C	58. A	77. D
2. E	21. E	40. B	59. B	78. D
3. A	22. E	41. E	60. D	79. A
4. D	23. B	42. D	61. C	80. C
5. E	24. C	43. A	62. E	81. A
6. B	25. A	44. B	63. A	82. E
7. A	26. A	45. C	64. D	83. C
8. E	27. E	46. B	65. C	84. B
9. D	28. B	47. C	66. E	85. D
10. A	29. D	48. C	67. B	86. E
11. D	30. C	49. D	68. D	87. E
12. C	31. B	50. A	69. E	88. A
13. B	32. E	51. C	70. B	89. C
14. C	33. A	52. E	71. D	90. B
15. B	34. D	53. B	72. C	91. D
16. A	35. E	54. B	73. A	92. A
17. E	36. C	55. B	74. B	93. E
18. D	37. E	56. D	75. E	94. A
19. A	38. A	57. E	76. D	

All the references in parentheses below are to the basic integration formulas on pages 125 and 126.

1. C.   Use, first, formula (2), then (3), replacing u by x.

2. E.   **Hint: Expand. The correct answer is** $\dfrac{x^3}{3} - x - \dfrac{1}{4x} + C$.

3. A.   By formula (3), with $u = 4 - 2t$ and $n = \dfrac{1}{2}$,

$$\int \sqrt{4 - 2t}\, dx = -\frac{1}{2}\int \sqrt{4 - 2t} \cdot (-2)\, dt = -\frac{1}{2}\frac{(4 - 2t)^{3/2}}{3/2} + C.$$

4. D.   Use (3) with $u = 2 - 3x$, noting that $du = -3\, dx$.

5. E.   Rewrite:

$$\int (2y - 3y^2)^{-1/2}(1 - 3y)\, dy = \frac{1}{2}\int (2y - 3y^2)^{-1/2}(2 - 6y)\, dy.$$

Use (3).

6. B.   Rewrite:

$$\frac{1}{3}\int (2x - 1)^{-2}\, dx = \frac{1}{3} \cdot \frac{1}{2}\int (2x - 1)^{-2} \cdot 2\, dx.$$

Using (3) yields $-\dfrac{1}{6(2x - 1)} + C$.

7. A.   This is equivalent to $\dfrac{2}{3}\displaystyle\int \dfrac{dv}{v}$. Use (4).

8. E.   Rewrite and modify to $\dfrac{1}{4}\displaystyle\int (2t^2 - 1)^{-1/2} \cdot 4t\, dt$. Use (3).

9. D.   Use (5) with $u = 3x$; $du = 3\, dx$.

10. A.   Use (4). If $u = 1 + 4x^2$, $du = 8x\, dx$.

11. D.   Use (18). Let $u = 2x$; then $du = 2\, dx$.

12. C.   Rewrite and modify to $\dfrac{1}{8}\displaystyle\int (1 + 4x^2)^{-2} \cdot 8x\, dx$. Use (3) with $n = -2$.

13. B.   Rewrite and modify to $\dfrac{1}{8}\displaystyle\int (1 + 4x^2)^{-1/2} \cdot 8x\, dx$. Use (3) with $n = -\dfrac{1}{2}$.

Note carefully the differences in the integrands in questions 10 through 13.

14. C.   Use (17). Note that $u = y$, $a = 2$.

15. B.   Rewrite and modify to $-\dfrac{1}{2}\displaystyle\int (4 - y^2)^{-1/2} \cdot -2y\, dy$. Use (3).

Compare the integrands in questions 14 and 15, noting the difference.

16. A.   Divide to obtain $\int \left(1 + \frac{1}{2} \cdot \frac{1}{x} \right) dx$. Use (2), (3), and (4). Remember that $\int k \, dx = kx + C$ whenever $k \neq 0$.

17. E. $\int \frac{(x-2)^3}{x^2} \, dx = \int \left(x - 6 + \frac{12}{x} - \frac{8}{x^2} \right) dx = \frac{x^2}{2} - 6x + 12 \ln |x| + \frac{8}{x} + C$. We used the binomial theorem with $n = 3$ on page 481 to expand $(x - 2)^3$.

18. D.   The integral is equivalent to $\int \left(t - 2 + \frac{1}{t} \right) dt$. Integrate term by term.

19. A.   Integrate term by term.

20. D.   Long division yields

$$\int \left((x - 2) + \frac{2x + 1}{x^2 + 2x + 1} \right) dx = \frac{1}{2}(x - 2)^2 + \int \frac{2x + 2 - 1}{x^2 + 2x + 1} \, dx.$$

The integral equals $\int \frac{2x + 2}{x^2 + 2x + 1} \, dx - \int \frac{1}{(x + 1)^2} \, dx$. Use formula (4) for the first integral, (3) for the second with $u = x + 1$ and $n = -2$. Note that $\ln |x + 1|^2 = 2 \ln |x + 1|$.

21. E.   Use formula (4) with $u = 1 - \sqrt{y} = 1 - y^{1/2}$. Then $du = -\frac{1}{2\sqrt{y}} \, dy$. Note that the integral can be written as $-2 \int \frac{1}{(1 - \sqrt{y})} \left(-\frac{1}{2\sqrt{y}} \right) dy$.

22. E.   Use formula (3) after multiplying outside the integral by $-\frac{1}{18}$, inside by -18.

23. B.   The integral is equal to $\frac{1}{2} \int \sin 2\theta \, d\theta$. Use formula (6) with $u = 2\theta$; $du = 2 \, d\theta$.

24. C.   Use formula (6) with $u = \sqrt{x}$; $du = \frac{1}{2\sqrt{x}} \, dx$.

25. A.   Use formula (5) with $u = 4t^2$; $du = 8t \, dt$.

26. A.   Using the half-angle formula (23) on page 483 with $\alpha = 2x$ yields $\int \left(\frac{1}{2} + \frac{1}{2} \cos 4x \right) dx$.

27. E.   Use formula (6).

28. B.   Integrate by parts (page 137). Let $u = x$ and $dv = \cos x \, dx$. Then $du = dx$ and $v = \sin x$. The given integral equals $x \sin x - \int \sin x \, dx$.

29. D.   Replace $\frac{1}{\cos^2 3u}$ by $\sec^2 3u$; then use formula (9).

30. C.   The integral is of the form $\int \frac{du}{\sqrt{u}}$, where $u = 1 + \sin x$ and $du = \cos x \, dx$. Use formula (3) with $n = -\frac{1}{2}$.

31. B.   The integral is equivalent to $\int \csc(\theta - 1) \cot(\theta - 1)\, d\theta$. Use formula (12).

32. E.   Use formula (13) with $u = \frac{t}{2}$; $du = \frac{1}{2}\, dt$.

33. A.   If we replace $\sin 2x$ by $2 \sin x \cos x$, the integral is equivalent to

$$-\int \frac{-2 \sin x \cos x}{\sqrt{1 + \cos^2 x}}\, dx = -\int \frac{du}{\sqrt{u}},$$

where $u = 1 + \cos^2 x$ and $du = -2 \sin x \cos x\, dx$. Use formula (3).

34. D.   Rewriting in terms of sines and cosines yields

$$\int \frac{\sin x}{\cos^{5/2} x}\, dx = -\int \cos^{-5/2} x(-\sin x)\, dx = -\left(-\frac{2}{3}\right) \cos^{-3/2} x + C.$$

35. E.   Use formula (7).

36. C.   Replace $\frac{1}{\sin^2 2x}$ by $\csc^2 2x$ and use formula (10).

37. E.   If we let $u = \tan^{-1} y$, then we integrate $\int u\, du$. The correct answer is $\frac{1}{2}(\tan^{-1} y)^2 + C$.

38. A.   Rewrite:

$$\int \sin^3 \theta(1 - \sin^2 \theta) \cos \theta\, d\theta = \int (\sin^3 \theta - \sin^5 \theta) \cos \theta\, d\theta.$$

39. C.   The answer is equivalent to $\frac{1}{2} \ln |1 - \cos 2t| + C$.

40. B.   To use formula (8), multiply outside the integral by $\frac{1}{2}$, inside by 2.

41. E.   Use formula (4) with $u = e^x - 1$; $du = e^x\, dx$.

42. D.   Use partial fractions: find A and B such that

$$\frac{x - 1}{x(x - 2)} = \frac{A}{x} + \frac{B}{x - 2}.$$

Then $x - 1 = A(x - 2) + Bx$.

Set $x = 0$: $-1 = -2A$ and $A = \frac{1}{2}$.

Set $x = 2$: $1 = 2B$ and $B = \frac{1}{2}$.

So the given integral equals

$$\int \left(\frac{1}{2x} + \frac{1}{2(x - 2)}\right) dx = \frac{1}{2} \ln |x| + \frac{1}{2} \ln |x - 2| + C$$

$$= \frac{1}{2} \ln |x(x - 2)| + C + \ln \sqrt{x^2 - 2x} + C.$$

43.   **A.**   Use formula (15) with $u = x^2$; $du = 2x\,dx$.

44.   **B.**   Use formula (15) with $u = \sin\theta$; $du = \cos\theta\,d\theta$.

45.   **C.**   Use formula (6) with $u = e^{2\theta}$; $du = 2e^{2\theta}\,d\theta$.

46.   **B.**   Use formula (15) with $u = \sqrt{x} = x^{1/2}$; $du = \dfrac{1}{2\sqrt{x}}\,dx$.

47.   **C.**   Use the Parts Formula. Let $u = x$, $dv = e^{-x}\,dx$; $du = dx$, $v = -e^{-x}$. We

$$\text{get } xe^{-x} + \int e^{-x}\,dx = -xe^{-x} - e^{-x} + C.$$

48.   **C.**   See Example 53, page 137.

49.   **D.**   The integral is of the form $\displaystyle\int \frac{du}{u}$; use (4).

50.   **A.**   This integral has the form $\displaystyle\int \frac{du}{1 + u^2}$. Use formula (18), with $u = e^x$,

               $du = e^x\,dx$, and $a = 1$.

51.   **C.**   Let $u = \ln v$; then $du = \dfrac{dv}{v}$. Use formula (3).

52.   **E.**   **Hint:** $\ln \sqrt{x} = \dfrac{1}{2}\ln x$.

53.   **B.**   Use parts, letting $u = \ln\xi$ and $dv = \xi^3\,d\xi$. Then $du = \dfrac{1}{\xi}\,d\xi$ and $v = \dfrac{\xi^4}{4}$.

               The integral equals $\dfrac{\xi^4}{4}\ln\xi - \dfrac{1}{4}\displaystyle\int \xi^3\,d\xi$.

54.   **B.**   Use parts, letting $u = \ln\eta$ and $dv = d\eta$. Then $du = \dfrac{1}{\eta}\,d\eta$ and $v = \eta$.

               The integral equals $\eta\ln\eta - \displaystyle\int d\eta$.

55.   **B.**   Rewrite $\ln x^3$ as $3\ln x$, and use the method in Answer 54, just above.

56.   **D.**   Use parts, letting $u = \ln y$ and $dv = y^{-2}\,dy$. Then $du = \dfrac{1}{y}\,dy$ and $v =$

               $-\dfrac{1}{y}$. The Parts Formula yields $\dfrac{-\ln y}{y} + \displaystyle\int \frac{1}{y^2}\,dy$.

57.   **E.**   The integral has the form $\displaystyle\int \frac{du}{u}$, where $u = \ln v$.

58.   **A.**   **Hint: The integrand is equivalent to** $1 - \dfrac{2}{y + 1}$.

59.   **B.**   **Hint: Let** $u = \sqrt{t + 1}$. **Then**

$$u^2 = t + 1, \qquad 2u\,du = dt, \qquad \text{and} \qquad t = u^2 - 1.$$

               The substitution yields $2\displaystyle\int (u^4 - u^2)\,du$.

60.   **D.**   **Hint: Multiply.**

61.   **C.**   See Example 54, page 137. Replace x by θ.

62.   E.   The integral equals $-\int (1 - \ln t)^2 \left(-\frac{1}{t}\, dt \right)$; it is equivalent to $-\int u^2\, du$, where $u = 1 - \ln t$.

63.   A.   Replace u by x in the given integral to avoid confusion in applying the Parts Formula. To integrate $\int x \sec^2 x\, dx$, let the variable u in the Parts Formula be x, and let dv be $\sec^2 x\, dx$. Then $du = dx$ and $v = \tan x$. So

$$\int x \sec^2 x\, dx = x \tan x - \int \tan x\, dx$$
$$= x \tan x + \ln |\cos x| + C.$$

64.   D.   The integral is equivalent to $\int \frac{2x}{4 + x^2}\, dx + \int \frac{1}{4 + x^2}\, dx$. Use formula (4) on the first integral and (18) on the second.

65.   C.   Rewrite:

$$\frac{1}{2} \int \frac{2x + 4}{x^2 + 2x + 10}\, dx = \frac{1}{2} \int \frac{2x + 2 + 2}{x^2 + 2x + 10}\, dx$$
$$= \frac{1}{2} \int \frac{2x + 2}{x^2 + 2x + 10}\, dx + \int \frac{dx}{(x + 1)^2 + 3^2}\, dx.$$

Use formulas (4) and (18).

66.   E.   Hint: Letting $u = 4x - 4x^2$, we see that we are integrating $-\frac{1}{4} \int u^{-1/2}\, du$.

67.   B.   Hint: Divide, getting $\int \left[e^x - \frac{e^x}{1 + e^x} \right] dx$.

68.   D.   Letting $u = \sin \theta$ yields the integral $\int \frac{du}{1 + u^2}$. Use formula (18).

69.   E.   Use integration by parts, letting $u = \tan^{-1} x$ and $dv = x\, dx$. Then

$$du = \frac{dx}{1 + x^2} \qquad \text{and} \qquad v = \frac{x^2}{2}.$$

The **Parts Formula** yields $\frac{x^2}{2} \tan^{-1} x - \frac{1}{2} \int \frac{x^2}{1 + x^2}\, dx$.

The integral is equivalent to $-\frac{1}{2} \int \left(1 - \frac{1}{1 + x^2} \right) dx$.

70.   B.   Hint: Note that

$$\frac{1}{1 - e^x} = \frac{1 - e^x + e^x}{1 - e^x} = 1 + \frac{e^x}{1 - e^x}.$$

Or multiply the integrand by $\frac{e^{-x}}{e^{-x}}$, recognizing that the correct answer is equivalent to $-\ln |e^{-x} - 1|$.

71. D.   **Hint: Expand the numerator and divide.** Then integrate term by term.

72. C.   **Hint: Observe that $e^{2 \ln u} = u^2$.**

73. A.   If we let $u = 1 + \ln y^2 = 1 + 2 \ln |y|$, we want to integrate $\frac{1}{2} \int \frac{du}{u}$.

74. B.   **Hint: Expand and note that**

$$\int(\tan^2 \theta - 2 \tan \theta + 1) \, d\theta = \int \sec^2 \theta \, d\theta - 2 \int \tan \theta \, d\theta.$$

Use formulas (9) and (7).

75. E.   **Multiply by $\frac{1 - \sin \theta}{1 - \sin \theta}$. The correct answer is $\tan \theta - \sec \theta + C$.**

76. D.   Note the initial conditions: when $t = 0$, $v = 0$ and $s = 0$. Integrate twice: $v = 6t^2$ and $s = 2t^3$. Let $t = 3$.

77. D.   Since $y' = x^2 - 2$, $y = \frac{1}{3}x^3 - 2x + C$. Replacing x by 1 and y by -3 yields $C = -\frac{4}{3}$.

78. D.   When $t = 0$, $v = 3$ and $s = 2$. So

$$v = 2t + 3t^2 + 3 \qquad \text{and} \qquad s = t^2 + t^3 + 3t + 2.$$

Let $t = 1$.

79. A. $\frac{dy}{dx} = 2y$, so $\frac{dy}{y} = 2 \, dx$; therefore $\ln y = 2x + C$. Since $\ln 1 = 4 + C$, $C = -4$; so $\ln y = 2x - 4$ and $y = e^{2x-4}$.

80. C.   Let P be the population at time t, P_0 at time $t = 0$, and let k be the factor of proportionality. Then

$$\frac{dP}{dt} = kP \qquad \text{and} \qquad \ln P = kt + \ln P_0.$$

So $\frac{P}{P_0} = e^{kt}$. Since $P = 2P_0$ when $t = 20$,

$$2 = e^{k \cdot 20} \qquad \text{and} \qquad \ln 2 = 20k \quad \text{or} \quad k = \frac{1}{20} \ln 2.$$

81. A.   Integrating yields $\frac{y^2}{2} = \frac{x^2}{2} + C$ or $y^2 = x^2 + 2C$ or $y^2 = x^2 + C'$, where we have replaced the arbitrary constant $2C$ by C'. Note that (A) is just $y^2 = x^2 + C'$, where $C' = -4$.

82. E.   Separate variables. The particular solution is $\ln y = \sqrt{x} - 2$.

83. C.   Separate variables. The particular solution is $-e^{-y} = x - 2$.

84. B. The general solution is $y = \sqrt{9 - x^2} + C$; $y = 5$ when $x = 4$ yields $C = 0$.

85. D. We separate variables to get $\csc^2\left(\frac{\pi}{2} s\right) ds = dt$. We integrate: $-\frac{2}{\pi} \cot\left(\frac{\pi}{2} s\right) = t + C$. With $t = 0$ and $s = 1$, $C = 0$. When $s = \frac{3}{2}$, we get $-\frac{2}{\pi} \cot \frac{3\pi}{4} = t$.

86. E. Since $\int \frac{dy}{y} = \int \frac{dx}{x}$, it follows that

$$\ln y = \ln x + C \qquad \text{or} \qquad \ln y = \ln x + \ln k;$$

so $y = kx$.

87. E. The solution is $y = ke^x$, $k \neq 0$.

88. A. We rewrite and separate variables, getting $y\frac{dy}{dx} = x$. The general solution is

$$y^2 = x^2 + C \qquad \text{or} \qquad f(x) = \pm\sqrt{x^2 + C}.$$

89. C. We are given that $\frac{dy}{dx} = \frac{3y}{x}$. The general solution is $\ln|y| = 3\ln|x| + C$. Thus, $|y| = c\,|x^3|$; $y = \pm c\,|x^3|$. Since $y = 1$ when $x = 1$, we get $c = 1$.

90. B. Since $\frac{dR}{dt} = cR$, $\frac{dR}{R} = c\,dt$, and $\ln R = ct + C$. When $t = 0$, $R = R_0$; so $\ln R_0 = C$ or $\ln R = ct + \ln R_0$. Thus

$$\ln R - \ln R_0 = ct;\ \ln \frac{R}{R_0} = ct \qquad \text{or} \qquad \frac{R}{R_0} = e^{ct}.$$

91. D. The question gives rise to the differential equation $\frac{dP}{dt} = kP$, where $P = 2P_0$ when $t = 50$. We seek $\frac{P}{P_0}$ for $t = 75$. We get $\ln \frac{P}{P_0} = kt$ with $\ln 2 = 50k$; then

$$\ln \frac{P}{P_0} = \frac{t}{50} \ln 2 \qquad \text{or} \qquad \frac{P}{P_0} = 2^{t/50}$$

92. A. Let S equal the amount present at time t; using $S = 40$ when $t = 0$ yields $\ln \frac{S}{40} = kt$. Since, when $t = 2$, $S = 10$, we get

$$k = \frac{1}{2} \ln \frac{1}{4} \qquad \text{or} \qquad \ln \frac{1}{2} \qquad \text{or} \qquad -\ln 2.$$

93. E. The general solution is $y = k \ln |x| + C$, and the particular solution is $y = 2 \ln |x| + 2$.

94. A. We replace $g(x)$ by y and then solve the equation $\frac{dy}{dx} = \pm\sqrt{y}$. We use the constraints given to find the particular solution $2\sqrt{y} = x$ or $2\sqrt{g(x)} = x$.

Practice Section II Problems on Integration

*1. Integrate

$$\text{(a)} \int \frac{x-3}{\sqrt{6x-x^2}}\,dx; \qquad \text{(b)} \int \frac{dx}{\sqrt{6x-x^2}};$$

and show how the above can be combined to integrate

$$\text{(c)} \int \frac{2x+3}{\sqrt{6x-x^2}}\,dx.$$

2. Integrate

$$\text{*(a)} \int \frac{dx}{x^3-x}; \qquad \text{(b)} \int x\cos 3x\,dx.$$

*3. Integrate:

$$\text{(a)} \int \frac{dx}{\sqrt{6x-x^2}}; \qquad \text{(b)} \int x^3 e^{-x^2}\,dx.$$

4. Find the equation of the curve whose slope at any point (x, y) equals $\frac{2y}{x}$ and which passes through the point $(1, 1)$.

*5. Integrate $\displaystyle\int \frac{2dx}{(x-1)(x^2+1)}$.

6. If $f'(x) = (\sin x)\cdot f(x)$ and $f(0) = 3$, find $f(x)$.

7. Find a function y which satisfies the conditions:

$$\text{(1)} \ \frac{y'}{y} = 2x; \qquad \text{(2)} \ y(1) = e.$$

8. (a) Evaluate $\displaystyle\int_1^e x \ln x\,dx$.

$\quad$ *(b) Integrate $\displaystyle\int \frac{x-3}{x^2-1}\,dx$.

---

*Questions preceded by an asterisk are likely to appear only on the Calculus BC Examination.

Solutions for Section II Problems: Integration

1. (a) Let $A = \int \dfrac{x-3}{\sqrt{6x-x^2}}\, dx$. Then

$$A = -\frac{1}{2}\int \frac{6-2x}{\sqrt{6x-x^2}}\, dx = -\frac{1}{2}\int (6x-x^2)^{-1/2}(6-2x)\, dx = -\frac{1}{2}\int u^{-1/2}\, du,$$

where $u = 6x - x^2$. Then

$$A = -\frac{1}{2}\frac{u^{1/2}}{\frac{1}{2}} + C_1 = -\sqrt{6x-x^2} + C_1.$$

(b) Let $B = \int \dfrac{dx}{\sqrt{6x-x^2}}$. This equals

$$\int \frac{dx}{\sqrt{9-(x^2-6x+9)}} = \int \frac{dx}{\sqrt{3^2-(x-3)^2}} = \int \frac{du}{\sqrt{a^2-u^2}},$$

where $a = 3$ and $u = x - 3$. Since

$$\int \frac{du}{\sqrt{a^2-u^2}} = \sin^{-1}\frac{u}{a} + C,$$

we see that $B = \sin^{-1}\dfrac{x-3}{3} + C_2$.

(c) Let $C = \int \dfrac{2x+3}{\sqrt{6x-x^2}}\, dx$ and note that this equals

$$\int \frac{2x-6+9}{\sqrt{6x-x^2}}\, dx = 2\int \frac{x-3}{\sqrt{6x-x^2}}\, dx + 9\int \frac{dx}{\sqrt{6x-x^2}} = 2A + 9B.$$

So, from (a) and (b) above, $C = -2\sqrt{6x-x^2} + 9\sin^{-1}\dfrac{x-3}{3} + C_3$.

2. (a) Hint: Use partial fractions. $\ln \frac{\sqrt{(x - 1)(x + 1)}}{x} + C$.

(b) Hint: Use integration by parts. $\frac{1}{3}x \sin 3x + \frac{1}{9}\cos 3x + C$.

3. (a) $\sin^{-1} \frac{x - 3}{3} + C$.

(b) Hint: Use the Parts Formula, letting $u = x^2$ and $dv = xe^{-x^2} dx$. The answer is $-\frac{1}{2}e^{-x^2}(x^2 + 1) + C$.

4. We get the differential equation $\frac{dy}{y} = 2\frac{dx}{x}$, whose general solution is $\ln y = 2 \ln x + C$. Since the point $(1, 1)$ is on the curve, $C = 0$. The curve is $y = x^2$.

5. Hint: Use partial fractions. If we let

$$\frac{2}{(x - 1)(x^2 + 1)} = \frac{A}{x - 1} + \frac{Bx + C}{x^2 + 1},$$

then $A = 1, B = -1, C = -1$. So

$$\int \frac{2dx}{(x - 1)(x^2 + 1)} = \ln \frac{|x - 1|}{\sqrt{x^2 + 1}} - \tan^{-1} x + C'.$$

6. Since the given equation is equivalent to $\frac{dy}{dx} = (\sin x) \cdot y$, with $f(x)$ replaced by y, we can separate to get $\frac{dy}{y} = \sin x \, dx$, then integrate, getting $\ln y = -\cos x + C$. The condition $f(0) = 3$ yields $\ln 3 = -1 + C$; so $C = 1 + \ln 3$. We then have

$$\ln y = -\cos x + 1 + \ln 3$$

$$\ln \frac{y}{3} = 1 - \cos x$$

$$\frac{y}{3} = e^{1 - \cos x}$$

$$y = 3e^{1 - \cos x} = f(x).$$

7. We get $\ln y = x^2 + C$ and use $y(1) = e$ to find that $C = 0$. Thus $y = e^{x^2}$.

8. (a)   Let $I = \int_1^e x \ln x \, dx$ and use the Parts Formula with $u = \ln x$, $dv = x \, dx$.

$$du = \frac{1}{x} \, dx \qquad v = \frac{x^2}{2}$$

Then

$$I = \frac{x^2}{2} \ln x \Big]_1^e - \frac{1}{2} \int_1^e x \, dx = \frac{e^2}{2} - \frac{1}{4} x^2 \Big]_1^e$$

$$= \frac{e^2}{2} - \left(\frac{e^2}{4} - \frac{1}{4}\right) = \frac{e^2}{4} + \frac{1}{4}$$

(b) Partial fractions yield

$$\frac{x - 3}{x^2 - 1} = \frac{A}{x - 1} + \frac{B}{x + 1}$$

$$x - 3 = A(x + 1) + B(x - 1)$$

When $x = 1$, we get $-2 = 2A$ and $A = -1$;
when $x = -1$, we get $-4 = -2B$ and $B = 2$.
So

$$\frac{x - 3}{x^2 - 1} = -\frac{1}{x - 1} + \frac{2}{x + 1}$$

The given integral is equal to $\ln |x + 1|^2 - \ln |x - 1|$

$$\text{or} \quad \ln \left| \frac{(x + 1)^2}{(x - 1)} \right|$$

Definite Integrals

Review of Definitions and Methods

A. Definition of Definite Integral_____

If f is continuous on the closed interval $[a, b]$ and $F' = f$, then according to the Fundamental Theorem of the Integral Calculus

$$\int_a^b f(x)\ dx = F(b) - F(a).$$

Here $\int_a^b f(x)\ dx$ is the *definite integral of f from a to b*; $f(x)$ is called the *integrand*; and a and b are called respectively the *lower* and *upper limits of integration*.
See, also, property (1) under §B below.

B. Properties of Definite Integrals_____

The following theorems about definite integrals are important. (The first is simply a restatement of the Fundamental Theorem.)

$$\frac{d}{dx} \int_a^x f(t)\ dt = f(x) \tag{1}$$

$$\int_a^b kf(x)\ dx = k \int_a^b f(x)\ dx \qquad (k \text{ a constant}) \tag{2}$$

$$\int_a^a f(x)\ dx = 0 \tag{3}$$

$$\int_a^b f(x)\ dx = - \int_b^a f(x)\ dx \tag{4}$$

$$\int_a^c f(x)\ dx + \int_c^b f(x)\ dx = \int_a^b f(x)\ dx \qquad (a < c < b) \tag{5}$$

If f and g are both integrable functions of x on $[a, b]$, then

$$\int_a^b [f(x) \pm g(x)] \, dx = \int_a^b f(x) \, dx \pm \int_a^b g(x) \, dx \tag{6}$$

The Mean Value Theorem for Integrals: If f is continuous on $[a, b]$ there exists at least one number c, $a < c < b$, such that

$$\int_a^b f(x) \, dx = f(c)(b - a) \tag{7}$$

By the *comparison property*, if f and g are integrable on $[a, b]$ and if $f(x) \geq g(x)$ for all x in $[a, b]$, then

$$\int_a^b f(x) \, dx \geq \int_a^b g(x) \, dx \tag{8}$$

The evaluation of a definite integral is illustrated in the following examples. A scientific calculator will be helpful for some numerical computations.

Examples

1. $\displaystyle\int_{-1}^2 (3x^2 - 2x) \, dx = x^3 - x^2 \Big|_{-1}^2 = (8 - 4) - (-1 - 1) = 6.$

2. $\displaystyle\int_1^2 \frac{x^2 + x - 2}{2x^2} \, dx = \frac{1}{2}\int_1^2 \left(1 + \frac{1}{x} - \frac{2}{x^2}\right) dx = \frac{1}{2}\left(x + \ln x + \frac{2}{x}\right)\Big|_1^2 =$
$\frac{1}{2}[(2 + \ln 2 + 1) - (1 + 2)] = \frac{1}{2} \ln 2$, or $\ln \sqrt{2}.$

3. $\displaystyle\int_5^8 \frac{dy}{\sqrt{9 - y}} = -\int_5^8 (9 - y)^{-1/2} \, (-dy) = -2\sqrt{9 - y} \Big|_5^8 =$
$-2(1 - 2) = 2.$

4. $\displaystyle\int_0^1 \frac{x \, dx}{(2 - x^2)^3} = -\frac{1}{2}\int_0^1 (2 - x^2)^{-3}(-2x \, dx) = -\frac{1}{2} \frac{(2 - x^2)^{-2}}{-2}\Big|_0^1 =$
$\frac{1}{4}\left(1 - \frac{1}{4}\right) = \frac{3}{16}.$

5. $\displaystyle\int_0^3 \frac{dt}{9 + t^2} = \frac{1}{3} \tan^{-1} \frac{t}{3}\Big|_0^3 = \frac{1}{3}(\tan^{-1} 1 - \tan^{-1} 0) = \frac{1}{3}\left(\frac{\pi}{4} - 0\right) = \frac{\pi}{12}.$

6. $\displaystyle\int_0^1 (3x - 2)^3 \, dx = \frac{1}{3}\int_0^1 (3x - 2)^3 (3 \, dx) = \frac{(3x - 2)^4}{12}\Big|_0^1 =$
$\frac{1}{12}(1 - 16) = -\frac{5}{4}.$

7. $\displaystyle\int_0^1 xe^{-x^2} \, dx = -\frac{1}{2} e^{-x^2}\Big|_0^1 = -\frac{1}{2}\left(\frac{1}{e} - 1\right) = \frac{e - 1}{2e}.$

8. $\displaystyle\int_{-\pi/4}^{\pi/4} \cos 2x \, dx = \frac{1}{2} \sin 2x \Big|_{-\pi/4}^{\pi/4} = \frac{1}{2}(1 + 1) = 1.$

9. $\int_{-1}^{1} xe^x \, dx = (xe^x - e^x)\Big|_{-1}^{1} = e - e - \left(-\frac{1}{e} - \frac{1}{e}\right) = \frac{2}{e}.$

10. $\int_{0}^{1/2} \frac{dx}{\sqrt{1-x^2}} = \sin^{-1} x\Big|_{0}^{1/2} = \frac{\pi}{6}.$

11. $\int_{0}^{e-1} \ln(x+1) \, dx = \Big[(x+1)\ln(x+1) - x\Big]_{0}^{e-1}$ (where the Parts Formula has been used) $= e \ln e - (e-1) - 0 = 1.$

*12. To evaluate $\int_{-1}^{1} \frac{dy}{y^2 - 4}$ we use the method of partial fractions and set

$$\frac{1}{y^2 - 4} = \frac{A}{y+2} + \frac{B}{y-2}.$$

Solving for A and B yields $A = -\frac{1}{4}$, $B = \frac{1}{4}$. Thus

$$\int_{-1}^{1} \frac{dy}{y^2 - 4} = \frac{1}{4} \ln \left|\frac{y-2}{y+2}\right|_{-1}^{1} = \frac{1}{4}\left(\ln\frac{1}{3} - \ln 3\right) = -\frac{1}{2}\ln 3.$$

13. $\int_{\pi/3}^{\pi/2} \tan\frac{\theta}{2} \sec^2\frac{\theta}{2} \, d\theta = \frac{2}{2}\tan^2\frac{\theta}{2}\Big|_{\pi/3}^{\pi/2} = 1 - \frac{1}{3} = \frac{2}{3}.$

*14. $\int_{0}^{\pi/2} \sin^2\frac{1}{2}x \, dx = \int_{0}^{\pi/2}\left(\frac{1}{2} - \frac{\cos x}{2}\right) dx = \frac{x}{2} - \frac{\sin x}{2}\Big|_{0}^{\pi/2} = \frac{\pi}{4} - \frac{1}{2}.$

15. $\frac{d}{dx}\int_{-1}^{x} \sqrt{1 + \sin^2 t} \, dt = \sqrt{1 + \sin^2 x}$ by theorem (1), page 172.

16. $\frac{d}{dx}\int_{x}^{1} e^{-t^2} \, dt = \frac{d}{dx}\left(-\int_{1}^{x} e^{-t^2} dt\right)$ by theorem (4), page 172, $= -\frac{d}{dx}\int_{1}^{x} e^{-t^2} \, dt = -e^{-x^2}$ by theorem (1).

17. If $F(x) = \int_{1}^{x^2} \frac{dt}{3+t}$, then

$$F'(x) = \frac{d}{dx}\int_{1}^{x^2} \frac{dt}{3+t}$$

$$= \frac{d}{dx}\int_{1}^{u} \frac{dt}{3+t} \quad \text{(where } u = x^2\text{)}$$

$$= \frac{d}{du}\int_{1}^{u} \frac{dt}{3+t} \cdot \frac{du}{dx} \quad \text{by the chain rule}$$

$$= \left(\frac{1}{3+u}\right)(2x) = \frac{2x}{3+x^2}.$$

18. If $F(x) = \int_0^{\cos x} \sqrt{1 - t^3}\, dt$, then to find $F'(x)$ we let $u = \cos x$. Thus

$$\frac{dF}{dx} = \frac{dF}{du} \cdot \frac{du}{dx} = \sqrt{1 - u^3}\,(- \sin x) = - \sin x \sqrt{1 - \cos^3 x}\,.$$

19. $\lim\limits_{h \to 0} \dfrac{1}{h} \int_x^{x+h} \sqrt{e^t - 1}\, dt = \sqrt{e^x - 1}$. Here we have let $f(t) = \sqrt{e^t - 1}$ and noted that

$$\lim_{h \to 0} \frac{1}{h} \int_x^{x+h} f(t)\, dt = \lim_{h \to 0} \frac{F(x + h) - F(x)}{h}$$

where

$$\frac{dF(x)}{dx} = f(x) = \sqrt{e^x - 1}.$$

The limit on the right, however, is, by definition, the derivative of $F(x)$, that is, $f(x)$.

20. Evaluate $\int_3^6 x \sqrt{x - 2}\, dx$.

Here we let $u = \sqrt{x - 2}$, $u^2 = x - 2$, and $2u\, du = dx$. The limits of the given integral are values of x. When we write the new integral in terms of the variable u, then the limits, if written, must be the values of u which correspond to the given limits. Thus, when $x = 3$, $u = 1$, and when $x = 6$, $u = 2$. So

$$\int_3^6 x \sqrt{x - 2}\, dx = 2 \int_1^2 (u^2 + 2)u^2\, du = 2 \int_1^2 (u^4 + 2u^2)\, du$$

$$= 2 \left(\frac{u^5}{5} + \frac{2u^3}{3} \right) \Big|_1^2 = 2 \left[\left(\frac{32}{5} + \frac{16}{3} \right) - \left(\frac{1}{5} + \frac{2}{3} \right) \right]$$

$$= \frac{326}{15}.$$

** **21.** Evaluate $\int_1^{\sqrt{2}} \dfrac{x^2}{\sqrt{4 - x^2}}\, dx.$

From Example 45, page 131, we see that, if we let $x = 2 \sin \theta$, then

$$\int \frac{x^2\, dx}{\sqrt{4 - x^2}} = 2 \int (1 - \cos 2\theta)\, d\theta.$$

Since $\theta = \dfrac{\pi}{6}$ when $x = 1$ and $\theta = \dfrac{\pi}{4}$ when $x = \sqrt{2}$, we have

$$\int_1^{\sqrt{2}} \frac{x^2}{\sqrt{4 - x^2}}\, dx = 2 \int_{\pi/6}^{\pi/4} (1 - \cos 2\theta)\, d\theta = (2\theta - \sin 2\theta) \Big|_{\pi/6}^{\pi/4}$$

$$= \left(\frac{\pi}{2} - 1 \right) - \left(\frac{\pi}{3} - \frac{\sqrt{3}}{2} \right) = \frac{\pi}{6} + \frac{\sqrt{3}}{2} - 1.$$

*C.   Integrals Involving Parametrically Defined Functions

The techniques are illustrated in Examples 22 and 23.

Example 22. To evaluate $\int_{-2}^{2} y\, dx$, where $x = 2 \sin \theta$ and $y = 2 \cos \theta$, we note that $dx = 2 \cos \theta\, d\theta$ and that

$$\theta = -\frac{\pi}{2} \quad \text{when } x = -2,$$

$$= \frac{\pi}{2} \quad \text{when } x = 2.$$

Then

$$\int_{-2}^{2} y\, dx = \int_{-\pi/2}^{\pi/2} 2 \cos \theta(2 \cos \theta)\, d\theta = 4 \int_{-\pi/2}^{\pi/2} \frac{1 + \cos 2\theta}{2}\, d\theta$$

$$= 2\left(\theta + \frac{\sin 2\theta}{2} \right)\Bigg|_{-\pi/2}^{\pi/2} = 2\pi.$$

When using parametric equations (as in the above integral) we must be sure to express all of the following in terms of the parameter: (1) the integrand; (2) the differential of the given variable; and (3) both limits. Note that the integral of Example 22 gives the area of the semicircle whose Cartesian equation is $x^2 + y^2 = 4$.

Example 23. Evaluate $\int_{0}^{2\pi} \sqrt{1 + \left(\frac{dy}{dx} \right)^2}\, dx$ where $\begin{array}{l} x = t - \sin t \\ y = 1 - \cos t \end{array}$

We see that the integral can be rewritten as $\int_{0}^{2\pi} \sqrt{dx^2 + dy^2}$. Since $dx = (1 - \cos t)\, dt$, $dy = \sin t\, dt$, $t = 0$ when $x = 0$ and $t = 2\pi$ when $x = 2\pi$, it follows that

$$\int_{0}^{2\pi} \sqrt{dx^2 + dy^2} = \int_{0}^{2\pi} \sqrt{(1 - \cos t)^2 + \sin^2 t}\, dt$$

$$= \sqrt{2} \int_{0}^{2\pi} \sqrt{1 - \cos t}\, dt = 2 \int_{0}^{2\pi} \sin \frac{t}{2}\, dt$$

$$= 4\left(-\cos \frac{t}{2} \right)\Bigg|_{0}^{2\pi} = 8.$$

*An asterisk denotes a topic covered only in Calculus BC.

D.   Definition of Definite Integral as the Limit of a Sum: The Fundamental Theorem_____

Most applications of integration are based on the Fundamental Theorem of the Integral Calculus. This theorem provides the tool for evaluating an infinite sum by means of a definite integral. Suppose that a function $f(x)$ is continuous on the closed interval $[a, b]$. Divide the interval into n equal* subintervals, of length $\Delta x = \dfrac{b - a}{n}$. Choose numbers, one in each subinterval, as follows: x_1 in the first, x_2 in the second, . . ., x_k in the kth, . . ., x_n in the nth. Then

$$\lim_{n \to \infty} \sum_{k=1}^{n} f(x_k) \, \Delta x = \int_a^b f(x) \, dx = F(b) - F(a) \quad \text{where} \quad \frac{dF(x)}{dx} = f(x).$$

If $f(x)$ is nonnegative on $[a, b]$, we see (Figure N6−1) that $f(x_k) \, \Delta x$ can be regarded as the area of a typical approximating rectangle, and that the area bounded by the x-axis, the curve, and the vertical lines $x = a$ and $x = b$ is given exactly by

$$\lim_{n \to \infty} \sum_{k=1}^{n} f(x_k) \, \Delta x \qquad \text{and hence by} \qquad \int_a^b f(x) \, dx.$$

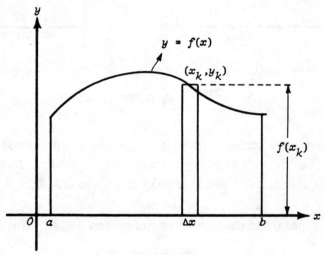

FIGURE N6−1

E.   Approximations to the Definite Integral_____

Often an integrand cannot be expressed in terms of elementary functions. It is always possible, however, to approximate the value of a definite integral. We interpret $\int_a^b f(x) \, dx$ as the area bounded above by $y = f(x)$, below by the x-axis, and vertically by the lines $x = a$ and $x = b$. The value of the definite integral is then approximated by dividing the area into n strips, approximating the area of each strip by a rectangle or other geometric figure, then summing these approxi-

*It is not in fact necessary that the subintervals be of equal length, but the formulation is simpler if one assumes they are.

mations. In this section we will divide the interval from a to b into n strips of equal width, Δx.

(1) USING RECTANGLES. Here we approximate $\int_a^b f(x)\, dx$ by

$$\sum_{k=1}^n f(x_k)\, \Delta x,$$

where $f(x_k)\, \Delta x$, as noted in §D above, is the area of a rectangle of altitude $f(x_k)$ and width Δx. If $f(x)$ is increasing (or decreasing) on $[a, b]$, we find an *upper sum* by using the ordinate at the right (left) endpoint of each strip as the altitude of the corresponding rectangle; for a lower sum we interchange right and left.

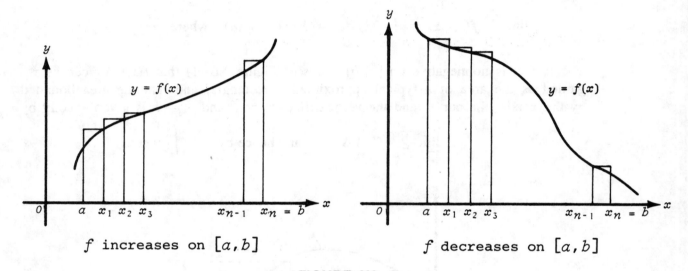

f increases on $[a, b]$ $\qquad\qquad$ f decreases on $[a, b]$

FIGURE N6–2

In each case we see in Figure N6–2 an upper-sum approximation:

$$\int_a^b f(x)\, dx \approx [f(x_1)\, \Delta x + f(x_2)\, \Delta x + \cdots + f(x_n)\, \Delta x].$$

If f both increases and decreases on $[a, b]$, we approximate $\int_a^b f(x)\, dx$ by

$$\sum_{k=1}^n f(c_k)\, \Delta x$$

for an upper (lower) sum, where $f(c_k)$ is the maximum (minimum) value of $f(x)$ in the kth subinterval.

Example 24. Approximate $\int_0^2 x^3\, dx$ by using four subintervals and calculating (a) an upper sum; (b) a lower sum.

Here

$$\Delta x = \frac{2-0}{4} = \frac{1}{2};$$

note that f increases on $[0, 2]$.

(a) For an upper sum we use right-hand altitudes at $x = \frac{1}{2}, 1, \frac{3}{2},$ and 2. The approximating sum is

$$\left(\frac{1}{2}\right)^3 \cdot \frac{1}{2} + (1)^3 \cdot \frac{1}{2} + \left(\frac{3}{2}\right)^3 \cdot \frac{1}{2} + (2)^3 \cdot \frac{1}{2} = \frac{25}{4}.$$

The upper sum uses the circumscribed rectangles shown in Figure N6-3.

(b) For a lower sum we use the left-hand altitudes at $x = 0, \frac{1}{2}, 1,$ and $\frac{3}{2}$. The approximating sum is

$$(0)^3 \cdot \frac{1}{2} + \left(\frac{1}{2}\right)^3 \cdot \frac{1}{2} + (1)^3 \cdot \frac{1}{2} + \left(\frac{3}{2}\right)^3 \cdot \frac{1}{2} = \frac{9}{4}.$$

The dotted lines show the inscribed rectangles used.

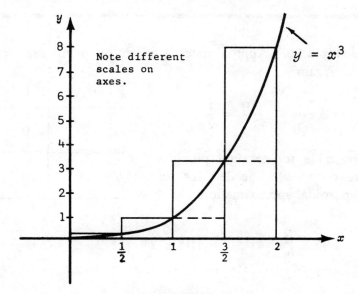

FIGURE N6-3

(2) USING TRAPEZOIDS. We now approximate the area of a strip by that of a trapezoid.

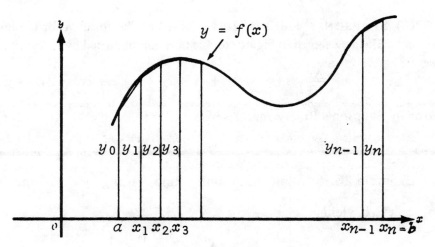

FIGURE N6-4

As Figure N6–4 shows, if $\Delta x = \frac{b-a}{n}$, then

$$\int_a^b f(x)\,dx \approx \frac{1}{2}(y_0 + y_1)\,\Delta x + \frac{1}{2}(y_1 + y_2)\,\Delta x$$

$$+ \frac{1}{2}(y_2 + y_3)\,\Delta x + \cdots + \frac{1}{2}(y_{n-1} + y_n)\,\Delta x$$

$$= \left[\frac{1}{2}y_0 + y_1 + y_2 + \cdots + y_{n-1} + \frac{1}{2}y_n\right]\Delta x.$$

$$= \frac{\Delta x}{2}(y_0 + 2y_1 + 2y_2 + \cdots + 2y_{n-1} + y_n).$$

This method of approximating area is called the Trapezoidal Rule.

Example 25. Approximate $\int_0^2 x^3\,dx$ by using four trapezoids.
Again

$$\Delta x = \frac{2-0}{4} = \frac{1}{2}.$$

The table to the right provides the coordinates needed for the trapezoidal approximation.

k	0	1	2	3	4
x_k	0	$\frac{1}{2}$	1	$\frac{3}{2}$	2
$y_k = (x_k)^3$	0	$\frac{1}{8}$	1	$\frac{27}{8}$	8

$$\int_0^2 x^3\,dx \approx \frac{1}{2}\cdot\frac{1}{2}\left(0 + 2\cdot\frac{1}{8} + 2\cdot 1 + 2\cdot\frac{27}{8} + 8\right) = \frac{17}{4}.$$

The exact value of the integral is $\left.\frac{x^4}{4}\right|_0^2 = 4$; the Trapezoidal Rule is clearly much better than either the upper or lower approximating sum in Example 24.

---

*(3) SIMPSON'S RULE. This is based on the formula for the area under the arc of a parabola. The area shown in Figure N6-4 is now approximated by

$$\frac{\Delta x}{3}(y_0 + 4y_1 + 2y_2 + 4y_3 + 2y_4 + \cdots + 2y_{n-2} + 4y_{n-1} + y_n).$$

To apply Simpson's Rule, n must be even.

Example 26. With Simpson's Rule, using $n = 6$, the approximation for $\int_0^2 x^3\,dx$ yields

$$\Delta x = \frac{2-0}{6} = \frac{1}{3},$$

and the integral equals

$$\frac{1}{3} \cdot \frac{1}{3} \left[1 \cdot \left(\frac{0}{6}\right)^3 + 4 \cdot \left(\frac{2}{6}\right)^3 + 2 \cdot \left(\frac{4}{6}\right)^3 + 4 \cdot \left(\frac{6}{6}\right)^3 + 2 \cdot \left(\frac{8}{6}\right)^3 + 4 \cdot \left(\frac{10}{6}\right)^3 + 1 \cdot \left(\frac{12}{6}\right)^3 \right]$$

$$= \frac{1}{9} \cdot \frac{1}{6^3} (7776) = 4.$$

This is the exact value of the definite integral.

*F.   Recognition of Limits of Sums as Definite Integrals

In this section we use the Fundamental Theorem of the Integral Calculus to evaluate infinite sums by means of definite integrals. If f is a continuous function on $[a, b]$, if the interval $a \leq x \leq b$ has been partitioned into n equal subintervals of length $\Delta x = \frac{b - a}{n}$, and if $x_k = a + k \Delta x$, then

$$\lim_{n \to \infty} \sum_{k=1}^{n} f(x_k) \, \Delta x = \lim_{n \to \infty} \sum_{k=0}^{n-1} f(x_k) \, \Delta x = \int_{a}^{b} f(x) \, dx. \tag{1}$$

For example,

$$\lim_{n \to \infty} \sum_{k=1}^{n} x_k^2 \, \Delta x = \int_{a}^{b} x^2 \, dx,$$

where the subdivisions are made on the interval $a \leq x \leq b$. Given an infinite sum, we need to identify Δx, x_k, $f(x_k)$, and the interval $[a, b]$. The technique is illustrated in the following examples.

Examples

27. To write

$$\lim_{n \to \infty} \left[f\left(\frac{1}{n}\right)\frac{1}{n} + f\left(\frac{2}{n}\right)\frac{1}{n} + \cdots + f\left(\frac{n}{n}\right)\frac{1}{n} \right] \tag{2}$$

as a definite integral, we see that (2) is equivalent to

$$\lim_{n \to \infty} \left[f\left(\frac{1}{n}\right) + f\left(\frac{2}{n}\right) + \cdots + f\left(\frac{n}{n}\right) \right]\frac{1}{n}.$$

This yields

$$\lim_{n \to \infty} \sum_{k=1}^{n} f\left(\frac{k}{n}\right) \Delta x = \lim_{n \to \infty} \sum_{k=1}^{n} f(x_k) \Delta x = \int_0^1 f(x) \, dx.$$

We have rewritten (2) precisely in the form of (1) with $\Delta x = \frac{1}{n}$; $x_k = 0 + \frac{k}{n}$ for $k = 1, 2, \ldots, n$; $a = 0$; and $b = 1$.

28. Evaluate

$$\lim_{n \to \infty} \frac{1}{n^4}(1^3 + 2^3 + \cdots + n^3) \tag{3}$$

by identifying it with an appropriate definite integral.

Equation (3) is equivalent to

$$\lim_{n \to \infty} \left[\left(\frac{1}{n}\right)^3 + \left(\frac{2}{n}\right)^3 + \cdots + \left(\frac{n}{n}\right)^3 \right] \frac{1}{n},$$

which equals

$$\lim_{n \to \infty} \sum_{k=1}^{n} \left(\frac{k}{n}\right)^3 \frac{1}{n} = \lim_{n \to \infty} \sum_{k=1}^{n} x_k^3 \, \Delta x = \int_0^1 x^3 \, dx = \frac{1}{4}.$$

Here

$$\Delta x = \frac{1}{n}; \quad x_k = 0 + \frac{k}{n} \quad \text{for } k = 1, 2, \ldots, n;$$

$$f(x_k) = x_k^3; \quad \text{and} \quad a = 0, b = 1.$$

29. Evaluate

$$\lim_{n \to \infty} \left(\frac{1}{n} + \frac{1}{n+1} + \cdots + \frac{1}{2n-1} \right). \tag{4}$$

Equation (4) is equivalent to

$$\lim_{n \to \infty} \left[\frac{1}{1} + \frac{1}{1 + \frac{1}{n}} + \cdots + \frac{1}{2 - \frac{1}{n}} \right] \cdot \frac{1}{n}, \tag{5}$$

where we have removed the factor $\frac{1}{n}$ to be identified as Δx. Equation (5), then, equals

$$\lim_{n \to \infty} \sum_{k=0}^{n-1} \frac{1}{1 + \frac{k}{n}} \cdot \frac{1}{n} = \lim_{n \to \infty} \sum_{k=0}^{n-1} \frac{1}{1 + x_k} \Delta x$$

$$= \int_0^1 \frac{1}{1 + x} dx = \ln(1 + x) \Big|_0^1 = \ln 2.$$

Note that

$$\Delta x = \frac{1}{n}; \qquad x_k = 0 + \frac{k}{n} \quad \text{for } k = 0, 1, 2, \ldots, n - 1;$$

$$f(x_k) = \frac{1}{1 + x_k}; \qquad \text{and} \qquad a = 0, b = 1.$$

We can arrive at the same answer by rewriting

$$\lim_{n \to \infty} \sum_{k=0}^{n-1} \frac{1}{1 + \frac{k}{n}} \cdot \frac{1}{n} = \lim_{n \to \infty} \sum_{k=0}^{n-1} \frac{1}{x_k} \Delta x = \int_1^2 \frac{1}{x} dx = \ln 2.$$

Again,

$$\Delta x = \frac{1}{n};$$

but now

$$x_k = 1 + \frac{k}{n} \quad \text{for } k = 0, 1, \ldots, n - 1;$$

$$f(x_k) = \frac{1}{x_k}; \qquad \text{and} \qquad a = 1, b = 2.$$

In Figures N6–5a and N6–5b the two equivalent definite integrals are interpreted as areas.

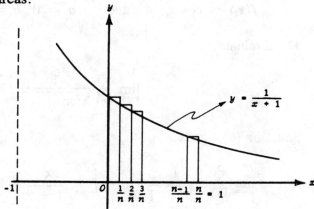

FIGURE N6–5a

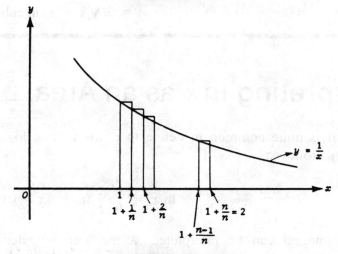

FIGURE N6–5b

30. $\displaystyle\lim_{n\to\infty}\sum_{k=0}^{n-1}\frac{\cos\dfrac{k\pi}{n}}{n} = \lim_{n\to\infty}\sum_{k=0}^{n-1}\left(\cos\pi\,\frac{k}{n}\right)\cdot\frac{1}{n}$

$$= \int_0^1 \cos\pi x\,dx = \frac{1}{\pi}\sin\pi x\,\bigg|_0^1 = 0,$$

where

$$\Delta x = \frac{1}{n};\qquad x_k = \frac{k}{n};\qquad f(x_k) = \cos\pi x_k;\qquad \text{and}\qquad a = 0,\ b = 1.$$

Alternatively,

$$\lim_{n\to\infty}\sum_{k=0}^{n-1}\left(\cos\frac{k\pi}{n}\right)\cdot\frac{1}{n} = \lim_{n\to\infty}\frac{1}{\pi}\sum_{k=0}^{n-1}\left(\cos k\,\frac{\pi}{n}\right)\cdot\frac{\pi}{n}$$

$$= \frac{1}{\pi}\int_0^\pi \cos x\,dx = \frac{1}{\pi}\sin x\,\bigg|_0^\pi = 0.$$

We see here that we can let

$$\Delta x = \frac{\pi}{n};\qquad x_k = \frac{k\pi}{n}\ \ \text{for}\ k = 0, 1, \ldots, n - 1;$$

$$f(x_k) = \cos x_k;\qquad \text{and}\qquad a = 0,\ b = \pi.$$

31. Evaluate

$$\lim_{n\to\infty}\sum_{k=1}^{n}\frac{1}{\sqrt{n^2 + kn}}. \tag{6}$$

Equation (6) equals

$$\lim_{n\to\infty}\sum_{k=1}^{n}\frac{1}{\sqrt{1 + \dfrac{k}{n}}}\cdot\frac{1}{n} = \int_0^1\frac{dx}{\sqrt{1 + x}}\qquad \text{or}\qquad \int_1^2\frac{dx}{\sqrt{x}}$$

$$= 2(\sqrt{2} - 1)\ \text{in either case.}$$

G.   Interpreting ln *x* as an Area

It is quite common to define ln x, the natural logarithm of x, as a definite integral, as follows:

$$\ln x = \int_1^x \frac{1}{t}\,dt \qquad (x > 0).$$

This integral can be interpreted as the area bounded above by the curve $y = \dfrac{1}{t}$ ($t > 0$), below by the t-axis, at the left by $t = 1$, and at the right by $t = x$ ($x > 1$).

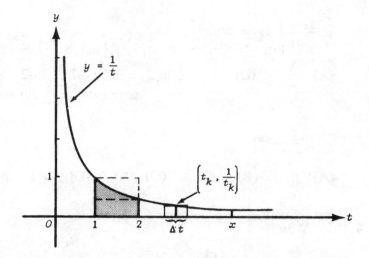

FIGURE N6–6

Note that if $x = 1$ the above definition yields $\ln 1 = 0$, and if $0 < x < 1$ we can rewrite as follows:

$$\ln x = -\int_x^1 \frac{1}{t}\, dt.$$

This shows that $\ln x < 0$ if $0 < x < 1$.

With this definition of $\ln x$ we can approximate $\ln x$ using rectangles or trapezoids.

Example 32.   Show that $\frac{1}{2} < \ln 2 < 1$.

Using the definition of $\ln x$ above yields $\ln 2 = \int_1^2 \frac{1}{t}\, dt$, which we interpret as the area under $y = \frac{1}{t}$, above the t-axis, and bounded at the left by $t = 1$ and at the right by $t = 2$ (the shaded region in Figure N6–6). Since $y = \frac{1}{t}$ is strictly decreasing, the area of the inscribed rectangle (height $\frac{1}{2}$, width 1) is less than $\ln 2$, which, in turn, is less than the area of the circumscribed rectangle (height 1, width 1). Thus $\frac{1}{2} \cdot 1 < \ln 2 < 1 \cdot 1$ or $\frac{1}{2} < \ln 2 < 1$.

Set 6: Multiple-Choice Questions on Definite Integrals

1. $\displaystyle\int_{-1}^{1} (x^2 - x - 1)\, dx =$

(A) $\dfrac{2}{3}$ (B) 0 (C) $-\dfrac{4}{3}$ (D) -2 (E) -1

2. $\int_1^2 \dfrac{3x-1}{3x} \, dx =$

(A) $\dfrac{3}{4}$ (B) $1 - \dfrac{1}{3} \ln 2$ (C) $1 - \ln 2$ (D) $-\dfrac{1}{3} \ln 2$ (E) 1

3. $\int_0^3 \dfrac{dt}{\sqrt{4-t}} =$

(A) 1 (B) -2 (C) 4 (D) -1 (E) 2

4. $\int_{-1}^0 \sqrt{3u+4} \, du =$

(A) 2 (B) $\dfrac{14}{9}$ (C) $\dfrac{14}{3}$ (D) 6 (E) $\dfrac{7}{2}$

5. $\int_2^3 \dfrac{dy}{2y-3} =$

(A) $\ln 3$ (B) $\dfrac{1}{2} \ln \dfrac{3}{2}$ (C) $\dfrac{16}{9}$ (D) $\ln \sqrt{3}$ (E) $\sqrt{3} - 1$

6. $\int_0^{\sqrt{3}} \dfrac{x}{\sqrt{4-x^2}} \, dx =$

(A) 1 (B) $\dfrac{\pi}{6}$ (C) $\dfrac{\pi}{3}$ (D) -1 (E) 2

7. $\int_0^1 (2t-1)^3 \, dt =$

(A) $\dfrac{1}{4}$ (B) 6 (C) $\dfrac{1}{2}$ (D) 0 (E) 4

8. $\int_0^1 \dfrac{dx}{\sqrt{4-x^2}} =$

(A) $\dfrac{\pi}{3}$ (B) $2 - \sqrt{3}$ (C) $\dfrac{\pi}{12}$ (D) $2(\sqrt{3} - 2)$ (E) $\dfrac{\pi}{6}$

9. $\int_4^9 \dfrac{2+x}{2\sqrt{x}} \, dx =$

(A) $\dfrac{25}{3}$ (B) $\dfrac{41}{3}$ (C) $\dfrac{100}{3}$ (D) $\dfrac{5}{3}$ (E) $\dfrac{1}{3}$

10. $\int_{-3}^3 \dfrac{dx}{9+x^2} =$

(A) $\dfrac{\pi}{2}$ (B) 0 (C) $\dfrac{\pi}{6}$ (D) $-\dfrac{\pi}{2}$ (E) $\dfrac{\pi}{3}$

11. $\int_0^1 e^{-x} \, dx =$

(A) $\dfrac{1}{e} - 1$ (B) $1 - e$ (C) $-\dfrac{1}{e}$ (D) $1 - \dfrac{1}{e}$ (E) $\dfrac{1}{e}$

12. $\int_0^1 xe^{x^2}\,dx =$

   (A) $e - 1$ **(B)** $\frac{1}{2}(e - 1)$ **(C)** $2(e - 1)$ **(D)** $\frac{e}{2}$ **(E)** $\frac{e}{2} - 1$

13. $\int_0^{\pi/4} \sin 2\theta\,d\theta =$

   (A) 2 **(B)** $\frac{1}{2}$ **(C)** -1 **(D)** $-\frac{1}{2}$ **(E)** -2

14. $\int_1^2 \frac{dz}{3 - z} =$

   (A) $-\ln 2$ **(B)** $\frac{3}{4}$ **(C)** $2(\sqrt{2} - 1)$ **(D)** $\frac{1}{2}\ln 2$ **(E)** $\ln 2$

15. $\int_1^e \ln y\,dy =$

   (A) $2e + 1$ **(B)** $\frac{1}{2}$ **(C)** 1 **(D)** $e - 1$ **(E)** -1

*16. $\int_{-4}^4 \sqrt{16 - x^2}\,dx =$

   (A) 8π **(B)** 4π **(C)** 4 **(D)** 8 **(E)** none of these

17. $\int_0^\pi \cos^2 \theta \sin \theta\,d\theta =$

   (A) $-\frac{2}{3}$ **(B)** $\frac{1}{3}$ **(C)** 1 **(D)** $\frac{2}{3}$ **(E)** 0

18. $\int_1^e \frac{\ln x}{x}\,dx =$

   (A) $\frac{1}{2}$ **(B)** $\frac{1}{2}(e^2 - 1)$ **(C)** 0 **(D)** 1 **(E)** $e - 1$

19. $\int_0^1 xe^x\,dx =$

   (A) -1 **(B)** $e + 1$ **(C)** 1 **(D)** $e - 1$ **(E)** $\frac{1}{2}(e - 1)$

20. $\int_0^{\pi/6} \frac{\cos \theta}{1 + 2 \sin \theta}\,d\theta =$

   (A) $\ln 2$ **(B)** $\frac{3}{8}$ **(C)** $-\frac{1}{2}\ln 2$ **(D)** $\frac{3}{2}$ **(E)** $\ln \sqrt{2}$

21. $\int_0^{\pi/4} \sqrt{1 - \cos 2\alpha}\,d\alpha =$

   (A) 1 **(B)** $\sqrt{2}$ **(C)** $\frac{1}{4}$ **(D)** $\sqrt{2} - 1$ **(E)** 2

---

*Questions preceded by an asterisk are likely to appear only on the Calculus BC Examination.

22. $\displaystyle\int_{\sqrt{2}}^{2} \frac{u}{u^2 - 1} \, du =$

(A) $\ln\sqrt{3}$ (B) $\dfrac{8}{9}$ (C) $\ln\dfrac{3}{2}$ (D) $\ln 3$ (E) $1 - \sqrt{3}$

23. $\displaystyle\int_{\sqrt{2}}^{2} \frac{u \, du}{(u^2 - 1)^2} =$

(A) $-\dfrac{1}{3}$ (B) $-\dfrac{2}{3}$ (C) $\dfrac{2}{3}$ (D) -1 (E) $\dfrac{1}{3}$

*24. $\displaystyle\int_{0}^{\pi/4} \cos^2\theta \, d\theta =$

(A) $\dfrac{1}{2}$ (B) $\dfrac{\pi}{8}$ (C) $\dfrac{\pi}{8} + \dfrac{1}{4}$ (D) $\dfrac{\pi}{8} + \dfrac{1}{2}$ (E) $\dfrac{\pi}{8} - \dfrac{1}{4}$

25. $\displaystyle\int_{\pi/12}^{\pi/4} \frac{\cos 2x \, dx}{\sin^2 2x} =$

(A) $-\dfrac{1}{4}$ (B) 1 (C) $\dfrac{1}{2}$ (D) $-\dfrac{1}{2}$ (E) -1

26. $\displaystyle\int_{0}^{1} \frac{e^{-x} + 1}{e^{-x}} \, dx =$

(A) e (B) $2 + e$ (C) $\dfrac{1}{e}$ (D) $1 + e$ (E) $e - 1$

27. $\displaystyle\int_{0}^{1} \frac{e^x}{e^x + 1} \, dx =$

(A) $\ln 2$ (B) e (C) $1 + e$ (D) $-\ln 2$ (E) $\ln\dfrac{e + 1}{2}$

28. If $f(x)$ is continuous on the interval $a \le x \le b$ and $a < c < b$, then $\displaystyle\int_{c}^{b} f(x) \, dx$ is equal to

(A) $\displaystyle\int_{a}^{c} f(x) \, dx + \int_{c}^{b} f(x) \, dx$ (B) $\displaystyle\int_{a}^{c} f(x) \, dx - \int_{a}^{b} f(x) \, dx$

(C) $\displaystyle\int_{c}^{a} f(x) \, dx + \int_{b}^{a} f(x) \, dx$ (D) $\displaystyle\int_{a}^{b} f(x)\, dx - \int_{a}^{c} f(x) \, dx$

(E) $\displaystyle\int_{a}^{c} f(x) \, dx - \int_{b}^{c} f(x) \, dx$

29. If $f(x)$ is continuous on $a \le x \le b$, then

(A) $\displaystyle\int_{a}^{b} f(x) \, dx = f(b) - f(a)$ (B) $\displaystyle\int_{a}^{b} f(x) \, dx = - \int_{b}^{a} f(x) \, dx$

(C) $\displaystyle\int_{a}^{b} f(x) \, dx \ge 0$ (D) $\displaystyle\frac{d}{dx}\int_{a}^{x} f(t) \, dt = f'(x)$

(E) $\displaystyle\frac{d}{dx}\int_{a}^{x} f(t) \, dt = f(x) - f(a)$

30. If $f(x)$ is continuous on the interval $a \leqq x \leqq b$, if this interval is partitioned into n equal subintervals of length Δx, and if x_k is a number in the kth subinterval, then $\lim\limits_{n \to \infty} \sum\limits_{1}^{n} f(x_k)\, \Delta x$ is equal to

(A) $f(b) - f(a)$

(B) $F(x) + C$, where $\dfrac{dF(x)}{dx} = f(x)$ and C is an arbitrary constant

(C) $\displaystyle\int_a^b f(x)\, dx$ (D) $F(b - a)$, where $\dfrac{dF(x)}{dx} = f(x)$

(E) none of these

31. If $F'(x) = G'(x)$ for all x, then

(A) $\displaystyle\int_a^b F'(x)\, dx = \int_a^b G'(x)\, dx$ (B) $\int F(x)\, dx = \int G(x)\, dx$

(C) $\displaystyle\int_a^b F(x)\, dx = \int_a^b G(x)\, dx$ (D) $\int F(x)\, dx = \int G(x)\, dx + C$

(E) none of the preceding is necessarily true

32. If $f(x)$ is continuous on the closed interval $[a, b]$, then there exists at least one number c, $a < c < b$, such that $\displaystyle\int_a^b f(x)\, dx$ is equal to

(A) $\dfrac{f(c)}{b - a}$ (B) $f'(c)(b - a)$ (C) $f(c)(b - a)$ (D) $\dfrac{f'(c)}{b - a}$

(E) $f(c)[f(b) - f(a)]$

33. If $f(x)$ is continuous on the closed interval $[a, b]$ and k is a constant, then $\displaystyle\int_a^b k f(x)\, dx$ is equal to

(A) $k(b - a)$ (B) $k[f(b) - f(a)]$

(C) $kF(b - a)$, where $\dfrac{dF(x)}{dx} = f(x)$ (D) $k\displaystyle\int_a^b f(x)\, dx$

(E) $\dfrac{[k f(x)]^2}{2} \Big]_a^b$

34. $\dfrac{d}{dt} \displaystyle\int_0^t \sqrt{x^3 + 1}\, dx =$

(A) $\sqrt{t^3 + 1}$ (B) $\dfrac{\sqrt{t^3 + 1}}{3t^2}$ (C) $\dfrac{2}{3}(t^3 + 1)(\sqrt{t^3 + 1} - 1)$

(D) $3x^2\sqrt{x^3 + 1}$ (E) none of these

35. If $F(u) = \displaystyle\int_1^u (2 - x^2)^3\, dx$, then $F'(u)$ is equal to

(A) $-6u(2 - u^2)^2$ (B) $\dfrac{(2 - u^2)^4}{4} - \dfrac{1}{4}$ (C) $(2 - u^2)^3 - 1$

(D) $(2 - u^2)^3$ (E) $-2u(2 - u^2)^3$

36. $\dfrac{d}{dx}\displaystyle\int_{\pi/2}^{x^2}\sqrt{\sin t}\ dt =$

(A) $\sqrt{\sin t^2}$ (B) $2x\sqrt{\sin x^2}-1$ (C) $\dfrac{2}{3}(\sin^{3/2} x^2 - 1)$

(D) $\sqrt{\sin x^2}-1$ (E) $2x\sqrt{\sin x^2}$

*37. If we let $x = \tan\theta$, then $\displaystyle\int_{1}^{\sqrt{3}}\sqrt{1+x^2}\ dx$ is equivalent to

(A) $\displaystyle\int_{\pi/4}^{\pi/3}\sec\theta\ d\theta$ (B) $\displaystyle\int_{1}^{\sqrt{3}}\sec^3\theta\ d\theta$ (C) $\displaystyle\int_{\pi/4}^{\pi/3}\sec^3\theta\ d\theta$

(D) $\displaystyle\int_{\pi/4}^{\pi/3}\sec^2\theta\tan\theta\ d\theta$ (E) $\displaystyle\int_{1}^{\sqrt{3}}\sec\theta\ d\theta$

38. If the substitution $u = \sqrt{x+1}$ is used, then $\displaystyle\int_{0}^{3}\dfrac{dx}{x\sqrt{x+1}}$ is equivalent to

(A) $\displaystyle\int_{1}^{2}\dfrac{du}{u^2-1}$ (B) $\displaystyle\int_{1}^{2}\dfrac{2\,du}{u^2-1}$ (C) $2\displaystyle\int_{0}^{3}\dfrac{du}{(u-1)(u+1)}$

(D) $2\displaystyle\int_{1}^{2}\dfrac{du}{u(u^2-1)}$ (E) $2\displaystyle\int_{0}^{3}\dfrac{du}{u(u-1)}$

*39. If $x = 4\cos\theta$ and $y = 3\sin\theta$, then $\displaystyle\int_{2}^{4}xy\ dx$ is equivalent to

(A) $48\displaystyle\int_{\pi/3}^{0}\sin\theta\cos^2\theta\ d\theta$ (B) $48\displaystyle\int_{2}^{4}\sin^2\theta\cos\theta\ d\theta$ (C) $36\displaystyle\int_{2}^{4}\sin\theta\cos^2\theta\ d\theta$

(D) $-48\displaystyle\int_{0}^{\pi/3}\sin\theta\cos^2\theta\ d\theta$ (E) $48\displaystyle\int_{0}^{\pi/3}\sin^2\theta\cos\theta\ d\theta$

*40. A curve is defined by the parametric equations $y = 2a\cos^2\theta$ and $x = 2a\tan\theta$, where $0 \leqq \theta \leqq \pi$. Then the definite integral $\pi\displaystyle\int_{0}^{2a}y^2\ dx$ is equivalent to

(A) $4\pi a^2\displaystyle\int_{0}^{\pi/4}\cos^4\theta\ d\theta$ (B) $8\pi a^3\displaystyle\int_{\pi/2}^{\pi}\cos^2\theta\ d\theta$ (C) $8\pi a^3\displaystyle\int_{0}^{\pi/4}\cos^2\theta\ d\theta$

(D) $8\pi a^3\displaystyle\int_{0}^{2a}\cos^2\theta\ d\theta$ (E) $8\pi a^3\displaystyle\int_{0}^{\pi/4}\sin\theta\cos^2\theta\ d\theta$

*41. A curve is given parametrically by $x = 1 - \cos t$ and $y = t - \sin t$, where $0 \leqq t \leqq \pi$. Then $\displaystyle\int_{0}^{3/2}y\ dx$ is equivalent to

(A) $\displaystyle\int_{0}^{3/2}\sin t(t - \sin t)\ dt$ (B) $\displaystyle\int_{2\pi/3}^{\pi}\sin t(t - \sin t)\ dt$

(C) $\displaystyle\int_{0}^{2\pi/3}(t - \sin t)\ dt$ (D) $\displaystyle\int_{0}^{2\pi/3}\sin t(t - \sin t)\ dt$

(E) $\displaystyle\int_{0}^{3/2}(t - \sin t)\ dt$

*42. $\lim\limits_{n\to\infty}\left[\dfrac{1}{n}\left(\dfrac{1}{1} + \dfrac{1}{1 + \dfrac{1}{n}} + \dfrac{1}{1 + \dfrac{2}{n}} + \cdots + \dfrac{1}{1 + \dfrac{n-1}{n}}\right)\right]$ is equal to the definite integral

(A) $\displaystyle\int_1^2 \ln x\, dx$ (B) $\displaystyle\int_1^2 \dfrac{1}{x}\, dx$ (C) $\displaystyle\int_1^2 \dfrac{1}{1+x}\, dx$ (D) $\displaystyle\int_1^2 x\, dx$

(E)   none of these

*43. $\lim\limits_{n\to\infty}\left(\dfrac{1}{n+1} + \dfrac{1}{n+2} + \cdots + \dfrac{1}{n+n}\right)$ is equal to

(A) $\ln 2$ (B) $\dfrac{3}{8}$ (C) $\ln\dfrac{3}{2}$ (D) $\dfrac{3}{2}$ (E)   none of these

*44. $\lim\limits_{n\to\infty}\sum\limits_{k=0}^{n-1}\dfrac{e^{k/n}}{n}$ is equal to the definite integral

(A) $\displaystyle\int_0^1 \dfrac{e^x}{x}\, dx$ (B) $\displaystyle\int_1^2 e^x\, dx$ (C) $\displaystyle\int_0^1 e^x\, dx$ (D) $\dfrac{1}{x}\displaystyle\int_0^1 e^x\, dx$

(E)   none of these

*45. $\lim\limits_{n\to\infty}\sum\limits_{k=1}^{n}\dfrac{1}{\sqrt{kn}}$ is equal to

(A)   2   (B) $\dfrac{2}{3}$ (C) $\dfrac{3}{2}$ (D)   1   (E)   none of these

*46. $\lim\limits_{n\to\infty}\sum\limits_{k=0}^{n-1}\left(\sin\dfrac{\pi k}{n}\right)\dfrac{1}{n}$ is equal to the definite integral

(A) $\displaystyle\int_0^1 \sin x\, dx$ (B) $\displaystyle\int_0^\pi \sin x\, dx$ (C) $\displaystyle\int_0^\pi \sin \pi x\, dx$

(D) $\displaystyle\int_0^1 \sin \pi x\, dx$ (E) $\dfrac{1}{\pi}\displaystyle\int_0^1 \sin \pi x\, dx$

47. $\displaystyle\int_{-1}^{3}|x|\, dx =$

(A) $\dfrac{7}{2}$ (B)   4   (C) $\dfrac{9}{2}$ (D)   5   (E) $\dfrac{11}{2}$

48. $\displaystyle\int_{-3}^{2}|x + 1|\, dx =$

(A) $\dfrac{5}{2}$ (B) $\dfrac{7}{2}$ (C)   5   (D) $\dfrac{11}{2}$ (E) $\dfrac{13}{2}$

49. Using the Trapezoidal Rule with a division at $x = \frac{5}{2}$, we can say that the area of the shaded region below is approximated by which of the following fractions?

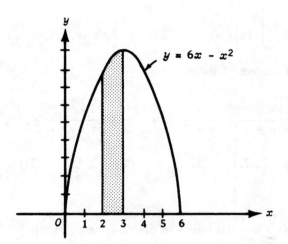

(A) $\dfrac{103}{16}$ (B) $\dfrac{67}{8}$ (C) $\dfrac{69}{8}$ (D) $\dfrac{20}{3}$ (E) $\dfrac{69}{4}$

50. The area of the following shaded region is equal exactly to ln 3. If we approximate ln 3 using an upper and lower sum and the two rectangles whose bases are from $x = 1$ to $x = 2$ and from $x = 2$ to $x = 3$, which of these inequalities follows?

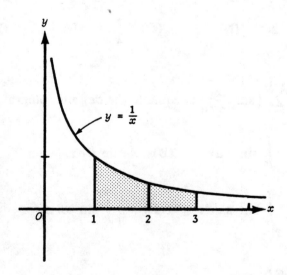

(A) $\dfrac{1}{2} < \displaystyle\int_1^2 \dfrac{1}{x}\,dx < 1$ (B) $\dfrac{1}{3} < \displaystyle\int_1^3 \dfrac{1}{x}\,dx < 2$ (C) $\dfrac{1}{2} < \displaystyle\int_0^2 \dfrac{1}{x}\,dx < 2$

(D) $\dfrac{1}{3} < \displaystyle\int_2^3 \dfrac{1}{x}\,dx < \dfrac{1}{2}$ (E) $\dfrac{5}{6} < \displaystyle\int_1^3 \dfrac{1}{x}\,dx < \dfrac{3}{2}$

51. $\displaystyle\int_0^2 \sqrt{3x + 1}\,dx$ is equal approximately to

(A) 3.646   (B) 3.893   (C) 4.116   (D) 4.646   (E) 11.680

*52. $\displaystyle\int_1^2 \frac{dx}{\sqrt{4-x^2}}$ is equal, to three decimal places, approximately to

    (A) 0.322   **(B)** 0.524   **(C)** 1.047   **(D)** 1.570   **(E)** none of these

53. If the Trapezoidal Rule is used with $n = 5$, then $\displaystyle\int_0^1 \frac{dx}{1+x^2}$ is equal, to three decimal places, to

    (A) 0.784   **(B)** 1.567   **(C)** 1.959   **(D)** 3.142   **(E)** 7.837

*54. If Simpson's Rule is used with $n = 4$, then $\displaystyle\int_0^1 \sqrt{1+x^3}\,dx$ is equal, to two decimal places, to

    (A) 1.00   **(B)** 1.11   **(C)** 1.20   **(D)** 2.22   **(E)** 3.33

55. $\displaystyle\int_2^5 \frac{1}{3x}\,dx$ is best approximated, to three decimal places, by

    (A) 0.268   **(B)** 0.286   **(C)** 0.305   **(D)** 0.916   **(E)** 2.749

Answers for Set 6: Definite Integrals

1. C	12. B	23. E	34. A	45. A
2. B	13. B	24. C	35. D	46. D
3. E	14. E	25. C	36. E	47. D
4. B	15. C	26. A	37. C	48. E
5. D	16. A	27. E	38. B	49. C
6. A	17. D	28. D	39. E	50. E
7. D	18. A	29. B	40. C	51. B
8. E	19. C	30. C	41. D	52. C
9. A	20. E	31. A	42. B	53. A
10. C	21. D	32. C	43. A	54. B
11. D	22. A	33. D	44. C	55. C

1. C. The integral is equal to

$$\left(\frac{1}{3}x^3 - \frac{1}{2}x^2 - x\right)\Bigg|_{-1}^{1} = -\frac{7}{6} - \frac{1}{6}.$$

2. B. Rewrite as $\displaystyle\int_1^2 \left(1 - \frac{1}{3}\cdot\frac{1}{x}\right)dx$. This equals

$$\left(x - \frac{1}{3}\ln x\right)\Bigg|_{1}^{2} = 2 - \frac{1}{3}\ln 2 - 1.$$

3. **E.**   Rewrite as

$$-\int_0^3 (4 - t)^{-1/2}(-1\ dt) \;=\; -2\sqrt{4 - t}\,\Big|_0^3 \;=\; -2(1 - 2).$$

4. **B.**   This one equals

$$\frac{1}{3}\int_{-1}^0 (3u + 4)^{1/2} \cdot 3\ du \;=\; \frac{1}{3} \cdot \frac{2}{3}(3u + 4)^{3/2}\,\Big|_{-1}^0$$

$$=\; \frac{2}{9}(4^{3/2} - 1^{3/2}).$$

5. **D.**   We have:

$$\frac{1}{2}\int_2^3 \frac{2\ dy}{2y - 3} \;=\; \frac{1}{2}\ln(2y - 3)\,\Big|_2^3 \;=\; \frac{1}{2}(\ln 3 - \ln 1).$$

6. **A.**   Rewrite:

$$-\frac{1}{2}\int_0^{\sqrt{3}} (4 - x^2)^{-1/2}(-2x\ dx) \;=\; -\frac{1}{2} \cdot 2\sqrt{4 - x^2}\,\Big|_0^{\sqrt{3}} \;=\; -(1 - 2).$$

7. **D.**   We expand, then integrate term by term:

$$(8t^3 - 12t^2 + 6t - 1)\,\Big|_0^1 \;=\; (2t^4 - 4t^3 + 3t^2 - t)\,\Big|_0^1 \;=\; (2 - 4 + 3 - 1) = 0.$$

8. **E.**   Use formula (17) on page 126.

$$\sin^{-1}\frac{x}{2}\,\Big|_0^1 \;=\; \sin^{-1}\frac{1}{2} - \sin^{-1} 0.$$

Compare with Question 6 above.

9. **A.**   We divide:

$$\int_4^9 \left(x^{-1/2} + \frac{1}{2}x^{1/2}\right) dx \;=\; \left(2x^{1/2} + \frac{1}{2} \cdot \frac{2}{3}x^{3/2}\right)\Big|_4^9$$

$$=\; \left(2 \cdot 3 + \frac{1}{3} \cdot 27\right) - \left(2 \cdot 2 + \frac{1}{3} \cdot 8\right).$$

10. **C.**   The integral equals

$$\frac{1}{3}\tan^{-1}\frac{x}{3}\,\Big|_{-3}^3 \;=\; \frac{1}{3}\left(\frac{\pi}{4} - \left(-\frac{\pi}{4}\right)\right).$$

11. **D.**   We get $-e^{-x}\,\Big|_0^1 = -(e^{-1} - 1).$

12. **B.**   The integral equals $\frac{1}{2}e^x\,\Big|_0^1 = \frac{1}{2}(e^1 - 1).$

13. **B.**   We evaluate $-\frac{1}{2}\cos 2\theta\,\Big|_0^{\pi/4}$, which equals $-\frac{1}{2}(0 - 1).$

14. **E.**   We evaluate $-\ln(3 - z)\,\Big|_1^2$ and get $-(\ln 1 - \ln 2).$

15. C.   Use parts as explained in Answer 54 on page 164 and evaluate
$(y \ln y - y)\Big|_1^e$. This equals $(e \ln e - e) - (0 - 1)$.

16. A.   We let $x = 4 \sin \theta$. Then $dx = 4 \cos \theta \, d\theta$, and the integral is equivalent to

$$\int_{-\pi/2}^{\pi/2} 16 \cos^2 \theta \, d\theta = 16 \int_{-\pi/2}^{\pi/2} \frac{1 + \cos 2\theta}{2} \, d\theta$$

$$= 8\left(\theta + \frac{\sin 2\theta}{2}\right)\Big|_{-\pi/2}^{\pi/2}$$

Note that the answer can be obtained immediately by recognizing the given integral as the area of a semicircle of radius 4.

17. D.   Evaluate $-\int_0^\pi \cos^2 \theta(-\sin \theta) \, d\theta$. This equals $-\frac{1}{3} \cos^3 \theta \Big|_0^\pi =$
$-\frac{1}{3}(-1 - 1)$.

18. A.   The integral equals $\frac{1}{2} \ln^2 x \Big|_1^e = \frac{1}{2}(1 - 0)$.

19. C.   We use the parts formula with $u = x$ and $dv = e^x \, dx$. Then $du = dx$ and $v = e^x$. We get

$$(xe^x - \int e^x \, dx)\Big|_0^1 = (xe^x - e^x)\Big|_0^1 = (e - e) - (0 - 1).$$

20. E.   We evaluate $\frac{1}{2} \ln (1 + 2 \sin \theta) \Big|_0^{\pi/6}$ and get $\frac{1}{2}(\ln (1 + 1) - \ln 1)$.

21. D.   Using formula (22), page 483, yields

$$\sqrt{2} \int_0^{\pi/4} \sin \alpha \, d\alpha = -\sqrt{2} \cos \alpha \Big|_0^{\pi/4} = -\sqrt{2}\left(\frac{\sqrt{2}}{2} - 1\right).$$

22. A.   Evaluate the integral $\frac{1}{2} \int_{\sqrt{2}}^2 \frac{2u}{u^2 - 1} \, du$. It equals

$$\frac{1}{2} \ln (u^2 - 1) \Big|_{\sqrt{2}}^2 \qquad \text{or} \qquad \frac{1}{2}(\ln 3 - \ln 1).$$

23. E.   We evaluate $\frac{1}{2} \int_{\sqrt{2}}^2 (u^2 - 1)^{-2} \cdot 2u \, du$ and get

$$-\frac{1}{2(u^2 - 1)} \Big|_{\sqrt{2}}^2 \qquad \text{or} \qquad -\frac{1}{2}\left(\frac{1}{3} - \frac{1}{1}\right).$$

Compare with question 22 above.

24. C.   Use formula (23), page 483, to obtain

$$\frac{1}{2} \int_0^{\pi/4} (1 + \cos 2\theta) \, d\theta = \frac{1}{2}\left(\theta + \frac{1}{2} \sin 2\theta\right)\Big|_0^{\pi/4} = \frac{1}{2}\left(\left(\frac{\pi}{4} + \frac{1}{2}\right) - 0\right).$$

25. **C.** Rewrite:

$$\frac{1}{2}\int_{\pi/12}^{\pi/4} \sin^{-2} 2x \cos 2x \,(2\,dx) = -\frac{1}{2}\cdot\frac{1}{\sin 2x}\bigg|_{\pi/12}^{\pi/4}$$

$$= -\frac{1}{2}\left(\frac{1}{1} - \frac{1}{1/2}\right).$$

26. **A.** The integral is equivalent to

$$\int_0^1 (1 + e^x)\,dx = (x + e^x)\bigg|_0^1 = (1 + e) - 1.$$

27. **E.** We evaluate $\ln(e^x + 1)\big|_0^1$, getting $\ln(e + 1) - \ln 2$.

28. **D.** Note that the integral from a to b is the sum of the two integrals from a to c and from c to b.

29. **B.** Find the errors in (A), (C), (D), and (E).

30. **C.** This is the definition of the definite integral given on page 177.

31. **A.** Find examples of functions F and G that show that (B), (C), and (D) are false.

32. **C.** This is the Mean Value Theorem for integrals (page 173). Draw some sketches that illustrate the theorem.

33. **D.** This is theorem (2) on page 172. Prove by counterexamples that (A), (B), (C), and (E) are false.

34. **A.** This is a restatement of the Fundamental Theorem. In theorem (1) on page 172, interchange t and x.

35. **D.** Apply (1) on page 172, noting that

$$F'(u) = \frac{d}{du}\int_a^u f(x)\,dx = f(u).$$

36. **E.** If we let $y = \int_{\pi/2}^{x^2} \sqrt{\sin t}\,dt$ and $u = x^2$, then

$$y = \int_{\pi/2}^{u} \sqrt{\sin t}\,dt.$$

By the chain rule, $\dfrac{dy}{dx} = \dfrac{dy}{du}\cdot\dfrac{du}{dx} = \sqrt{\sin u}\cdot 2x$, where we used theorem (1) on page 172 to find $\dfrac{dy}{du}$. Replace u by x^2.

37. **C.** Note that $dx = \sec^2\theta\,d\theta$ and that $\sqrt{1 + \tan^2\theta} = \sec\theta$. Be sure to express the limits as values of θ: $1 = \tan\theta$ yields $\theta = \dfrac{\pi}{4}$; $\sqrt{3} = \tan\theta$ yields $\theta = \dfrac{\pi}{3}$.

38. **B.** If $u = \sqrt{x + 1}$, then $u^2 = x + 1$, and $2u\,du = dx$. When we substitute for the limits we get $2\int_1^2 \dfrac{u\,du}{u(u^2 - 1)}$. Since $u \neq 0$ on its interval of integration, we may divide numerator and denominator by it.

39. E.   Since $dx = -4 \sin \theta \, d\theta$, we get the new integral $-48 \int_{\pi/3}^{0} \sin^2 \theta \cos \theta \, d\theta$. Use theorem (4) on page 172 to get the correct answer.

40. C.   Since $dx = 2a \sec^2 \theta \, d\theta$, we get $8\pi a^3 \int_{0}^{\pi/4} \cos^4 \theta \sec^2 \theta \, d\theta$. Use the fact that $\cos^2 \theta \sec^2 \theta = 1$.

41. D.   Use the facts that $dx = \sin t \, dt$, that $t = 0$ when $x = 0$, and that $t = \dfrac{2\pi}{3}$ when $x = \dfrac{3}{2}$.

42. B.   See the second solution of Example 29 on page 183.

43. A.   Note that the given limit can be rewritten as

$$\lim_{n \to \infty} \frac{1}{n}\left(\frac{1}{1 + \dfrac{1}{n}} + \frac{1}{1 + \dfrac{2}{n}} + \cdots + \frac{1}{1 + \dfrac{n}{n}} \right) \quad \text{or} \quad \lim_{n \to \infty} \sum_{k=1}^{n} f(x_k)\, \Delta x,$$

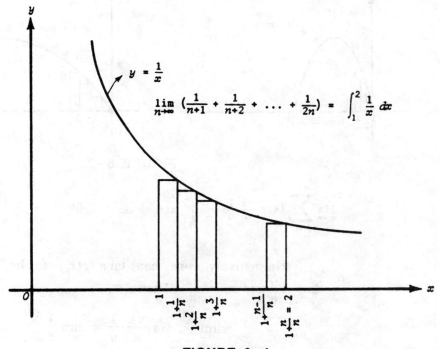

FIGURE 6–1

where $\Delta x = \dfrac{1}{n}$, $x_k = 1 + \dfrac{k}{n}$, $f(x_k) = \dfrac{1}{x_k}$, and the subdivisions are made on the interval $1 \leqq x \leqq 2$. See Figure 6–1, where the given limit, and the equivalent definite integral $\displaystyle\int_{1}^{2} \frac{dx}{x}$, are interpreted as the area under $y = \dfrac{1}{x}$ from $x = 1$ to $x = 2$.

44. C.   This is precisely in the form of equation (1) on page 181, where $x_k = \dfrac{k}{n}$, $f(x_k) = e^{x_k}$, and $\Delta x = \dfrac{1}{n}$. The interval $[a, b]$ is $[0, 1]$.

45. A.   Here

$$\lim_{n \to \infty} \sum_{k=1}^{n} \frac{1}{\sqrt{kn}} = \lim_{n \to \infty} \sum_{k=1}^{n} \frac{1}{\sqrt{\frac{k}{n}}} \frac{1}{n} \qquad \text{or} \qquad \lim_{n \to \infty} \sum_{k=1}^{n} f(x_k) \, \Delta x,$$

where $\Delta x = \frac{1}{n}$, $x_k = \frac{k}{n}$, $f(x_k) = \frac{1}{\sqrt{x_k}}$, on the interval $[0, 1]$. The equivalent definite integral $\int_0^1 \frac{1}{\sqrt{x}} \, dx$ is improper but does converge. (See Example 26, page 258).

46. D.   This limit can again be recognized as $\lim_{n \to \infty} \sum_{k=0}^{n-1} f(x_k) \, \Delta x$, where $\Delta x = \frac{1}{n}$, $x_k = \frac{k}{n}$, and $f(x_k) = \sin \pi x_k$, on the interval $[0, 1]$. The equivalent definite integral is thus $\int_0^1 \sin \pi x \, dx$, which equals $\frac{2}{\pi}$.

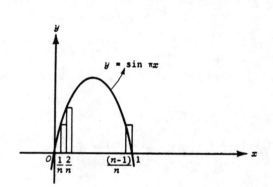 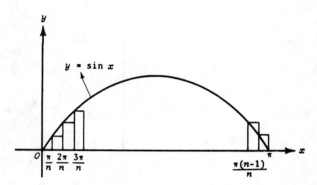

FIGURE 6–2

$$\lim_{n \to \infty} \sum_{k=0}^{n-1} \left(\sin \frac{\pi k}{n}\right)\frac{1}{n} = \int_0^1 \sin \pi x \, dx \qquad \text{OR} \qquad = \frac{1}{\pi} \int_0^\pi \sin x \, dx$$

Alternatively, one can take $f(x_k)$ to be $\sin x_k$ on $[0, \pi]$ with $\Delta x = \frac{\pi}{n}$ and $x_k = \frac{\pi k}{n}$. Then

$$\lim_{n \to \infty} \sum_{k=0}^{n-1} \sin \frac{\pi k}{n} \cdot \frac{1}{n} = \lim_{n \to \infty} \frac{1}{\pi} \sum_{k=0}^{n-1} \sin \frac{\pi k}{n} \cdot \frac{\pi}{n},$$

which is equivalent to $\frac{1}{\pi} \int_0^\pi \sin x \, dx$. The value of this integral is also $\frac{2}{\pi}$. In Figure 6–2 these two definite integrals have been interpreted as areas. Note that the area of the bounded region at the left is equal to the area of the bounded region at the right divided by π.

47. D.   We must rewrite the integral to evaluate it, using the fact that x changes sign at 0. We get

$$\int_{-1}^0 (-x) \, dx + \int_0^3 x \, dx = -\frac{x^2}{2}\Big|_{-1}^0 + \frac{x^2}{2}\Big|_0^3.$$

Draw a sketch of $y = |x|$ and verify that the area over $-1 \leq x \leq 3$ equals 5.

48. E.   Since $x + 1$ changes sign at $x = -1$, $|x + 1| = -(x + 1)$ if $x < -1$ but equals $x + 1$ if $x \geq -1$. The given integral is therefore equivalent to

$$\int_{-3}^{-1} -(x + 1)\, dx + \int_{-1}^{2} (x + 1)\, dx = -\frac{(x + 1)^2}{2}\Big|_{-3}^{-1} + \frac{(x + 1)^2}{2}\Big|_{-1}^{2}$$

$$= -\frac{1}{2}(0 - 4) + \frac{1}{2}(9 - 0).$$

Draw a sketch of $y = |x + 1|$, and verify that the area over $-3 \leq x \leq 2$ is $\frac{13}{2}$.

49. C.   See page 179. We compute $\left(\frac{1}{2} y_0 + y_1 + \frac{1}{2} y_2\right)\frac{1}{2}$, where y_0, y_1, and y_2 are evaluated at 2, $\frac{5}{2}$, and 3, respectively.

50. E.   For the lower sum we add the areas of the two inscribed rectangles, namely, $\frac{1}{2} \cdot 1 + \frac{1}{3} \cdot 1$; for the upper sum we use the circumscribed rectangles, $1 \cdot 1 + \frac{1}{2} \cdot 1$.

51. B.   Evaluate $\frac{2}{9}(3x + 1)^{3/2}\Big|_0^2$ using a scientific calculator.

52. C.   Evaluate $\sin^{-1}\frac{x}{2}\Big|_1^2$ using a scientific calculator and radian measure.

53. A. $\int_0^1 \frac{dx}{1 + x^2}$, using a trapezoidal approximation with $n = 5$, equals

$$\frac{1}{2}(0.2)\left(\frac{1}{1 + 0} + 2 \cdot \frac{1}{1.04} + 2 \cdot \frac{1}{1.16} + 2 \cdot \frac{1}{1.36} + 2 \cdot \frac{1}{1.64} + \frac{1}{1 + 1}\right).$$

54. B.   With $n = 4$, Simpson's Rule yields

$$\int_0^1 \sqrt{1 + x^3}\, dx = \frac{1}{3} \cdot \frac{1}{4}\left(\sqrt{1} + 4\sqrt{1 + 0.25^3} + 2\sqrt{1 + 0.5^3} + 4\sqrt{1 + 0.75^3} + \sqrt{2}\right).$$

55. C. $\int_2^5 \frac{1}{3x}\, dx = \frac{1}{3}\ln x\Big|_2^5 = 0.305$, with a scientific calculator.

Practice Section II Problems on Definite Integrals

1. (a) Integrate $\int \frac{x^3}{x^{n-2}}\, dx$, where n may be any real number.

(b) Integrate $\int_0^1 \frac{2}{2 - e^{-x}}\, dx$.

(c) If $y = \ln\frac{1 + \sin x}{\cos x}$, find $\frac{dy}{dx}$ and use this result to evaluate $\int_0^{\pi/3} \sec x\, dx$.

2. Find a nonzero function, $f(x)$, differentiable for all x and such that, if

$$t \geqq 0, \quad \int_0^t x f(x)\, dx = f^2(t).$$

3. Let n be an integer greater than 1.
   (a) Define $\ln n$ as an integral, and interpret $\ln n$ as an area.
   (b) Show that

$$\frac{1}{2} + \frac{1}{3} + \cdots + \frac{1}{n} < \ln n < 1 + \frac{1}{2} + \frac{1}{3} + \cdots + \frac{1}{n-1}.$$

4. Find $\dfrac{1}{x}\dfrac{d^2x}{dy^2}$ if $y = \displaystyle\int_0^x \frac{1}{\sqrt{3 + 2t^2}}\, dt$.

5. (a) Evaluate, if it exists, $\displaystyle\lim_{h \to 0} \frac{1}{h} \int_{\pi/4}^{(\pi/4)+h} \frac{\sin x}{x}\, dx$.

   (b) By identifying the following with an appropriate definite integral, evaluate $\displaystyle\lim_{n \to \infty} \frac{1}{n} \sum_{k=1}^{n} \cos^2 \frac{\pi k}{n}$.

Solutions for Section II Problems: Definite Integrals

1. (a) For $n \neq 6$,

$$\int \frac{x^3}{x^{n-2}}\, dx = \int x^{5-n}\, dx = \frac{x^{6-n}}{6 - n} + C.$$

But if $n = 6$, it equals $\displaystyle\int \frac{dx}{x} = \ln |x| + C$.

(b) $\displaystyle\int \frac{2}{2 - e^{-x}}\, dx = \int \frac{2e^x}{2e^x - 1}\, dx = \int \frac{du}{u}$, where $u = 2e^x - 1$. The definite integral then equals

$$\ln |2e^x - 1|_0^1 = \ln (2e - 1) - \ln 1 = \ln (2e - 1).$$

(c) If $y = \ln \dfrac{1 + \sin x}{\cos x}$, then it equals $\ln (1 + \sin x) - \ln \cos x$, and

$$\frac{dy}{dx} = \frac{\cos x}{1 + \sin x} + \frac{\sin x}{\cos x} = \frac{\cos^2 x + \sin x + \sin^2 x}{(\cos x)(1 + \sin x)}$$

$$= \frac{1 + \sin x}{(\cos x)(1 + \sin x)} = \sec x \quad \text{if } \sin x \neq -1.$$

So

$$\int_0^{\pi/3} \sec x\, dx = \ln \frac{1 + \sin x}{\cos x}\Bigg|_0^{\pi/3} = \ln \frac{1 + \sqrt{3}/2}{\frac{1}{2}} = \ln (2 + \sqrt{3}).$$

2. Differentiating both sides with respect to t yields $t \cdot f(t) = 2f(t)f'(t)$. Then $f'(t) = \frac{t}{2}$ and $f(t) = \frac{t^2}{4} + C$. So $f(x) = \frac{x^2}{4} + C$.

3. (a) $\ln n = \int_1^n \frac{1}{t}\, dt$. Here, $\ln n$ may be interpreted as the area under the curve of $y = \frac{1}{t}$ (and above the t-axis) from $t = 1$ to $t = n$.

(b) $\frac{1}{2} + \frac{1}{3} + \cdots + \frac{1}{n}$ can be regarded as the sum of the areas of inscribed rectangles, each of width 1 and of heights $\frac{1}{2}, \frac{1}{3}, \ldots, \frac{1}{n}$, as indicated by the broken lines in Figure S6–3a. $1 + \frac{1}{2} + \frac{1}{3} + \cdots + \frac{1}{n-1}$ can be regarded as the sum of

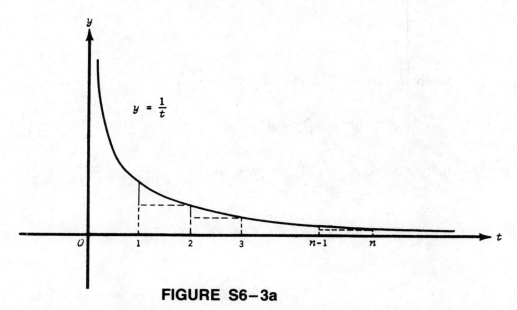

FIGURE S6–3a

the areas of circumscribed rectangles, each of width 1 and of heights $1, \frac{1}{2}, \frac{1}{3}, \ldots, \frac{1}{n-1}$, as shown in Figure S6–3b. Clearly, the inequality given in part (b) of this question holds.

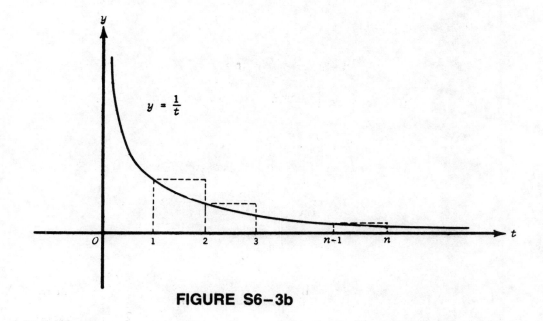

FIGURE S6–3b

4. Since $\dfrac{dy}{dx} = \dfrac{1}{\sqrt{3 + 2x^2}}$, $\dfrac{dx}{dy} = \sqrt{3 + 2x^2}$.

Then $\dfrac{d^2x}{dy^2} = \dfrac{d}{dy}\left(\dfrac{dx}{dy}\right) = \dfrac{d}{dx}\left(\dfrac{dx}{dy}\right) \cdot \dfrac{dx}{dy} = \dfrac{4x}{2\sqrt{3 + 2x^2}} \cdot \sqrt{3 + 2x^2} = 2x$.

Therefore $\dfrac{1}{x} \cdot \dfrac{d^2x}{dy^2} = 2$

5. (a) The given limit is equivalent to

$$\lim_{h \to 0} \frac{F\left(\dfrac{\pi}{4} + h\right) - F\left(\dfrac{\pi}{4}\right)}{h} = F'\left(\dfrac{\pi}{4}\right),$$

where $F'(x) = \dfrac{\sin x}{x}$. The answer is $\dfrac{2\sqrt{2}}{\pi}$.

(b) $\displaystyle\lim_{n \to \infty} \frac{1}{n} \sum_{k=1}^{n} \cos^2 \frac{\pi k}{n} = \frac{1}{\pi} \int_0^\pi \cos^2 x \; dx = \frac{1}{\pi} \int_0^\pi \left(\frac{1}{2} + \frac{\cos 2x}{2}\right) dx$, using

half-angle formula (23) on page 483, $= \dfrac{1}{\pi}\left[\dfrac{x}{2} + \dfrac{\sin 2x}{4}\right]_0^\pi = \dfrac{1}{2}$.

Chapter 7

Applications of Integration to Area

Review of Definitions and Methods

A. Calculating Areas

To find an area, we

(1) draw a sketch of the given region and of a typical element;

(2) write the expression for the area of a typical rectangle; and

(3) set up the definite integral that is the limit of the sum of the n areas as $n \to \infty$.

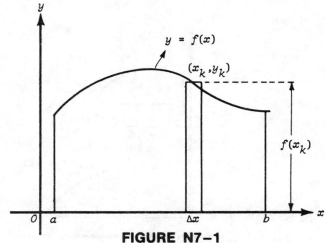

FIGURE N7–1

If $f(x)$ is nonnegative on $[a, b]$, as in Figure N7–1, then $f(x_k) \, \Delta x$ can be regarded as the area of a typical approximating rectangle, and the area bounded by the x-axis, the curve, and the vertical lines $x = a$ and $x = b$ is given exactly by

$$\lim_{n \to \infty} \sum_{k=1}^{n} f(x_k) \, \Delta x \qquad \text{and hence by} \qquad \int_a^b f(x) \, dx.$$

See Questions 1, 6, and 11 of Set 7.

If $f(x)$ changes sign on the interval (Figure N7–2), we find the values of x for which $f(x) = 0$ and note where the function is positive, where it is negative. The total area bounded by the x-axis, the curve, $x = a$, and $x = b$ is here given

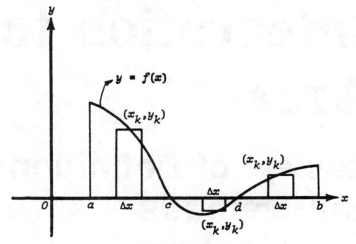

FIGURE N7–2

exactly by

$$\int_a^c f(x) \, dx - \int_c^d f(x) \, dx + \int_d^b f(x) \, dx,$$

where we have taken into account that $f(x_k) \, \Delta x$ is a negative number if $c < x < d$.

See Question 12 of Set 7.

If x is given as a function of y, say $x = g(y)$, then (Figure N7–3) the sub-

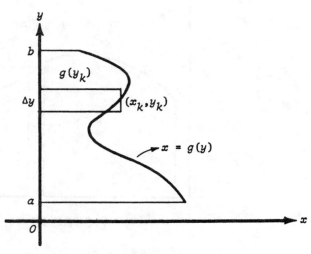

FIGURE N7–3

divisions are made along the y-axis, and the area bounded by the y-axis, the curve, and the horizontal lines $y = a$ and $y = b$ is given exactly by

$$\lim_{n \to \infty} \sum_{k=1}^{n} g(y_k) \, \Delta y = \int_a^b g(y) \, dy.$$

See Questions 3 and 14 of Set 7.

B.   Area between Curves

To find the area between curves (Figure N7–4), we first find where they intersect and then write the area of a typical element for each region between the

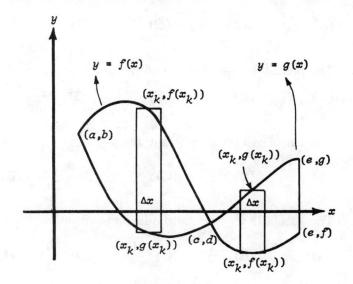

FIGURE N7–4

points of intersection. To find the total area bounded by the curves $y = f(x)$ and $y = g(x)$ between $x = a$ and $x = e$ we see that, if they intersect at $[c, d]$, the total area is given exactly by

$$\int_a^c [f(x) - g(x)] \, dx + \int_c^e [g(x) - f(x)] \, dx.$$

See Questions 4, 7, 8, and 10 of Set 7.

C.   Using Symmetry

Frequently we seek the area of a region that is symmetric to the x- or y-axis (or both) or to the origin. In such cases it is almost always simpler to make use of this symmetry when integrating. For example:

in Question 2 on page 209, the area sought, which is symmetric to the y-axis, is twice that to the right of the y-axis;

in Question 5 on page 210, the area, which is symmetric to the x-axis, is twice that above the x-axis;

in problem 21 on page 122, the area between the curve and the line over the interval $-2\pi \le x \le 2\pi$, which is symmetric to the origin, is twice the area in the first quadrant on $0 \le x \le 2\pi$.

Set 7: Multiple-Choice Questions on Applications of Integration to Area

In Questions 1–12, choose the alternative that gives the area of the region whose boundaries are given.

1. The curve of $y = x^2$, $y = 0$, $x = -1$, and $x = 2$.

   (A) $\dfrac{11}{3}$ (B) $\dfrac{7}{3}$ (C) 3   (D) 5   (E) none of these

2. The parabola $y = x^2 - 3$ and the line $y = 1$.

   (A) $\dfrac{8}{3}$ (B) 32   (C) $\dfrac{32}{3}$ (D) $\dfrac{16}{3}$ (E) none of these

3. The curve of $x = y^2 - 1$ and the y-axis.

   (A) $\dfrac{4}{3}$ (B) $\dfrac{2}{3}$ (C) $\dfrac{8}{3}$ (D) $\dfrac{1}{2}$ (E) none of these

4. The parabola $y^2 = x$ and the line $x + y = 2$.

   (A) $\dfrac{5}{2}$ (B) $\dfrac{3}{2}$ (C) $\dfrac{11}{6}$ (D) $\dfrac{9}{2}$ (E) $\dfrac{29}{6}$

5. The curve of $x^2 = y^2(4 - y^2)$.

   (A) $\dfrac{16}{3}$ (B) $\dfrac{32}{3}$ (C) $\dfrac{8}{3}$ (D) $\dfrac{64}{15}$ (E) 4

6. The curve of $y = \dfrac{4}{x^2 + 4}$, the x-axis, and the vertical lines $x = -2$ and $x = 2$.

   (A) $\dfrac{\pi}{4}$ (B) $\dfrac{\pi}{2}$ (C) 2π (D) π (E) none of these

7. The parabolas $x = y^2 - 5y$ and $x = 3y - y^2$.

   (A) $\dfrac{32}{3}$ (B) $\dfrac{139}{6}$ (C) $\dfrac{64}{3}$ (D) $\dfrac{128}{3}$ (E) none of these

8. The curve of $y = \dfrac{2}{x}$ and $x + y = 3$.

   (A) $\dfrac{1}{2} - 2\ln 2$ (B) $\dfrac{3}{2}$ (C) $\dfrac{1}{2} - \ln 4$ (D) $\dfrac{5}{2}$ (E) $\dfrac{3}{2} - \ln 4$

9. In the first quadrant, bounded below by the x-axis and above by the curves of $y = \sin x$ and $y = \cos x$.

   (A) $2 - \sqrt{2}$ (B) $2 + \sqrt{2}$ (C) 2   (D) $\sqrt{2}$ (E) $2\sqrt{2}$

10. Above by the curve $y = \sin x$ and below by $y = \cos x$ from $x = \frac{\pi}{4}$ to $x = \frac{5\pi}{4}$.

   (A) $2\sqrt{2}$ (B) $\frac{2}{\sqrt{2}}$ (C) $\frac{1}{2\sqrt{2}}$ (D) $2(\sqrt{2} - 1)$ (E) $2(\sqrt{2} + 1)$

11. The curve $y = \cot x$, the lines $x = \frac{\pi}{4}$ and $x = \frac{\pi}{2}$, and the x-axis.

   (A) $\ln 2$ (B) $+\ln \frac{1}{2}$ (C) 1   (D) $\frac{1}{2} \ln 2$ (E) 2

12. The curve of $y = x^3 - 2x^2 - 3x$ and the x-axis.

   (A) $\frac{28}{3}$ (B) $\frac{79}{6}$ (C) $\frac{45}{4}$ (D) $\frac{71}{6}$ (E) none of these

13. The total area bounded by the cubic $x = y^3 - y$ and the line $x = 3y$ is equal to

   (A) 4   (B) $\frac{16}{3}$ (C) 8   (D) $\frac{32}{3}$ (E) 16

14. The area bounded by $y = e^x$, $y = 1$, $y = 2$, and $x = 3$ is equal to

   (A) $3 + \ln 2$ (B) $3 - 3 \ln 3$ (C) $4 + \ln 2$

   (D) $3 - \frac{1}{2} \ln^2 2$ (E) $4 - \ln 4$

*15. The area enclosed by the ellipse with parametric equations $x = 2 \cos \theta$ and $y = 3 \sin \theta$ equals

   (A) 6π (B) $\frac{9}{2}\pi$ (C) 3π (D) $\frac{3}{2}\pi$ (E) none of these

*16. The area enclosed by one loop of the cycloid with parametric equations $x = \theta - \sin \theta$ and $y = 1 - \cos \theta$ equals

   (A) $\frac{3\pi}{2}$ (B) 3π (C) 2π (D) 6π (E) none of these

17. The area enclosed by the curve $y^2 = x(1 - x)$ is given by

   (A) $2\int_0^1 x\sqrt{1 - x}\, dx$ (B) $2\int_0^1 \sqrt{x - x^2}\, dx$ (C) $4\int_0^1 \sqrt{x - x^2}\, dx$
   (D) π (E) 2π

18. The area bounded by the parabola $y = 2 - x^2$ and the line $y = x - 4$ is given by

   (A) $\int_{-2}^3 (6 - x - x^2)\, dx$ (B) $\int_{-2}^1 (2 + x + x^2)\, dx$

   (C) $\int_{-3}^2 (6 - x - x^2)\, dx$ (D) $2\int_0^{\sqrt{2}} (2 - x^2)\, dx + \int_{-3}^2 (4 - x)\, dx$

   (E) none of these

*Questions preceded by an asterisk are likely to appear only on the Calculus BC Examination.

*19. The area enclosed by the hypocycloid with parametric equations $x = \cos^3 t$ and $y = \sin^3 t$ is given by

(A) $3\displaystyle\int_{\pi/2}^{0} \sin^4 t \cos^2 t \, dt$ (B) $4\displaystyle\int_{0}^{1} \sin^3 t \, dt$ (C) $-4\displaystyle\int_{\pi/2}^{0} \sin^6 t \, dt$

(D) $12\displaystyle\int_{0}^{\pi/2} \sin^4 t \cos^2 t \, dt$ (E) none of these

20. The figure below shows part of the curve of $y = x^3$ and a rectangle with two vertices at $(0, 0)$ and $(c, 0)$. What is the ratio of the area of the rectangle to the shaded part of it above the cubic?

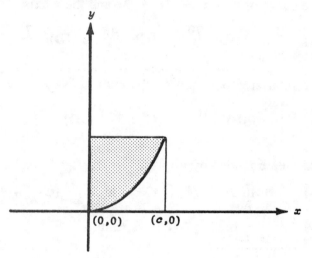

(A) $4 : 3$ (B) $4 : 1$ (C) $3 : 2$ (D) $3 : 1$ (E) $5 : 4$

*21. Suppose the following is a table of ordinates for $y = f(x)$, given that f is continuous on $[1, 5]$:

x	1	2	3	4	5
y	1.62	4.15	7.5	9.0	12.13

If Simpson's Rule is used, with $n = 4$, then the area under the curve, from $x = 1$ to $x = 5$, is equal, to two decimal places, to

(A) 16.2   (B) 20.3   (C) 27.12   (D) 40.68   (E) 54.23

Answers for Set 7: Applications of Integration to Area

1.	C	**5.**	B	**9.**	A	**13.**	C	**17.**	B
2.	C	**6.**	D	**10.**	A	**14.**	E	**18.**	C
3.	A	**7.**	C	**11.**	D	**15.**	A	**19.**	D
4.	D	**8.**	E	**12.**	D	**16.**	B	**20.**	A
								21.	C

We give below, for each question, a sketch of the region, and indicate a typical element of area. The area of the region is given by the definite integral. We exploit symmetry wherever possible.

1. **C.**

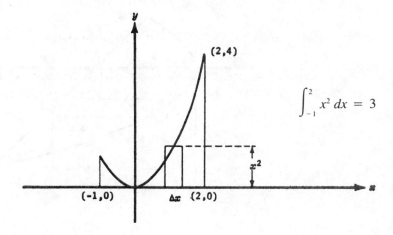

$$\int_{-1}^{2} x^2 \, dx = 3$$

2. **C.**

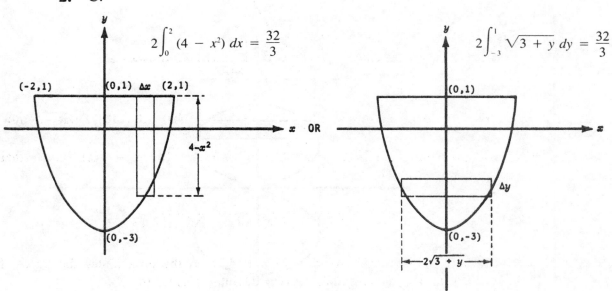

$$2\int_{0}^{2} (4 - x^2) \, dx = \frac{32}{3}$$

OR

$$2\int_{-3}^{1} \sqrt{3 + y} \, dy = \frac{32}{3}$$

3. **A.**

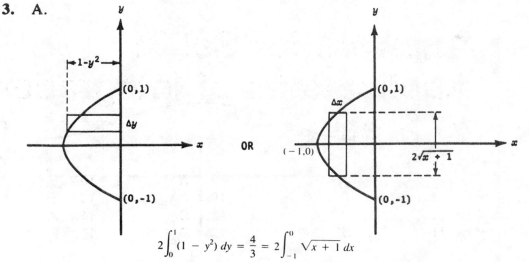

OR

$$2\int_{0}^{1} (1 - y^2) \, dy = \frac{4}{3} = 2\int_{-1}^{0} \sqrt{x + 1} \, dx$$

4.   D.

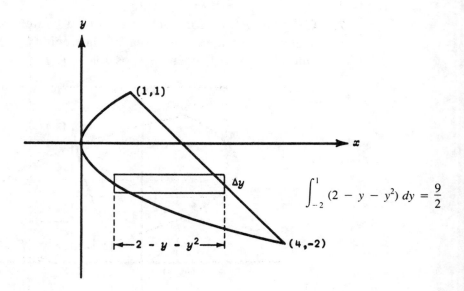

$$\int_{-2}^{1} (2 - y - y^2)\, dy = \frac{9}{2}$$

5.   B.

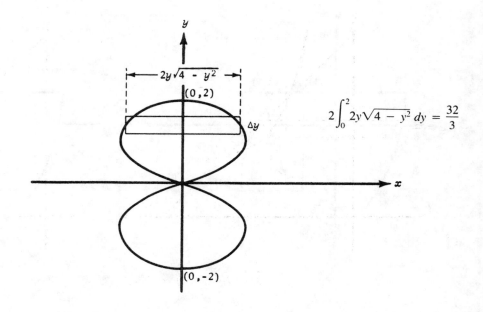

$$2\int_{0}^{2} 2y\sqrt{4 - y^2}\, dy = \frac{32}{3}$$

6.   D.

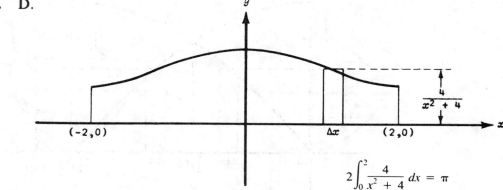

$$2\int_{0}^{2} \frac{4}{x^2 + 4}\, dx = \pi$$

7. C.

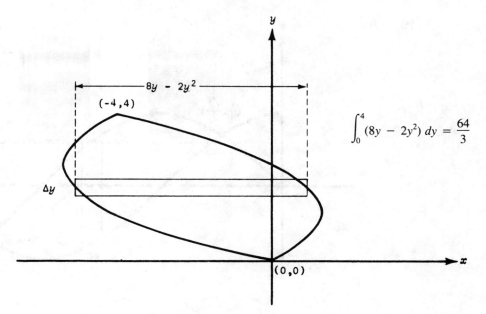

$$\int_0^4 (8y - 2y^2)\, dy = \frac{64}{3}$$

8. E.

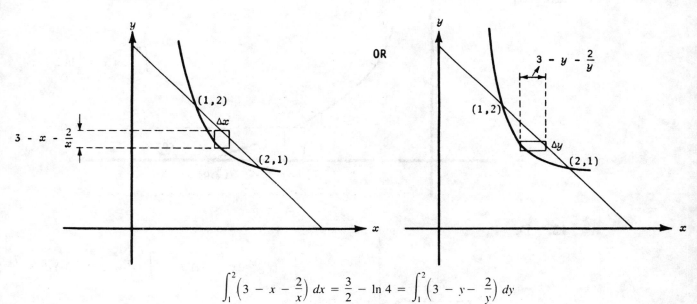

OR

$$\int_1^2 \left(3 - x - \frac{2}{x}\right) dx = \frac{3}{2} - \ln 4 = \int_1^2 \left(3 - y - \frac{2}{y}\right) dy$$

9. A.

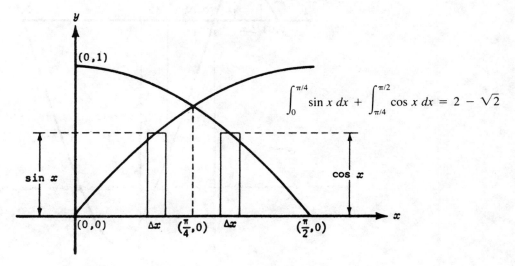

$$\int_0^{\pi/4} \sin x\, dx + \int_{\pi/4}^{\pi/2} \cos x\, dx = 2 - \sqrt{2}$$

10. A.

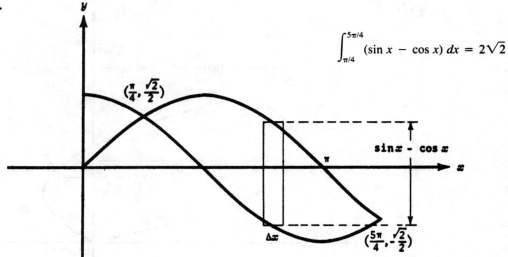

$$\int_{\pi/4}^{5\pi/4} (\sin x - \cos x)\, dx = 2\sqrt{2}$$

11. D.

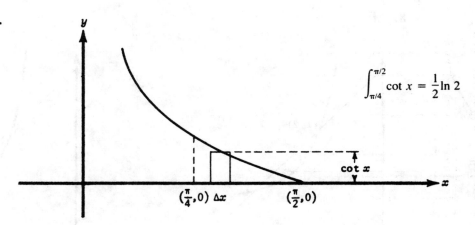

$$\int_{\pi/4}^{\pi/2} \cot x = \frac{1}{2}\ln 2$$

12. D.

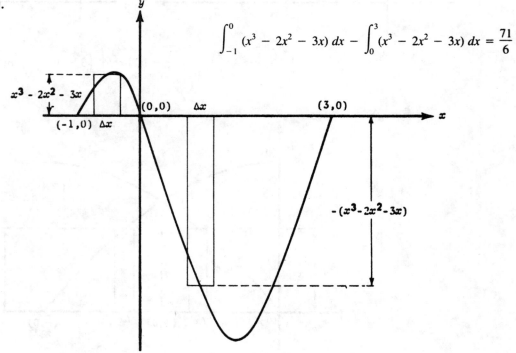

$$\int_{-1}^{0} (x^3 - 2x^2 - 3x)\, dx - \int_{0}^{3} (x^3 - 2x^2 - 3x)\, dx = \frac{71}{6}$$

13. C.

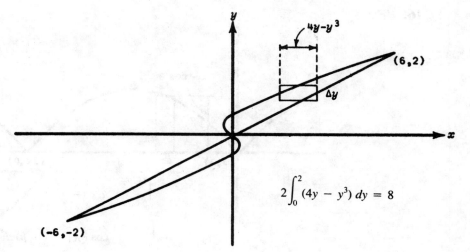

$$2\int_0^2 (4y - y^3)\, dy = 8$$

14. E.

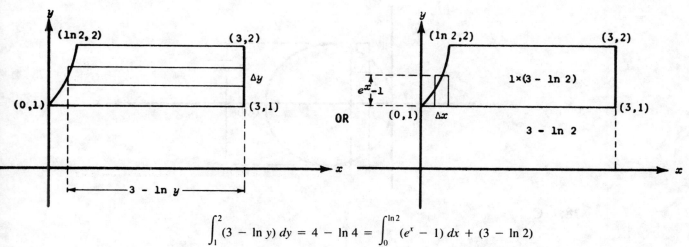

$$\int_1^2 (3 - \ln y)\, dy = 4 - \ln 4 = \int_0^{\ln 2} (e^x - 1)\, dx + (3 - \ln 2)$$

15. A.

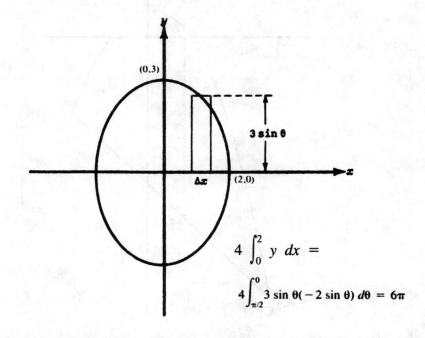

$$4\int_0^2 y\, dx =$$

$$4\int_{\pi/2}^0 3\sin\theta(-2\sin\theta)\, d\theta = 6\pi$$

16. B.

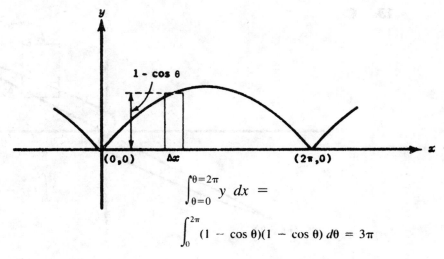

$$\int_{\theta=0}^{\theta=2\pi} y \ dx =$$

$$\int_0^{2\pi} (1 - \cos\theta)(1 - \cos\theta) \ d\theta = 3\pi$$

17. B.

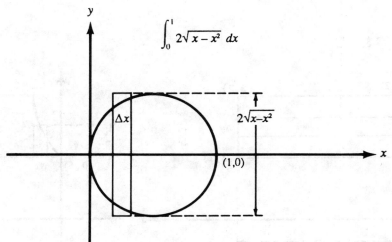

$$\int_0^1 2\sqrt{x - x^2} \ dx$$

18. C.

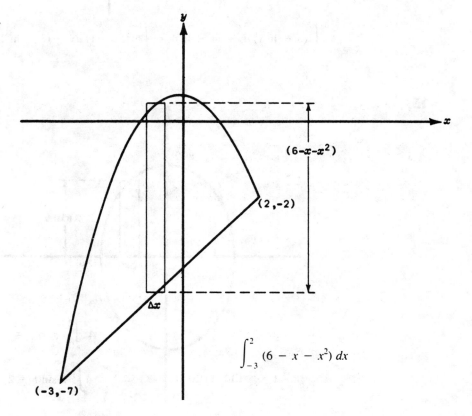

$$\int_{-3}^2 (6 - x - x^2) \ dx$$

19.   D.

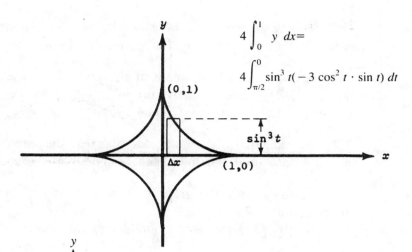

$$4 \int_0^1 y \, dx =$$

$$4 \int_{\pi/2}^0 \sin^3 t (-3 \cos^2 t \cdot \sin t) \, dt$$

(0,1)

$\sin^3 t$

Δx (1,0)

20.   A.

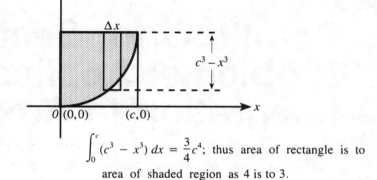

Δx

$c^3 - x^3$

$0 | (0,0)$ $(c,0)$

$$\int_0^c (c^3 - x^3) \, dx = \frac{3}{4} c^4; \text{ thus area of rectangle is to}$$

area of shaded region as 4 is to 3.

21.   C.   If Simpson's Rule is used, with $\Delta x = 1$, then the area equals

$$\frac{1}{3} \cdot 1 \cdot [1.62 + 4(4.15) + 2(7.5) + 4(9.0) + 12.13].$$

Practice Section II Problems on Applications of Integration to Area

1. (a) Sketch the region in the first quadrant bounded above by the line $y = x + 4$, below by the line $y = 4 - x$, and to the right by the parabola $y = x^2 + 2$.
(b) Find the area of this region.

2. (a) Find the coordinates of any relative maximum or minimum points of the curve of $y = x^4 - 4x^2$.
(b) Sketch the curve.
(c) Find the area bounded by the curve and the x-axis.

3. Find the area of the region bounded by the curves $y = x^3 - 2x^2$ and $y = x^2$.

4. Sketch the curve of $y = x^3 - 4x^2 + 3x$, and find the area bounded by the curve and the line joining the points $(0, 0)$ and $(4, 12)$.

5. Find the x-coordinate of the point R on the curve of $y = x - x^2$ such that the line OR (where O is the origin) divides the area bounded by the curve and the x-axis into two regions of equal area.

6. Sketch, on the same set of axes, the curves of $y = x^3 - 4x$ and of $y = x^2 - 4$, and find the total area bounded by these curves.

7. (a) Sketch the graph of $y = x^4 - 6x^2$, showing any maximum, minimum, or inflection points.
   (b) Find the area bounded by this curve and the x-axis.

Solutions for Section II Problems: Applications of Integration to Area

1. (a) The region is sketched in Figure S7–1. The pertinent points of intersection are labeled.

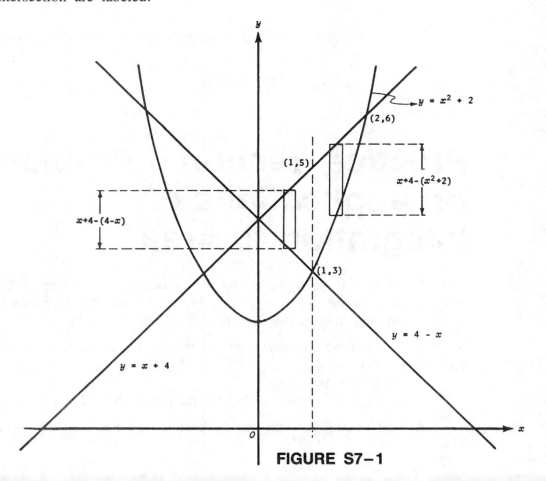

FIGURE S7–1

(b) The required area consists of two parts. The area of the triangle is represented by $\int_0^1 [(x + 4) - (4 - x)]\, dx$ and is equal to 1, while the area of the region bounded at the left by $x = 1$, above by $y = x + 4$, and at the right by the parabola is represented by $\int_1^2 [(x + 4) - (x^2 + 2)]\, dx$. This equals

$$\int_1^2 (x + 2 - x^2)\, dx = \frac{x^2}{2} + 2x - \frac{x^3}{3}\Big|_1^2 = \frac{7}{6}.$$

The required area, thus, equals $2\frac{1}{6}$ or $\frac{13}{6}$.

2. (a) Since $y = x^4 - 4x^2$,

$$y' = 4x^3 - 8x \quad\text{and}\quad y'' = 12x^2 - 8.$$

Then $y' = 4x(x^2 - 2)$, and $0,\ \pm\sqrt{2}$ are critical values. Since $y''(0) < 0$ while $y''(\pm\sqrt{2}) > 0$, the graph has a relative maximum point at $(0, 0)$ and minima at $(\pm\sqrt{2}, -4)$.

(b) The curve is sketched in Figure S7–2.

(c) The area bounded by the curve and the x-axis is given by

$$A = 2\int_0^2 -(x^4 - 4x^2)\, dx,$$

where we use the symmetry about the y-axis. Thus

$$A = 2\int_0^2 (4x^2 - x^4)\, dx = \frac{128}{15}.$$

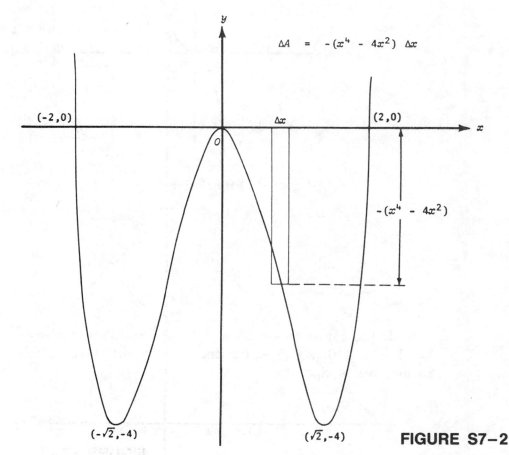

FIGURE S7–2

3. The region is sketched in Figure S7–3. We solve $y = x^3 - 2x^2$ and $y = x^2$ simultaneously:

$$x^3 - 2x^2 = x^2 \quad \text{or} \quad x^2(x - 3) = 0.$$

The curves intersect, then, for $x = 0$ and $x = 3$. For the area A we have

$$A = \int_0^3 (y_1 - y_2)\, dx = \int_0^3 x^2 - (x^3 - 2x^2)\, dx$$

$$= \int_0^3 (3x^2 - x^3)\, dx = x^3 - \frac{x^4}{4}\Big|_0^3 = \frac{27}{4}.$$

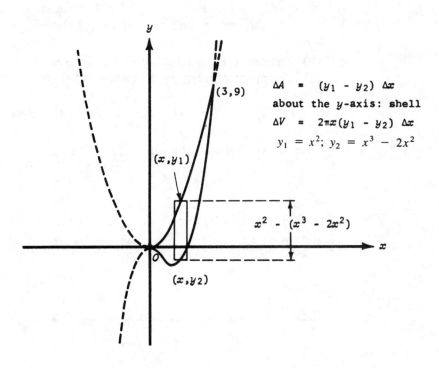

$$\Delta A = (y_1 - y_2)\,\Delta x$$
about the y-axis: shell
$$\Delta V = 2\pi x(y_1 - y_2)\,\Delta x$$
$$y_1 = x^2;\ y_2 = x^3 - 2x^2$$

FIGURE S7–3

4. See Figure S7–4. Since $3x - (x^3 - 4x^2 + 3x) = 4x^2 - x^3 = x^2(4 - x)$, and $4 - x \geqq 0$ if $x \leqq 4$, the line is above the curve on the interval from 0 to 4. So the area is equal to

$$\int_0^4 [3x - (x^3 - 4x^2 + 3x)]\, dx = \frac{64}{3}.$$

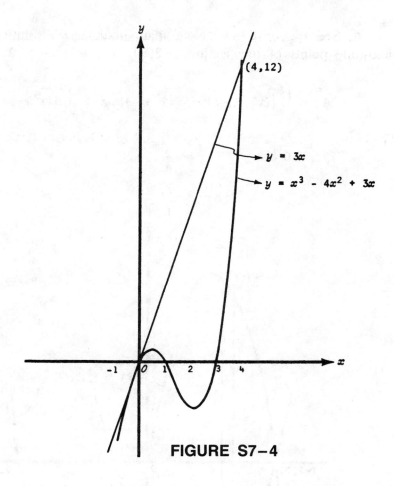

FIGURE S7–4

5. See Figure S7–5, where R is the point (a, b). We seek a such that

$$\int_0^a \left(x - x^2 - \frac{b}{a} \cdot x\right) dx = \frac{1}{2} \int_0^1 (x - x^2) \, dx$$

This yields

$$\frac{a^2}{2} - \frac{a^3}{3} - \frac{ab}{2} = \frac{1}{4} - \frac{1}{6}$$

Replacing b by $a - a^2$, we get

$$\frac{a^2}{2} - \frac{a^3}{3} - \frac{a}{2}(a - a^2) = \frac{1}{12}$$

$$a^3 = \frac{1}{2} \text{ and } a = \frac{1}{\sqrt[3]{2}} = \frac{\sqrt[3]{4}}{2}$$

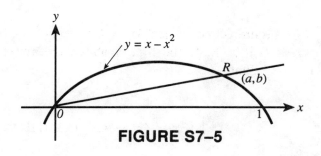

FIGURE S7–5

6. See Figure S7–6. Note that simultaneous solution of the two equations yields the points of intersection $(-2, 0)$, $(1, -3)$, and $(2, 0)$. The total area

$$A = \int_{-2}^{1}[(x^3 - 4x) - (x^2 - 4)]\,dx + \int_{1}^{2}[(x^2 - 4) - (x^3 - 4x)]\,dx$$

$$= \frac{71}{6}.$$

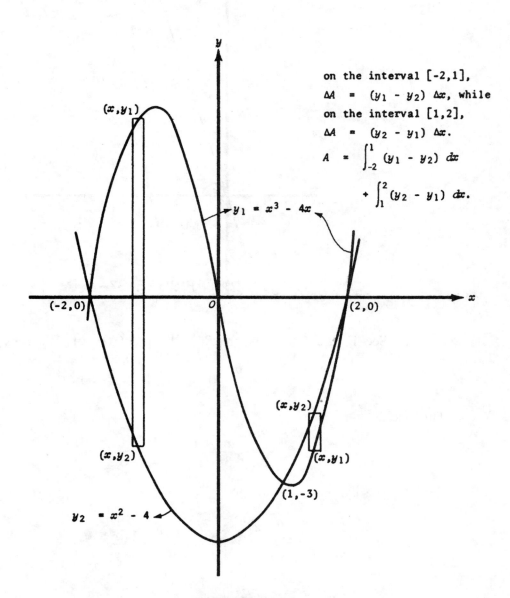

on the interval $[-2,1]$,
$\Delta A = (y_1 - y_2)\,\Delta x$, while
on the interval $[1,2]$,
$\Delta A = (y_2 - y_1)\,\Delta x$.
$$A = \int_{-2}^{1}(y_1 - y_2)\,dx$$
$$+ \int_{1}^{2}(y_2 - y_1)\,dx.$$

$y_1 = x^3 - 4x$

$y_2 = x^2 - 4$

(x, y_1)

(x, y_2)

$(-2, 0)$

O

$(2, 0)$

(x, y_2)

(x, y_1)

$(1, -3)$

FIGURE S7–6

7. (a) The curve has a relative minimum at $(0, 0)$, relative maxima at $(\pm \sqrt{3}, -9)$, and inflection points at $(\pm 1, -5)$. It is sketched in Figure S7-7.

(b) The area bounded by the curve and the x-axis equals $\dfrac{48\sqrt{6}}{5}$.

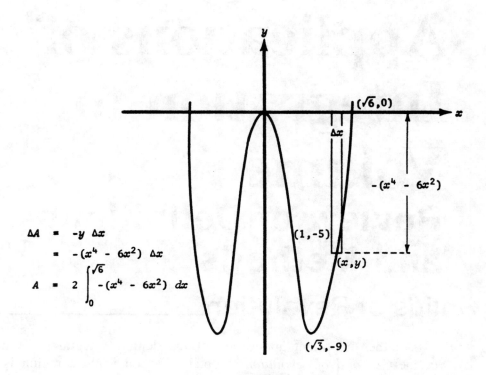

$$\Delta A = -y \, \Delta x$$
$$= -(x^4 - 6x^2) \, \Delta x$$
$$A = 2 \int_0^{\sqrt{6}} -(x^4 - 6x^2) \, dx$$

FIGURE S7-7

Chapter 8

Applications of Integration to Volume
Review of Definitions and Methods

A. Solids of Revolution_____

Another important application of the definite integral is its use in determining volume. A *solid of revolution* is obtained when a plane region is revolved about a fixed line, called the *axis of revolution*. There are three major methods of obtaining the volume of a solid of revolution.

DISKS.   For example, if the region bounded by a semicircle (with center O and radius r) and its diameter is revolved about the x-axis, the solid of revolution obtained is a sphere of radius r. We think of the "rectangular strip" of the semicircular region at the left below as generating the solid disk or slice of the sphere shown at the right.

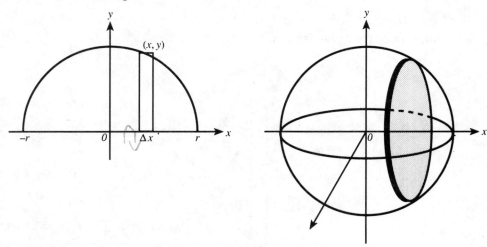

> The volume of a disk (cylinder) is $\pi R^2 H$,
> where R is the radius and H is the height.

In our example, the volume ΔV of a typical disk is given by $\Delta V = \pi y^2 / \Delta x$. The equation of the circle is $x^2 + y^2 = r^2$. To find the volume of the sphere, we let the number of slices become infinite, then apply the Fundamental Theorem; that is, we integrate:

$$V = \int_{-r}^{r} \pi y^2 \, dx = \pi \int_{-r}^{r} (r^2 - x^2) \, dx = \pi \left[r^2 x - \frac{x^3}{3} \right]_{-r}^{r} = \frac{4}{3} \pi r^3.$$

See Questions 1, 10, 17, 20, and 21 of Set 8 for examples of finding volumes by disks.

SHELLS.   A cylindrical shell is a solid bounded by two concentric circular cylinders with the same height. If r_1 and r_2 are the radii of the inner and outer cylinders, respectively, and the common height is H, then the volume V of the shell equals

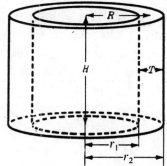

$$\pi r_2^2 H - \pi r_1^2 H = \pi (r_2^2 - r_1^2) H$$

$$= \pi (r_2 + r_1)(r_2 - r_1) H$$

$$= 2\pi \frac{(r_2 + r_1)}{2} (r_2 - r_1) H.$$

With $r_2 - r_1 = T$, the *thickness* of the shell, and $\frac{r_2 + r_1}{2} = R$, the *midradius*, we see that:

> The volume of a shell is $2\pi RHT$,
> where R is the midradius, H is the height,
> and T is the thickness.

To find the volume of a sphere by the shell method, we can revolve the semicircular region shown at the left below about the y-axis.

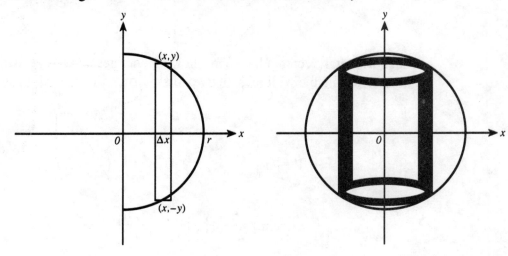

The typical vertical rectangular strip in the region generates the cylindrical shell shown at the right, with midradius x, height $2y$, and thickness Δx. The volume of the shell (think of it as a hollow, thin soup can with no top or bottom) is

$$\Delta V = 2\pi x \cdot 2y \cdot \Delta x.$$

The sphere therefore has volume

$$V = 4\pi \int_0^r xy \, dx = 4\pi \int_0^r x\sqrt{r^2 - x^2} \, dx$$

$$= \frac{4\pi}{-2} \int_0^r (r^2 - x^2)^{1/2}(-2x) \, dx$$

$$= -2\pi\left(\frac{2}{3}\right) (r^2 - x^2)^{3/2} \Big|_0^r = -\frac{4\pi}{3}(0 - r^3) = \frac{4}{3}\pi r^3.$$

See Questions 2, 3, 9, 12, and 25 of Set 8 for examples of finding volumes by shells.

WASHERS.   A washer is a disk with a hole in it. The volume may be regarded as the difference in the volumes of two concentric disks. As an example, we find the volume obtained when the region bounded by $y = x^2$ and $y = 2x$ is revolved about the y-axis. The curves intersect at the origin and at $(2, 4)$. Note that we distinguish between the functions by letting (x_1, y) be a point on the line, (x_2, y) a point on the parabola.

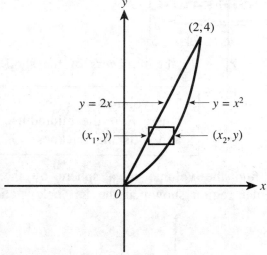

The horizontal rectangular strip shown generates a washer whose volume, ΔV, is the difference in the volumes of two disks, of radii x_1 and x_2 and height Δy. So

$$\Delta V = \pi x_2^2 \, \Delta y - \pi x_1^2 \, \Delta y;$$

$$V = \pi \int_0^4 (x_2^2 - x_1^2) \, dy = \pi \int_0^4 \left(y - \frac{y^2}{4}\right) dy$$

$$= \pi \left[\frac{y^2}{2} - \frac{y^3}{12}\right]_0^4 = \pi\left(\frac{16}{2} - \frac{64}{12}\right) = \frac{8}{3}\pi.$$

The volume of a solid of revolution using washers can also be found by regarding a typical washer as a shell.

As difference of two disks:

WASHER: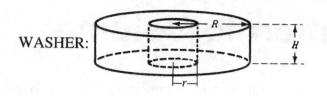

$V = \pi R^2 H - \pi r^2 H$, where

R = radius of larger disk,

r = radius of smaller disk;

OR as shell: $V = 2\pi\left(\dfrac{R + r}{2}\right)(R - r)H,$

which yields, as above, $\pi(R^2 - r^2)H$.

See Questions 2, 5, 8, 11, and 25 of Set 8 for examples in which washers are regarded as the differences of two disks; see Questions 6 and 19 for examples of washers viewed as shells.

Occasionally when more than one method is satisfactory we try to use the most efficient. In the key for Set 8, a sketch is shown for each question and for each the type and volume of a typical element are given. The required volume is then found by letting the number of elements become infinite and applying the Fundamental Theorem.

*B. Solids with Known Cross-Sections_____

If the *area of a cross-section* of the solid *is known* and can be expressed in terms of x, then the volume of a typical slice, ΔV, can be determined. The volume of the solid is obtained, as usual, by letting the number of slices increase indefinitely. In Figure N8–1, the slices are taken perpendicular to the x-axis so that $\Delta V = A(x)\,\Delta x$, where $A(x)$ is the area of a cross-section and Δx is the thickness of the slice. When the cross-section is a circle, the solid can be generated by revolving a plane area.

Questions 22 and 23 of Set 8 illustrate solids with known cross-sections.

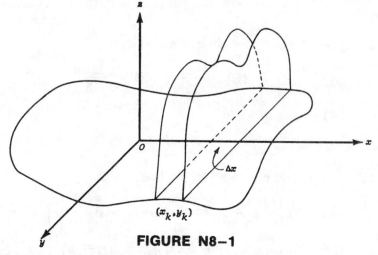

FIGURE N8–1

*An asterisk denotes a topic covered only in Calculus BC.

Set 8: Multiple-Choice Questions on Applications of Integration to Volume

In Questions 1–20, the region whose boundaries are given is rotated about the line indicated. Choose the alternative which gives the volume of the solid generated.

1. $y = x^2$, $x = 2$, and $y = 0$ about the x-axis.

   (A) $\dfrac{64\pi}{3}$ **(B)** $\dfrac{32\pi}{5}$ **(C)** $\dfrac{8\pi}{3}$ **(D)** $\dfrac{128\pi}{5}$ **(E)** 8π

2. $y = x^2$, $x = 2$, and $y = 0$ about the y-axis.

   (A) $\dfrac{16\pi}{3}$ **(B)** 4π **(C)** $\dfrac{32\pi}{5}$ **(D)** 8π **(E)** $\dfrac{8\pi}{3}$

3. $y = x^2$, $x = 2$, and $y = 0$ about the line $x = 2$.

   (A) 4π **(B)** $\dfrac{4\pi}{3}$ **(C)** $\dfrac{8\pi}{3}$ **(D)** $\dfrac{16\pi}{3}$ **(E)** 8π

4. The first quadrant region bounded by $y = x^2$, the y-axis, and $y = 4$; about the y-axis.

   (A) 8π **(B)** 4π **(C)** $\dfrac{64\pi}{3}$ **(D)** $\dfrac{32\pi}{3}$ **(E)** $\dfrac{16\pi}{3}$

5. $y = x^2$ and $y = 4$ about the x-axis.

   (A) $\dfrac{64\pi}{5}$ **(B)** $\dfrac{512\pi}{15}$ **(C)** $\dfrac{256\pi}{5}$ **(D)** $\dfrac{128\pi}{5}$

   (E) none of these

6. $y = x^2$ and $y = 4$ about the line $x = 2$.

   (A) $2\pi\displaystyle\int_{-2}^{2} (2 - x)(4 - x^2)\,dx$ **(B)** $4\pi\displaystyle\int_{0}^{2} (2 - x)(4 - x^2)\,dx$

   (C) $4\pi\displaystyle\int_{0}^{2} (4 - x)(2 - x^2)\,dx$ **(D)** $4\pi\displaystyle\int_{0}^{4} \sqrt{y}\,dy$ **(E)** $8\pi\displaystyle\int_{0}^{4} y\,dy$

7. $y = x^2$ and $y = 4$ about the line $y = 4$.

   (A) $\dfrac{256\pi}{15}$ (B) $\dfrac{256\pi}{5}$ (C) $\dfrac{512\pi}{5}$ (D) $\dfrac{512\pi}{15}$ (E) $\dfrac{64\pi}{3}$

8. $y = x^2$ and $y = 4$ about the line $y = -1$.

   (A) $4\pi \displaystyle\int_{-1}^{4} (y + 1)\sqrt{y}\ dy$ (B) $2\pi \displaystyle\int_{0}^{2} (4 - x^2)^2\ dx$

   (C) $\pi \displaystyle\int_{-2}^{2} (16 - x^4)\ dx$ (D) $4\pi \displaystyle\int_{0}^{4} (y + 1)\sqrt{y}\ dy$ (E) none of these

9. $y = 3x - x^2$ and $y = 0$ about the y-axis.

   (A) 27π (B) $\dfrac{81\pi}{10}$ (C) $\dfrac{648\pi}{15}$ (D) $\dfrac{27\pi}{4}$ (E) $\dfrac{27\pi}{2}$

10. $y = 3x - x^2$ and $y = 0$ about the x-axis.

   (A) $\pi \displaystyle\int_{0}^{3} (9x^2 + x^4)\ dx$ (B) $\pi \displaystyle\int_{0}^{3} (3x - x^2)^2\ dx$ (C) $\pi \displaystyle\int_{0}^{\sqrt{3}} (3x - x^2)\ dx$

   (D) $2\pi \displaystyle\int_{0}^{3} y\sqrt{9 - 4y}\ dy$ (E) $\pi \displaystyle\int_{0}^{9/4} y^2\ dy$

11. $y = 3x - x^2$ and $y = x$ about the x-axis.

   (A) $\pi \displaystyle\int_{0}^{3/2} [(3x - x^2)^2 - x^2]\ dx$ (B) $\pi \displaystyle\int_{0}^{2} (9x^2 - 6x^3)\ dx$

   (C) $\pi \displaystyle\int_{0}^{2} [(3x - x^2)^2 - x^2]\ dx$ (D) $\pi \displaystyle\int_{0}^{3} [(3x - x^2)^2 - x^4]\ dx$

   (E) $\pi \displaystyle\int_{0}^{3} (2x - x^2)^2\ dx$

12. $y = 3x - x^2$ and $y = x$ about the y-axis.

   (A) $\dfrac{8\pi}{3}$ (B) 4π (C) $\dfrac{40\pi}{3}$ (D) $\dfrac{32\pi}{3}$ (E) 8π

13. An arch of $y = \sin x$ and the x-axis about the x-axis.

   (A) $\dfrac{\pi}{2}\left(\pi - \dfrac{1}{2}\right)$ (B) $\dfrac{\pi^2}{2}$ (C) $\dfrac{\pi^2}{4}$ (D) π^2 (E) $\pi(\pi - 1)$

14. $y = \sin x,\ x = 0,\ x = \pi$, and $y = 0$ about the y-axis.

   (A) $2\pi^2$ (B) $2\pi^2 - 2$ (C) 4π (D) π^2 (E) $\dfrac{\pi^2}{2}$

15. The right branch of the hyperbola $x^2 - y^2 = 3$ and $x = 2$ about the y-axis.

   (A) $\dfrac{8\pi}{3}$ (B) 4π (C) $\dfrac{4\pi}{3}$ (D) $\dfrac{2\pi}{3}(3\sqrt{3} - 5)$

   (E) none of these

16. The trapezoid with vertices at $(2, 0)$, $(2, 2)$, $(4, 0)$, and $(4, 4)$ about the y-axis.

(A) $\dfrac{56\pi}{3}$ (B) $\dfrac{128\pi}{3}$ (C) $\dfrac{92\pi}{3}$ (D) $\dfrac{112\pi}{3}$

(E) none of these

17. $y = \ln x$, $y = 0$, $x = e$ about the line $x = e$.

(A) $2\pi \displaystyle\int_1^e (e - x) \ln x \, dx$ (B) $\pi \displaystyle\int_0^1 (e - e^{2y}) \, dy$

(C) $2\pi \displaystyle\int_1^e (e - \ln x) \, dx$ (D) $\pi \displaystyle\int_0^e (e^2 - 2e^{y+1} + e^{2y}) \, dy$

(E) none of these

18. The circle $x^2 - 2x + y^2 = 0$ about the y-axis.

(A) $8\pi \displaystyle\int_0^1 x\sqrt{2x - x^2} \, dx$ (B) $4\pi \displaystyle\int_0^2 x\sqrt{2x - x^2} \, dx$

(C) $4\pi \displaystyle\int_{-2}^2 \sqrt{y^2 - 1} \, dy$ (D) $2\pi \displaystyle\int_0^2 x\sqrt{2x - x^2} \, dx$

(E) none of these

*19. The circle with parametric equations $x = a \cos \theta$, $y = a \sin \theta$ $(a > 0)$, about the line $x = 2a$.

(A) $4\pi a^3 \displaystyle\int_0^\pi (2 \sin \theta - \sin^2 \theta) \cos \theta \, d\theta$

(B) $4\pi a^3 \displaystyle\int_\pi^0 (2 \sin \theta - \cos \theta) \, d\theta$

(C) $-4\pi a^3 \displaystyle\int_\pi^0 (2 \sin \theta \cos \theta - \sin \theta \cos^2 \theta) \, d\theta$

(D) $16\pi a^3 \displaystyle\int_0^{\pi/2} \cos^2 \theta \, d\theta$

(E) $8\pi a^3 \displaystyle\int_0^{\pi/2} (2 \sin^2 \theta - \sin^2 \theta \cos \theta) \, d\theta$

*20. The curve with parametric equations $x = \tan \theta$, $y = \cos^2 \theta$, and the lines $x = 0$, $x = 1$, and $y = 0$ about the x-axis.

(A) $\pi \displaystyle\int_0^{\pi/4} \cos^4 \theta \, d\theta$ (B) $\pi \displaystyle\int_0^{\pi/4} \cos^2 \theta \sin \theta \, d\theta$ (C) $\pi \displaystyle\int_0^{\pi/4} \cos^2 \theta \, d\theta$

(D) $\pi \displaystyle\int_0^1 \cos^2 \theta \, d\theta$ (E) $\pi \displaystyle\int_0^1 \cos^4 \theta \, d\theta$

*Questions preceded by an asterisk are likely to appear only on the Calculus BC Examination.

21. A sphere of radius r is divided into two parts by a plane at distance h $(0 < h < r)$ from the center. The volume of the smaller part equals

(A) $\frac{\pi}{3}(2r^3 + h^3 - 3r^2h)$ (B) $\frac{\pi h}{3}(3r^2 - h^2)$ (C) $\frac{4}{3}\pi r^3 + \frac{h^3}{3} - r^2h$

(D) $\frac{\pi}{3}(2r^3 + 3r^2h - h^3)$ (E) none of these

*22. The base of a solid is a circle of radius a, and every plane section perpendicular to a diameter is a square. The solid has volume

(A) $\frac{8}{3}a^3$ (B) $2\pi a^3$ (C) $4\pi a^3$ (D) $\frac{16}{3}a^3$ (E) $\frac{8\pi}{3}a^3$

*23. The base of a solid is the region bounded by the parabola $x^2 = 8y$ and the line $y = 4$, and each plane section perpendicular to the y-axis is an equilateral triangle. The volume of the solid is

(A) $\frac{64\sqrt{3}}{3}$ (B) $64\sqrt{3}$ (C) $32\sqrt{3}$ (D) 32

(E) none of these

24. $f(x)$ is positive on $[a, b]$, where $0 < a < b$. If the region bounded by $y = 0$, $x = a$, $x = b$ and the curve of $y = f(x)$ is rotated about the y-axis, the volume generated is given by

(A) $2\pi \int_a^b (x - a) f(x)\, dx$ (B) $2\pi \int_0^b x f(x)\, dx$ (C) $2\pi \int_a^b x f(x)\, dx$

(D) $2\pi \int_0^b (x - a) f(x)\, dx$ (E) none of these

25. If the curves of $f(x)$ and $g(x)$ intersect for $x = a$ and $x = b$ and if $f(x) > g(x) > 0$ for all x on (a, b), then the volume obtained when the region bounded by the curves is rotated about the x-axis is equal to

(A) $\pi \int_a^b f^2(x)\, dx - \int_a^b g^2(x)\, dx$

(B) $\pi \int_a^b [f(x) - g(x)]^2\, dx$

(C) $2\pi \int_a^b x[f(x) - g(x)]\, dx$

(D) $\pi \int_a^b [f^2(x) - g^2(x)]\, dx$

(E) none of these

Answers for Set 8: Applications of Integration to Volume

1.	B	**6.**	A	**11.**	C	**16.**	D	**21.**	A
2.	D	**7.**	D	**12.**	A	**17.**	A	**22.**	D
3.	C	**8.**	D	**13.**	B	**18.**	B	**23.**	B
4.	A	**9.**	E	**14.**	A	**19.**	D	**24.**	C
5.	C	**10.**	B	**15.**	C	**20.**	C	**25.**	D

One or more sketches are given below for each question, in addition to the definite integral for each volume.

1. B.

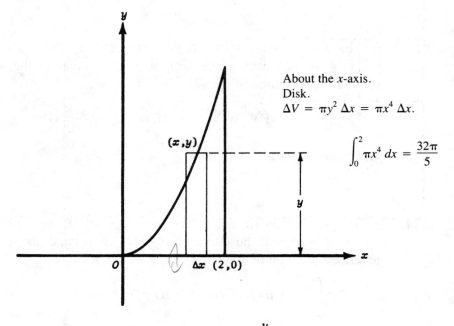

About the *x*-axis.
Disk.
$\Delta V = \pi y^2 \, \Delta x = \pi x^4 \, \Delta x.$

$$\int_0^2 \pi x^4 \, dx = \frac{32\pi}{5}$$

2. D.

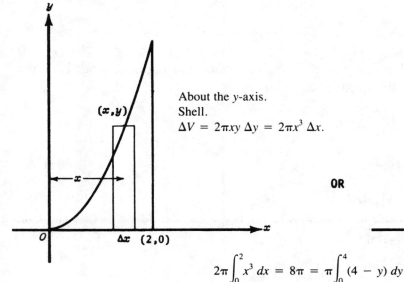

About the *y*-axis.
Shell.
$\Delta V = 2\pi xy \, \Delta y = 2\pi x^3 \, \Delta x.$

OR

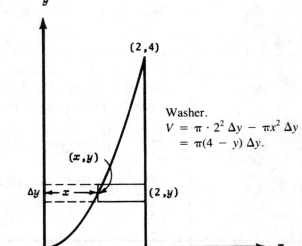

Washer.
$V = \pi \cdot 2^2 \, \Delta y - \pi x^2 \, \Delta y$
$= \pi(4 - y) \, \Delta y.$

$$2\pi \int_0^2 x^3 \, dx = 8\pi = \pi \int_0^4 (4 - y) \, dy$$

3. C.

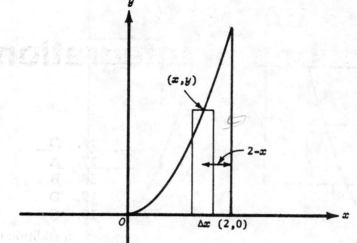

About $x = 2$.
Shell.
$\Delta V = 2\pi(2 - x)y\,\Delta x$
$\quad = 2\pi(2 - x)x^2\,\Delta x.$

$2\pi\displaystyle\int_0^2 (2 - x)x^2\,dx = \dfrac{8\pi}{3}$

4. A.

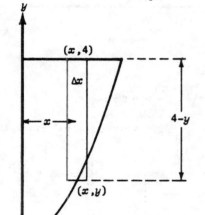

OR

About the y-axis.

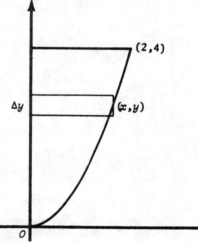

Shell.
$\Delta V = 2\pi x(4 - y)\,\Delta x$
$\quad = 2\pi x(4 - x^2)\,\Delta x.$

Disk.
$\Delta V = \pi x^2\,\Delta y$
$\quad = \pi y\,\Delta y.$

$2\pi\displaystyle\int_0^2 x(4 - x^2)\,dx = 8\pi = \pi\int_0^4 y\,dy$

5. C.

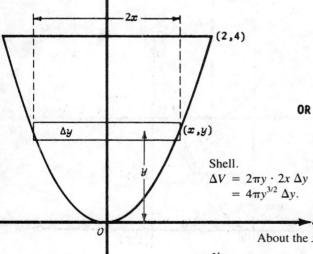

OR

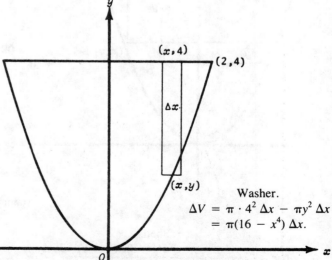

Shell.
$\Delta V = 2\pi y \cdot 2x\,\Delta y$
$\quad = 4\pi y^{3/2}\,\Delta y.$

Washer.
$\Delta V = \pi \cdot 4^2\,\Delta x - \pi y^2\,\Delta x$
$\quad = \pi(16 - x^4)\,\Delta x.$

About the x-axis.

$4\pi\displaystyle\int_0^4 y^{3/2}\,dy = \dfrac{256\pi}{5} = 2\pi\int_0^2 (16 - x^4)\,dx$

6. A.

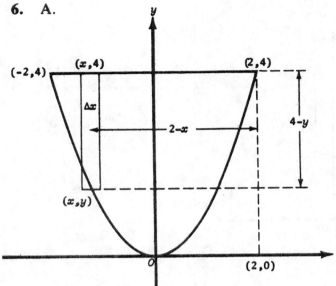 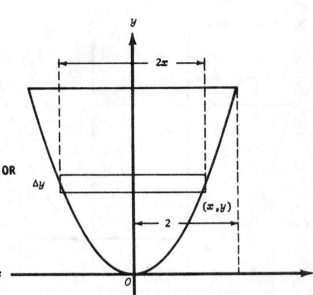

OR

About $x = 2$.

Shell.

$\Delta V = 2\pi(2 - x)(4 - y) \, \Delta x$

$= 2\pi(2 - x)(4 - x^2) \, \Delta x$.

Washer (shell).

$\Delta V = 2\pi \cdot 2 \cdot 2x \, \Delta y$

$= 8\pi y^{1/2} \, \Delta y$.

$$2\pi \int_{-2}^{2} (2 - x)(4 - x^2) \, dx = 8\pi \int_{0}^{4} y^{1/2} \, dy$$

7. D.

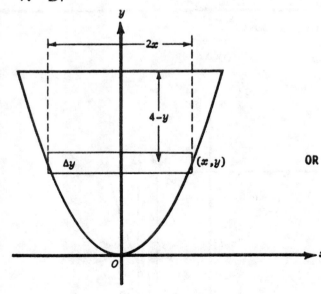 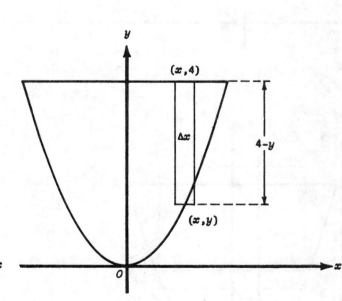

OR

About $y = 4$.

Shell.

$\Delta V = 2\pi(4 - y) \cdot 2x \, \Delta y$

$= 4\pi(4 - y)y^{1/2} \, \Delta y$.

Disk.

$\Delta V = \pi(4 - y)^2 \, \Delta x$

$= \pi(4 - x^2)^2 \, \Delta x$.

$$4\pi \int_{0}^{4} (4y^{1/2} - y^{3/2}) \, dy = \frac{512\pi}{15} = 2\pi \int_{0}^{2} (4 - x^2)^2 \, dx$$

8. D.

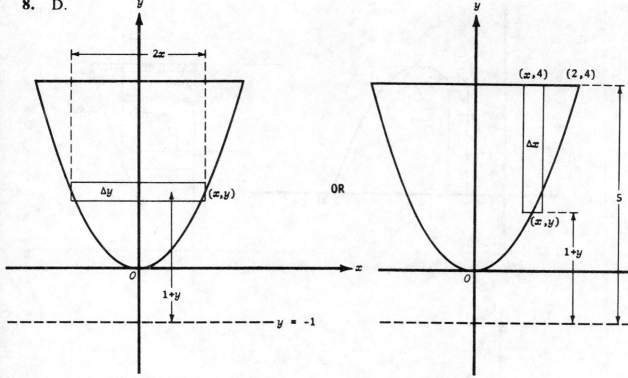

About $y = -1$.

Shell.
$$\Delta V = 2\pi(1 + y) \cdot 2x \, \Delta y$$
$$= 4\pi(1 + y)y^{1/2} \, \Delta y.$$

Washer.
$$\Delta V = \pi \cdot 5^2 \, \Delta x - \pi(1 + y)^2 \, \Delta x$$
$$= \pi[25 - (1 + y)^2] \, \Delta x.$$

$$4\pi \int_0^4 (1 + y)y^{1/2} \, dy = 2\pi \int_0^2 (24 - 2x^2 - x^4) \, dx$$

9. E.

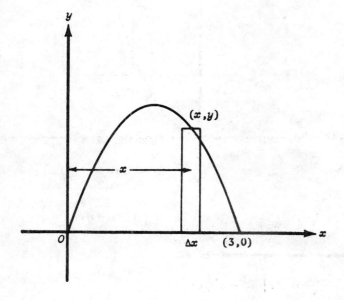

About the y-axis.
Shell.
$$\Delta V = 2\pi xy \, \Delta x$$
$$= 2\pi x(3x - x^2) \, \Delta x.$$

$$2\pi \int_0^3 x(3x - x^2) \, dx = \frac{27\pi}{2}$$

10. B.

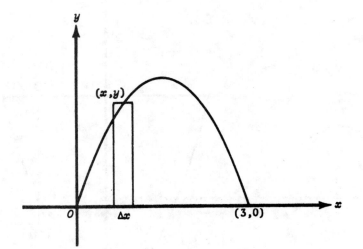

About the x-axis.
Disk.
$$\Delta V = \pi y^2\, \Delta x$$
$$\quad = \pi(3x - x^2)^2\, \Delta x.$$

$$\pi \int_0^3 (3x - x^2)^2\, dx$$

11. C.

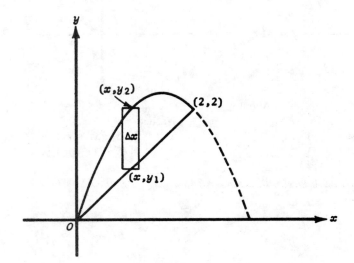

About the x-axis.
Washer.
$$\Delta V = \pi y_2^2\, \Delta x - \pi y_1^2\, \Delta x$$
$$\quad = \pi[(3x - x^2)^2 - x^2]\, \Delta x.$$

$$\pi \int_0^2 [(3x - x^2)^2 - x^2]\, dx$$

12. A.

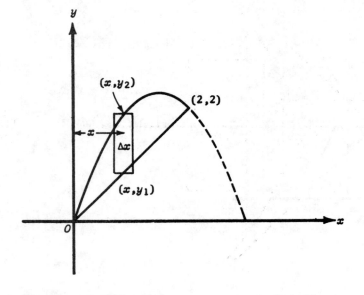

About the y-axis.
Shell.
$$\Delta V = 2\pi x(y_2 - y_1)\, \Delta x$$
$$\quad = 2\pi x(2x - x^2)\, \Delta x.$$

$$2\pi \int_0^2 (2x^2 - x^3)\, dx = \frac{8\pi}{3}$$

13. B.

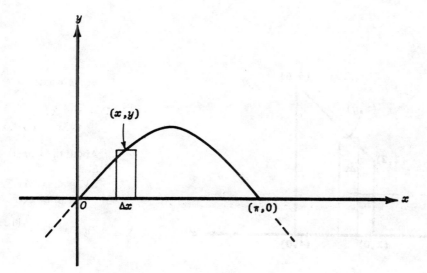

About the x-axis.
Disk.
$$\Delta V = \pi y^2\,\Delta x$$
$$\quad\; = \pi \sin^2 x\,\Delta x.$$

$$\pi\int_0^\pi \sin^2 x\,dx = \frac{\pi^2}{2}$$

14. A.

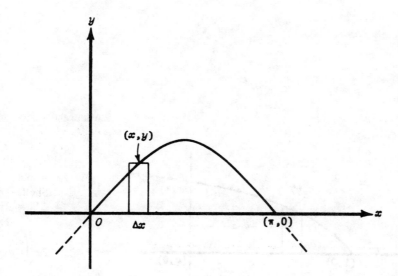

About the y-axis.
Shell.
$$\Delta V = 2\pi xy\,\Delta x$$
$$\quad\; = 2\pi x \sin x\,\Delta x.$$

$$2\pi\int_0^\pi x \sin x\,dx = 2\pi^2$$

15. C.

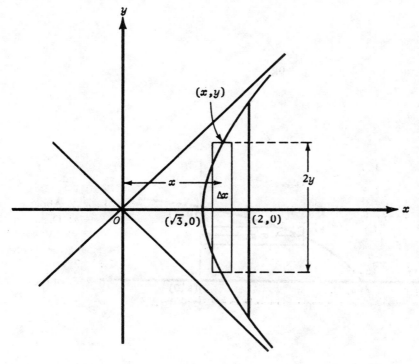

About the y-axis.
Shell.
$$\Delta V = 2\pi x(2y)\,\Delta x$$
$$\quad\; = 4\pi x\sqrt{x^2 - 3}\,\Delta x.$$

$$4\pi\int_{\sqrt3}^2 x\sqrt{x^2 - 3}\,dx = \frac{4\pi}{3}$$

16. D.

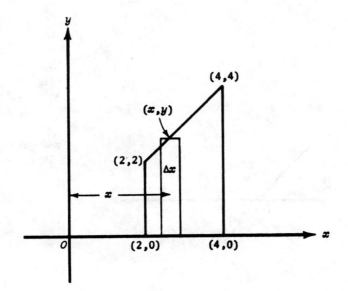

About the y-axis.
Shell.
$$\Delta V = 2\pi xy\,\Delta x$$
$$= 2\pi x^2\,\Delta x.$$

$$2\pi \int_2^4 x^2\,dx = \frac{112\pi}{3}$$

17. A.

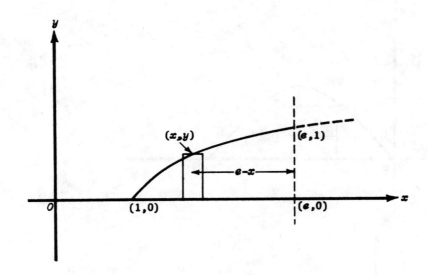

About $x = e$.
Shell.
$$\Delta V = 2\pi(e - x)y\,\Delta x$$
$$= 2\pi(e - x)\ln x\,\Delta x.$$

$$2\pi \int_1^e (e - x)\ln x\,dx.$$

OR

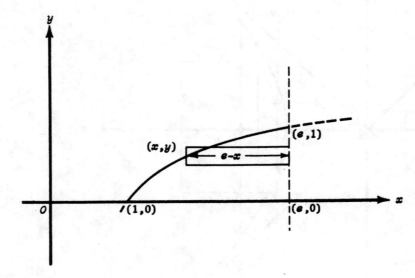

About $x = e$.
Disk.
$$\Delta V = \pi(e - x)^2\,\Delta y$$
$$= \pi(e - e^y)^2\,\Delta y.$$

$$\pi \int_0^1 (e - e^y)^2\,dy$$

18. B.

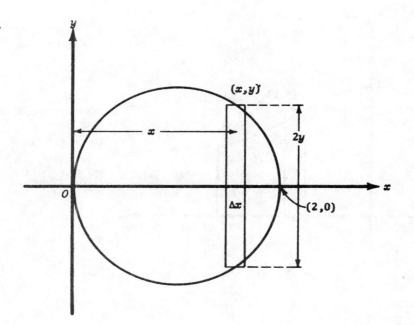

About the y-axis.
Shell.
$$\Delta V = 2\pi x \cdot 2y\ \Delta x$$
$$= 4\pi x\sqrt{2x - x^2}\ \Delta x.$$

$$4\pi \int_0^2 x\sqrt{2x - x^2}\ dx$$

19. D.

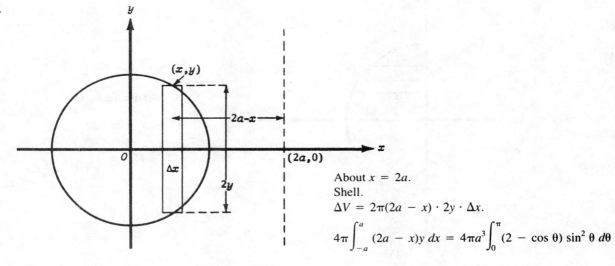

About $x = 2a$.
Shell.
$$\Delta V = 2\pi(2a - x) \cdot 2y \cdot \Delta x.$$

$$4\pi \int_{-a}^a (2a - x)y\ dx = 4\pi a^3 \int_0^\pi (2 - \cos\theta)\sin^2\theta\ d\theta$$

OR

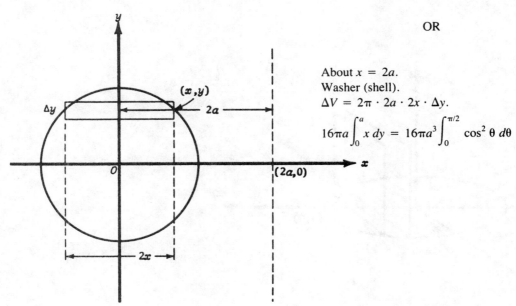

About $x = 2a$.
Washer (shell).
$$\Delta V = 2\pi \cdot 2a \cdot 2x \cdot \Delta y.$$

$$16\pi a \int_0^a x\ dy = 16\pi a^3 \int_0^{\pi/2} \cos^2\theta\ d\theta$$

20. C.

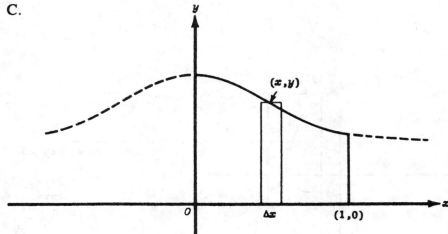

About the x-axis.
Disk.
$\Delta V = \pi y^2 \, \Delta x$.

$$\pi \int_0^1 y^2 \, dx \;=\; \pi \int_0^{\pi/4} \cos^2 \theta \, d\theta$$

21. A.

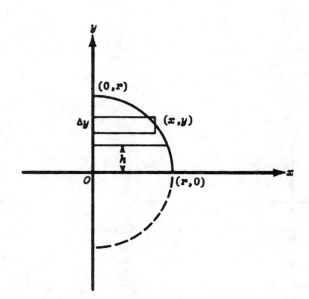

About the y-axis.
Disk.
$\Delta V = \pi x^2 \, \Delta y$
$\quad\;\; = \pi (r^2 - y^2) \, \Delta y$.

$$\pi \int_h^r (r^2 - y^2) \, dy = \frac{\pi}{3}(2r^3 + h^3 - 3r^2 h)$$

22. D.

$\Delta V = (2y)^2 \, \Delta x$
$\quad\;\; = 4(a^2 - x^2) \, \Delta x$.

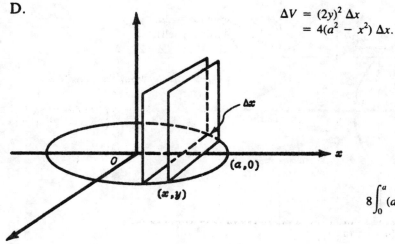

$$8 \int_0^a (a^2 - x^2) \, dx = \frac{16a^3}{3}$$

23. B.

$$\Delta V = x^2\sqrt{3}\ \Delta y = 8\sqrt{3}\ y\ \Delta y.$$

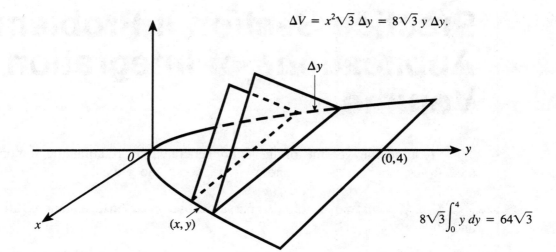

$$8\sqrt{3}\int_0^4 y\ dy = 64\sqrt{3}$$

24. C.

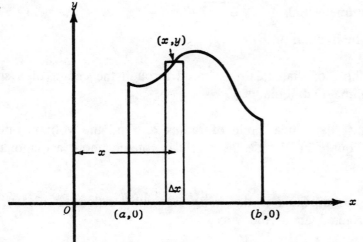

About the y-axis.
Shell.
$$\Delta V = 2\pi xy\ \Delta x$$
$$= 2\pi x \cdot f(x)\ \Delta x.$$

$$2\pi\int_a^b x \cdot f(x)\ dx$$

25. D.

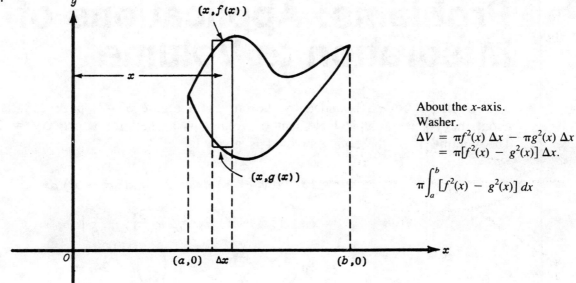

About the x-axis.
Washer.
$$\Delta V = \pi f^2(x)\ \Delta x - \pi g^2(x)\ \Delta x$$
$$= \pi[f^2(x) - g^2(x)]\ \Delta x.$$

$$\pi\int_a^b [f^2(x) - g^2(x)]\ dx$$

Practice Section II Problems on Applications of Integration to Volume

1. Find the volume of the solid generated by rotating the region bounded by the curves $y = x^3 - 2x^2$ and $y = x^2$ about the y-axis.

*2. A point moves on the curve whose parametric equations are $x = \sqrt{t-2}$ and $2y \neq \sqrt{6} = t$.

(a) Find a single equation in x and y for the path of the point, and sketch the part of the curve defined by the parametric equations.

(b) The region bounded by this curve and the axes is rotated about the y-axis. Show that its volume equals $\frac{\pi}{2} \int_2^6 \sqrt{6-t}\, dt$.

(c) Evaluate the integral in (b).

3. A cylindrical hole of diameter a is bored through the center of a sphere of radius a. Find the volume that remains.

4. A square is inscribed in a circle of radius a. Find the volume obtained if the region outside the square but inside the circle is rotated about a diagonal of the square.

5. Find the volume obtained if the trapezoid with vertices at $(2, 0)$, $(2, 2)$, $(5, 0)$, and $(5, 5)$ is rotated about the y-axis.

Solutions for Section II Problems: Applications of Integration to Volume

1. If the region bounded by the curves in Figure S7–3, page 218, is rotated about the y-axis, a typical element of volume is a shell for which $\Delta V = 2\pi RHT$. The volume V of the solid obtained is given by

$$V = 2\pi \int_0^3 x(y_1 - y_2)\, dx = 2\pi \int_0^3 x(3x^2 - x^3)\, dx$$

$$= 2\pi \int_0^3 (3x^3 - x^4)\, dx = 2\pi \left(\frac{3x^4}{4} - \frac{x^5}{5} \right)_0^3$$

$$= 2\pi \left(\frac{3^5}{4} - \frac{3^5}{5} \right) = 2\pi \cdot 3^5 \left(\frac{1}{20} \right) = 24.3\pi.$$

2. (a) $x^2 + 4y^2 = 4$.   The curve is sketched in Figure S8–2; note that $2 \leqq t \leqq 6$.

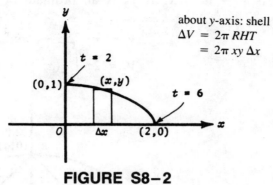

about y-axis: shell
$\Delta V = 2\pi RHT$
$\quad\quad = 2\pi xy \, \Delta x$

FIGURE S8–2

(b) Since a typical element of volume about the y-axis is a shell, $V = 2\pi RHT$ and

$$V = 2\pi \int_{x=0}^{2} xy \, dx = 2\pi \int_{t=2}^{6} \sqrt{t-2} \cdot \frac{\sqrt{6-t}}{2} \cdot \frac{dt}{2\sqrt{t-2}}$$

$$= \frac{\pi}{2} \int_{2}^{6} \sqrt{6-t} \, dt.$$

(c) $V = \frac{8\pi}{3}$.

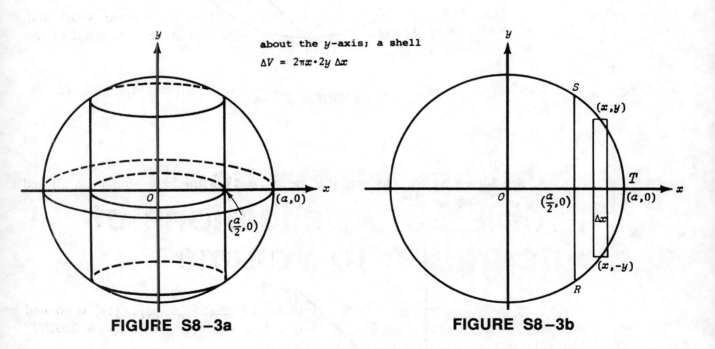

about the y-axis; a shell
$\Delta V = 2\pi x \cdot 2y \, \Delta x$

FIGURE S8–3a   **FIGURE S8–3b**

3. The solids are indicated in Figure S8–3a and a plane cross-section in Figure S8–3b. The required volume can be obtained by rotating the region of the circle labeled RST about the y-axis. If shells are used, then

$$V = 4\pi \int_{a/2}^{a} xy \, dx = 4\pi \int_{a/2}^{a} x\sqrt{a^2 - x^2} \, dx$$

$$= \frac{\sqrt{3}}{2} \pi a^3.$$

4. See Figure S8−4. The equation of the circle is $x^2 + y^2 = a^2$; the equation of RS is $y = a - x$. If y_2 is an ordinate of the circle and y_1 of the line, then

$$\Delta V = \pi y_2^2 \Delta x - \pi y_1^2 \Delta x,$$

$$V = 2\pi \int_0^a \left[(a^2 - x^2) - (a - x)^2 \right] dx = \frac{\pi a^3}{3}.$$

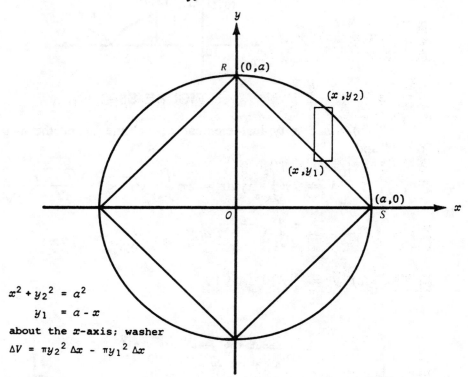

$x^2 + y_2^2 = a^2$

$y_1 = a - x$

about the x-axis; washer

$\Delta V = \pi y_2^2 \, \Delta x - \pi y_1^2 \, \Delta x$

FIGURE S8−4

5. See Figure S8−5, and note that the equation of the line through (2, 2) and (5, 5) is $y = x$. $V = \frac{234\pi}{3}$.

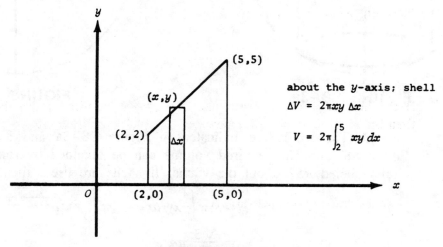

about the y-axis; shell

$\Delta V = 2\pi x y \, \Delta x$

$V = 2\pi \int_2^5 x y \, dx$

FIGURE S8−5

Chapter 9

Further Applications of Integration

Review of Definitions and Methods

In Chapter 7 we used definite integrals to find areas, and in Chapter 8 to find volumes. In this chapter we review applications of integration to the following topics: length of a curve (arc length); average (mean) value of a function; motion in a straight line and along a curve; area in polar coordinates; improper integrals; and work.

*A. Arc Length

If the derivative of a function $y = f(x)$ is continuous on the interval $a \leqq x \leqq b$, then the length s of the arc of the curve of $y = f(x)$ from the point where $x = a$ to the point where $x = b$ is given by

$$s = \int_a^b \sqrt{1 + \left(\frac{dy}{dx}\right)^2}\ dx. \tag{1}$$

If the derivative of the function $x = g(y)$ is continuous on the interval $c \leqq y \leqq d$, then the length s of the arc from $y = c$ to $y = d$ is given by

$$s = \int_c^d \sqrt{1 + \left(\frac{dx}{dy}\right)^2}\ dy. \tag{2}$$

*An asterisk denotes a topic covered only in Calculus BC.

If a curve is defined parametrically by the equations $x = x(t)$ and $y = y(t)$, if the derivatives of the functions $x(t)$ and $y(t)$ are continuous on $[t_a, t_b]$, (and if the curve does not intersect itself), then the length of the arc from $t = t_a$ to $t = t_b$ is given by

$$s = \int_{t_a}^{t_b} \sqrt{\left(\frac{dx}{dt}\right)^2 + \left(\frac{dy}{dt}\right)^2} \, dt. \tag{3}$$

The parenthetical clause above is equivalent to the requirement that the curve is traced out just once as t varies from t_a to t_b.

As indicated in Equation (4), page 81, the formulas in (1), (2), and (3) above can all be derived easily from the very simple relation

$$ds^2 = dx^2 + dy^2. \tag{4}$$

Examples

1. Find the length of the arc of $y = x^{3/2}$ from $x = 1$ to $x = 8$.

Here $\dfrac{dy}{dx} = \dfrac{3}{2} x^{1/2}$, so, by (1),

$$s = \int_1^8 \sqrt{1 + \frac{9}{4} x} \, dx = \frac{4}{9} \int_1^8 \left(1 + \frac{9}{4} x\right)^{1/2} \frac{9}{4} \, dx$$

$$= \frac{2}{3} \cdot \frac{4}{9} \left(1 + \frac{9}{4} x\right)^{3/2} \Big|_1^8 = \frac{8}{27} \left(19^{3/2} - \frac{13^{3/2}}{8}\right).$$

2. Find the length of the curve $(x - 2)^2 = 4y^3$ from $y = 0$ to $y = 1$. Since

$$x - 2 = 2y^{3/2} \qquad \text{and} \qquad \frac{dx}{dy} = 3y^{1/2},$$

(2) yields

$$s = \int_0^1 \sqrt{1 + 9y} \, dy = \frac{2}{27} (1 + 9y)^{3/2} \Big|_0^1 = \frac{2}{27} (10^{3/2} - 1).$$

3. The position (x, y) of a particle at time t is given parametrically by $x = t^2$ and $y = \dfrac{t^3}{3} - t$. Find the distance it travels between $t = 1$ and $t = 2$.

We can use (4): $ds^2 = dx^2 + dy^2$, where $dx = 2t \, dt$ and $dy = (t^2 - 1) \, dt$. Thus

$$ds = \sqrt{4t^2 + t^4 - 2t^2 + 1} \, dt,$$

and

$$s = \int_1^2 \sqrt{(t^2 + 1)^2} \, dt = \int_1^2 (t^2 + 1) \, dt$$

$$= \left. \frac{t^3}{3} + t \right|_1^2 = \frac{10}{3}.$$

4. Find the length of the arc of $y = \ln \sec x$ from $x = 0$ to $x = \frac{\pi}{3}$.
Here

$$\frac{dy}{dx} = \frac{\sec x \, \tan x}{\sec x},$$

so

$$s = \int_0^{\pi/3} \sqrt{1 + \tan^2 x} \, dx = \int_0^{\pi/3} \sec x \, dx$$

$$= \left. \ln (\sec x + \tan x) \right|_0^{\pi/3} = \ln (2 + \sqrt{3}).$$

B. Average (Mean) Value

If the function $y = f(x)$ is continuous on the interval $a \leqq x \leqq b$, then the average or mean value of y with respect to x over the interval $[a, b]$ is denoted by $(y_{av})_x$ and is defined by

$$(y_{av})_x = \frac{1}{b - a} \int_a^b f(x) \, dx. \tag{1}$$

If $a < b$, then (1) is equivalent to

$$(b - a)(y_{av})_x = \int_a^b f(x) \, dx. \tag{2}$$

If $f(x) \geqq 0$ for all x on $[a, b]$, we can interpret (2) in terms of areas as follows: The right-hand member represents the area under the curve of $y = f(x)$, above the x-axis, and bounded by the vertical lines $x = a$ and $x = b$. The left-hand member of (2) represents the area of a rectangle with base $(b - a)$ and height $(y_{av})_x$. See Figure N9–1.

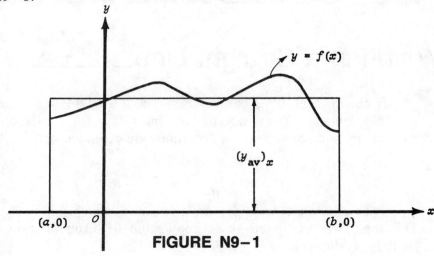

FIGURE N9–1

Examples

5. The average (mean) value of $f(x) = \ln x$ with respect to x on the interval $[1, 4]$ is

$$\frac{1}{4 - 1}\int_1^4 \ln x \, dx = \frac{1}{3}(x \ln x - x)\Big|_1^4 = \frac{4 \ln 4 - 3}{3}.$$

6. The mean value of ordinates of the semicircle $y = \sqrt{4 - x^2}$ with respect to x on $[-2, 2]$ is given by

$$(y_{av})_x = \frac{1}{2 - (-2)}\int_{-2}^2 \sqrt{4 - x^2} \, dx = \frac{1}{4}\frac{\pi(2^2)}{2} = \frac{\pi}{2}. \tag{3}$$

In (3) we have used the fact that the definite integral equals exactly the area of a semicircle of radius 2.

7. Find the average ordinate of the semicircle $y = \sqrt{4 - x^2}$ over the arc of the semicircle.

Whereas in Example 6 we may think of dividing the interval *along the x-axis* into equal subintervals, here we divide the *semicircle* into arcs of equal length. We then seek the mean value of the ordinates at these points of subdivision; that is, we want $(y_{av})_s$, where s denotes arc length. Thus,

$$(y_{av})_x = \frac{1}{2\pi}\int_{s_1}^{s_2} y \, ds, \tag{4}$$

where 2π is the length of the semicircle. Since

$$ds^2 = dx^2 + dy^2 = dx^2 + \frac{x^2}{4 - x^2} \, dx^2 = \frac{4}{y^2} \, dx^2,$$

(4) yields

$$(y_{av})_s = \frac{1}{2\pi}\int_{-2}^2 y \cdot \frac{2}{y} \, dx = \frac{1}{\pi} x \Big|_{-2}^2 = \frac{4}{\pi}.$$

C. Motion in a Straight Line

If the motion of a particle P along a straight line is given by the equation $s = F(t)$, where s is the distance at time t of P from a fixed point on the line, then the velocity and acceleration of P at time t are given respectively by

$$v = \frac{ds}{dt} \quad \text{and} \quad a = \frac{dv}{dt} = \frac{d^2s}{dt^2}.$$

This topic was discussed as an application of differentiation on page 78. Here we shall apply integration.

If we know that the particle P has velocity $v = f(t)$, where v is a continuous function, then the distance traveled by the particle during the time interval from $t = a$ to $t = b$ is

$$s = \int_a^b |f(t)|\ dt. \tag{1}$$

If $f(t) \geqq 0$ for all t on $[a, b]$ (i.e., P moves only in the positive direction), then (1) is equivalent to $s = \int_a^b f(t)\ dt$; similarly, if $f(t) \leqq 0$ on $[a, b]$ (P moves only in the negative direction), then $s = -\int_a^b f(t)\ dt$. If $f(t)$ changes sign on $[a, b]$ (i.e., the direction of motion changes), then (1) gives the *total* distance traveled. Suppose, for example, that the situation is as follows:

$$a \leqq t \leqq c \qquad f(t) \geqq 0;$$
$$c \leqq t \leqq d \qquad f(t) \leqq 0;$$
$$d \leqq t \leqq b \qquad f(t) \geqq 0.$$

Then the total distance s traveled during the time interval from $t = a$ to $t = b$ is exactly

$$s = \int_a^c f(t)\ dt - \int_c^d f(t)\ dt + \int_d^b f(t)\ dt.$$

Examples

8. If a particle moves along a straight line with velocity $v = t^3 + 3t^2$, then the distance traveled between $t = 1$ and $t = 4$ is given by

$$s = \int_1^4 (t^3 + 3t^2)\ dt = \left(\frac{t^4}{4} + t^3 \right) \Big|_1^4 = \frac{507}{4}.$$

Note that $v > 0$ for all t on $[1, 4]$.

9. A particle moves along the x-axis so that its position at time t is given by $x(t) = 2t^3 - 9t^2 + 12t - 4$. Find the total distance covered between $t = 0$ and $t = 4$.

Since $v(t) = 6t^2 - 18t + 12 = 6(t - 1)(t - 2)$, we see that:

$$\text{if} \qquad t < 1, \qquad \text{then } v > 0;$$
$$\text{if } 1 < t < 2, \qquad \text{then } v < 0;$$
$$\text{if } 2 < t, \qquad \text{then } v > 0.$$

Thus

$$s = \int_0^1 v(t)\ dt - \int_1^2 v(t)\ dt + \int_2^4 v(t)\ dt. \tag{2}$$

If we substitute for $v(t)$ in (2), we get

$$s = x(t)\Big|_0^1 - x(t)\Big|_1^2 + x(t)\Big|_2^4$$

$$= x(1) - x(0) - x(2) + x(1) + x(4) - x(2)$$

$$= x(4) - 2x(2) + 2x(1) - x(0)$$

$$= 28 - 2 \cdot 0 + 2 \cdot 1 - (-4) = 34.$$

This example is identical to Example 24, page 79, in which the required distance is computed by another method.

10. The acceleration of a particle moving on a line is given at time t by $a = \sin t$; when $t = 0$ the particle is at rest. Find the distance it travels from $t = 0$ to $t = \frac{5\pi}{6}$.

Since $a = \frac{d^2s}{dt^2} = \frac{dv}{dt} = \sin t$, it follows that

$$v(t) = \frac{ds}{dt} = \int \sin t\, dt; \qquad v(t) = -\cos t + C.$$

Also, $v(0) = 0$ yields $C = 1$. Thus $v(t) = 1 - \cos t$; and since $\cos t \leqq 1$ for all t we see that $v(t) \geqq 0$ for all t. So

$$s = \int_0^{5\pi/6} (1 - \cos t)\, dt = (t - \sin t)\Big|_0^{5\pi/6} = \frac{5\pi}{6} - \frac{1}{2}.$$

11. Find k if a deceleration of k ft/sec^2 is needed to bring a particle moving with a velocity of 75 ft/sec to a stop in 5 sec.

Let $a = \frac{dv}{dt} = -k$; then

$$v = -kt + C. \tag{3}$$

Since $v = 75$ when $t = 0$, we get $C = 75$. Then (3) becomes

$$v = -kt + 75$$

so

$$0 = -5k + 75 \qquad \text{and} \qquad k = 15.$$

*D. Motion along a Plane Curve

In Chapter 4, §J2, it was pointed out that, if the motion of a particle P along a curve is given parametrically by the equations $x = x(t)$ and $y = y(t)$, then at time t the position vector $\mathbf{R}$, the velocity vector $\mathbf{v}$, and the acceleration vector $\mathbf{a}$ have the following values:

$$\mathbf{R} = x\mathbf{i} + y\mathbf{j};$$

$$\mathbf{v} = \frac{d\mathbf{R}}{dt} = \frac{dx}{dt}\mathbf{i} + \frac{dy}{dt}\mathbf{j} = \dot{x}\mathbf{i} + \dot{y}\mathbf{j} = v_x\mathbf{i} + v_y\mathbf{j};$$

$$\mathbf{a} = \frac{d^2\mathbf{R}}{dt^2} = \frac{d\mathbf{V}}{dt} = \frac{d^2x}{dt^2}\mathbf{i} + \frac{d^2y}{dt^2}\mathbf{j} = \ddot{x}\mathbf{i} + \ddot{y}\mathbf{j} = a_x\mathbf{i} + a_y\mathbf{j}.$$

* An asterisk denotes a topic covered only in Calculus BC.

The components in the horizontal and vertical directions of **R**, **v**, and **a** are given respectively by the coefficients of **i** and **j** in the corresponding vector. The slope of **v** is $\dfrac{dy}{dx}$; its magnitude,

$$|\mathbf{v}| = \sqrt{\left(\frac{dx}{dt}\right)^2 + \left(\frac{dy}{dt}\right)^2} = \frac{ds}{dt},$$

is the speed of the particle, and the velocity vector is tangent to the path. The slope of **a** is $\dfrac{d^2y}{dt^2} \Big/ \dfrac{d^2x}{dt^2}$; its magnitude is $|\mathbf{a}| = \sqrt{a_x^2 + a_y^2}$.

How integration may be used to solve problems of curvilinear motion is illustrated in the following examples.

Examples

12. The motion of a particle satisfies the equations

$$\frac{d^2x}{dt^2} = 0, \qquad \frac{d^2y}{dt^2} = -g.$$

Find parametric equations for the motion if the initial conditions are $x = 0$, $y = 0$, $\dfrac{dx}{dt} = v_0 \cos \alpha$, and $\dfrac{dy}{dt} = v_0 \sin \alpha$, where v_0 and α are constants.

We integrate each of the given equations twice and determine the constants as indicated:

$$\frac{dx}{dt} = C_1 = v_0 \cos \alpha; \qquad\qquad \frac{dy}{dt} = -gt + C_2;$$

$$v_0 \sin \alpha = C_2;$$

$$x = (v_0 \cos \alpha)t + C_3; \qquad\qquad y = -\frac{1}{2} gt^2 + (v_0 \sin \alpha)t + C_4;$$

$$x(0) = 0 \text{ yields } C_3 = 0. \qquad y(0) = 0 \text{ yields } C_4 = 0.$$

Finally, then, we have

$$x = (v_0 \cos \alpha)t; \qquad\qquad y = -\frac{1}{2} gt^2 + (v_0 \sin \alpha)t.$$

These are the equations for the path of a projectile that starts at the origin with initial velocity v_0 and at an angle of elevation α. If desired, t can be eliminated from this pair of equations to yield a parabola in rectangular coordinates.

13. A particle $P(x, y)$ moves along a curve so that

$$\frac{dx}{dt} = 2\sqrt{x} \qquad \text{and} \qquad \frac{dy}{dt} = \frac{1}{x} \quad \text{at any time } t \geq 0.$$

At $t = 0$, $x = 1$ and $y = 0$. Find the parametric equations of motion.

Since $\dfrac{dx}{\sqrt{x}} = 2\,dt$, we integrate to get $2\sqrt{x} = 2t + C$, and use $x(0) = 1$ to find that $C = 2$. So $\sqrt{x} = t + 1$ and

$$x = (t + 1)^2. \tag{1}$$

Then $\dfrac{dy}{dt} = \dfrac{1}{x} = \dfrac{1}{(t + 1)^2}$ by (1), so $dy = \dfrac{dt}{(t + 1)^2}$ and

$$y = -\frac{1}{t + 1} + C'. \tag{2}$$

Since $y(0) = 0$, this yields $C' = 1$. So (2) becomes

$$y = 1 - \frac{1}{t + 1} = \frac{t}{t + 1}.$$

Thus the parametric equations are

$$x = (t + 1)^2 \qquad \text{and} \qquad y = \frac{t}{t + 1}.$$

14. If a particle moves on a curve so that the acceleration vector is always perpendicular to the position vector, prove that the speed of the particle is constant.

The hypothesis is equivalent to the statement

$$\frac{\dfrac{d^2 y}{dt^2}}{\dfrac{d^2 x}{dt^2}} \cdot \frac{\dfrac{dy}{dt}}{\dfrac{dx}{dt}} = -1$$

or

$$\frac{\ddot{y}}{\ddot{x}} \cdot \frac{\dot{y}}{\dot{x}} = -1. \tag{3}$$

Equation (3) is equivalent to

$$\dot{y}\ddot{y} = -\dot{x}\ddot{x}, \tag{4}$$

and (4) is simply

$$\frac{d}{dt}\left(\frac{1}{2}\dot{y}^2\right) = \frac{d}{dt}\left(-\frac{1}{2}\dot{x}^2\right).$$

So $\dfrac{1}{2}\dot{y}^2 = -\dfrac{1}{2}\dot{x}^2 + C'$, or

$$\dot{y}^2 + \dot{x}^2 = C. \tag{5}$$

But since the left-hand member of (5) is precisely $\left(\dfrac{ds}{dt}\right)^2$ or $|\mathbf{v}|^2$, (5) says that the square of the speed (and consequently the speed) is constant.

15. A particle $P(x, y)$ moves along a curve so that its acceleration is given by

$$\mathbf{a} = -4 \cos 2t\,\mathbf{i} - 2 \sin t\,\mathbf{j} \qquad \left(-\frac{\pi}{2} \leq t \leq \frac{\pi}{2}\right);$$

when $t = 0$, then $x = 1$, $y = 0$, $\dot{x} = 0$, and $\dot{y} = 2$. (a) Find the position vector $\mathbf{R}$ at any time t. (b) Find a Cartesian equation for the path of the particle and identify the conic on which P moves.

(a) $\mathbf{v} = (-2 \sin 2t + c_1)\mathbf{i} + (2 \cos t + c_2)\mathbf{j}$, and since $\mathbf{v} = 2\mathbf{j}$ when $t = 0$, it follows that $c_1 = c_2 = 0$. So $\mathbf{v} = -2 \sin 2t\,\mathbf{i} + 2 \cos t\,\mathbf{j}$. Also $\mathbf{R} = (\cos 2t + c_3)\mathbf{i} + (2 \sin t + c_4)\mathbf{j}$; and since $\mathbf{R} = \mathbf{i}$ when $t = 0$, we see that $c_3 = c_4 = 0$. Finally, then,

$$\mathbf{R} = \cos 2t\,\mathbf{i} + 2 \sin t\,\mathbf{j}.$$

(b) From (a) the parametric equations of motion are

$$x = \cos 2t, \qquad y = 2 \sin t.$$

By a trigonometric identity,

$$x = 1 - 2 \sin^2 t = 1 - \frac{y^2}{2}.$$

P travels along *part* of a parabola that has its vertex at $(1, 0)$ and opens to the left. The path of the particle is sketched in Figure N9–2; note that $-1 \leq x \leq 1$, $-2 \leq y \leq 2$.

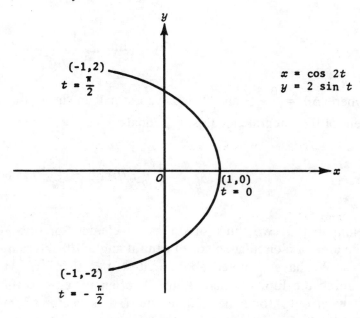

FIGURE N9–2

*E. Area Bounded by Polar Curves

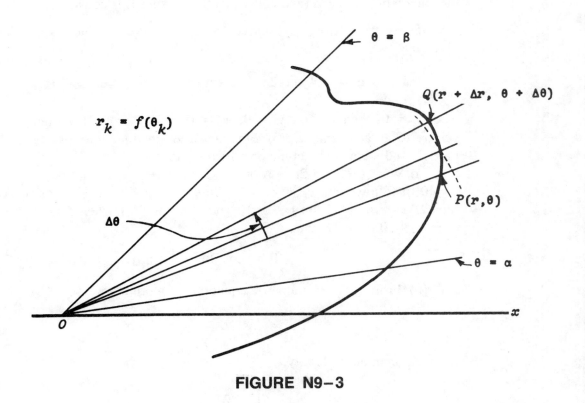

FIGURE N9–3

To find the area A bounded by the polar curve $r = f(\theta)$ and the rays $\theta = \alpha$, $\theta = \beta$ (see Figure N9–3) we divide the region AOB into n sectors as shown. Then the area of the region is given by

$$A = \lim_{n \to \infty} \sum_{k=1}^{n} \frac{1}{2} f^2(\theta_k) \, \Delta\theta, \tag{1}$$

where $\Delta\theta = \dfrac{\beta - \alpha}{n}$ and θ_k is a value of θ in the kth sector. By the fundamental theorem of the integral calculus, (1) equals

$$\int_{\alpha}^{\beta} \frac{1}{2} f^2(\theta) \, d\theta = \int_{\alpha}^{\beta} \frac{1}{2} r^2 \, d\theta. \tag{2}$$

Note that, if we think of dA as an element of area, then $dA = \frac{1}{2} r^2 \, d\theta$, and this is the area of a circular sector of central angle $d\theta$ and radius r.

We have assumed above that $f(\theta) \geqq 0$ on $[\alpha, \beta]$. We must be careful in determining the limits α and β in (2); often it helps to think of the required area as that "swept out" (or generated) as the radius vector (from the pole) rotates from $\theta = \alpha$ to $\theta = \beta$. It is also useful to exploit symmetry of the curve wherever possible.

The relations between rectangular and polar coordinates, some common polar equations, and graphs of polar curves are given in the Appendix, starting on page 486.

Examples

16. Find the area enclosed by the cardioid $r = 2(1 + \cos \theta)$.

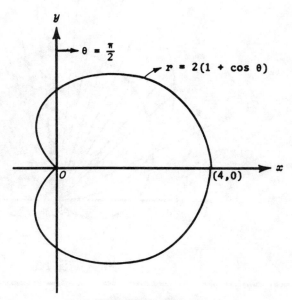

FIGURE N9-4

See Figure N9-4. We use the symmetry of the curve with respect to the polar axis and write

$$A = 2 \frac{1}{2} \int_0^\pi r^2 \, d\theta = 4 \int_0^\pi (1 + \cos \theta)^2 \, d\theta$$

$$= 4 \int_0^\pi (1 + 2 \cos \theta + \cos^2 \theta) \, d\theta$$

$$= 4 \int_0^\pi \left(1 + 2 \cos \theta + \frac{1}{2} + \frac{\cos 2\theta}{2} \right) \, d\theta$$

$$= 4 \left[\theta + 2 \sin \theta + \frac{\theta}{2} + \frac{\sin 2\theta}{4} \right] \Big|_0^\pi = 6\pi.$$

17. Find the area inside both the circle $r = 3 \sin \theta$ and the cardioid $r = 1 + \sin \theta$.

See Figure N9-5, where one half of the required area is shaded. Since $3 \sin \theta = 1 + \sin \theta$ when $\theta = \frac{\pi}{6}$ or $\frac{5\pi}{6}$, we see that the desired area is twice the sum of two parts: the area of the circle swept out by θ as it varies from 0 to $\frac{\pi}{6}$ plus the area of the cardioid swept out by a radius vector as θ varies from $\frac{\pi}{6}$ to $\frac{\pi}{2}$. Consequently

$$A = 2 \left[\int_0^{\pi/6} \frac{9}{2} \sin^2 \theta \, d\theta + \int_{\pi/6}^{\pi/2} \frac{1}{2}(1 + \sin \theta)^2 \, d\theta \right] = \frac{5\pi}{4}.$$

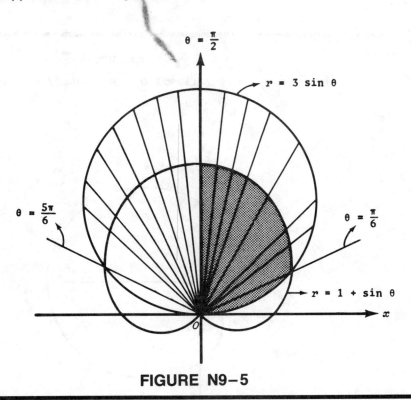

FIGURE N9–5

*F. Improper Integrals

There are two classes of improper integrals: (1) those in which at least one of the limits of integration is infinite; and (2) those of the type $\int_a^b f(x)\,dx$, where $f(x)$ has a point of discontinuity (becoming infinite) at $x = c$, $a \leqq c \leqq b$.

Illustrations of improper integrals of class (1) are:

$$\int_0^\infty \frac{dx}{\sqrt[3]{x+1}}; \qquad \int_1^\infty \frac{dx}{x}; \qquad \int_{-\infty}^\infty \frac{dx}{a^2+x^2}; \qquad \int_{-\infty}^0 e^{-x}\,dx;$$

$$\int_{-\infty}^{-1} \frac{dx}{x^n} \quad (n \text{ a real number}); \qquad \int_{-\infty}^\infty \frac{dx}{e^x + e^{-x}};$$

$$\int_0^\infty \frac{dx}{(4+x)^2}; \qquad \int_{-\infty}^\infty e^{-x^2}\,dx; \qquad \int_1^\infty \frac{e^{-x^2}}{x^2}\,dx.$$

The following improper integrals of type $\int_a^b f(x)\,dx$ are of class (2):

$$\int_0^1 \frac{dx}{x}; \qquad \int_1^2 \frac{dx}{(x-1)^n} \quad (n \text{ a real number}); \qquad \int_{-1}^1 \frac{dx}{1-x^2};$$

$$\int_0^2 \frac{x}{\sqrt{4-x^2}}\,dx; \qquad \int_\pi^{2\pi} \frac{dx}{1+\sin x}; \qquad \int_{-1}^2 \frac{dx}{x(x-1)^2};$$

$$\int_0^{2\pi} \frac{\sin x\,dx}{\cos x + 1}; \qquad \int_a^b \frac{dx}{(x-c)^n} \quad (n \text{ real}; a \leqq c \leqq b);$$

$$\int_0^1 \frac{dx}{\sqrt{x+x^4}}; \qquad \int_{-2}^2 \sqrt{\frac{2+x}{2-x}}\,dx.$$

Sometimes an improper integral belongs to both classes. Consider, for example:

$$\int_0^\infty \frac{dx}{x}; \qquad \int_0^\infty \frac{dx}{\sqrt{x + x^4}}; \qquad \int_{-\infty}^1 \frac{dx}{\sqrt{1 - x}}.$$

Each of the integrands in this set fails to exist at some point on the interval of integration.

Note, however, that each integral of the following set is *proper*:

$$\int_{-1}^3 \frac{dx}{\sqrt{x + 2}}; \qquad \int_{-2}^2 \frac{dx}{x^2 + 4}; \qquad \int_0^{\pi/6} \frac{dx}{\cos x};$$

$$\int_0^e \ln (x + 1)\ dx; \qquad \int_{-3}^3 \frac{dx}{e^x + 1}.$$

In each of the above the integrand is defined at each number on the interval of integration.

Improper integrals of class (1) are handled as follows:

$$\int_a^\infty f(x)\ dx = \lim_{b \to \infty} \int_a^b f(x)\ dx,$$

where f is continuous on $[a, b]$. If the limit on the right exists, the improper integral on the left is said to *converge* to this limit; if the limit on the right fails to exist, we say that the improper integral *diverges* (or is *meaningless*).

The evaluation of improper integrals of class (1) is illustrated in Examples 18–24.

Examples

18. $\int_1^\infty \frac{dx}{x^2} = \lim\limits_{b \to \infty} \int_1^b x^{-2}\ dx = \lim\limits_{b \to \infty} -\frac{1}{x} \Big|_1^b = \lim\limits_{b \to \infty} -\left(\frac{1}{b} - 1\right) = 1.$ The given integral thus converges to 1. In Figure N9–6 we interpret $\int_1^\infty \frac{dx}{x^2}$ as the area above the x-axis, under the curve of $y = \frac{1}{x^2}$, and bounded at the left by the vertical line $x = 1$.

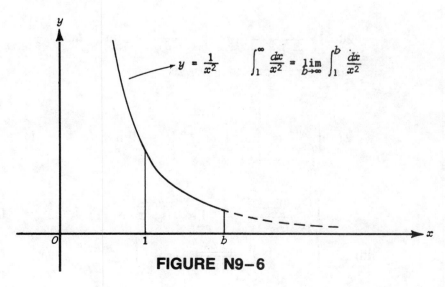

FIGURE N9–6

19. $\displaystyle\int_1^\infty \frac{dx}{\sqrt{x}} = \lim_{b\to\infty} \int_1^b x^{-1/2}\,dx = \lim_{b\to\infty} 2\sqrt{x}\,\Big|_1^b = \lim_{b\to\infty} 2(\sqrt{b} - 1) =$

$+\infty.$ So $\displaystyle\int_1^\infty \frac{dx}{\sqrt{x}}$ diverges. It can be proved that $\displaystyle\int_1^\infty \frac{dx}{x^p}$ converges if $p > 1$ but

diverges if $p \leqq 1$. Figure N9–7 gives a geometric interpretation in terms of

area of $\displaystyle\int_1^\infty \frac{dx}{x^p}$ for $p = \frac{1}{2}$, 1, 2. Only the first-quadrant area under $y = \frac{1}{x^2}$

bounded at the left by $x = 1$ exists. Note that

$$\int_1^\infty \frac{dx}{x} = \lim_{b\to\infty} \ln x\,\Big|_1^b = +\infty.$$

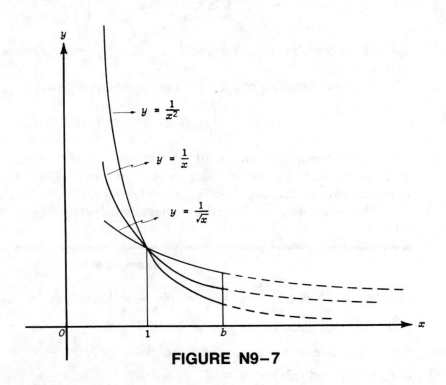

FIGURE N9–7

20. $\displaystyle\int_0^\infty \frac{dx}{x^2 + 9} = \lim_{b\to\infty} \int_0^b \frac{dx}{x^2 + 9} = \lim_{b\to\infty} \frac{1}{3} \tan^{-1} \frac{x}{3}\,\Big|_0^b = \lim_{b\to\infty} \frac{1}{3} \tan^{-1} \frac{b}{3} =$

$\dfrac{1}{3} \cdot \dfrac{\pi}{2} = \dfrac{\pi}{6}.$

21. $\displaystyle\int_0^\infty \frac{dy}{e^y} = \lim_{b\to\infty} \int_0^b e^{-y}\,dy = \lim_{b\to\infty} -(e^{-b} - 1) = 1.$

22. $\displaystyle\int_{-\infty}^0 \frac{dz}{(z - 1)^2} = \lim_{b\to-\infty} \int_b^0 (z - 1)^{-2}\,dz = \lim_{b\to-\infty} -\frac{1}{z - 1}\,\Big|_b^0 =$

$\lim_{b\to-\infty} -\left(-1 - \frac{1}{b - 1}\right) = 1.$

23. $\displaystyle\int_{-\infty}^{0} e^{-x}\, dx = \lim_{b\to-\infty} -e^{-x}\Big|_{b}^{0} = \lim_{b\to-\infty} -(1 - e^{-b}) = +\infty.$

Thus this improper integral diverges.

24. $\displaystyle\int_{0}^{\infty} \cos x\, dx = \lim_{b\to\infty} \sin x\Big|_{0}^{b} = \lim_{b\to\infty} \sin b.$ Since this limit does not

exist ($\sin b$ takes on values between -1 and 1 as $b \to \infty$), it follows that the given integral diverges. Note, however, that it does not become infinite; rather, it diverges by oscillation.

Improper integrals of class (2) are handled as follows.

To investigate $\displaystyle\int_{a}^{b} f(x)\, dx$ where f becomes infinite at $x = a$, we define $\displaystyle\int_{a}^{b} f(x)\, dx$

to be $\displaystyle\lim_{h\to 0^{+}} \int_{a+h}^{b} f(x)\, dx.$ The given integral then converges or diverges according as

the limit on the right does or does not exist. If f has its discontinuity at b, we de-

fine $\displaystyle\int_{a}^{b} f(x)\, dx$ to be $\displaystyle\lim_{h\to 0^{+}} \int_{a}^{b-h} f(x)\, dx$; again, the given integral converges or di-

verges as the limit does or does not exist. When, finally, the integrand has a dis-continuity at an interior point c on the interval of integration ($a < c < b$), we let

$$\int_{a}^{b} f(x)\, dx = \lim_{h\to 0^{+}} \int_{a}^{c-h} f(x)\, dx + \lim_{h\to 0^{+}} \int_{c+h}^{b} f(x)\, dx.$$

Now the improper integral converges only if both of the limits exist. If *either* limit does not exist, the improper integral diverges.

The evaluation of improper integrals of class (2) is illustrated in Examples 25–32.

Examples

25. $\displaystyle\int_{0}^{1} \frac{dx}{\sqrt[3]{x}} = \lim_{h\to 0^{+}} \int_{h}^{1} x^{-1/3}\, dx = \lim_{h\to 0^{+}} \frac{3}{2} x^{2/3}\Big|_{h}^{1} = \lim_{h\to 0^{+}} \frac{3}{2}(1 - h^{2/3}) = \frac{3}{2}.$ In

Figure N9–8 we interpret this integral as the first-quadrant area under $y = \dfrac{1}{\sqrt[3]{x}}$ and to the left of $x = 1$.

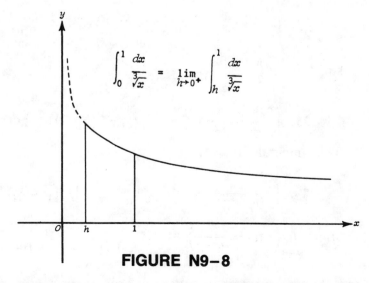

FIGURE N9–8

26. $\displaystyle\int_0^1 \frac{dx}{x^3} = \lim_{h\to 0^+} \int_h^1 x^{-3}\, dx = \lim_{h\to 0^+} -\frac{1}{2x^2}\Big|_h^1 = \lim_{h\to 0^+} -\frac{1}{2}\left(1 - \frac{1}{h^2}\right) = \infty.$

So this integral diverges.

It can be shown that $\displaystyle\int_0^a \frac{dx}{x^p}\ (a > 0)$ converges if $p < 1$ but diverges if $p \geqq 1$. Figure N9–9 shows an interpretation of $\displaystyle\int_0^1 \frac{dx}{x^p}$ in terms of areas where $p = \frac{1}{3}$, 1, and 3. Only the first-quadrant area under $y = \frac{1}{\sqrt[3]{x}}$ to the left of $x = 1$ exists. Note that

$$\int_0^1 \frac{dx}{x} = \lim_{h\to 0^+} \ln x \Big|_h^1 = \lim_{h\to 0^+} \ln 1 - \ln h = +\infty.$$

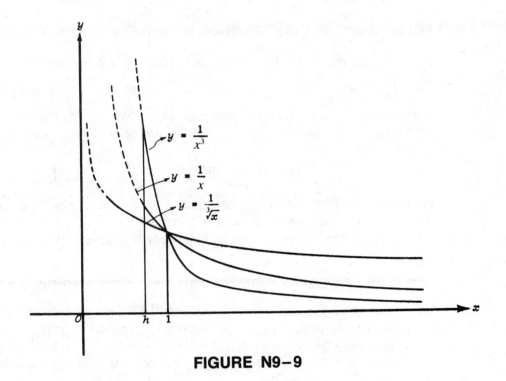

FIGURE N9–9

27. $\displaystyle\int_0^2 \frac{dy}{\sqrt{4 - y^2}} = \lim_{h\to 0^+} \int_0^{2-h} \frac{dy}{\sqrt{4 - y^2}} = \lim_{h\to 0^+} \sin^{-1} \frac{y}{2}\Big|_0^{2-h} = \sin^{-1} 1 -$

$\sin^{-1} 0 = \dfrac{\pi}{2}.$

28. $\displaystyle\int_2^3 \frac{dt}{(3 - t)^2} = \lim_{h\to 0^+} - \int_2^{3-h} (3 - t)^{-2}(-dt) = \lim_{h\to 0^+} \frac{1}{3 - t}\Big|_2^{3-h} = +\infty.$

This integral diverges.

29. $\displaystyle\int_0^2 \frac{dx}{(x - 1)^{2/3}} = \lim_{h\to 0^+} \int_0^{1-h} (x-1)^{-2/3}\, dx + \lim_{h\to 0^+} \int_{1+h}^2 (x - 1)^{-2/3}$

$dx = \lim_{h\to 0^+} 3(x - 1)^{1/3}\Big|_0^{1-h} + \lim_{h\to 0^+} 3(x - 1)^{1/3}\Big|_{1+h}^2 = 3(0 + 1) + 3(1 - 0) = 6.$

30. $\displaystyle\int_{-2}^{2} \frac{dx}{x^2} = \lim_{h \to 0^+} \int_{-2}^{0-h} x^{-2}\,dx + \lim_{h \to 0^+} \int_{0+h}^{2} x^{-2}\,dx = \lim_{h \to 0^+} -\frac{1}{x}\Big|_{-2}^{-h} +$

$\displaystyle\lim_{h \to 0^+} -\frac{1}{x}\Big|_{h}^{2}$. Neither of these two limits exists; the given integral diverges.

This example demonstrates how careful one must be to notice a discontinuity at an interior point. If it were overlooked, one might proceed as follows:

$$\int_{-2}^{2} \frac{dx}{x^2} = -\frac{1}{x}\Big|_{-2}^{2} = -\left(\frac{1}{2} + \frac{1}{2}\right) = -1.$$

Since this integrand is positive except at zero the result obtained is clearly meaningless. Figure N9–10 shows the impossiblity of this answer.

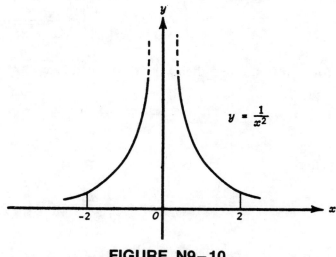

FIGURE N9–10

31. Determine whether or not $\displaystyle\int_{1}^{\infty} e^{-x^2}\,dx$ converges.

Although there is no elementary function whose derivative is e^{-x^2}, we can still show that the given improper integral converges. Note, first, that if $x \geqq 1$ then $x^2 \geqq x$, so that $-x^2 \leqq -x$ and $e^{-x^2} \leqq e^{-x}$. Figure N9–11 en-

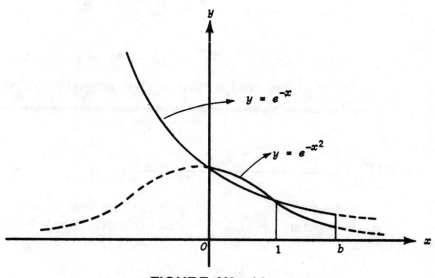

FIGURE N9–11

ables us to compare the areas under the curves of $y = e^{-x^2}$ and $y = e^{-x}$ from $x = 1$ to $x = b$. Thus, if $b > 1$, then

$$\int_1^b e^{-x^2}\, dx \leq \int_1^b e^{-x}\, dx \leq \int_1^\infty e^{-x}\, dx = \frac{1}{e}.$$

Note also that $\int_1^b e^{-x^2}\, dx$ increases with b. Since the latter is true and since $\int_1^b e^{-x^2}\, dx$ is bounded, it follows that $\lim\limits_{b \to \infty} \int_1^b e^{-x^2}\, dx$ exists $\left(\text{and is} \leq \frac{1}{e}\right)$. Therefore $\int_1^\infty e^{-x^2}\, dx$ converges.

32. Show that $\int_0^\infty \dfrac{dx}{\sqrt{x + x^4}}$ converges.

$$\int_0^\infty \frac{dx}{\sqrt{x + x^4}} = \int_0^1 \frac{dx}{\sqrt{x + x^4}} + \int_1^\infty \frac{dx}{\sqrt{x + x^4}}.$$

Since if $0 < x \leq 1$ then $x + x^4 > x$ and $\sqrt{x + x^4} > \sqrt{x}$, it follows that

$$\frac{1}{\sqrt{x + x^4}} < \frac{1}{\sqrt{x}} \qquad (0 < x \leq 1).$$

Thus

$$\int_0^1 \frac{dx}{\sqrt{x + x^4}} \leq \int_0^1 \frac{dx}{\sqrt{x}} = 2.$$

Further, if $x \geq 1$ then $x + x^4 \geq x^4$ and $\sqrt{x + x^4} \geq \sqrt{x^4} = x^2$, so that

$$\frac{1}{\sqrt{x + x^4}} \leq \frac{1}{x^2} \qquad (x \geq 1)$$

and

$$\int_1^\infty \frac{dx}{\sqrt{x + x^4}} \leq \int_1^\infty \frac{dx}{x^2} = -\frac{1}{x}\Big|_1^\infty = 1.$$

Thus, since $\int_0^\infty \dfrac{dx}{\sqrt{x + x^4}} \leq (2 + 1)$, it is bounded. And since $\int_0^b \dfrac{dx}{\sqrt{x + x^4}}$ increases with b, it follows that the given integral converges.

*G. Work

If the function F is continuous on the interval $a \leq x \leq b$, then the work W done by the force F on an object as it moves along the x-axis from a to b is given by

$$W = \int_a^b F(x)\, dx. \tag{1}$$

In the special case that $F(x)$ is constant and equal to F, we see that

$$W = F \cdot (b - a), \tag{2}$$

that is, force times distance.

When the force is variable we use (2) to approximate an element of work, ΔW, done over a short distance Δx, by $\Delta W = F(x_k) \, \Delta x$, where $F(x_k)$ is the force acting at some point in the kth subinterval. We use the Fundamental Theorem to get

$$W = \lim_{n \to \infty} \sum_{k=0}^{n} F(x_k) \, \Delta x = \int_a^b F(x) \, dx.$$

If the force is given in pounds and the distance in feet, then the work is given in foot-pounds (ft-lb). Problems typical of those involving computation of work are given in the following examples.

Examples

33. Find the work done by a force F, in pounds, which moves a particle along the x-axis from $x = 4$ ft to $x = 9$ ft, if $F(x) = \dfrac{1}{2\sqrt{x}}$.

$$W = \int_4^9 \frac{dx}{2\sqrt{x}} = \sqrt{x} \Big|_4^9 = 3 - 2 = 1 \text{ ft-lb.}$$

34. If the natural length of a spring is 1 ft and it takes a force of 5 lb to extend it 3 in., find the work done in stretching the spring (a) from its natural length to 16 in.; (b) from 14 in. to 18 in.

By Hooke's law, the force, $F(x)$, required to keep a spring extended (compressed) x units beyond (short of) its natural length is given by

$$F(x) = kx. \tag{3}$$

Since here $F(3)$ is 5, the "spring constant" k in (3) is determined by $5 = k \cdot 3$. Consequently,

$$k = \frac{5}{3} \qquad \text{and} \qquad F(x) = \frac{5}{3} x.$$

(a) When the spring is 1 ft long, then $x = 0$; when it is 16 in. long, then $x = 4$. So

$$W = \int_0^4 \frac{5}{3} x \, dx = \frac{5}{6} x^2 \Big|_0^4 = \frac{40}{3} \text{ in.-lb.}$$

(b) When the spring is 14 in. long, then $x = 2$; when it is 18 in. long, then $x = 6$. So, here,

$$W = \int_2^6 \frac{5}{3} x \, dx = \frac{5}{6} (36 - 4) = \frac{80}{3} \text{ in.-lb.}$$

35. A conical cistern is full of water. If it has a radius of 6 ft and a height of 15 ft, find the work done (a) in pumping all the water out of the cistern; (b) in pumping the water to a height of 5 ft above the top of the cistern.

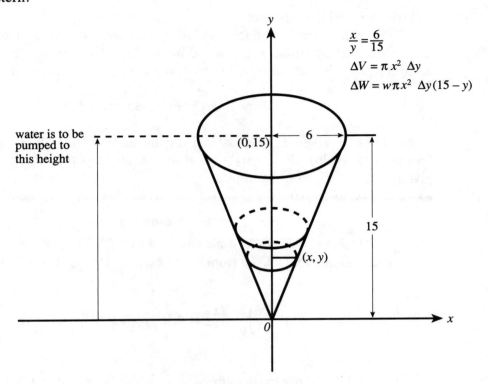

$$\frac{x}{y} = \frac{6}{15}$$

$$\Delta V = \pi x^2 \, \Delta y$$

$$\Delta W = w\pi x^2 \, \Delta y (15 - y)$$

water is to be pumped to this height

$(0, 15)$ 6

15

(x, y)

FIGURE N9–12a

(a) See Figure N9–12a, where we have set up the coordinate axes so that the origin is at the bottom of the tank. We partition the interval along the y-axis from 0 to 15 and find the work, ΔW, done in lifting a typical slice (disk) of water. ΔW is the force (the weight of the slice) times the distance through which the force acts; the weight of the slice is its volume times w (the weight per cubic unit). We have, then,

$$\Delta W = \Delta V \cdot w(15 - y) = \pi x^2 \, \Delta y \cdot w(15 - y);$$

and

$$W = \pi w \int_0^{15} x^2(15 - y) \, dy. \tag{4}$$

To express x in terms of y in (4), we use similar triangles and note that $\frac{x}{y} = \frac{6}{15} = \frac{2}{5}$; so $x^2 = \frac{4}{25} y^2$, and (4) becomes

$$W = \frac{4}{25} \pi w \int_0^{15} y^2(15 - y) \, dy$$

$$= \frac{4}{25} \pi w \int_0^{15} (15y^2 - y^3) \, dy = 675\pi w \text{ ft-lb.}$$

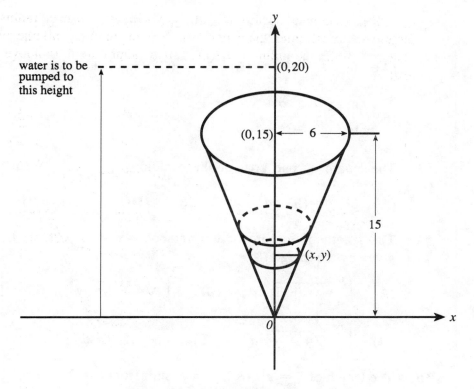

FIGURE N9−12b

(b) See Figure N9−12b. This part differs from (a) only in the distance through which a slice moves. Here a slice is lifted $(20 - y)$ ft and

$$W = \frac{4}{25}\,\pi w \int_0^{15} y^2(20 - y)\,dy = 1575\pi w \text{ ft-lb.}$$

Note that the limits are the same as in part (a).

Set 9: Multiple-Choice Questions on Further Applications of Integration

*1. The length of the arc of the curve $y^2 = x^3$ cut off by the line $x = 4$ is

(A) $\frac{4}{3}(10\sqrt{10} - 1)$ (B) $\frac{8}{27}(10^{3/2} - 1)$ (C) $\frac{16}{27}(10^{3/2} - 1)$

(D) $\frac{16}{27}10\sqrt{10}$ (E) none of these

*Questions preceded by an asterisk are likely to appear only on the Calculus BC Examination.

*2. The length of the arc of $y = \ln \cos x$ from $x = \frac{\pi}{4}$ to $x = \frac{\pi}{3}$ equals

(A) $\ln \dfrac{\sqrt{3} + 2}{\sqrt{2} + 1}$ (B) 2 (C) $\ln (1 + \sqrt{3} - \sqrt{2})$ (D) $\sqrt{3} - 2$

(E) $\dfrac{\ln (\sqrt{3} + 2)}{\ln (\sqrt{2} + 1)}$

*3. The length of one arch of the cycloid $\begin{array}{l} x = t - \sin t \\ y = 1 - \cos t \end{array}$ equals

(A) 3π (B) 4 (C) 16 (D) 8 (E) 2π

*4. The length of the arc of the parabola $4x = y^2$ cut off by the line $x = 2$ is given by the integral

(A) $\displaystyle\int_{-1}^{1} \sqrt{x^2 + 1}\ dx$ (B) $\dfrac{1}{2}\displaystyle\int_{0}^{2} \sqrt{4 + y^2}\ dy$ (C) $\displaystyle\int_{-1}^{1} \sqrt{1 + x}\ dx$

(D) $\displaystyle\int_{0}^{2\sqrt{2}} \sqrt{4 + y^2}\ dy$ (E) none of these

*5. The length of $x = e^t \cos t$, $y = e^t \sin t$ from $t = 2$ to $t = 3$ is equal to

(A) $\sqrt{2}e^2 \sqrt{e^2 - 1}$ (B) $\sqrt{2}(e^3 - e^2)$ (C) $2(e^3 - e^2)$
(D) $e^3(\cos 3 + \sin 3) - e^2(\cos 2 + \sin 2)$ (E) none of these

6. The average (mean) value of $\cos x$ over the interval $\frac{\pi}{3} \leq x \leq \frac{\pi}{2}$ is

(A) $\dfrac{3}{\pi}$ (B) $\dfrac{1}{2}$ (C) $\dfrac{3(2 - \sqrt{3})}{\pi}$ (D) $\dfrac{3}{2\pi}$ (E) $\dfrac{2}{3\pi}$

7. The average (mean) value of $\csc^2 x$ over the interval from $x = \frac{\pi}{6}$ to $x = \frac{\pi}{4}$ is

(A) $\dfrac{3\sqrt{3}}{\pi}$ (B) $\dfrac{\sqrt{3}}{\pi}$ (C) $\dfrac{12}{\pi}(\sqrt{3} - 1)$ (D) $3\sqrt{3}$

(E) $3(\sqrt{3} - 1)$

8. Let the motion of a freely falling body be given by $s(t) = 16t^2$ (where s is the distance in feet that the body has fallen in t seconds). The average velocity with respect to s of the body from $t = 0$ to $t = 1$ is equal, in feet per second, to

(A) $\dfrac{32}{3}$ (B) 16 (C) $\dfrac{64}{3}$ (D) 32 (E) none of these

9. A body moves along a straight line so that its velocity v at time t is given by $v = 4t^3 + 3t^2 + 5$. The distance it covers from $t = 0$ to $t = 2$ equals

(A) 34 (B) 55 (C) 24 (D) 44 (E) none of these

10. A particle moves along a line with velocity $v = 3t^2 - 6t$. The total distance traveled from $t = 0$ to $t = 3$ equals

(A) 9 (B) 4 (C) 2 (D) 16 (E) none of these

11. The acceleration of a particle moving on a straight line is given by $a = \cos t$, and when $t = 0$ the article is at rest. The distance it covers from $t = 0$ to $t = 2$ equals

(A) $\sin 2$ (B) $1 - \cos 2$ (C) $\cos 2$ (D) $\sin 2 - 1$
(E) $-\cos 2$

*12. The area enclosed by the four-leaved rose $r = \cos 2\theta$ equals

(A) $\dfrac{\pi}{4}$ (B) $\dfrac{\pi}{2}$ (C) π (D) 2π (E) $\dfrac{\pi}{2} + \dfrac{1}{2}$

*13. The area bounded by the small loop of the limaçon $r = 1 - 2 \sin \theta$ is given by the definite integral

(A) $\displaystyle\int_{\pi/3}^{5\pi/3} \left[\dfrac{1}{2}(1 - 2 \sin \theta)\right]^2 d\theta$

(B) $\displaystyle\int_{7\pi/6}^{3\pi/2} (1 - 2 \sin \theta)^2\, d\theta$

(C) $\displaystyle\int_{\pi/6}^{\pi/2} (1 - 2 \sin \theta)^2\, d\theta$

(D) $\displaystyle\int_{0}^{\pi/6} \left[\dfrac{1}{2}(1 - 2 \sin \theta)\right]^2 d\theta + \int_{5\pi/6}^{\pi} \left[\dfrac{1}{2}(1 - 2 \sin \theta)\right]^2 d\theta$

(E) $\displaystyle\int_{0}^{\pi/3} (1 - 2 \sin \theta)^2\, d\theta$

*14. Which one of the following is an improper integral?

(A) $\displaystyle\int_{0}^{2} \dfrac{dx}{\sqrt{x + 1}}$ (B) $\displaystyle\int_{-1}^{1} \dfrac{dx}{1 + x^2}$ (C) $\displaystyle\int_{0}^{2} \dfrac{x\, dx}{1 - x^2}$ (D) $\displaystyle\int_{0}^{\pi/3} \dfrac{\sin x\, dx}{\cos^2 x}$

(E)   none of these

*15. $\displaystyle\int_{0}^{\infty} e^{-x}\, dx =$

(A) 1 (B) $\dfrac{1}{e}$ (C) -1 (D) $-\dfrac{1}{e}$ (E)   none of these

*16. $\displaystyle\int_{0}^{e} \dfrac{du}{u} =$

(A) 1 (B) $\dfrac{1}{e}$ (C) $-\dfrac{1}{e^2}$ (D) -1 (E)   none of these

*17. $\displaystyle\int_{1}^{2} \dfrac{dt}{\sqrt[3]{t - 1}} =$

(A) $\dfrac{2}{3}$ (B) $\dfrac{3}{2}$ (C) 3 (D) 1 (E)   none of these

*18. $\displaystyle\int_2^4 \frac{dx}{(x-3)^{2/3}} =$

    (A) 6   **(B)** $\dfrac{6}{5}$ **(C)** $\dfrac{2}{3}$ **(D)** 0   **(E)** none of these

*19. $\displaystyle\int_2^4 \frac{dx}{(x-3)^2} =$

    (A) 2   **(B)** -2 **(C)** 0   **(D)** $\dfrac{2}{3}$ **(E)** none of these

*20. $\displaystyle\int_0^{\pi/2} \frac{\sin x}{\sqrt{1 - \cos x}}\, dx =$

    (A) -2 **(B)** $\dfrac{2}{3}$ **(C)** 2   **(D)** $\dfrac{1}{2}$ **(E)** none of these

*21. Which one of the following improper integrals diverges?

    (A) $\displaystyle\int_1^\infty \frac{dx}{x^2}$ **(B)** $\displaystyle\int_0^\infty \frac{dx}{e^x}$ **(C)** $\displaystyle\int_{-1}^1 \frac{dx}{\sqrt[3]{x}}$

    (D) $\displaystyle\int_{-1}^1 \frac{dx}{x^2}$ **(E)** none of these

*22. Which one of the following improper integrals diverges?

    (A) $\displaystyle\int_0^\infty \frac{dx}{1+x^2}$ **(B)** $\displaystyle\int_0^1 \frac{dx}{x^{1/3}}$ **(C)** $\displaystyle\int_0^\infty \frac{dx}{x^3+1}$ **(D)** $\displaystyle\int_0^\infty \frac{dx}{e^x+2}$

    (E) $\displaystyle\int_1^\infty \frac{dx}{x^{1/3}}$

In Questions 29–33, choose the alternative which gives the area, if it exists, of the region described.

*23. In the first quadrant under the curve of $y = e^{-x}$.

    (A) 1   **(B)** e **(C)** $\dfrac{1}{e}$ **(D)** 2   **(E)** none of these

*24. In the first quadrant under the curve of $y = xe^{-x^2}$.

    (A) 2   **(B)** $\dfrac{2}{e}$ **(C)** $\dfrac{1}{2}$ **(D)** $\dfrac{1}{2e}$ **(E)** none of these

*25. In the first quadrant above $y = 1$ and bounded by the y-axis and the curve $xy = 1$.

    (A) 1   **(B)** 2   **(C)** $\dfrac{1}{2}$ **(D)** 4   **(E)** none of these

*26. Between the curve $y = \dfrac{4}{1+x^2}$ and its asymptote.

    (A) 2π **(B)** 4π **(C)** 8π **(D)** π **(E)** none of these

*27. Between the curve $y = \dfrac{4}{\sqrt{1-x^2}}$ and its asymptotes.

(A) $\dfrac{\pi}{2}$ (B) π (C) 2π (D) 4π (E) none of these

In Questions 28–32, choose the alternative whch gives the volume, if it exists, of the solid generated.

*28. $y = \dfrac{1}{x}$, at the left by $x = 1$, and below by $y = 0$, about the x-axis.

(A) $\dfrac{\pi}{2}$ (B) π (C) 2π (D) 4π (E) none of these

*29. The region in Question 28, about the y-axis.

(A) π (B) $\dfrac{\pi}{2}$ (C) 2π (D) 4π (E) none of these

*30. The first-quadrant region under $y = e^{-x}$ about the x-axis.

(A) π (B) 2π (C) 4π (D) $\dfrac{\pi}{2}$ (E) none of these

*31. The first-quadrant region under $y = e^{-x}$ about the y-axis.

(A) $\dfrac{\pi}{2}$ (B) π (C) 2π (D) 4π (E) none of these

*32. The first-quadrant region under $y = e^{-x^2}$ about the y-axis.

(A) π (B) $\dfrac{\pi}{2}$ (C) 2π (D) 4π (E) none of these

*33. If the force F, in pounds, acting on a particle on the x-axis is given by $F = \dfrac{1}{x^2}$, then the work done, in foot-pounds, in moving it from $x = 1$ to $x = 3$, in feet, is equal to

(A) 2 (B) $\dfrac{2}{3}$ (C) $\dfrac{26}{27}$ (D) 1 (F) $\dfrac{3}{2}$

*34. A force of 40 lb is required to stretch a spring from its natural length, 10 ft, to 11 ft. The work done in stretching the string from a length of 12 ft to a length of 14 ft is equal, in foot-pounds, to

(A) 360   (B) 80   (C) 320   (D) 160   (E) 240

*35. A cylindrical reservoir of diameter 4 ft and height 6 ft is half full of water weighing w lb/ft^3. The work done, in foot-pounds, in emptying the water over the top is equal to

(A) $216\pi w$ (B) $18w$ (C) $72\pi w$ (D) $54\pi w$ (E) $108\ \pi w$

*36. A force of 20 lb compresses a spring from its natural length of 30 in. to 26 in. The work done, in inch-pounds, in compressing the spring from 26 in. to 22 in. is equal to

(A) 20   (B) 40   (C) 80   (D) 120   (E) 240

*37. A hemispherical tank of radius 4 ft is filled with a liquid weighing w lb/ft^3. The work done in pumping all the liquid just to the top of the tank, expressed in foot-pounds, is

(A) 128 πw (B) $\dfrac{256}{3} \pi w$ (C) $\dfrac{320}{3} \pi w$ (D) 96 πw

(E) 64 πw

Answers for Set 9: Further Applications of Integration

1. C	7. C	13. C	19. E	25. E	31. C
2. A	8. C	14. C	20. C	26. B	32. A
3. D	9. A	15. A	21. D	27. D	33. B
4. D	10. E	16. E	22. E	28. B	34. E
5. B	11. B	17. B	23. A	29. E	35. D
6. C	12. B	18. A	24. C	30. D	36. D
					37. E

1. C.   Note that the curve is symmetric to the x-axis. The arc length equals

$$2 \int_0^4 \sqrt{1 + \frac{9}{4} x} \, dx.$$

2. A.   Integrate $\displaystyle\int_{\pi/4}^{\pi/3} \sqrt{1 + \tan^2 x} \, dx$. Replace the integrand by sec x, and use

formula (13) on page 125 to get $\ln \left| \sec x + \tan x \right| \Big|_{\pi/4}^{\pi/3}$.

3. D.   From (3) on page 244, we obtain the length:

$$\int_0^{2\pi} \sqrt{(1 - \cos t)^2 + (\sin t)^2} \, dt = 2 \int_0^{2\pi} \sqrt{\frac{1 - \cos t}{2}} \, dt = 2 \int_0^{2\pi} \sin \frac{t}{2} \, dt$$

by identity (22) on page 483.

4. D.   Note that the curve is symmetric to the x-axis. Use (2) on page 243.

5. B.   Use (3) on page 244 to get the integral:

$$\int_2^3 \sqrt{(-e^t \sin t + e^t \cos t)^2 + (e^t \cos t + e^t \sin t)^2} \, dt.$$

The integrand simplifies to $\sqrt{2} \, e^t$.

6. C.   Equation (1) on page 245 yields

$$(y_{av})_x = \frac{1}{\dfrac{\pi}{2} - \dfrac{\pi}{3}} \int_{\pi/3}^{\pi/2} \cos x\, dx.$$

7. C.   The mean value is equal to $\dfrac{1}{\dfrac{\pi}{4} - \dfrac{\pi}{6}} \displaystyle\int_{\pi/6}^{\pi/4} \csc^2 x\, dx.$

8. C.   If we let $\bar{v}$ be the average velocity with respect to s, then

$$\bar{v} = \frac{1}{s(1) - s(0)} \int_{s(0)}^{s(1)} v\, ds = \frac{1}{16} \int_0^{16} 32t\, ds = 2 \int_0^{16} \frac{\sqrt{s}}{4}\, ds = \frac{64}{3}.$$

9. A.   Since $v > 0$ for $0 \leq t \leq 2$, the distance is equal to $\displaystyle\int_0^2 (4t^3 + 3t^2 + 5)\, dt.$

10. E.   The answer is 8. Since the particle reverses direction when $t = 2$, and $v > 0$ for $t > 2$ but $v < 0$ for $t < 2$, therefore

$$s = -\int_0^2 (3t^2 - 6t)\, dt + \int_2^3 (3t^2 - 6t)\, dt.$$

11. B.   Since $v = \sin t$ is positive on $0 < t \leq 2$,

$$s = \int_0^2 \sin t\, dt = 1 - \cos 2.$$

12. B. $A = 8 \displaystyle\int_0^{\pi/4} \frac{1}{2} \cos^2 2\theta\, d\theta = \frac{\pi}{2}.$ Use (2) on page 252.

13. C.   The small loop is generated as θ varies from $\dfrac{\pi}{6}$ to $\dfrac{5\pi}{5}.$ Use (2) on page 252.

14. C.   The integrand is discontinuous at $x = 1$, which is on the interval of integration.

15. A.   The integral equals $\displaystyle\lim_{b \to \infty} \left. -\frac{1}{e^x} \right|_0^b = -(0 - 1).$

16. E. $\displaystyle\int_0^e \frac{du}{u} = \lim_{h \to 0^+} \int_h^e \frac{du}{u} = \lim_{h \to 0^+} \left. \ln |u| \right|_h^e = \lim_{h \to 0^+} (\ln e - \ln h).$ So the integral diverges to infinity.

17. B.   Redefine as $\displaystyle\lim_{h \to 0^+} \int_{1+h}^2 (t - 1)^{-1/3}\, dt.$

18. A.   Rewrite as $\displaystyle\lim_{h \to 0^+} \int_2^{3-h} (x - 3)^{-2/3}\, dx + \lim_{h \to 0^+} \int_{3+h}^4 (x - 3)^{-2/3}\, dx.$ Each integral converges to 3.

19. E. $\displaystyle\int_2^4 \frac{dx}{(x - 3)^2} = \int_2^3 \frac{dx}{(x - 3)^2} + \int_3^4 \frac{dx}{(x - 3)^2}.$ Neither of the latter integrals converges; therefore the original integral diverges.

20. C.   Evaluate $\lim\limits_{h \to 0^+} 2\sqrt{1 - \cos x}\Big|_0^{(\pi/2)-h}$

21. D.   The integral in (D) is the sum of two integrals from −1 to 0 and from 0 to 1. Both diverge (see Example 30, page 259). Note that (A), (B), and (C) all converge.

22. E.   Note, first, that

$$\int_0^\infty \frac{dx}{1 + x^2} = \lim_{b \to \infty} \tan^{-1} x \Big|_0^b = \frac{\pi}{2}.$$

Since $Q(b) = \int_0^b \frac{dx}{x^3 + 1}$ increases with b and, further,

$$Q(b) \leq \int_0^b \frac{dx}{x^2 + 1},$$

$Q(b)$ converges as $b \to \infty$. Similarly, since $\int_0^\infty \frac{dx}{e^x} = 1$ and since, further, $S(b) = \int_0^b \frac{dx}{e^x + 2}$ increases with b and

$$\int_0^b \frac{dx}{e^x + 2} \leq \int_0^b \frac{dx}{e^x},$$

$S(b)$ also converges as $b \to \infty$.

The convergence of (B) is shown in Example 25 on page 257.

23. A.

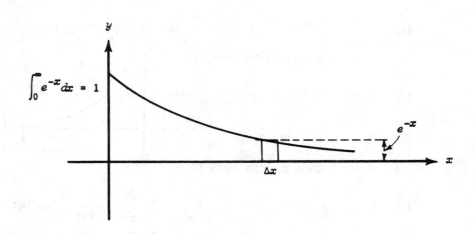

24. C.

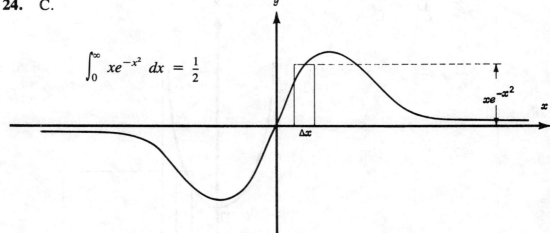

$$\int_0^\infty xe^{-x^2}\, dx = \frac{1}{2}$$

25. E.

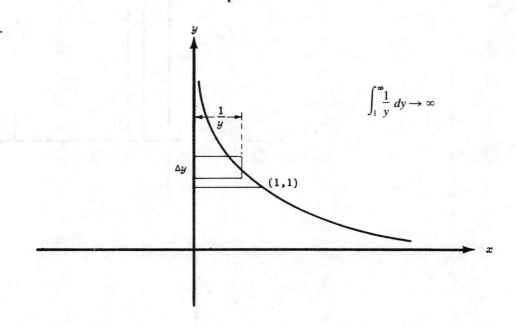

$$\int_1^\infty \frac{1}{y}\, dy \to \infty$$

26. B.

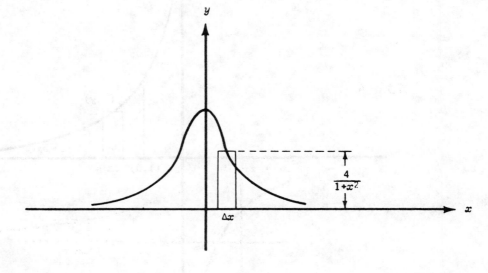

$$\int_{-\infty}^\infty \frac{4}{1 + x^2}\, dx = \lim_{b \to \infty} 4 \tan^{-1} x \Big|_{-b}^{b} = 4\pi.$$

27. D.

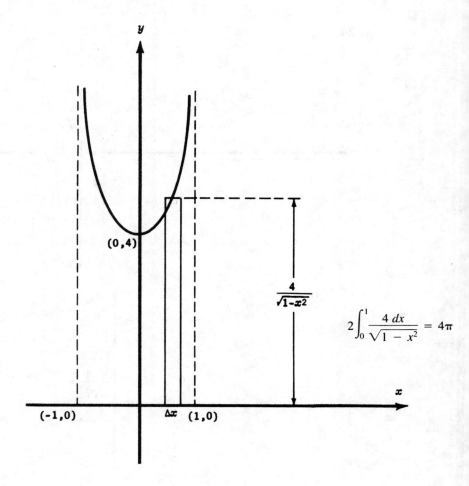

$$2\int_0^1 \frac{4\, dx}{\sqrt{1 - x^2}} = 4\pi$$

28. B.

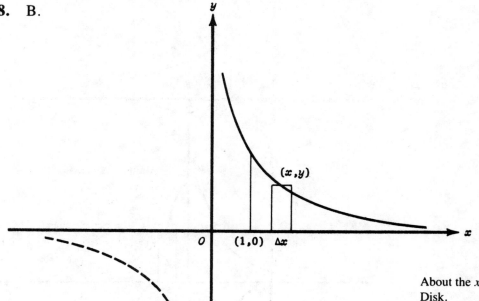

About the x-axis.
Disk.

$$\Delta V = \pi y^2\, \Delta x = \frac{\pi}{x^2}\, \Delta x.$$

$$\int_1^\infty \frac{\pi}{x^2}\, dx = \pi$$

29. E.

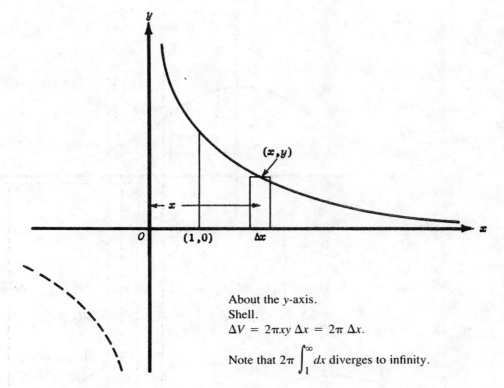

About the y-axis.
Shell.
$\Delta V = 2\pi xy\ \Delta x = 2\pi\ \Delta x.$

Note that $2\pi \displaystyle\int_{1}^{\infty} dx$ diverges to infinity.

30. D.

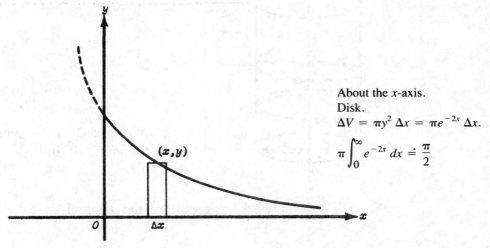

About the x-axis.
Disk.
$\Delta V = \pi y^2\ \Delta x = \pi e^{-2x}\ \Delta x.$

$\pi \displaystyle\int_{0}^{\infty} e^{-2x}\ dx \doteq \dfrac{\pi}{2}$

31. C.

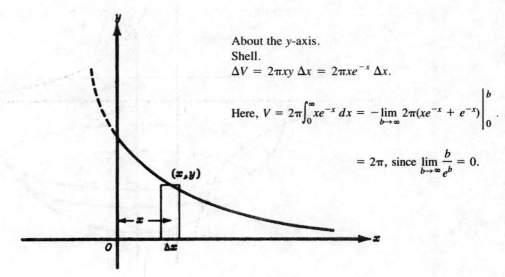

About the y-axis.
Shell.
$\Delta V = 2\pi xy\ \Delta x = 2\pi xe^{-x}\ \Delta x.$

Here, $V = 2\pi \displaystyle\int_{0}^{\infty} xe^{-x}\ dx = -\lim_{b\to\infty} 2\pi (xe^{-x} + e^{-x})\Big|_{0}^{b}$.

$= 2\pi$, since $\displaystyle\lim_{b\to\infty}\dfrac{b}{e^{b}} = 0.$

32.   A.

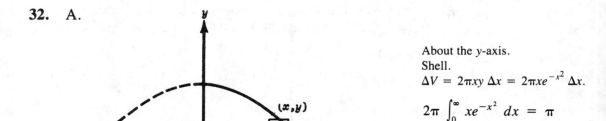

About the y-axis.
Shell.
$\Delta V = 2\pi xy\ \Delta x = 2\pi xe^{-x^2}\ \Delta x.$

$2\pi \displaystyle\int_0^\infty xe^{-x^2}\ dx = \pi$

33.   B. $W = \displaystyle\int_1^3 \frac{1}{x^2}\ dx = -\left.\frac{1}{x}\right|_1^3 = \frac{2}{3}$ ft-lb.

34.   E.   Using $F(x) = kx$, we have that $F(1) = k \cdot 1 = 40$; so $k = 40$ and

$F(x) = 40x.\ W = \displaystyle\int_2^4 40x\ dx = 240$ ft-lb.

35.   D.

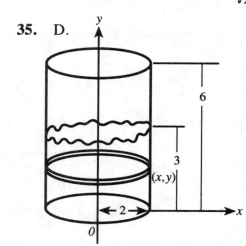

The volume of a slice of water is $\Delta V = \pi x^2\ \Delta y$, where $x = 2$. A slice at height y is lifted $(6 - y)$ ft.

$$\Delta W = w \cdot \pi \cdot 4\Delta y(6 - y);$$

$$W = 4\pi w \int_0^3 (6 - y)\ dy = 54\pi w \text{ ft-lb.}$$

We used 3 as the upper limit since the reservoir is only half full.

36.   D.   The work is given by the integral $\displaystyle\int_4^8 5x\ dx$, where 5 is the spring "constant."

37.   E.   In the figure, the generating circle has equation $x^2 + y^2 = 16$. Note

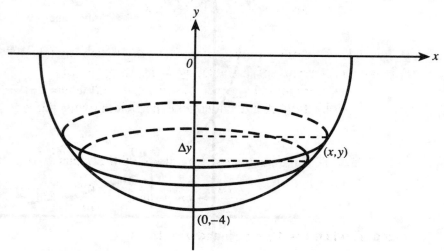

that over the interval of integration y is negative, and that a slice must be lifted a distance of $(-y)$ feet. Then for the work, W, we have

$$W = \pi w \int_{-4}^{0} (-y)x^2 \, dy = -\pi w \int_{-4}^{0} y(16 - y^2) \, dy = 64\pi w.$$

Practice Section II Problems on Further Applications of Integration

1. A car is traveling on a straight road at 45 mi/hr when the brakes are applied. The car then slows down at the constant rate of k ft/sec^2 and comes to a stop after traveling 60 ft. Find k.

2. (a) Show, by long division, if $t \geqq 0$, that

$$1 - t + t^2 \geqq \frac{1}{1 + t} \geqq 1 - t + t^2 - t^3.$$

(b) Use the result in (a) to show, if $0 < x \leqq 1$, that

$$x - \frac{x^2}{2} + \frac{x^3}{3} \geqq \ln (1 + x) \geqq x - \frac{x^2}{2} + \frac{x^3}{3} - \frac{x^4}{4}.$$

(c) Show that it follows from (b) that $0.0954 > \ln 1.1 > 0.0953$.

3. (a) Evaluate $\int_{1}^{m} \frac{1}{x^2} \, dx$.

(b) Show that $\sum_{k=2}^{m} \frac{1}{k^2} < 1 - \frac{1}{m}$ (m an integer $\geqq 2$).

*(c) Show that $\lim_{m \to \infty} \sum_{k=2}^{m} \frac{1}{k^2} < 1$.

4. A particle moves along a straight line so that its acceleration at any time t is given in terms of its velocity v by $a = -2v$.
(a) Find v in terms of t if $v = 20$ when $t = 0$.
(b) Find the distance the particle travels while v changes from $v = 20$ to $v = 5$.

*5. (a) Find parametric equations of the locus of $P(x, y)$ if, at any time $t \geqq 0$,

$$\frac{dx}{dt} = y \quad \text{and} \quad \frac{dy}{dt} = \sqrt{1 + 2y}$$

and if, when $t = 0$, $x = 0$ and $y = 0$.

(b) Find the x- and y-components of the acceleration in terms of y.

(c) Find the speed of P and the magnitude of its acceleration at the instant when $t = 1$.

*6. (a) Sketch the curve whose polar equation is $r = \cos \theta + \sin \theta$.

(b) Find the area enclosed by both the polar curves $r = 4 \sin \theta$ and $r = 4 \cos \theta$.

7. The velocity $\mathbf{v}$ of a particle moving on a curve is given, at time t, by $\mathbf{v} = t\mathbf{i} - (1 - t)\mathbf{j}$. When $t = 0$, the particle is at the point $(0, 1)$.

(a) Find the position vector $\mathbf{R}$ at time t.

(b) Find the acceleration vector $\mathbf{a}$ at time $t = 2$.

(c) For what positive t is the speed of the particle a minimum?

*8. (a) Integrate $\displaystyle\int \frac{dx}{e^x + 1}$

(b) Evaluate, if possible, $\displaystyle\lim_{b \to \infty} \int_0^b \frac{dx}{e^x + 1}$

*9. Find the area which the polar curves of $r = 2 - \cos \theta$ and $r = 3 \cos \theta$ have in common.

*10. Evaluate, if possible, $\displaystyle\int_0^\infty xe^{-x^2} \, dx$.

*11. A particle moves along a curve so that its position vector and velocity vector are perpendicular at all times. Prove that the particle moves along a circle whose center is at the origin.

*12. The acceleration of a particle is given by the vector $\mathbf{a} = e^{-t}\mathbf{i} + e^t\mathbf{j}$. When $t = 0$, the particle is at the point $(1, 2)$ and has velocity $\mathbf{v} = 2\mathbf{i}$. Find the position vector $\mathbf{R}$ at any time t.

*13. Evaluate, if possible, $\displaystyle\int_0^\infty \frac{dx}{x^2 + 4}$.

*14. A reservoir in the shape of a sphere of radius 4 ft has its center 8 ft above the surface of a pond, and water is pumped into the bottom of the reservoir from the pond. Determine the work done in filling the reservoir to a depth of 6 ft.

15. If a freely falling body starts from rest, then the distance s it falls is given in terms of time t by $s = \frac{1}{2} gt^2$, where g is constant.

(a) Prove that its velocity $v = \sqrt{2gs}$.

(b) Prove that the average velocity *with respect to* t from $t = 0$ to $t = t_1$ equals one-half v_1 (the velocity when $t = t_1$).

(c) Prove that the average velocity *with respect to* s from $t = 0$ to $t = t_1$ equals two-thirds v_1.

Solutions for Section II Problems: Further Applications of Integration

1. Since 45 mi/hr is equivalent to 66 ft/sec, we know that the acceleration $a = -k$ and that, when $t = 0$, the velocity $v = 66$ and the distance $s = 0$. Since $a = \frac{dv}{dt}$, we integrate to get $v = -kt + C$, and use the initial velocity to get $C = 66$, so that

$$v = -kt + 66. \tag{1}$$

Since $v = \frac{ds}{dt}$ we integrate again, getting $s = -\frac{1}{2}kt^2 + 66t + C'$, and use the initial distance to get

$$s = -\frac{1}{2}kt^2 + 66t. \tag{2}$$

To find k, note that v, in (1), equals zero when the car comes to a stop, that is, when $t = \frac{66}{k}$, and that $s = 60$ when $v = 0$. Thus, in (2),

$$60 = -\frac{1}{2}\left(\frac{66}{k}\right)^2 + 66\left(\frac{66}{k}\right) = \frac{1}{2}\frac{(66)^2}{k},$$

and $k = \frac{363}{10}$ or 36.3 ft/sec².

2. (a) Long division of 1 by $(1 + t)$ begins as follows:

$$
\begin{array}{r}
1 - t + t^2 - t^3 \\
1 + t \overline{\smash{\big)}\ 1} \\
\underline{1 + t} \\
- t \\
\underline{- t - t^2} \\
+ t^2 \\
\underline{+ t^2 + t^3} \\
- t^3 \\
\underline{- t^3 - t^4} \\
+ t^4
\end{array}
$$

Observe that

$$1 - t + t^2 - \frac{t^3}{1 + t} = \frac{1}{1 + t} = 1 - t + t^2 - t^3 + \frac{t^4}{1 + t}$$

and that, if $t \geq 0$, $-\frac{t^3}{1 + t} \leq 0$ while $\frac{t^4}{1 + t} \geq 0$. Thus, if $t \geq 0$, we have

$$1 - t + t^2 \geq \frac{1}{1 + t} \geq 1 - t + t^2 - t^3.$$

(b) If $0 \leq t \leq 1$, each member of the inequality in (a) is positive, and it follows, if $0 \leq x \leq 1$, that

$$\int_0^x (1 - t + t^2) \, dt \geq \int_0^x \frac{dt}{1 + t} \geq \int_0^x (1 - t + t^2 - t^3) \, dt.$$

Integrating gives

$$t - \frac{t^2}{2} + \frac{t^3}{3} \Big|_0^x \geq \ln (1 + t) \Big|_0^x \geq t - \frac{t^2}{2} + \frac{t^3}{3} - \frac{t^4}{4} \Big|_0^x$$

or

$$x - \frac{x^2}{2} + \frac{x^3}{3} \geq \ln (1 + x) \geq x - \frac{x^2}{2} + \frac{x^3}{3} - \frac{x^4}{4}.$$

(c) We let $x = 0.1$ in (b) approximate $\ln 1.1$, and see that

$$0.1 - \frac{0.01}{2} + \frac{0.001}{3} \geq \ln 1.1 \geq 0.1 - \frac{0.01}{2} + \frac{0.001}{3} - \frac{0.0001}{4}.$$

Since 0.0954 is slightly greater than the left member here, while 0.0953 is slightly less than the right, it follows that $0.0954 \geq \ln 1.1 \geq 0.0953$.

3. (a) $\int_1^m \frac{1}{x^2} \, dx = -\frac{1}{x} \Big|_1^m = -\left(\frac{1}{m} - 1 \right) = 1 - \frac{1}{m}.$

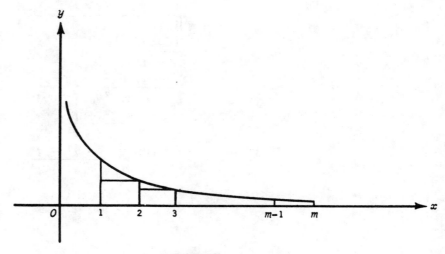

FIGURE S9–3

(b) If m is an integer $\geqq 2$, $\sum_{k=2}^{m} \frac{1}{k^2}$ can be interpreted as the sum of the areas of inscribed rectangles of width 1 and height $\frac{1}{k^2}$ (see Figure S9–3) for the region under the curve and above the x-axis from $x = 1$ to $x = m$. Since this sum is clearly less than the actual area under the curve of $y = \frac{1}{x^2}$ from $x = 1$ to $x = m$, it follows that

$$\sum_{k=2}^{m} \frac{1}{k^2} < \int_{1}^{m} \frac{1}{x^2}\, dx = 1 - \frac{1}{m}.$$

(c) The inequality in (b) is true for all $m \geqq 2$, and so

$$\lim_{m \to \infty} \sum_{k=2}^{m} \frac{1}{k^2} < \lim_{m \to \infty} \int_{1}^{m} \frac{1}{x^2}\, dx.$$

The latter defines an improper integral which converges to 1. Therefore

$$\lim_{m \to \infty} \sum_{k=2}^{m} \frac{1}{k^2} < 1.$$

4. (a) Since $a = \frac{dv}{dt} = -2v$, we have, separating variables, $\frac{dv}{v} = -2\, dt$. Integrating, we get

$$\ln v = -2t + C, \tag{1}$$

and since $v = 20$ when $t = 0$ we get $C = \ln 20$. Then (1) becomes $\ln \frac{v}{20} = -2t$ or, solving for v,

$$v = 20e^{-2t}. \tag{2}$$

(b) The second part of the problem can be done in any of the following three ways:

METHOD 1. Since, from (2), $v = \frac{ds}{dt} = 20e^{-2t}$, we can integrate to get

$$s = -10e^{-2t} + C_1. \tag{3}$$

If we let $s = 0$ when $t = 0$ (i.e., when $v = 20$), then $C_1 = 10$. So (3) becomes

$$s = -10e^{-2t} + 10. \tag{4}$$

When $v = 5$, we see from (2) that $e^{-2t} = \frac{1}{4}$. Then, in (4), $s = -10\left(\frac{1}{4}\right) + 10 = \frac{15}{2}$.

METHOD 2.   The above method is entirely equivalent to the following. Let s be the required distance traveled (as v decreases from 20 to 5); then

$$s = \int_{v=20}^{v=5} 20e^{-2t}\, dt = \int_{t=0}^{\ln 2} 20e^{-2t}\, dt, \tag{5}$$

where, when $v = 5$, we get, using (2), $\frac{1}{4} = e^{-2t}$ or $-\ln 4 = -2t$. Evaluating s in (5) gives

$$s = -10e^{-2t}\Big|_{0}^{\ln 2} = -10\left(\frac{1}{4} - 1\right) = \frac{15}{2}.$$

METHOD 3.   Since $a = \dfrac{dv}{dt} = \dfrac{ds}{dt}\dfrac{dv}{ds}$, we have $v\dfrac{dv}{ds} = -2v$, which yields $dv = -2ds$. We integrate to get $v = -2s + C_2$, and let $s = 0$ when $v = 20$. Then $C_2 = 20$ and $v = -2s + 20$. When $v = 5$, s, again, equals $\frac{15}{2}$.

Method 3 is especially neat because s is given directly in terms of v and the data of the second part of the problem involve only these two variables. However, either of the first two methods for solving part (b) follows naturally from the solution of part (a).

5. (a) Since $\dfrac{dx}{dt} = y$ and $\dfrac{dy}{dt} = \sqrt{1 + 2y}$, we can separate variables in the latter to get $\dfrac{dy}{\sqrt{1 + 2y}} = dt$. Integrating yields

$$\sqrt{1 + 2y} = t + C_1.$$

Since $y = 0$ when $t = 0$, this gives $1 = 0 + C_1$, and $C_1 = 1$. So $\sqrt{1 + 2y} = t + 1$, and we can square this to get $1 + 2y = (t + 1)^2$. Then

$$y = \frac{t^2}{2} + t. \tag{1}$$

Using (1) in $\dfrac{dx}{dt} = y$, it follows that $\dfrac{dx}{dt} = \dfrac{t^2}{2} + t$, and we can again separate and integrate to get $x = \dfrac{t^3}{6} + \dfrac{t^2}{2} + C_2$. Using $x = 0$ when $t = 0$ shows that $C_2 = 0$. Thus

$$x = \frac{t^3}{6} + \frac{t^2}{2} \quad \text{and} \quad y = \frac{t^2}{2} + t. \tag{2}$$

(b) For this we describe two alternatives.

Alternative 1. We can find the x- and y-components of the acceleration in terms of y by differentiating the given equations with respect to t. Thus

$$\frac{d^2x}{dt^2} = \frac{dy}{dt} \quad \text{and} \quad \frac{d^2y}{dt^2} = \frac{2}{2\sqrt{1 + 2y}}\frac{dy}{dt}.$$

Since it is given that $\dfrac{dy}{dt} = \sqrt{1 + 2y}$, it follows that

$$\frac{d^2x}{dt^2} = \sqrt{1 + 2y} \quad \text{and} \quad \frac{d^2y}{dt^2} = 1.$$

Alternative 2. Or we can differentiate twice in (2) to get the components of acceleration in terms of t:

$$\frac{dx}{dt} = \frac{t^2}{2} + t, \qquad \frac{dy}{dt} = t + 1,$$

$$\frac{d^2x}{dt^2} = t + 1; \qquad \frac{d^2y}{dt^2} = 1.$$

Note that these are equivalent to those obtained in Alternative 1, since, from (a), $t + 1 = \sqrt{1 + 2y}$.

      (c) Alternative 2 of (b) can be used to obtain the speed $|\mathbf{v}|$ and magnitude of acceleration when $t = 1$ immediately. Thus

$$|\mathbf{v}| = \sqrt{\left(\frac{dx}{dt}\right)^2 + \left(\frac{dy}{dt}\right)^2},$$

and when $t = 1$ this is

$$\sqrt{\left(\frac{3}{2}\right)^2 + 2^2} = \frac{5}{2}.$$

The magnitude of the acceleration equals

$$\sqrt{\left(\frac{d^2x}{dt^2}\right)^2 + \left(\frac{d^2y}{dt^2}\right)^2},$$

and when $t = 1$ this is

$$\sqrt{2^2 + 1^2} = \sqrt{5}.$$

      Or we may note, from (1), that when $t = 1$, $y = \frac{3}{2}$. Then we can use the given equations to get, when $t = 1$, $\frac{dx}{dt} = \frac{3}{2}$ and $\frac{dy}{dt} = \sqrt{4} = 2$, and proceed to use Alternative 1 of (b) to get

$$\frac{d^2x}{dt^2} = \sqrt{4} = 2 \qquad \text{and} \qquad \frac{d^2y}{dt^2} = 1.$$

Again, then, when $t = 1$ the speed of P is $\frac{5}{2}$ and its acceleration has magnitude $\sqrt{5}$.

6. (a) See Figure S9–6.

The curve is a circle.

Its center is at $\left(\frac{\sqrt{2}}{2}, \frac{\pi}{4}\right)$.

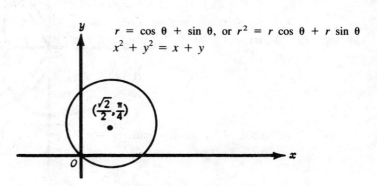

$r = \cos\theta + \sin\theta$, or $r^2 = r\cos\theta + r\sin\theta$
$x^2 + y^2 = x + y$

$\left(\frac{\sqrt{2}}{2}, \frac{\pi}{4}\right)$

FIGURE S9–6

(b) Both curves are circles with centers at, respectively, $(2, 0)$ and at $\left(2, \frac{\pi}{2}\right)$; they intersect at $\left(2\sqrt{2}, \frac{\pi}{4}\right)$. The common area is given by

$$2\int_0^{\pi/4} (4 \sin \theta)^2 \, d\theta \qquad \text{or} \qquad 2\int_{\pi/4}^{\pi/2} (4 \cos \theta)^2 \, d\theta.$$

The answer is $2(\pi - 2)$.

7. (a) $\mathbf{R} = \dfrac{t^2}{2}\mathbf{i} + \dfrac{t^2 - 2t + 2}{2}\mathbf{j}$.

(b) $\mathbf{a} = \mathbf{i} + \mathbf{j}$ for all t.

(c) $\mathbf{v} = t\mathbf{i} + (t - 1)\mathbf{j}$. $|\mathbf{v}| = \sqrt{t^2 + (t - 1)^2}$. $\dfrac{d|\mathbf{v}|}{dt} = \dfrac{2t - 1}{|\mathbf{v}|}$.

The speed is a minimum when $t = \dfrac{1}{2}$.

8. (a) Hint: $\displaystyle\int \frac{dx}{e^x + 1} = \int \frac{e^x + 1 - e^x}{e^x + 1} \, dx = x - \ln(e^x + 1) + C$.

(b) We see $\displaystyle\lim_{b \to \infty} [b - \ln(e^b + 1) + \ln 2]$. Since

$$\lim_{b \to \infty} [\ln e^b - \ln(e^b + 1)] = \lim_{b \to \infty} \ln \frac{e^b}{e^b + 1} = \ln \lim_{b \to \infty} \frac{e^b}{e^b + 1} = \ln 1 = 0,$$

the answer to this part is $\ln 2$.

9. See Figure S9–9. The required area A is twice the sum of the following areas: that of the limaçon from 0 to $\frac{\pi}{3}$, and that of the circle from $\frac{\pi}{3}$ to $\frac{\pi}{2}$. Thus

$$A = 2\left[\frac{1}{2}\int_0^{\pi/3} (2 - \cos \theta)^2 \, d\theta + \frac{1}{2}\int_{\pi/3}^{\pi/2} (3 \cos \theta)^2 \, d\theta\right]$$

$$= \frac{9\pi}{4} - 3\sqrt{3}.$$

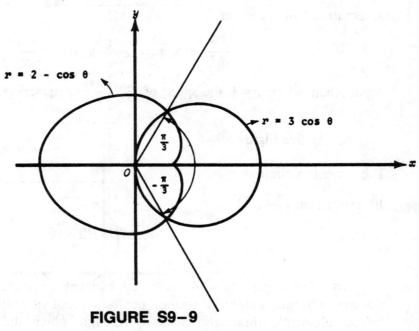

$r = 2 - \cos \theta$

$r = 3 \cos \theta$

FIGURE S9–9

10. $\int_0^\infty xe^{-x^2}\,dx = \lim_{b\to\infty} \left. -\frac{1}{2}e^{-x^2}\right|_0^b = \frac{1}{2}.$

11. Since $\mathbf{R} = x\mathbf{i} + y\mathbf{j}$, its slope is $\frac{y}{x}$; since $\mathbf{v} = \frac{dx}{dt}\mathbf{i} + \frac{dy}{dt}\mathbf{j}$, its slope is $\frac{dy}{dx}$.

If $\mathbf{R}$ is perpendicular to $\mathbf{v}$, then $\frac{y}{x}\cdot\frac{dy}{dx} = -1$. So

$$\frac{y^2}{2} = -\frac{x^2}{2} + C \quad \text{and} \quad x^2 + y^2 = k \quad (k > 0).$$

12. We have $\mathbf{v} = (3 - e^{-t})\mathbf{i} + (e^t - 1)\mathbf{j}$. So

$$\mathbf{R} = (3t + e^{-t})\mathbf{i} + (e^t - t + 1)\mathbf{j}.$$

(Hint: When $t = 0$, $\mathbf{v} = 2\mathbf{i} + 0\mathbf{j}$, $\mathbf{R} = \mathbf{i} + 2\mathbf{j}$.)

13. $\int_0^\infty \frac{dx}{x^2 + 4} = \lim_{b\to\infty}\int_0^b \frac{dx}{x^2 + 4} = \lim_{b\to\infty}\left.\frac{1}{2}\tan^{-1}\frac{x}{2}\right|_0^b = \lim_{b\to\infty}\frac{1}{2}\left(\tan^{-1}\frac{b}{2} - 0\right) = \frac{\pi}{4}.$

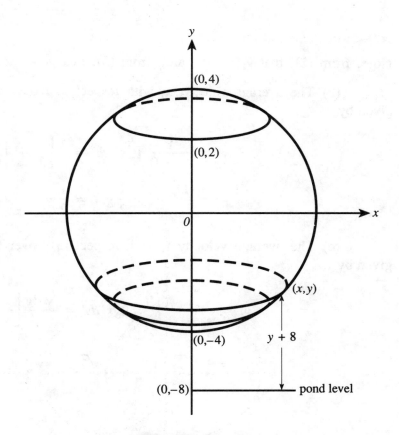

FIGURE S9–14

14. See Figure S9–14. ΔW, an element of work, equals the weight, $w\,\Delta V$ (where w is the weight in pounds per cubic foot of volume, and ΔV is the volume of an element), times the distance through which the element is to be moved.

Thus

$$\Delta W = w \cdot \pi x^2 \, \Delta y \cdot (y + 8);$$

$$W = \pi w \int_{-4}^{2} x^2(y + 8) \, dy,$$

where $x^2 = 16 - y^2$. The work W equals $540\pi w$ ft-lb.

15. (a) We know that

$$s = \frac{1}{2} g t^2 \qquad (1)$$

and thus that

$$v = \frac{ds}{dt} = gt. \qquad (2)$$

Solving (1) for t yields $\sqrt{\dfrac{2s}{g}} = t \ (t \geq 0)$, and (2) becomes

$$v = g \sqrt{\frac{2s}{g}} = \sqrt{2gs}. \qquad (3)$$

Note, from (2), that $v_1 = gt_1$, and, from (3), that $v_1 = \sqrt{2gs_1}$.

(b) The average velocity v_t with respect to t over the time interval $[0, t_1]$ is given by

$$v_t = \frac{1}{t_1 - 0} \int_0^{t_1} gt \, dt = \frac{1}{t_1} \left[\frac{1}{2} g t^2 \right]_0^{t_1}$$

$$= \frac{1}{t_1} \frac{g t_1^2}{2} = \frac{1}{2} g t_1 = \frac{1}{2} v_1.$$

(c) The average velocity v_s with respect to s over the time interval $[0, t_1]$ is given by

$$v_s = \frac{1}{s_1 - 0} \int_0^{s_1} \sqrt{2gs} \, ds = \frac{\sqrt{2g}}{s_1} \left[\frac{2}{3} s^{3/2} \right]_0^{s_1}$$

$$= \frac{2}{3} \sqrt{2g} \cdot s_1^{1/2} = \frac{2}{3} \sqrt{2gs_1} = \frac{2}{3} v_1.$$

Chapter 10

Sequences and Series

Review of Principles

A. Sequences of Real Numbers

A1. Definitions.

An *infinite sequence* is a function whose domain is the set of positive integers. The sequence $\{n, f(n) | n = 1, 2, 3, \ldots\}$ is often denoted simply by $\{f(n)\}$. The sequence defined, for example, by $f(n) = \frac{1}{n}$ is the set of numbers $1, \frac{1}{2}, \frac{1}{3}, \ldots,$ $\frac{1}{n}, \ldots$. The elements in this set are called the *terms* of the sequence, and the *nth* or *general* term of this sequence is $\frac{1}{n}$. Frequently the sequence is denoted by its *nth* term: $\{s_n\}$ is the sequence whose *nth* term is s_n. Thus

$$\left\{\frac{n}{n^2 + 1}\right\} = \frac{1}{2}, \frac{2}{5}, \frac{3}{10}, \ldots;$$

$$\left\{\frac{2^n}{n!}\right\} = \frac{2}{1}, \frac{2^2}{2!}, \frac{2^3}{3!}, \ldots.$$

[Recall that for every nonnegative integer the definition of $n!$ (read "n factorial") is given by $0! = 1$, $(n + 1)! = (n + 1)n!$. So, for example, $1! = 1$, $2! = 2 \cdot 1$, $7! = 7 \cdot 6 \cdot 5 \cdot 4 \cdot 3 \cdot 2 \cdot 1$.]

$$\{1 + (-1)^n\} = 0, 2, 0, 2, \ldots;$$

$$\left\{1 + \frac{(-1)^n}{n}\right\} = 0, \frac{3}{2}, \frac{2}{3}, \frac{5}{4}, \frac{4}{5}, \frac{7}{6}, \frac{6}{7}, \ldots.$$

*An asterisk denotes a topic covered only in Calculus BC.

A sequence $\{s_n\}$ is said to have the *limit L* if, given any positive ϵ, we can find an integer N such that all terms beyond the Nth in the sequence differ from L by less than ϵ. If $\{s_n\}$ has the limit L, we say it *converges to L*, and write

$$\lim_{n \to \infty} s_n = L.$$

If $\lim_{n \to \infty} s_n = L$, then for every $\epsilon > 0$ there exists an integer N such that $|s_n - L| < \epsilon$ for all $n > N$. The sequence is said to be *convergent* in this case. If a sequence fails to have a (finite) limit, it is said to be *divergent*.

A2. Theorems.

Several theorems follow from the definition of convergence:

THEOREM 2a.   The limit of a convergent sequence is unique.

THEOREM 2b. $\lim_{n \to \infty} (c \cdot s_n) = c \lim_{n \to \infty} s_n$ (c a number).

THEOREM 2c.   · If two sequences converge, so do their sum, product, and quotient (with division by zero to be avoided).

Examples

1. $\lim_{n \to \infty} \dfrac{1}{n} = 0.$

2. $\lim_{n \to \infty} 1 + \dfrac{(-1)^n}{n} = 1.$

3. $\lim_{n \to \infty} \dfrac{2n^2}{3n^3 - 1} = 0$ (by the Rational Function Theorem, page 23).

4. $\lim_{n \to \infty} \dfrac{3n^4 + 5}{4n^4 - 7n^2 + 9} = \dfrac{3}{4}.$

5. $\left\{\dfrac{n^2 - 1}{n}\right\}$ diverges (to infinity), since $\lim_{n \to \infty} \dfrac{n^2 - 1}{n} = \infty.$

6. $\{\sin n\}$ diverges since the sequence fails to have a limit, but note that it does not diverge to infinity.

7. $\{(-1)^{n+1}\} = 1, -1, 1, -1, \ldots$ diverges because it oscillates.

8. $\lim_{n \to \infty} \dfrac{\ln n}{n} = \lim_{n \to \infty} \dfrac{1/n}{1} = 0$ by L'Hôpital's rule, page 43.

9. $\lim_{n \to \infty} \dfrac{e^n}{n^2} = \lim_{n \to \infty} \dfrac{e^n}{2n} = \lim_{n \to \infty} \dfrac{e^n}{2} = \infty$ by repeated application of L'Hôpital's rule; the sequence diverges.

DEFINITIONS.   The sequence $\{s_n\}$ is said to be *increasing* if, for all n, $s_n \leqq s_{n+1}$, and to be *decreasing* if, for all n, $s_{n+1} \leqq s_n$; a *monotonic* sequence is either increasing or decreasing. If $|s_n| \leqq M$ for all n, then $\{s_n\}$ is said to be *bounded*. The following theorem is the basic one on sequences:

THEOREM 2d.   A monotonic sequence converges if and only if it is bounded. In particular, if $\{s_n\}$ is increasing and $s_n \leqq M$ for all n, then $\{s_n\}$ converges to a limit which is less than or equal to M.

THEOREM 2e. Every unbounded sequence diverges. (Of course, a bounded sequence may also diverge; for example $1, -1, 1, -1, \ldots$, or $\{\cos n\}$.)

Examples

10. Use Theorem 2d to prove that $\left\{1 - \frac{2}{n}\right\}$ converges.

If $s_n = 1 - \frac{2}{n}$, then $-1 \leqq s_n < 1$ for all n; also $s_{n+1} > s_n$, since

$$s_{n+1} - s_n = \left(1 - \frac{2}{n+1}\right) - \left(1 - \frac{2}{n}\right) = \frac{2}{n} - \frac{2}{n+1} = \frac{2}{n(n+1)},$$

which is positive for all n. Note that $\{s_n\}$ converges to 1.

11. For what r's does $\{r^n\}$ converge?

If $|r| > 1$, then $\{r^n\}$ is unbounded, so diverges by Theorem 2e.

If $|r| < 1$, then $\lim_{n \to \infty} r^n = 0$.

If $r = 1$, then $\{r^n\}$ converges to 1.

If $r = -1$, then $\{r^n\}$ diverges by oscillation.

Finally, then, $\{r^n\}$ converges if $-1 < r \leqq 1$.

B. Series of Constants

B1. Definitions.

If $\{u_n\}$ is a sequence of real numbers, then an *infinite series* is an expression of the form

$$\sum_{k=1}^{\infty} u_k = u_1 + u_2 + u_3 + \cdots + u_n + \cdots . \qquad (1)$$

The elements in the sum (1) are called *terms*; u_n is the *nth* or *general* term of series (1). Associated with series (1) is the sequence $\{s_n\}$, where

$$s_n = u_1 + u_2 + u_3 + \cdots + u_n; \qquad (2)$$

that is,

$$s_n = \sum_{k=1}^{n} u_k.$$

The sequence $\{s_n\}$ is called the *sequence of partial sums* of series (1). Note that

$$\{s_n\} = s_1, s_2, s_3, \ldots , s_n, \ldots \qquad (3)$$

$$= u_1, u_1 + u_2, u_1 + u_2 + u_3, \ldots , u_1 + u_2 + \cdots + u_n, \ldots .$$

If there is a finite number S such that

$$\lim_{n \to \infty} s_n = S,$$

then we say that series (1) is *convergent*, or *converges to S*, or *has the sum S*. We write, in this case,

$$\sum_{k=1}^{\infty} u_k = S, \quad \text{where} \quad \sum_{k=1}^{\infty} u_k = \lim_{n \to \infty} \sum_{k=1}^{n} u_k .$$

When there is no source of confusion, the infinite series (1) may be indicated simply by

$$\Sigma \mu_k \quad \text{or} \quad \Sigma \mu_n .$$

Note that, if the sequence $\{s_n\}$ of partial sums given by (3) converges, then so does series (1); if the sequence $\{s_n\}$ diverges, then so does the series.

Examples

12. Show that the geometric series

$$\frac{1}{2} + \frac{1}{4} + \cdots + \frac{1}{2^n} + \cdots \tag{4}$$

converges to 1.

The series can be rewritten as

$$\left(1 - \frac{1}{2}\right) + \left(\frac{1}{2} - \frac{1}{4}\right) + \cdots + \left(\frac{1}{2^{n-1}} - \frac{1}{2^n}\right) + \cdots; \tag{5}$$

hence

$$s_n = \left(1 - \frac{1}{2}\right) + \left(\frac{1}{2} - \frac{1}{4}\right) + \cdots + \left(\frac{1}{2^{n-1}} - \frac{1}{2^n}\right)$$

$$= 1 - \frac{1}{2^n}.$$

Since $\lim_{n \to \infty} s_n = 1$,

$$\sum_{k=1}^{\infty} \frac{1}{2^k} = 1.$$

The trick above is equivalent to the following:

$$s_n = \frac{1}{2} + \frac{1}{4} + \frac{1}{8} + \cdots + \frac{1}{2^n}; \tag{6}$$

$$\frac{1}{2} s_n = \frac{1}{4} + \frac{1}{8} + \cdots + \frac{1}{2^n} + \frac{1}{2^{n+1}}. \tag{7}$$

Subtracting (7) from (6) yields

$$\frac{1}{2} s_n = \frac{1}{2} - \frac{1}{2^{n+1}} \qquad \text{or} \qquad s_n = 1 - \frac{1}{2^n}.$$

So, as before, $\lim\limits_{n \to \infty} s_n = 1$.

13. Show that the *harmonic series*

$$1 + \frac{1}{2} + \frac{1}{3} + \frac{1}{4} + \cdots + \frac{1}{n} + \cdots \tag{8}$$

diverges.

The terms in (8) can be grouped as follows:

$$1 + \frac{1}{2} + \left(\frac{1}{3} + \frac{1}{4}\right) + \left(\frac{1}{5} + \frac{1}{6} + \frac{1}{7} + \frac{1}{8}\right) + \left(\frac{1}{9} + \frac{1}{10} + \cdots + \frac{1}{16}\right)$$

$$+ \left(\frac{1}{17} + \cdots + \frac{1}{32}\right) + \cdots. \tag{9}$$

The sum in (9) clearly exceeds

$$1 + \frac{1}{2} + 2\left(\frac{1}{4}\right) + 4\left(\frac{1}{8}\right) + 8\left(\frac{1}{16}\right) + 16\left(\frac{1}{32}\right) + \cdots,$$

which equals

$$1 + \frac{1}{2} + \frac{1}{2} + \frac{1}{2} + \frac{1}{2} + \frac{1}{2} + \cdots. \tag{10}$$

Since the sum in (10) can be made arbitrarily large, it follows that $\sum \frac{1}{n}$ diverges.

14. Show that the *geometric series*

$$\sum_{k=1}^{\infty} ar^{k-1} = a + ar + ar^2 + \cdots + ar^{n-1} + \cdots \qquad (a \neq 0) \tag{11}$$

converges if $|r| < 1$ but diverges if $|r| \geqq 1$.

Series (4) in Example 12 is a special case of (11) with $a = r = \frac{1}{2}$; the technique applied there is used here. For (11),

$$s_n = a + ar + ar^2 + \cdots + ar^{n-1};$$

$$rs_n = \qquad ar + ar^2 + \cdots + ar^{n-1} + ar^n;$$

$$s_n(1 - r) = a - ar^n = a(1 - r^n);$$

and

$$s_n = \begin{cases} \dfrac{a(1 - r^n)}{1 - r} & \text{if } r \neq 1, \\[2mm] a + a + \cdots + a = na & \text{if } r = 1. \end{cases} \tag{12}$$

We now investigate $\lim\limits_{n\to\infty} s_n$ to determine the behavior of series (11). We see from (12) that there are four cases, respectively for $|r| < 1$, $|r| > 1$, $r = 1$, and $r = -1$.

If $|r| < 1$, then, since $\lim\limits_{n\to\infty} r^n = 0$, $\lim\limits_{n\to\infty} s_n = \dfrac{a}{1-r}$; thus

$$\Sigma\, ar^{k-1} = \frac{a}{1-r}. \tag{13}$$

If $|r| > 1$, then $\lim\limits_{n\to\infty} s_n$ does not exist, because $\lim\limits_{n\to\infty} r^n = \infty$.

If $r = 1$, then the given series diverges.

If $r = -1$, then $\Sigma\, ar^{k-1} = a - a + a - a + \cdots$; thus $s_1 = a$, $s_2 = 0$, $s_3 = a$, $s_4 = 0$. In this case $\{s_n\}$ oscillates and thus diverges; and so does the series.

The geometric series therefore converges if $|r| < 1$ but diverges otherwise.

B2. Theorems about Convergence or Divergence of Series of Constants.

The following theorems are important.

THEOREM 2a. If Σu_k converges, then $\lim\limits_{n\to\infty} u_n = 0$.

This provides a convenient and useful test for divergence, since it is equivalent to the statement: If u_n does not approach zero, then the series Σu_n diverges. Note, however, particularly that the converse of Theorem 2a is *not* true. The condition that u_n approach zero is *necessary but not sufficient* for the convergence of the series. The harmonic series $\Sigma \dfrac{1}{n}$ is an excellent example of a series whose nth term goes to zero but which diverges (see Example 13 above). The series $\Sigma \dfrac{n}{n+1}$ diverges because $\lim\limits_{n\to\infty} u_n = 1$, not zero; the series $\Sigma \dfrac{n}{n^2+1}$ does not converge even though $\lim\limits_{n\to\infty} u_n = 0$.

THEOREM 2b. A finite number of terms may be added to or deleted from a series without affecting its convergence or divergence; thus

$$\sum_{k=1}^{\infty} u_k \quad \text{and} \quad \sum_{k=m}^{\infty} u_k$$

(where m is any positive integer) both converge or both diverge. (Of course, the sums may differ.)

THEOREM 2c. The terms of a series may be multiplied by a nonzero constant without affecting the convergence or divergence; thus

$$\sum_{k=1}^{\infty} a_k \quad \text{and} \quad \sum_{k=1}^{\infty} ca_k \quad (c \neq 0)$$

both converge or both diverge. (Again, the sums may differ.)

THEOREM 2d. If Σa_n and Σb_n both converge, so does $\Sigma(a_n + b_n)$.

THEOREM 2e. If the terms of a convergent series are regrouped, the new series converges.

B3. Tests for Convergence of Positive Series.

The series Σu_n is called a *positive* series if $u_n > 0$ for all n. Note for such a series that the associated sequence of partial sums $\{s_n\}$ is increasing. It then follows from Theorem 2d, page 286, that a positive series converges if and only if $\{s_n\}$ has an upper bound.

Examples

15. $\sum \frac{1}{k!} = 1 + \frac{1}{2!} + \frac{1}{3!} + \cdot \cdot \cdot + \frac{1}{n!} + \cdot \cdot \cdot$ converges, since

$$s_n = 1 + \frac{1}{2!} + \frac{1}{3!} + \frac{1}{4!} + \cdot \cdot \cdot + \frac{1}{n!}$$

$$< 1 + \frac{1}{2} + \frac{1}{2\cdot 2} + \frac{1}{2\cdot 2\cdot 2} + \cdot \cdot \cdot + \frac{1}{2^n}$$

$$< 2$$

(using Equation (13), page 290, with $a = 1$ and $r = \frac{1}{2}$). Thus $\{s_n\}$ has an upper bound and the given series converges.

16. $\sum \frac{1}{n}$ diverges since the sequence of sums $\{s_n\}$ is unbounded (see Example 13).

TESTS TO APPLY TO $\{u_n\}$ IF $u_n > 0$.

TEST 3a. If $\lim\limits_{n\to\infty} u_n \neq 0$, then Σu_n diverges. This is the contrapositive of Theorem 2a, page 290 A statement and its contrapositive are equivalent.

TEST 3b. THE INTEGRAL TEST. If $f(x)$ is a continuous, positive, decreasing function for which $f(n)$ is the nth term u_n of the series, then Σu_n converges if and only if the improper integral $\int_1^{\infty} f(x)\, dx$ converges.

Examples

17. Does $\sum \frac{n}{n^2 + 1}$ converge?
The associated improper integral is

$$\int_1^{\infty} \frac{x\, dx}{x^2 + 1},$$

which equals

$$\lim_{b\to\infty} \frac{1}{2} \ln (x^2 + 1) \Big|_1^b = \infty.$$

The improper integral and the infinite series both diverge.

18. Test the series $\sum \frac{n}{e^n}$ for convergence.

$$\int_1^\infty \frac{x}{e^x}\,dx = \lim_{b\to\infty} \int_1^b xe^{-x}\,dx = \lim_{b\to\infty} -e^{-x}(1+x)\Big|_1^b$$

$$= -\lim_{b\to\infty} \left(\frac{1+b}{e^b} - \frac{2}{e}\right) = \frac{2}{e}$$

by an application of L'Hôpital's rule. Thus $\sum \frac{n}{e^n}$ converges.

19. Show that the *p-series* $\sum \frac{1}{n^p}$ converges if $p > 1$ but diverges if $p \leq 1$.

Let $f(x) = \frac{1}{x^p}$ and investigate $\int_1^\infty \frac{dx}{x^p}$.

(a) If $p > 1$, then

$$\int_1^\infty x^{-p}\,dx = \lim_{b\to\infty} \frac{1}{1-p}\cdot x^{1-p}\Big|_1^b = \lim_{b\to\infty} \frac{1}{1-p}(b^{1-p} - 1)$$

$$= \frac{1}{p-1} \lim_{b\to\infty} \left(1 - \frac{1}{b^{p-1}}\right) = \frac{1}{p-1}.$$

Consequently, $\sum \frac{1}{n^p}$ converges if $p > 1$.

(b) If $p = 1$, the improper integral is

$$\int_1^\infty \frac{dx}{x} = \lim_{b\to\infty} \ln b = \infty,$$

so $\sum \frac{1}{n}$ diverges, a result previously well established.

(c) If $p < 1$, then

$$\int_1^\infty x^{-p}\,dx = \lim_{b\to\infty} \frac{1}{1-p}(b^{1-p} - 1) = \infty,$$

so $\sum \frac{1}{n^p}$ diverges.

TEST 3c. THE COMPARISON TEST. We compare the general term of Σu_n, the positive series we are investigating, with the general term of a series known to converge or diverge.

(1)   If Σa_n converges and $u_n \leq a_n$, then Σu_n converges.

(2)   If Σb_n diverges and $u_n \geq b_n$, then Σu_n diverges.

Any known series can be used for comparison; particularly useful are the *p*-series, which converges if $p > 1$ but diverges if $p \leq 1$ (see Example 19), and the geometric series, which converges if $|r| < 1$ but diverges if $|r| \geq 1$ (see Example 14, page 289).

Examples

20. $1 + \frac{1}{2!} + \frac{1}{3!} + \cdots + \frac{1}{n!} + \cdots$ converges, since $\frac{1}{n!} < \frac{1}{2^{n-1}}$, and the latter is the general term of the geometric series $\sum_1^\infty r^{n-1}$ with $r = \frac{1}{2}$.

21. $\frac{1}{\sqrt{2}} + \frac{1}{\sqrt{5}} + \frac{1}{\sqrt{8}} + \cdots + \frac{1}{\sqrt{3n-1}} + \cdots$ diverges, since

$$\frac{1}{\sqrt{3n-1}} > \frac{1}{\sqrt{3n}} = \frac{1}{\sqrt{3}\cdot n^{1/2}};$$

the latter is the general term of the divergent p-series $\sum \frac{c}{n^p}$, where $c = \frac{1}{\sqrt{3}}$ and $p = \frac{1}{2}$.

Remember in using the Comparison Test that we may either discard a finite number of terms or multiply each term by a nonzero constant without affecting the convergence of the series we are testing.

Example 22. $\sum \frac{1}{n^n} = 1 + \frac{1}{2^2} + \frac{1}{3^3} + \cdots + \frac{1}{n^n} + \cdots$ converges,

since $\frac{1}{n^n} < \frac{1}{2^{n-1}}$ and $\sum \frac{1}{2^{n-1}}$ converges.

A variant of the Comparison Test, referred to as the Limit Comparison Test, is more convenient when we are not certain exactly which series to use for ordinary comparison but think that the series in question behaves like a known one for large n.

Test 3d. The Limit Comparison Test. Let $a_n \geqq 0$, $b_n > 0$, and let $\lim_{n\to\infty} \frac{a_n}{b_n} = L$. If $0 < L < \infty$, then Σa_n and Σb_n both converge or both diverge. If $L = 0$ and Σb_n converges, then so does Σa_n.

Example 23. To test the series $\sum \frac{1}{\sqrt{n^2 + 10n}}$, we note that for (very) large n the nth term is essentially $\frac{1}{n}$. Using test 3d, we see that

$$\lim_{n\to\infty} \frac{a_n}{b_n} = \lim_{n\to\infty} \frac{\frac{1}{\sqrt{n^2+10n}}}{\frac{1}{n}} = \lim_{n\to\infty} \sqrt{\frac{n^2}{n^2+10n}} = 1$$

Since (the harmonic series) $\sum \frac{1}{n}$ diverges, so does the series we are checking.

Observe that, since $\dfrac{1}{\sqrt{n^2 + 10n}}$ is *less than* $\dfrac{1}{n}$, we cannot use $\Sigma \dfrac{1}{n}$ to test the given series by ordinary comparison. It is important to remember that we can conclude nothing about how the series Σu_n behaves from either of the following comparisons:

    (1)   The series Σa_n converges and $u_n \geqq a_n$.

    (2)   The series Σb_n diverges and $u_n \leqq b_n$.

TEST 3e. THE RATIO TEST. Let $\lim\limits_{n \to \infty} \dfrac{u_n + 1}{u_n} = L$, if it exists. Then Σu_n converges if $L < 1$ and diverges if $L > 1$. If $L = 1$, this test fails and we must apply one of the other tests.

Examples

24. For $\Sigma \dfrac{1}{n!}$,

$$\lim_{n \to \infty} \frac{u_{n+1}}{u_n} = \lim_{n \to \infty} \frac{\dfrac{1}{(n + 1)!}}{\dfrac{1}{n!}} = \lim_{n \to \infty} \frac{n!}{(n + 1)!} = \lim_{n \to \infty} \frac{1}{n + 1} = 0.$$

Therefore this series converges by the Ratio Test. (Compare Examples 15, page 291, and 20, page 293.

25. For $\Sigma \dfrac{n^n}{n!}$, we again use the Ratio Test:

$$\frac{u_{n+1}}{u_n} = \frac{(n + 1)^{n+1}}{(n + 1)!} \cdot \frac{n!}{n^n} = \frac{(n + 1)^n}{n^n}$$

and

$$\lim_{n \to \infty} \left(\frac{n + 1}{n}\right)^n = \lim_{n \to \infty} \left(1 + \frac{1}{n}\right)^n = e.$$

(See E2, page 25.) So the given series diverges since $e > 1$.

26. If the ratio test is applied to the *p*-series, $\Sigma \dfrac{1}{n^p}$, then

$$\frac{u_{n+1}}{u_n} = \frac{\dfrac{1}{(n + 1)^p}}{\dfrac{1}{n^p}} = \left(\frac{n}{n + 1}\right)^p \quad \text{and} \quad \lim_{n \to \infty} \left(\frac{n}{n + 1}\right)^p = 1 \quad \text{for all } p.$$

But if $p > 1$ then $\Sigma \dfrac{1}{n^p}$ converges, while if $p \leq 1$ then $\Sigma \dfrac{1}{n^p}$ diverges. This illustrates the failure of the ratio test to resolve the question of convergence when the limit of the ratio is 1.

B4. Alternating Series and Absolute Convergence.

Any test that can be applied to a positive series can be used for a series all of whose terms are negative. We consider here only one type of series with mixed signs, the so-called *alternating series*. This has the form:

$$\sum_{k=1}^{\infty} (-1)^{k+1} u_k = u_1 - u_2 + u_3 - u_4 + \cdots + (-1)^{k+1} u_k + \cdots, \qquad (1)$$

where each $u_k > 0$. The series

$$1 - \frac{1}{2} + \frac{1}{3} - \frac{1}{4} + \cdots + (-1)^{n+1} \cdot \frac{1}{n} + \cdots \qquad (2)$$

is the *alternating harmonic* series.

THEOREM 4a.   The alternating series (1) converges if $u_{n+1} < u_n$ for all n and if $\lim\limits_{n \to \infty} u_n = 0$.

Examples

27. The alternating harmonic series (2) converges, since $\frac{1}{n+1} < \frac{1}{n}$ for all n and since $\lim\limits_{n \to \infty} \frac{1}{n} = 0$.

28. The series $\frac{1}{2} - \frac{2}{3} + \frac{3}{4} - \cdots$ diverges since we see that $\lim\limits_{n \to \infty} u_n = \lim\limits_{n \to \infty} \frac{n}{n+1}$ is 1, not 0. (By Theorem 2a, page 290, if u_n does not approach 0, then Σu_n does not converge.

THEOREM 4b.   If $u_{n+1} < u_n$ and $\lim\limits_{n \to \infty} u_n = 0$ for the alternating series, then the numerical difference between the sum of the first n terms and the sum of the series is less than u_{n+1}; that is,

$$\left| \sum_{k=1}^{\infty} u_k - \sum_{k=1}^{n} u_k \right| < u_{n+1}.$$

Theorem 4b is exceedingly useful in computing the maximum possible error if a finite number of terms of a convergent series is used to approximate the sum of the series. If we stop with the nth term, the error is less than u_{n+1}, the first term omitted.

Example 29.   The sum $\sum_{k=1}^{\infty} \frac{(-1)^{k+1}}{k}$ differs from the sum

$$\left(1 - \frac{1}{2} + \frac{1}{3} - \frac{1}{4} + \frac{1}{5} - \frac{1}{6} \right)$$

by less than $\frac{1}{7}$.

A series with mixed signs is said to *converge absolutely* (or to be *absolutely convergent*) if the series obtained by taking the absolute values of its terms converges; that is, Σu_n converges absolutely if $\Sigma|u_n| = |u_1| + |u_2| + \cdots + |u_n| + \cdots$ converges. A series which converges but not absolutely is said to converge *conditionally* (or to be *conditionally convergent*). The alternating harmonic series (2) converges conditionally since it converges, but does not converge absolutely. (The harmonic series diverges.)

THEOREM 4c.   If a series converges absolutely, then it converges.

Examples

30. Determine whether $\sum \dfrac{\sin \frac{n\pi}{3}}{n^2}$ converges absolutely, converges conditionally, or diverges.

Note that, since $\left|\sin n\dfrac{\pi}{3}\right| \leqq 1$,

$$\left|\frac{\sin n \frac{\pi}{3}}{n^2}\right| \leqq \frac{1}{n^2} \quad \text{for all } n.$$

But $\dfrac{1}{n^2}$ is the general term of a convergent p-series, so by the Comparison Test the given series converges absolutely (therefore, by Theorem 4c above, it converges).

31. The series

$$1 + \frac{1}{2} - \frac{1}{4} - \frac{1}{8} + \frac{1}{16} + \frac{1}{32} - \cdots (\pm) \frac{1}{2^n} \pm \cdots, \qquad (3)$$

whose signs are $+ \ + \ - \ - \ + \ + \ - \ - \ \ldots$, converges absolutely, since $\sum \dfrac{1}{2^n}$ is a convergent geometric series. Consequently, (3) converges.

B5. Computations with Series of Constants.

The nth partial sum of a convergent series, given by $s_n = \displaystyle\sum_{k=1}^{n} u_k$, is an approximation to the sum of the whole series, $S = \displaystyle\sum_{k=1}^{\infty} u_k$. The error introduced by using s_n for S, referred to as the remainder *after n* terms, is denoted by R_n. Note that

$$R_n = u_{n+1} + u_{n+2} + u_{n+3} + \cdots.$$

We summarize below three useful theorems about the remainder for *convergent* series:

THEOREM 5a.   For the convergent geometric series $\sum_{n=1}^{\infty} ar^{n-1}$, with $|r| < 1$,

$$|R_n| = \left| \frac{ar^n}{1-r} \right|.$$

This follows immediately from the results of Example 14, page 289.

$$s_n = \frac{a(1-r^n)}{1-r} \quad \text{and} \quad S = \frac{a}{1-r}.$$

THEOREM 5b.   For the convergent alternating series

$$u_1 - u_2 + u_3 - u_4 + \cdots + (-1)^{n+1} u_n + \cdots,$$

where $u_n > 0$ for all n, $R_n < u_{n+1}$. Furthermore, $R_n > 0$ if n is even but $R_n < 0$ if n is odd. Thus the error is less than the magnitude of the first term dropped. Note that Theorem 5b is a restatement of Theorem 4b, page 295.

THEOREM 5c.   Suppose the Integral Test (page 291) shows that the positive series $\sum_{n=1}^{\infty} u_n$ converges. Then

$$R_n < \int_n^{\infty} f(x)\, dx \quad \left(\text{where } \int_1^{\infty} f(x)\, dx \text{ converges}\right).$$

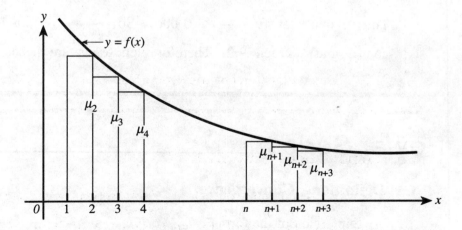

In the figure, $f(n) = u_n$, where $f(x)$ is a continuous, positive, decreasing function. Since the base of each rectangle is one unit, the rectangles, starting at the left, have areas $u_2, u_3, u_4, \ldots, u_{n+1}, u_{n+2}, u_{n+3}, \ldots$. We see that

$$R_n = u_{n+1} + u_{n+2} + u_{n+3} + \cdots < \int_n^{\infty} f(x)\, dx.$$

Examples

32. Estimate the error made in approximating $\sum_{n=1}^{\infty} \frac{n}{e^n}$ by the sum of its first 10 terms.

The series was shown to converge using the integral test in Example 18, page 292. Thus,

$$R_{10} < \int_{10}^{\infty} xe^{-x} \, dx = \lim_{b \to \infty} -e^{-x}(1 + x) \Big|_{10}^{b}$$

$$= -\lim_{b \to \infty} \left(\frac{1 + b}{e^b} - \frac{11}{e^{10}} \right) < 0.0005.$$

33. (a) Estimate the error introduced in approximating $\sum_{n=1}^{\infty} \frac{5}{3^n}$ by the sum of its first four terms. (b) How many terms are needed for the error to be less than 0.0001?

(a) The series is a geometric one with $a = \frac{5}{3}$ and $r = \frac{1}{3}$. Theorem 5a, page 297, yields

$$R_4 = \frac{ar^4}{1 - r} = \frac{\frac{5}{3} \cdot \left(\frac{1}{3}\right)^4}{\frac{2}{3}} = \frac{5}{2 \cdot 3^4} = 0.03 \text{ to two decimal places.}$$

(b) We seek n such that

$$R_n = \frac{ar^n}{1 - r} = \frac{\frac{5}{3} \cdot \left(\frac{1}{3}\right)^n}{\frac{2}{3}} < 0.0001.$$

Thus n must satisfy $\frac{5}{2 \cdot 3^n} < 0.0001$. So $\frac{5}{0.0002} < 3^n$, $n \ln 3 > \ln 25{,}000$, and n must exceed 9.2. Therefore ten terms are needed for the required accuracy.

C. Power Series

C1. Definitions; Convergence.

An expression of the form

$$\sum_{k=0}^{\infty} a_k x^k = a_0 + a_1 x + a_2 x^2 + \cdots + a_n x^n + \cdots, \tag{1}$$

where the a's are constants, is called a *power series in x*; and

$$\sum_{k=0}^{\infty} a_k (x - a)^k = a_0 + a_1(x - a) + a_2(x - a)^2 + \cdots + a_n(x - a)^n + \cdots \tag{2}$$

is called a *power series in (x − a)*.

If in (1) or (2) x is replaced by a specific real number, then the power series becomes a series of constants that either converges or diverges. Note that series (1) converges if $x = 0$ and series (2) converges if $x = a$.

The set of values of x for which a power series converges is called its *interval of convergence*. If the interval of convergence of the power series (1) contains values of x other than 0, then 0 will be the midpoint of this interval; similarly, a is the midpoint of the interval of convergence of series (2).

The interval of convergence of a power series may be found by applying the ratio test to the series of absolute values.

<hr>

Examples

34. Find the interval of convergence of

$$1 + x + x^2 + \cdots + x^n + \cdots. \tag{3}$$

We find

$$\lim_{n \to \infty} \left| \frac{u_{n+1}}{u_n} \right| = \lim_{n \to \infty} \left| \frac{x^{n+1}}{x^n} \right| = \lim_{n \to \infty} |x| = |x|.$$

The series clearly converges absolutely (and therefore converges) if $|x| < 1$; it diverges if $|x| > 1$. The endpoints must be tested separately since the ratio test fails when the limit equals 1. When $x = 1$, (3) becomes $1 + 1 + 1 + \cdots$ and diverges; when $x = -1$, (3) becomes $1 - 1 + 1 - 1 + \cdots$ and diverges. So the interval of convergence of (3) is $-1 < x < 1$.

35. For what x's does $\displaystyle\sum_{n=1}^{\infty} \frac{(-1)^{n-1} x^{n-1}}{n + 1}$ converge?

$$\lim_{n \to \infty} \left| \frac{u_{n+1}}{u_n} \right| = \lim_{n \to \infty} \left| \frac{x^n}{n + 2} \cdot \frac{n + 1}{x^{n-1}} \right| = \lim_{n \to \infty} |x| = |x|.$$

The series converges if $|x| < 1$ and diverges if $|x| > 1$. When $x = 1$ we have $\frac{1}{2} - \frac{1}{3} + \frac{1}{4} - \frac{1}{5} + \cdots$, an alternating convergent series; when $x = -1$ the series is $\frac{1}{2} + \frac{1}{3} + \frac{1}{4} + \cdots$, which diverges. So the series converges if $-1 < x \leqq 1$.

36. $\displaystyle\sum_{n=1}^{\infty} \frac{x^n}{n!}$ converges for all x, since

$$\lim_{n \to \infty} \left| \frac{u_{n+1}}{u_n} \right| = \lim_{n \to \infty} \left| \frac{x^{n+1}}{(n + 1)!} \cdot \frac{n!}{x^n} \right| = \lim_{n \to \infty} \frac{|x|}{n + 1} = 0,$$

for all x. Thus the interval of convergence is $-\infty < x < \infty$.

37. Find the interval of convergence of

$$1 + \frac{x - 2}{2^1} + \frac{(x - 2)^2}{2^2} + \cdots + \frac{(x - 2)^{n-1}}{2^{n-1}} + \cdots. \tag{4}$$

$$\lim_{n \to \infty} \left| \frac{u_{n+1}}{u_n} \right| = \lim_{n \to \infty} \left| \frac{(x-2)^n}{2^n} \cdot \frac{2^{n-1}}{(x-2)^{n-1}} \right| = \lim_{n \to \infty} \frac{|x-2|}{2} = \frac{|x-2|}{2},$$

which is less than 1 if $|x - 2| < 2$, that is, if $0 < x < 4$. Series (4) converges on this interval and diverges if $|x - 2| > 2$, that is, if $x < 0$ or $x > 4$.

When $x = 0$, (4) is $1 - 1 + 1 - 1 + \cdots$ and diverges. When $x = 4$, (4) is $1 + 1 + 1 + \cdots$ and diverges. So (4) converges if $0 < x < 4$.

38. $\sum_{n=1}^{\infty} n! x^n$ converges only at $x = 0$, since

$$\lim_{n \to \infty} \frac{u_{n+1}}{u_n} = \lim_{n \to \infty} (n + 1)x = \infty$$

unless $x = 0$.

RADIUS OF CONVERGENCE. If the power series (1) converges when $|x| < r$ and diverges when $|x| > r$, then r is called the *radius of convergence*. Similarly, r is the radius of convergence of series (2) when (2) converges if $|x - a| < r$ but diverges if $|x - a| > r$. Note the radii of convergence for the series investigated in Examples 34–38:

EXAMPLE	INTERVAL OF CONVERGENCE	RADIUS OF CONVERGENCE
34	$-1 < x < 1$	1
35	$-1 < x \leq 1$	1
36	$-\infty < x < \infty$	∞
37	$0 < x < 4$	2
38	$x = 0$ only	0

THEOREMS ABOUT CONVERGENCE. The following theorems about convergence of power series are useful:

THEOREM 1a. If $\Sigma a_n x^n$ converges for a nonzero number c, then it converges absolutely for all x such that $|x| < |c|$.

THEOREM 1b. If $\Sigma a_n x^n$ diverges for $x = d$, then it diverges for all x such that $|x| > |d|$.

C2. Functions Defined by Power Series.

If $x = x_0$ is a number within the interval of convergence of the power series $\sum_{k=0}^{\infty} a_k(x - a)^k$, then the sum at $x = x_0$ is unique. We let the function f be defined by

$$f(x) = \sum_{k=0}^{\infty} a_k(x - a)^k$$

$$= a_0 + a_1(x - a) + \cdots + a_n(x - a)^n + \cdots; \qquad (1)$$

its domain is the interval of convergence of the series; and it follows that

$$f(x_0) = \sum_{k=0}^{\infty} a_k(x_0 - a)^k.$$

Functions defined by power series behave very much like polynomials, as indicated by the following *properties*:

PROPERTY 2a.   The function defined by (1) is continuous for each x in the interval of convergence of the series.

PROPERTY 2b.   The series formed by differentiating the terms of series (1) converges to $f'(x)$ for each x within the interval of convergence of (1); that is,

$$f'(x) = \sum_{1}^{\infty} ka_k(x - a)^{k-1}$$
$$= a_1 + 2a_2(x - a) + \cdots + na_n(x - a)^{n-1} + \cdots. \qquad (2)$$

Note that we can conclude from 2b that the power series (1) and its derived series (2) have the same radius of convergence but not necessarily the same interval of convergence.

Example 39.   Let

$$f(x) = \sum_{k=1}^{\infty} \frac{x^k}{k(k + 1)} = \frac{x}{1\cdot2} + \frac{x^2}{2\cdot3} + \cdots + \frac{x^n}{n(n + 1)} + \cdots; \qquad (3)$$

then

$$f'(x) = \sum_{k=1}^{\infty} \frac{x^{k-1}}{k + 1} = \frac{1}{2} + \frac{x}{3} + \frac{x^2}{4} + \cdots + \frac{x^{n-1}}{n + 1} + \cdots. \qquad (4)$$

From (3) we see that

$$\lim_{n\to\infty} \left| \frac{x^{n+1}}{(n + 1)(n + 2)} \cdot \frac{n(n + 1)}{x^n} \right| = |x|;$$

that

$$f(1) = \frac{1}{1\cdot2} + \frac{1}{2\cdot3} + \cdots + \frac{1}{n(n + 1)} + \cdots;$$

and that

$$f(-1) = -\frac{1}{1\cdot2} + \frac{1}{2\cdot3} - \cdots + \frac{(-1)^n}{n(n + 1)} + \cdots;$$

and conclude that (3) converges if $-1 \leqq x \leqq 1$.
    From (4) we see that

$$\lim_{n\to\infty} \left| \frac{x^n}{n + 2} \cdot \frac{n + 1}{x^{n-1}} \right| = |x|;$$

that

$$f'(1) = \frac{1}{2} + \frac{1}{3} + \frac{1}{4} + \cdots;$$

and that

$$f'(-1) = \frac{1}{2} - \frac{1}{3} + \frac{1}{4} - \cdots;$$

and conclude that series (4) converges if $-1 \leqq x < 1$.

So the series given for $f(x)$ and $f'(x)$ do have the same radius of convergence but not the same interval.

PROPERTY 2c.   The series obtained by integrating the terms of the given series (1) converges to $\int_a^x f(t)\, dt$ for each x within the interval of convergence of (1); that is,

$$\int_a^x f(t)\, dt = a_0(x - a) + \frac{a_1(x - a)^2}{2} + \frac{a_2(x - a)^3}{3}$$

$$+ \cdots + \frac{a_n(x - a)^{n+1}}{n + 1} + \cdots \qquad (5)$$

$$= \sum_{k=0}^{\infty} \frac{a_k(x - a)^{k+1}}{k + 1}.$$

Example 40.   Obtain a series for $\dfrac{1}{(1 - x)^2}$ by long division.

$$
\begin{array}{r}
1 + 2x + 3x^2 + 4x^3 + \cdots \\
1 - 2x + x^2\, \overline{)\, 1 } \\
\underline{1 - 2x + x^2} \\
2x - x^2 \\
\underline{2x - 4x^2 + 2x^3} \\
+ 3x^2 - 2x^3 \\
\underline{3x^2 - 6x^3 + 3x^4} \\
4x^3 - 3x^4
\end{array}
$$

So

$$\frac{1}{(1 - x)^2} = 1 + 2x + 3x^2 + \cdots + (n + 1)x^n + \cdots. \qquad (6)$$

Let us assume that the power series on the right in (6) converges to

$$f(x) = \frac{1}{(1 - x)^2} \quad \text{if } -1 < x < 1.$$

Then by property 2c

$$\int_0^x \frac{1}{(1-t)^2}\,dt = \int_0^x \left[1 + 2t + 3t^2 + \cdots + (n+1)t^n + \cdots\right]dt$$

$$= \frac{1}{1-t}\bigg|_0^x = \frac{1}{1-x} - 1$$

$$= x + x^2 + x^3 + \cdots + x^{n+1} + \cdots.$$

So

$$\frac{1}{1-x} = 1 + x + x^2 + x^3 + \cdots + x^n + \cdots. \tag{7}$$

Note that the series on the right in (7) is exactly the one obtained for the function $\frac{1}{1-x}$ by long division:

$$
\begin{array}{r}
1 + x + x^2 + x^3 + \cdots \\
1 - x \overline{)\,1} \\
\underline{1 - x} \\
+ x \\
\underline{x - x^2} \\
x^2 \\
\underline{x^2 - x^3} \\
+ x^3
\end{array}
$$

Furthermore, the right-hand member of (7) is a geometric series with ratio $r = x$ and with $a = 1$; if $|x| < 1$, its sum is $\frac{a}{1-r} = \frac{1}{1-x}$.

PROPERTY 2d. Two series may be added, subtracted, multiplied, or divided (with division by zero to be avoided) for x's which lie within the intervals of convergence of both.

Example 41. If

$$\frac{1}{1-x} = 1 + x + x^2 + \cdots + x^n + \cdots \qquad (|x| < 1), \tag{8}$$

then $\frac{1}{(1-x)^2}$ can be obtained by multiplying (8) by itself.

$$
\begin{array}{r}
1 + x + x^2 + x^3 + \cdots \\
1 + x + x^2 + x^3 + \cdots \\
\hline
1 + x + x^2 + x^3 + \cdots \\
+ x + x^2 + x^3 + \cdots \\
+ x^2 + x^3 + \cdots \\
+ x^3 + \cdots \\
\hline
\end{array}
$$

$$\frac{1}{(1-x)^2} = 1 + 2x + 3x^2 + 4x^3 + \cdots \qquad (|x| < 1) \tag{9}$$

The series on the right in (9) is precisely the one obtained for $\dfrac{1}{(1-x)^2}$ by long division in Example 40.

Also, since the derivative of $\dfrac{1}{1-x}$ is exactly $\dfrac{1}{(1-x)^2}$, we can differentiate both sides of (8) above to get (9) once again.

C3. Finding a Power Series for a Function; Taylor Series.

If a function $f(x)$ is representable by a power series of the form

$$a_0 + a_1(x - a) + a_2(x - a)^2 + \cdots + a_n(x - a)^n + \cdots$$

on an interval $|x - a| < r$, then

$$a_0 = f(a), \ a_1 = f'(a), \ a_2 = \frac{f''(a)}{2!}, \cdots, \ a_n = \frac{f^{(n)}(a)}{n!}, \cdots, \tag{1}$$

so that

$$f(x) = f(a) + f'(a)(x - a) + \frac{f''(a)}{2!}(x - a)^2 + \cdots + \frac{f^{(n)}(a)}{n!}(x - a)^n + \cdots. \tag{2}$$

Series (2) is called the *Taylor series* of the function f about the number a. There is never more than one power series in $(x - a)$ for $f(x)$. Implicit in (2) is the requirement that the function and all its derivatives exist at $x = a$ if the function $f(x)$ is to generate a Taylor series expansion.

When $a = 0$ we have a special case of (2) given by

$$f(x) = f(0) + f'(0)x + \frac{f''(0)}{2!}x^2 + \cdots + \frac{f^{(n)}(0)}{n!}x^n + \cdots. \tag{3}$$

Series (3) is called the *Maclaurin series* of the function f; this is the expansion of f about $x = 0$.

Examples

42. If $f(x)$ and all its derivatives exist at $x = 0$ and if $f(x)$ is representable by a power series in x, show that $f(x)$ will have precisely the form given by (3).

Let $f(x) = a_0 + a_1x + a_2x^2 + a_3x^3 + \cdots + a_nx^n + \cdots$ for all x in the interval of convergence. Then

$$f'(x) = a_1 + 2a_2x + 3a_3x^2 + \cdots + na_nx^{n-1} + \cdots,$$

$$f''(x) = 2a_2 + 3\cdot2a_3x + \cdots + n(n-1)a_nx^{n-2} + \cdots,$$

$$f'''(x) = 3\cdot2a_3 + 4\cdot3\cdot2a_4x + \cdots + n(n-1)(n-2)a_nx^{n-3} + \cdots,$$

$$f^{iv}(x) = 4\cdot3\cdot2a_4 + \cdots + n(n-1)(n-2)(n-3)a_nx^{n-4} + \cdots,$$

$$\vdots$$

$$f^{(n)}(x) = n!a_n + \cdots.$$

If we let $x = 0$ in each of the preceding equations, we get

$$f(0) = a_0 \qquad \text{so that} \qquad a_0 = f(0);$$

$$f'(0) = a_1 \qquad\qquad\qquad a_1 = f'(0);$$

$$f''(0) = 2a_2 \qquad\qquad\qquad a_2 = \frac{f''(0)}{2};$$

$$f'''(0) = 3!a_3 \qquad\qquad\qquad a_3 = \frac{f'''(0)}{3!};$$

$$f^{iv}(0) = 4!a_4 \qquad\qquad\qquad a_4 = \frac{f^{iv}(0)}{4!};$$

$$\vdots \qquad\qquad\qquad\qquad \vdots$$

$$f^{(n)}(0) = n!a_n \qquad\qquad\qquad a_n = \frac{f^{(n)}(0)}{n!}.$$

43. Find the Maclaurin series for $f(x) = e^x$.

Here $f'(x) = e^x, \ldots, f^{(n)}(x) = e^x, \ldots,$ for all n. So $f'(0) = 1, \ldots,$ $f^{(n)}(0) = 1, \ldots,$ for all n. By (3), then,

$$e^x = 1 + x + \frac{x^2}{2!} + \frac{x^3}{3!} + \cdots + \frac{x^n}{n!} + \cdots.$$

44. Find the Maclaurin expansion for $f(x) = \sin x$.

$$f(x) = \sin x; \qquad\qquad f(0) = 0;$$
$$f'(x) = \cos x; \qquad\qquad f'(0) = 1;$$
$$f''(x) = -\sin x; \qquad\qquad f''(0) = 0;$$
$$f'''(x) = -\cos x; \qquad\qquad f'''(0) = -1;$$
$$f^{iv}(x) = \sin x; \qquad\qquad f^{iv}(0) = 0.$$

Thus

$$\sin x = x - \frac{x^3}{3!} + \frac{x^5}{5!} - \cdots + (-1)^{n-1}\frac{x^{2n-1}}{(2n-1)!} + \cdots.$$

45. Find the Maclaurin series for $f(x) = \dfrac{1}{1-x}$.

$$f(x) = (1-x)^{-1}; \qquad\qquad f(0) = 1;$$
$$f'(x) = (1-x)^{-2}; \qquad\qquad f'(0) = 1;$$
$$f''(x) = 2(1-x)^{-3}; \qquad\qquad f''(0) = 2;$$
$$f'''(x) = 3!(1-x)^{-4}; \qquad\qquad f'''(0) = 3!;$$
$$\vdots \qquad\qquad\qquad\qquad \vdots$$
$$f^{(n)}(x) = n!(1-x)^{-(n+1)}; \qquad\qquad f^{(n)}(0) = n!.$$

So

$$\frac{1}{1-x} = 1 + x + x^2 + x^3 + \cdots + x^n + \cdots .$$

Note that this agrees exactly with the power series in x obtained by three different methods in Examples 40 and 41.

46. Find the Taylor series for the function $f(x) = \ln x$ about $x = 1$.

$$f(x) = \ln x; \qquad\qquad f(1) = \ln 1 = 0;$$

$$f'(x) = \frac{1}{x}; \qquad\qquad f'(1) = 1;$$

$$f''(x) = -\frac{1}{x^2}; \qquad\qquad f''(1) = -1;$$

$$f'''(x) = \frac{2}{x^3}; \qquad\qquad f'''(1) = 2;$$

$$f^{iv}(x) = \frac{-3!}{x^4}; \qquad\qquad f^{iv}(1) = -3!;$$

$$\vdots \qquad\qquad\qquad\qquad \vdots$$

$$f^{(n)}(x) = \frac{(-1)^{n-1}(n-1)!}{x^n}; \qquad f^{(n)}(1) = (-1)^{n-1}(n-1)!.$$

So

$$\ln x = (x - 1) - \frac{(x-1)^2}{2} + \frac{(x-1)^3}{3} - \frac{(x-1)^4}{4}$$
$$+ \cdots + \frac{(-1)^{n-1}(x-1)^n}{n} + \cdots .$$

FUNCTIONS THAT GENERATE NO SERIES. Note that the following functions are among those that fail to generate a specific series in $(x - a)$ because the function and/or one or more derivatives do not exist at $x = a$:

FUNCTION	SERIES IT FAILS TO GENERATE
$\ln x$	about 0
$\ln(x - 1)$	about 1
$\sqrt{x - 2}$	about 2
$\sqrt{x - 2}$	about 0
$\tan x$	about $\frac{\pi}{2}$
$\sqrt{1 + x}$	about -1

C4. Taylor's Polynomials.

The function $f(x)$ at the point $x = a$ is approximated as follows by the *Taylor polynomial of order n*:

$$f(x) \approx f(a) + f'(a)(x - a) + \frac{f''(a)}{2!}(x - a)^2 + \cdots + \frac{f^{(n)}(a)}{n!}(x - a)^n. \qquad (1)$$

The Taylor polynomial at the right agrees with the function and its first n derivatives at $x = a$.

If $a = 0$, we refer to the Taylor polynomial as the *Maclaurin polynomial of order n,* to wit:

$$f(x) \approx f(0) + f'(0)x + \frac{f''(0)}{2!}x^2 + \cdots + \frac{f^{(n)}(0)}{n!}x^n. \qquad (2)$$

Example 47.   Find the Maclaurin polynomial of order 4 for $f(x) = e^{-x}$. Use this to approximate $f(0.25)$.

The first four derivatives are $-e^{-x}$, e^{-x}, $-e^{-x}$, and e^{-x}; at $a = 0$, these equal -1, 1, -1, and 1, respectively. The approximating Maclaurin polynomial of order 4 is therefore

$$e^{-x} \approx 1 - x + \frac{1}{2!}x^2 - \frac{1}{3!}x^3 + \frac{1}{4!}x^4.$$

With $x = 0.25$ we have

$$e^{-0.25} \approx 1 - 0.25 + \frac{1}{2!}(0.25)^2 - \frac{1}{3!}(0.25)^3 + \frac{1}{4!}(0.25)^4$$

$$\approx 0.7788.$$

This is the precise value of $e^{-0.25}$ correct to four places.

C5. Taylor's Formula with Remainder.

The following theorems resolve the problem of determining *when* a function is representable by a power series.

THEOREM 5a (TAYLOR'S THEOREM).   If a function f and its first $(n + 1)$ derivatives are continuous on the interval $|x - a| < r$, then for each x in this interval

$$f(x) = f(a) + f'(a)(x - a) + \frac{f''(a)}{2!}(x - a)^2 + \cdots + \frac{f^{(n)}(a)}{n!}(x - a)^n + R_n(x), \qquad (1)$$

where

$$R_n(x) = \frac{1}{n!}\int_a^x (x - t)^n f^{(n+1)}(t)\, dt.$$

Note that $R_n(x)$ is the remainder after $(n + 1)$ terms of the Taylor series for $f(x)$.

Two forms of the remainder other than the integral form (1) are given here because they are usually easier to use:

LAGRANGE'S FORM OF THE REMAINDER is

$$R_n(x) = \frac{f^{(n+1)}(\xi)(x - a)^{n+1}}{(n + 1)!}, \tag{2}$$

where ξ is some number between a and x.

CAUCHY'S FORM OF THE REMAINDER is

$$R_n(x) = \frac{f^{(n+1)}(\xi^*)(x - \xi^*)(x - a)}{n!}, \tag{3}$$

where ξ^* is some number between a and x.

When we truncate a series after the $(n + 1)$st term, we can compute R_n, according to Lagrange, for example, if we know what to substitute for ξ. In practice we do not find R_n exactly but only an upper bound for it by assigning to ξ the value between a and x which makes R_n as large as possible.

THEOREM 5b. If the function f has derivatives of all orders in some interval about $x = a$, then the power series

$$f(a) + f'(a)(x - a) + \frac{f''(a)}{2!}(x - a)^2 + \cdots + \frac{f^{(n)}(a)(x - a)^n}{n!} + \cdots$$

represents the function f for those x's, and only those, for which $\lim\limits_{n \to \infty} R_n(x) = 0$; here $R_n(x)$ is the remainder given above by (1), (2), or (3).

Examples

48. Prove that the Maclaurin series generated by e^x represents the function e^x for all real numbers.

From Example 43 we know that $f(x) = e^x$ generates the Maclaurin series

$$e^x = 1 + x + \frac{x^2}{2!} + \cdots + \frac{x^n}{n!} + \cdots. \tag{4}$$

The Lagrange remainder, by (2), is

$$R_n(x) = \frac{e^\xi (x)^{n+1}}{(n + 1)!} \qquad (0 < \xi < x).$$

If x is a positive fixed number, then $0 < e^{\xi} < e^x = C$, where C is a constant, and

$$|R_n(x)| < \frac{Cx^{n+1}}{(n+1)!}.$$

We showed in Example 36, page 299, that $\sum_1^{\infty} \frac{x^k}{k!}$ converges for all x; consequently, by Theorem 2a of §B2 it follows that the general term must approach zero. So, if $x > 0$ then $R_n(x) \to 0$. By Theorem 5b just above, then, (4) represents e^x if $x > 0$.

If $x = 0$, then (4) reduces to 1.

.If $x < 0$, then $x < \xi < 0$ and $e^{\xi} < 1$. So

$$|R_n(x)| < \frac{|x|^{n+1}}{(n+1)!},$$

which approaches zero.

Consequently, (4) represents e^x for all real x.

49. Find the Maclaurin expansion for $\ln(1 + x)$ with the integral form of the remainder.

$$f(x) = \ln(1 + x); \qquad\qquad f(0) = 0;$$

$$f'(x) = \frac{1}{1 + x}; \qquad\qquad f'(0) = 1;$$

$$f''(x) = -\frac{1}{(1 + x)^2}; \qquad\qquad f''(0) = -1;$$

$$f'''(x) = \frac{2}{(1 + x)^3}; \qquad\qquad f'''(0) = 2!;$$

$$\vdots \qquad\qquad\qquad\qquad \vdots$$

$$f^{(n)}(x) = \frac{(-1)^{n-1}(n-1)!}{(1 + x)^n}; \qquad f^{(n)}(0) = (-1)^{n-1}(n-1)!$$

$$f^{(n+1)}(x) = \frac{(-1)^n \cdot n!}{(1 + x)^{n+1}}.$$

So

$$\ln(1 + x) = x - \frac{x^2}{2} + \frac{x^3}{3} - \frac{x^4}{4} + \cdots + (-1)^{n-1} \cdot \frac{x^n}{n} + R_n(x),$$

where

$$R_n(x) = \int_0^x \frac{(-1)^n(x - t)^n}{(1 + t)^{n+1}} \, dt.$$

50. For what x may $\cos x$ be represented by its power series in $\left(x - \frac{\pi}{3}\right)$?

$$f(x) = \cos x; \qquad\qquad f\left(\frac{\pi}{3}\right) = \frac{1}{2};$$

$$f'(x) = -\sin x; \qquad\qquad f'\left(\frac{\pi}{3}\right) = -\frac{\sqrt{3}}{2};$$

$$f''(x) = -\cos x; \qquad\qquad f''\left(\frac{\pi}{3}\right) = -\frac{1}{2};$$

$$f'''(x) = \sin x; \qquad\qquad f'''\left(\frac{\pi}{3}\right) = \frac{\sqrt{3}}{2};$$

$$f^{iv}(x) = \cos x. \qquad\qquad f^{iv}\left(\frac{\pi}{3}\right) = \frac{1}{2}.$$

Note that $f^{(n)}(x) = \pm\sin x$ or $\pm\cos x$.

$$\cos x = \frac{1}{2} - \frac{\sqrt{3}}{2}\left(x - \frac{\pi}{3}\right) - \frac{1}{2}\cdot\frac{\left(x - \frac{\pi}{3}\right)^2}{2} + \frac{\sqrt{3}}{2}\cdot\frac{\left(x - \frac{\pi}{3}\right)^3}{3!} + \cdots$$

The Lagrange remainder (2) is

$$R_n(x) = \frac{f^{(n+1)}(\xi)\left(x - \frac{\pi}{3}\right)^{n+1}}{(n+1)!} \qquad \left(x < \xi < \frac{\pi}{3}\right).$$

Although we do not know exactly what ξ equals, we do know that

$$\left|f^{(n+1)}(\xi)\right| \leqq 1,$$

so that

$$\left|R_n(x)\right| \leqq \frac{\left|x - \frac{\pi}{3}\right|^{n+1}}{(n+1)!} \qquad \text{and} \qquad \lim_{n\to\infty} R_n(x) = 0$$

for all x.

C6. Computations with Power Series.

We list here for reference some frequently used series expansions together with their intervals of convergence:

FUNCTION	SERIES EXPANSION	INTERVAL OF CONVERGENCE	
$\sin x$	$x - \dfrac{x^3}{3!} + \dfrac{x^5}{5!} - \cdots + \dfrac{(-1)^{n-1}x^{2n-1}}{(2n-1)!} + \cdots$	$-\infty < x < \infty$	(1)
$\sin x$	$\sin a + (x-a)\cos a - \dfrac{(x-a)^2}{2!}\sin a$ $\qquad - \dfrac{(x-a)^3}{3!}\cos a + \cdots$	$-\infty < x < \infty$	(2)
$\cos x$	$1 - \dfrac{x^2}{2!} + \dfrac{x^4}{4!} - \cdots + \dfrac{(-1)^{n-1}x^{2n-2}}{(2n-2)!} + \cdots$	$-\infty < x < \infty$	(3)
$\cos x$	$\cos a - (x-a)\sin a - \dfrac{(x-a)^2}{2!}\cos a$ $\qquad + \dfrac{(x-a)^3}{3!}\sin a + \cdots$	$-\infty < x < \infty$	(4)
e^x	$1 + x + \dfrac{x^2}{2!} + \dfrac{x^3}{3!} + \cdots + \dfrac{x^n}{n!} + \cdots$	$-\infty < x < \infty$	(5)
$\ln(1+x)$	$-\dfrac{x^2}{2} + \dfrac{x^3}{3} - \dfrac{x^4}{4} + \cdots + \dfrac{(-1)^{n-1}x^n}{n} + \cdots$	$-1 < x \leqq 1$	(6)
$\ln x$	$(x-1) - \dfrac{(x-1)^2}{2} + \dfrac{(x-1)^3}{3} - \cdots$ $\qquad + \dfrac{(-1)^{n-1}(x-1)^n}{n} + \cdots$	$0 < x \leqq 2$	(7)
$\tan^{-1}x$	$x - \dfrac{x^3}{3} + \dfrac{x^5}{5} - \dfrac{x^7}{7} + \cdots + \dfrac{(-1)^{n-1}x^{2n-1}}{2n-1} + \cdots$	$-1 \leqq x \leqq 1$	(8)

If we use the nth partial sum of a convergent power series for $f(x_0)$ as an approximation to $f(x_0)$, we need information about the magnitude of the error involved. We know that, if the first $(n+1)$ terms of the Taylor series are used for $f(x_0)$, the error is $R_n(x_0)$ as given by (1), (2), or (3) in §C5 above. We have also indicated that it is necessary, not that we know exactly how large the error is, but only that we find an upper bound for it. If a convergent geometric series is involved, then

$$\sum_{n=1}^{\infty} ar^n = a + ar + \cdots + ar^{n-1} + R_n,$$

where

$$|R_n| = \left| \frac{ar^n}{1-r} \right|.$$

Recall, also, that for a convergent alternating series the error is less absolutely than the first term dropped; that is, if

$$S = a_1 - a_2 + a_3 - \cdots + (-1)^{n-1}a_n + R_n,$$

then

$$|R_n| < a_{n+1};$$

in fact, R_n has the same sign as the $(n+1)$st term.

Examples[†]

51. Compute $\dfrac{1}{\sqrt{e}}$ to four decimal places.

We can use the Maclaurin series (5),

$$e^x = 1 + x + \frac{x^2}{2!} + \frac{x^3}{3!} + \frac{x^4}{4!} + \cdots,$$

and let $x = -\dfrac{1}{2}$ to get

$$
\begin{aligned}
e^{-1/2} &= 1 - \frac{1}{2} + \frac{1}{4\cdot2} - \frac{1}{8\cdot3!} + \frac{1}{16\cdot4!} - \frac{1}{32\cdot5!} + R_5 \\
&= 1 - 0.50000 + 0.12500 - 0.02083 + 0.00260 - 0.00026 + R_5 \\
&= 0.60651 + R_5.
\end{aligned}
$$

Note that, since we have a convergent alternating series, R_5 is less than the first term dropped:

$$R_5 < \frac{1}{64\cdot6!} < 0.00003.$$

so $\dfrac{1}{\sqrt{e}} = 0.6065$, correct to four decimal places.

52. Compute $\cos 32°$ correct to four decimal places.

The Taylor series in powers of $\left(x - \dfrac{\pi}{6}\right)$ is obtained immediately from series (4) with $a = \dfrac{\pi}{6}$:

$$
\begin{aligned}
\cos x = \cos \frac{\pi}{6} &- \left(x - \frac{\pi}{6}\right) \sin \frac{\pi}{6} - \frac{\left(x - \dfrac{\pi}{6}\right)^2}{2!} \cos \frac{\pi}{6} \\
&+ \frac{\left(x - \dfrac{\pi}{6}\right)^3}{3!} \sin \frac{\pi}{6} + \cdots.
\end{aligned}
$$

[†]The series referred to in parentheses in this set of examples are listed on page 311.

Since $x = 32°$, $\left(x - \dfrac{\pi}{6}\right) = 2° = 2(0.01745) = 0.03490$, and

$$
\begin{aligned}
\cos 32° &= \frac{\sqrt{3}}{2} - (0.03490)\frac{1}{2} - \frac{(0.03490)^2}{2} \cdot \frac{\sqrt{3}}{2} + R_2 \\
&= 0.86602 - 0.01745 - 0.00053 + R_2 \\
&= 0.84804 + R_2.
\end{aligned}
$$

If we use the Lagrange remainder here, then

$$
R_2 = \frac{f'''(\xi)\left(x - \dfrac{\pi}{3}\right)^3}{3!} = \frac{(\sin \xi)(0.03490)^3}{6},
$$

where $30° < \xi < 32°$. Although we do not know exactly what ξ is, we do know that $\sin \xi < 1$; so $R_2 < 0.00001$ and $\cos 32° = 0.8480$ to four decimal places.

53. For what values of x is the approximate formula

$$
\ln (1 + x) = x - \frac{x^2}{2} \tag{9}
$$

correct to three decimal places?

We can use series (6):

$$
\ln (1 + x) = x - \frac{x^2}{2} + \frac{x^3}{3} - \cdots .
$$

Since this is a convergent alternating series, the error committed by using the first two terms is less than $\dfrac{|x|^3}{3}$. If $\dfrac{|x|^3}{3} < 0.0005$, then the approximate formula (9) will yield accuracy to three decimal places. We therefore require that $|x|^3 < 0.0015$ or that $|x| < 0.115$.

54. Estimate the error if the approximate formula

$$
\sqrt{1 + x} = 1 + \frac{x}{2} \tag{10}
$$

is used and $|x| < 0.02$.

We obtain the first few terms of the Maclaurin series generated by $f(x) = \sqrt{1 + x}$:

$$
\begin{aligned}
f(x) &= \sqrt{1 + x}; & f(0) &= 1; \\
f'(x) &= \frac{1}{2}(1 + x)^{-1/2}; & f'(0) &= \frac{1}{2}; \\
f''(x) &= -\frac{1}{4}(1 + x)^{-3/2}; & f''(0) &= -\frac{1}{4}; \\
f'''(x) &= \frac{3}{8}(1 + x)^{-5/2}. & f'''(0) &= \frac{3}{8}.
\end{aligned}
$$

So

$$\sqrt{1 + x} = 1 + \frac{x}{2} - \frac{1}{4}\cdot\frac{x^2}{2} + \frac{3}{8}\cdot\frac{x^3}{6} - \cdots.$$

If the first term is omitted, the series is strictly alternating; so if formula (10) is used, then $|R_1| < \frac{1}{4}\cdot\frac{x^2}{2}$. With $|x| < 0.02$, $|R_1| < 0.00005$.

Notice that the Lagrange remainder $|R_1|$ here is $\left|\frac{f''(\xi)x^2}{2!}\right|$, where $0 < \xi < 0.02$. Since $|\xi| < 0.02$, we see that

$$|R_1| < \frac{(0.02)^2}{8(1 + 0.02)^{3/2}} < 0.00005.$$

INDETERMINATE FORMS. Series may also be used to evaluate indeterminate forms, as indicated in Examples 55, 56, and 57.

Examples

55. Use series to evaluate $\lim\limits_{x\to 0} \frac{\sin x}{x}$.
From series (1),

$$\sin x = x - \frac{x^3}{3!} + \frac{x^5}{5!} - \cdots.$$

$$\lim_{x\to 0} \frac{\sin x}{x} = \lim_{x\to 0} \left(1 - \frac{x^2}{3!} + \frac{x^4}{5!} - \cdots\right) = 1,$$

a well-established result obtained previously.

56. Use series to evaluate $\lim\limits_{x\to 0} \frac{\ln(x + 1)}{3x}$.
We can use series (6) and write

$$\lim_{x\to 0} \frac{x - \frac{x^2}{2} + \frac{x^3}{3} - \frac{x^4}{4} + \cdots}{3x} = \lim_{x\to 0} \frac{1}{3} - \frac{x}{6} + \frac{x^2}{9} - \cdots$$

$$= \frac{1}{3}.$$

57. $\lim\limits_{x\to 0} \frac{e^{-x^2} - 1}{x^2}$

$$= \lim_{x\to 0} \frac{\left(1 - x^2 + \frac{x^4}{2!} - \frac{x^6}{3!} + \cdots\right) - 1}{x^2}$$

$$= \lim_{x\to 0} \frac{-x^2 + \frac{x^4}{2!} - \cdots}{x^2} = \lim_{x\to 0} -1 + \frac{x^2}{2!} - \frac{x^4}{4!} + \cdots$$

$$= -1.$$

58. Show how series may be used to evaluate π.

Since $\frac{\pi}{4} = \tan^{-1} 1$, a series for $\tan^{-1} x$ may prove helpful. Note that

$$\tan^{-1} x = \int_0^x \frac{dt}{1 + t^2}$$

and that a series for $\frac{1}{1 + t^2}$ is obtainable easily by long division to yield

$$\frac{1}{1 + t^2} = 1 - t^2 + t^4 - t^6 + \cdots . \qquad (11)$$

If we integrate (11) term by term and then evaluate the definite integral we get

$$\tan^{-1} x = x - \frac{x^3}{3} + \frac{x^5}{5} - \frac{x^7}{7} + \cdots + \frac{(-1)^{n-1} x^{2n-1}}{2n - 1} + \cdots . \qquad (12)$$

Compare with series (8) above, and note especially that (12) converges on $-1 \leqq x \leqq 1$. We replace x by 1 in (12) to get

$$\tan^{-1} 1 = 1 - \frac{1}{3} + \frac{1}{5} - \frac{1}{7} + \cdots .$$

So

$$\frac{\pi}{4} = 1 - \frac{1}{3} + \frac{1}{5} - \frac{1}{7} + \cdots \qquad (13)$$

and

$$\pi = 4\left(1 - \frac{1}{3} + \frac{1}{5} - \frac{1}{7} + \cdots\right).$$

It must be pointed out that it would take several hundred terms of (13) to get even two-place accuracy; there are series expressions for π that converge more rapidly than (13).

59. Use series to evaluate $\int_0^{0.1} e^{-x^2} \, dx$ to four decimal places.

Although $\int e^{-x^2} \, dx$ cannot be expressed in terms of elementary functions, we can write a series for e^u, replace u by $(-x^2)$, and integrate term by term. Thus

$$e^{-x^2} = 1 - x^2 + \frac{x^4}{2!} - \frac{x^6}{3!} + \cdots ,$$

so

$$\int_0^{0.1} e^{-x^2}\, dx = x - \frac{x^3}{3} + \frac{x^5}{5 \cdot 2!} - \frac{x^7}{7 \cdot 3!} + \cdots \Bigg|_0^{0.1}$$

$$= 0.1 - \frac{0.001}{3} + \frac{0.00001}{10} - \frac{0.0000001}{42} + \cdots \qquad (14)$$

$$= 0.1 - 0.00033 + 0.000001 + R_6$$

$$= 0.09967 + R_6.$$

Since (14) is a convergent alternating series, $|R_6| < \frac{10^{-7}}{42}$, which will not affect the fourth decimal place. So, correct to four decimal places,

$$\int_0^{0.1} e^{-x^2}\, dx = 0.0997.$$

---

†C7. Power Series over Complex Numbers.

A *complex number* is one of the form $a + bi$, where a and b are real and $i^2 = -1$. If we allow complex numbers as replacements for x in power series, we obtain some interesting results.

Consider, for instance, the series

$$e^x = 1 + x + \frac{x^2}{2!} + \frac{x^3}{3!} + \frac{x^4}{4!} + \cdots + \frac{x^n}{n!} \cdots . \qquad (1)$$

When $x = yi$, then (1) becomes

$$e^{yi} = 1 + yi + \frac{(yi)^2}{2!} + \frac{(yi)^3}{3!} + \frac{(yi)^4}{4!} + \cdots$$

$$= 1 + yi - \frac{y^2}{2!} - \frac{y^3 i}{3!} + \frac{y^4}{4!} + \cdots$$

$$= \left(1 - \frac{y^2}{2!} + \frac{y^4}{4!} + \cdots\right) + i\left(y - \frac{y^3}{3!} + \frac{y^5}{5!} - \cdots\right). \qquad (2)$$

So

$$e^{yi} = \cos y + i \sin y, \qquad (3)$$

since the series within the parentheses of equation (2) converge respectively to $\cos y$ and $\sin y$. Equation (3) is called *Euler's formula*. It follows from (3) that

$$e^{\pi i} = -1 \qquad \text{and} \qquad e^{2\pi i} = 1;$$

the latter is sometimes referred to as *Euler's magic formula*.

†This symbol denotes an optional topic not included in the BC Course Description.

*Set 10: Multiple-Choice Questions on Sequences and Series

(The asterisk applies to all the questions in this set. This topic is covered only in Calculus BC.)

1. If a sequence $\{s_n\}$ converges to L, then

   (A) L equals zero   (B) $|s_n - L| < \epsilon$ for all n
   (C) the difference between s_n and L may be made arbitrarily small
   (D) if $\epsilon > 0$, there is a number N such that $|s_n - L| < \epsilon$ when $n > N$
   (E) $s_n \leqq L$ for all n

2. If $\{s_n\} = \left\{1 + \dfrac{(-1)^n}{n}\right\}$, then

   (A) $\{s_n\}$ diverges by oscillation   (B) $\{s_n\}$ converges to zero
   (C) $\lim\limits_{n \to \infty} s_n = 1$ (D) $\{s_n\}$ diverges to infinity
   (E) none of the above is true

3. The sequence $\left\{\sin \dfrac{n\pi}{6}\right\}$

   (A) is unbounded   (B) is monotonic
   (C) converges to a number less than 1   (D) is bounded
   (E) diverges to infinity

4. Which of the following sequences diverges?

   (A) $\left\{\dfrac{1}{n}\right\}$ (B) $\left\{\dfrac{(-1)^{n+1}}{n}\right\}$ (C) $\left\{\dfrac{2^n}{e^n}\right\}$ (D) $\left\{\dfrac{n^2}{e^n}\right\}$ (E) $\left\{\dfrac{n}{\ln n}\right\}$

5. Which of the following statements about sequences is false?

   (A) If $\{s_n\}$ is bounded, then it is convergent.
   (B) If $\lim\limits_{n \to \infty} s_n = L$, then $|s_n - L| < 0.001$ except for at most a finite number of n's.
   (C) If $\{s_n\}$ converges, then $\{s_n\}$ is bounded.
   (D) If $\{s_n\}$ is unbounded, then it diverges.   (E) None of the above.

6. The sequence $\{r^n\}$ converges if and only if

   (A) $|r| < 1$ (B) $|r| \leqq 1$ (C) $-1 < r \leqq 1$ (D) $0 < r < 1$
   (E) $|r| > 1$

7. The sequence $\{s_n\}$, where $s_n = \frac{n}{n+1}$, converges to 1. It follows then, if $\epsilon > 0$, that there exists a positive integer N such that $|s_n - 1| < \epsilon$ when $n > N$. Let $\epsilon = 0.01$; then the least such N is

    (A) 10   **(B)** 90   **(C)** 99   **(D)** 100   **(E)** 101

8. Σu_n is a series of constants for which $\lim\limits_{n \to \infty} u_n = 0$. Which of the following statements is always true?

    (A) Σu_n converges to a finite sum.   **(B)** Σu_n equals zero.
    (C) Σu_n does not diverge to infinity.   **(D)** Σu_n is a positive series.
    (E) None of the preceding.

9. Note that $\frac{1}{n(n+1)} = \frac{1}{n} - \frac{1}{n+1} (n \geq 1)$. $\sum\limits_{n=1}^{\infty} \frac{1}{n(n+1)}$ equals

    (A) $\frac{4}{3}$ **(B)** 1   **(C)** $\frac{3}{2}$ **(D)** $\frac{3}{4}$ **(E)** ∞

10. The sum of the geometric series $\left(2 - 1 + \frac{1}{2} - \frac{1}{4} + \frac{1}{8} - \cdots \right)$ is

    (A) $\frac{4}{3}$ **(B)** $\frac{5}{4}$ **(C)** 1   **(D)** $\frac{3}{2}$ **(E)** $\frac{3}{4}$

11. Which of the following statements about series is true?

    (A) If $\lim\limits_{n \to \infty} u_n = 0$, then Σu_n converges.
    (B) If $\lim\limits_{n \to \infty} u_n \neq 0$, then Σu_n diverges.
    (C) If Σu_n diverges, then $\lim\limits_{n \to \infty} u_n \neq 0$.
    (D) Σu_n converges if and only if $\lim\limits_{n \to \infty} u_n = 0$.
    (E) None of the preceding.

12. Which of the following statements about series is false?

    (A) $\sum\limits_{k=1}^{\infty} u_k = \sum\limits_{k=m}^{\infty} u_k$, where m is any positive integer.
    (B) If Σu_n converges, so does $\Sigma c u_n$ if $c \neq 0$.
    (C) If Σa_n and Σb_n converge, so does $\Sigma(ca_n + b_n)$, where $c \neq 0$.
    (D) If 1000 terms are added to a convergent series, the new series also converges.
    (E) Rearranging the terms of a positive convergent series will not affect its convergence or its sum.

13. Which of the following series converges?

    (A) $\sum \frac{1}{\sqrt[3]{n}}$ **(B)** $\sum \frac{1}{\sqrt{n}}$ **(C)** $\sum \frac{1}{n}$ **(D)** $\sum \frac{1}{10n-1}$

    (E) $\sum \frac{2}{n^2 - 5}$

14. Which of the following series diverges?

(A) $\displaystyle\sum_{n=1}^{\infty} \frac{1}{n(n+1)}$ (B) $\displaystyle\sum_{n=1}^{\infty} \frac{n+1}{n!}$ (C) $\displaystyle\sum_{n=2}^{\infty} \frac{1}{n \ln n}$ (D) $\displaystyle\sum_{n=1}^{\infty} \frac{\ln n}{2^n}$

(E) $\displaystyle\sum_{n=1}^{\infty} \frac{n}{2^n}$

15. Which of the following series diverges?

(A) $\displaystyle\sum \frac{1}{n^2}$ (B) $\displaystyle\sum \frac{1}{n^2+n}$ (C) $\displaystyle\sum \frac{n}{n^3+1}$ (D) $\displaystyle\sum \frac{n}{\sqrt{4n^2-1}}$

(E) none of the preceding

16. For which of the following series does the Ratio Test fail?

(A) $\displaystyle\sum \frac{1}{n!}$ (B) $\displaystyle\sum \frac{n}{2^n}$ (C) $1 + \dfrac{1}{2^{3/2}} + \dfrac{1}{3^{3/2}} + \dfrac{1}{4^{3/2}} + \cdots$

(D) $\dfrac{\ln 2}{2^2} + \dfrac{\ln 3}{2^3} + \dfrac{\ln 4}{2^4} + \cdots$ (E) $\displaystyle\sum \frac{n^n}{n!}$

17. Which of the following alternating series diverges?

(A) $\displaystyle\sum \frac{(-1)^{n-1}}{n}$ (B) $\displaystyle\sum \frac{(-1)^{n+1}(n-1)}{n+1}$ (C) $\displaystyle\sum \frac{(-1)^{n+1}}{\ln(n+1)}$

(D) $\displaystyle\sum \frac{(-1)^{n-1}}{\sqrt{n}}$ (E) $\displaystyle\sum \frac{(-1)^{n-1}(n)}{n^2+1}$

18. Which of the following series converges conditionally?

(A) $3 - 1 + \dfrac{1}{9} - \dfrac{1}{27} + \cdots$ (B) $\dfrac{1}{\sqrt{2}} - \dfrac{1}{\sqrt{3}} + \dfrac{1}{\sqrt{4}} - \cdots$

(C) $\dfrac{1}{2^2} - \dfrac{1}{3^2} + \dfrac{1}{4^2} - \cdots$ (D) $1 - 1.1 + 1.21 - 1.331 + \cdots$

(E) $\dfrac{1}{1\cdot 2} - \dfrac{1}{2\cdot 3} + \dfrac{1}{3\cdot 4} - \dfrac{1}{4\cdot 5} + \cdots$

19. Let $S = \displaystyle\sum_{n=1}^{\infty} \left(\frac{2}{3}\right)^n$; then S equals

(A) 1   (B) $\dfrac{3}{2}$ (C) $\dfrac{4}{3}$ (D) 2   (E) 3

20. Which of the following statements is true?

(A) If a series converges, then it converges absolutely.
(B) If a series is truncated after the nth term, then the error is less than the first term omitted.
(C) If the terms of an alternating series decrease, then the series converges.
(D) If $r < 1$, then the series Σr^n converges.
(E) None of the preceding.

21. Which of the following expansions is impossible?

(A) $\sqrt{x-1}$ in powers of x (B) $\sqrt{x+1}$ in powers of x

(C) $\ln x$ in powers of $(x-1)$ (D) $\tan x$ in powers of $\left(x - \dfrac{\pi}{4}\right)$

(E) $\ln(1-x)$ in powers of x

22. The power series $x + \dfrac{x^2}{2} + \dfrac{x^3}{3} + \cdots + \dfrac{x^n}{n} + \cdots$ converges if and only if

(A) $-1 < x < 1$ (B) $-1 \leqq x \leqq 1$ (C) $-1 \leqq x < 1$
(D) $-1 < x \leqq 1$ (E) $x = 0$

23. The power series

$$(x+1) - \frac{(x+1)^2}{2!} + \frac{(x+1)^3}{3!} - \frac{(x+1)^4}{4!} + \cdots$$

diverges

(A) for no real x (B) if $-2 < x \leqq 0$ (C) if $x < -2$ or $x > 0$
(D) if $-2 \leqq x < 0$ (E) if $x \neq -1$

24. The series $\displaystyle\sum_{n=0}^{\infty} n!(x-3)^n$ converges if and only if

(A) $x = 0$ (B) $2 < x < 4$ (C) $x = 3$ (D) $2 \leqq x \leqq 4$
(E) $x < 2$ or $x > 4$

25. The interval of convergence of the series obtained by differentiating term by term the series

$$(x-2) + \frac{(x-2)^2}{4} + \frac{(x-2)^3}{9} + \frac{(x-2)^4}{16} + \cdots$$

is

(A) $1 \leqq x \leqq 3$ (B) $1 \leqq x < 3$ (C) $1 < x \leqq 3$ (D) $0 \leqq x \leqq 4$
(E) none of the preceding

26. Let $f(x) = \displaystyle\sum_{n=0}^{\infty} x^n$. The interval of convergence of $\displaystyle\int_0^x f(t)\, dt$ is

(A) $x = 0$ only   (B) $|x| \leqq 1$ (C) $-\infty < x < \infty$
(D) $-1 \leqq x < 1$ (E) $-1 < x < 1$

27. The coefficient of x^4 in the Maclaurin series for $f(x) = e^{-x/2}$ is

(A) $-\dfrac{1}{24}$ (B) $\dfrac{1}{24}$ (C) $\dfrac{1}{96}$ (D) $-\dfrac{1}{384}$ (E) $\dfrac{1}{384}$

28. The Maclaurin polynomial of order 3 for $f(x) = \sqrt{1 + x}$ is

(A) $1 + \dfrac{x}{2} - \dfrac{x^2}{4} + \dfrac{3x^3}{8}$ (B) $1 + \dfrac{x}{2} - \dfrac{x^2}{8} + \dfrac{x^3}{16}$

(C) $1 - \dfrac{x}{2} + \dfrac{x^2}{8} - \dfrac{x^3}{16}$ (D) $1 + \dfrac{x}{2} - \dfrac{x^2}{8} + \dfrac{x^3}{8}$

(E) $1 - \dfrac{x}{2} + \dfrac{x^2}{4} - \dfrac{3x^3}{8}$

29. The Taylor polynomial of order 3 at $x = 1$ for e^x is

(A) $1 + (x - 1) + \dfrac{(x - 1)^2}{2} + \dfrac{(x - 1)^3}{3}$

(B) $e\left[1 + (x - 1) + \dfrac{(x - 1)^2}{2} + \dfrac{(x - 1)^3}{3!} \right]$

(C) $e\left[1 + (x + 1) + \dfrac{(x + 1)^2}{2!} + \dfrac{(x + 1)^3}{3!} \right]$

(D) $e\left[1 + (x - 1) + \dfrac{(x - 1)^2}{2!} + \dfrac{(x - 1)^3}{3!} \right]$

(E) $e\left[1 - (x - 1) + \dfrac{(x - 1)^2}{2!} - \dfrac{(x - 1)^3}{3!} \right]$

30. The coefficient of $\left(x - \dfrac{\pi}{4}\right)^3$ in the Taylor series about $\dfrac{\pi}{4}$ of $f(x) = \cos x$ is

(A) $\dfrac{\sqrt{3}}{12}$ (B) $-\dfrac{1}{12}$ (C) $\dfrac{1}{12}$ (D) $\dfrac{1}{6\sqrt{2}}$ (E) $-\dfrac{1}{3\sqrt{2}}$

31. Which of the following series can be used to compute ln 0.8?

(A) ln $(x - 1)$ expanded about $x = 0$
(B) ln x about $x = 0$
(C) ln x in powers of $(x - 1)$
(D) ln $(x - 1)$ in powers of $(x - 1)$
(E) none of the preceding

32. If $e^{-0.1}$ is computed using series, then, correct to three decimal places, it equals

(A) 0.905   (B) 0.950   (C) 0.904   (D) 0.900   (E) 0.949

33. The coefficient of x^2 in the Maclaurin series for $e^{\sin x}$ is

(A) 0   (B) 1   (C) $\dfrac{1}{2!}$ (D) -1 (E) $\dfrac{1}{4}$

34. Let $f(x) = \sum\limits_{n=0}^{\infty} a_n x^n$, $g(x) = \sum\limits_{n=0}^{\infty} b_n x^n$. Suppose both series converge for $|x| < R$. Let x_0 be a number such that $|x_0| < R$. Which of the following statements is false?

(A) $\sum\limits_{n=0}^{\infty} (a_n + b_n)(x_0)^n$ converges to $f(x_0) + g(x_0)$.

(B) $\left[\sum\limits_{n=0}^{\infty} a_n(x_0)^n \right]\left[\sum\limits_{n=0}^{\infty} b_n(x_0)^n \right]$ converges to $f(x_0)g(x_0)$.

(C) $f(x) = \sum\limits_{n=0}^{\infty} a_n x^n$ is continuous at $x = x_0$.

(D) $\sum\limits_{n=1}^{\infty} n a_n x^{n-1}$ converges to $f'(x_0)$.

(E) None of the preceding.

35. The coefficient of $(x - 1)^5$ in the Taylor series for $x \ln x$ about $x = 1$ is

(A) $-\dfrac{1}{20}$ (B) $\dfrac{1}{5!}$ (C) $-\dfrac{1}{5!}$ (D) $\dfrac{1}{4!}$ (E) $-\dfrac{1}{4!}$

36. The radius of convergence of the series $\sum\limits_{n=1}^{\infty} \dfrac{x^n}{2^n} \cdot \dfrac{n^n}{n!}$ is

(A) 0 (B) 2 (C) $\dfrac{2}{e}$ (D) $\dfrac{e}{2}$ (E) ∞

37. If the approximate formula $\sin x = x - \dfrac{x^3}{3!}$ is used and $|x| < 1$ (radian), then the error is numerically less than

(A) 0.001 (B) 0.003 (C) 0.005 (D) 0.008 (E) 0.009

38. If an appropriate series is used to evaluate $\int_0^{0.3} x^2 e^{-x^2}\, dx$, then, correct to three decimal places, the definite integral equals

(A) 0.009 (B) 0.082 (C) 0.098 (D) 0.008 (E) 0.090

39. If a suitable series is used, then $\int_0^{0.2} \dfrac{e^{-x} - 1}{x}\, dx$, correct to three decimal places, is

(A) -0.200 (B) 0.180 (C) 0.190 (D) -0.190
(E) -0.990

40. The function $f(x) = \sum\limits_{n=0}^{\infty} a_n x^n$ and $f'(x) = -f(x)$ for all x. If $f(0) = 1$, then $f(0.2)$, correct to three decimal places, is

(A) 0.905 (B) 1.221 (C) 0.819 (D) 0.820 (E) 1.220

41. The sum of the series $\sum\limits_{n=1}^{\infty} \left(\frac{\pi^3}{3^\pi}\right)^n$ is equal to

(A) 0   (B) 1   (C) $\dfrac{3^\pi}{\pi^3 - 3^\pi}$ (D) $\dfrac{\pi^3}{3^\pi - \pi^3}$ (E) none of these

42. When $\sum\limits_{1}^{\infty} \dfrac{(-1)^{n-1}}{3n-1}$ is approximated by the sum of its first 300 terms, the error is closest to

(A) 0.001   (B) 0.002   (C) 0.005   (D) 0.01   (E) 0.02

Answers for Set 10: Sequences and Series

1.	D	**9.**	B	**17.**	B	**25.**	B	**34.**	E
2.	C	**10.**	A	**18.**	B	**26.**	D	**35.**	A
3.	D	**11.**	B	**19.**	D	**27.**	E	**36.**	C
4.	E	**12.**	A	**20.**	E	**28.**	B	**37.**	E
5.	A	**13.**	E	**21.**	A	**29.**	D	**38.**	A
6.	C	**14.**	C	**22.**	C	**30.**	D	**39.**	D
7.	C	**15.**	D	**23.**	A	**31.**	C	**40.**	C
8.	E	**16.**	C	**24.**	C	**32.**	A	**41.**	D
						33.	C	**42.**	A

1. D.   (D) is the definition of convergence of a sequence. Find examples of $\{s_n\}$ and L that show that (A), (B), (C), and (E) are false.

2. C.   Note that $\lim\limits_{n\to\infty} \dfrac{(-1)^n}{n} = 0$.

3. D.   The sine function varies continuously between -1 and 1 inclusive.

4. E.   Note that $\left\{\dfrac{2}{e}\right\}^n$ is a sequence of the type $\{r^n\}$ with $|r| < 1$; also that $\lim \dfrac{n^2}{e^n} = 0$ by repeated application of L'Hôpital's rule.

5. A.   The sequence $1, -1, 1, -1, \ldots$ is a good counterexample for statement (A).

6. C.   See Example 11, page 287.

7. C. $\left|\dfrac{n}{n+1} - 1\right| < 0.01$ if $\left|-\dfrac{1}{n+1}\right| = \dfrac{1}{n+1} < 0.01$; $n + 1 > \dfrac{1}{0.01}$ or $n > 100 - 1 = 99$.

8. E.   The harmonic series $\sum_{1}^{\infty} \frac{1}{k}$ is a counterexample for (A), (B), and (C).

$\sum_{1}^{\infty} \frac{(-1)^{k+1}}{k}$ shows that (D) does not follow.

9. B. $\sum_{n=1}^{\infty} \frac{1}{n(n+1)} = \frac{1}{1\cdot2} + \frac{1}{2\cdot3} + \frac{1}{3\cdot4} + \cdots + \frac{1}{n(n+1)} + \cdots;$ so

$$s_n = 1 - \frac{1}{2} + \frac{1}{2} - \frac{1}{3} + \frac{1}{3} - \cdots + \frac{1}{n} - \frac{1}{n+1} = 1 - \frac{1}{n+1},$$

and $\lim_{n\to\infty} s_n = 1$.

10. A. $S = \frac{a}{1-r} = \frac{2}{1-(-\frac{1}{2})} = \frac{4}{3}.$

11. B.   Find counterexamples for statements (A), (C), and (D).

12. A.   If $\sum_{k=1}^{\infty} u_k$ converges, so does $\sum_{k=m}^{\infty} u_k$, where m is any positive integer; but their *sums* are probably different.

13. E.   See Example 19, page 292. Each series given is essentially a p-series. Only in (E) is $p > 1$.

14. C.   Use the Integral Test on page 291.

15. D. $\frac{n}{\sqrt{4n^2-1}} > \frac{n}{\sqrt{4n^2}} = \frac{1}{2}$, the general term of a divergent series.

16. C.   The limit of the ratio for the series $\sum \frac{1}{n^{3/2}}$ is 1, so this test fails; note for (E) that

$$\lim_{n\to\infty} \frac{u_{n+1}}{u_n} = \lim_{n\to\infty} \left(\frac{n+1}{n}\right)^n = \lim_{n\to\infty} \left(1 + \frac{1}{n}\right)^n = e.$$

17. B. $\lim_{n\to\infty} \frac{(-1)^{n+1}(n-1)}{n+1}$ does not equal 0.

18. B.   (A), (C), and (E) all converge absolutely; (D) is the divergent geometric series with $r = -1.1$.

19. D. $S = \frac{a}{1-r} = \frac{\frac{2}{3}}{1-\frac{2}{3}} = \frac{\frac{2}{3}}{\frac{1}{3}} = 2.$

20. E.   Note the following counterexamples:

(A) $\sum \frac{(-1)^{n-1}}{n}$ **(B)** $\sum \frac{1}{n}$ **(C)** $\sum \frac{(-1)^{n-1}\cdot n}{2n-1}$

(D) $\sum \left(-\frac{3}{2}\right)^{n-1}.$

21. A.   If $f(x) = \sqrt{x-1}$, then $f(0)$ is not defined.

22. C. Since $\lim\limits_{n\to\infty}\left|\dfrac{u_{n+1}}{u_n}\right| = |x|$, the series converges if $|x| < 1$. We must test the endpoints: when $x = 1$, we get the divergent harmonic series; $x = -1$ yields the convergent alternating harmonic series.

23. A. $\lim\limits_{n\to\infty}\left|\dfrac{x+1}{n+1}\right| = 0$ for all $x \neq -1$; since the given series converges to 0 if $x = -1$, it therefore converges for *all* x.

24. C. $\lim\limits_{n\to\infty}(n+1)(x-3) = \infty$ unless $x = 3$.

25. B. The differentiated series is $\displaystyle\sum_{n=1}^{\infty} \dfrac{(x-2)^{n-1}}{n}$; so

$$\lim_{n\to\infty}\left|\dfrac{u_{n+1}}{u_n}\right| = |x-2|.$$

26. D. The integrated series is $\displaystyle\sum_{n=0}^{\infty} \dfrac{x^{n+1}}{n+1}$ or $\displaystyle\sum_{n=1}^{\infty} \dfrac{x^{n}}{n}$. See Question 22 above.

27. E. $e^{-x/2} = 1 + \left(-\dfrac{x}{2}\right) + \left(-\dfrac{x}{2}\right)^2 \cdot \dfrac{1}{2!} + \left(-\dfrac{x}{2}\right)^3 \cdot \dfrac{1}{3!} + \left(-\dfrac{x}{2}\right)^4 \cdot \dfrac{1}{4!} + \cdots.$

28. B. Use (2) on page 307.

29. D. Note that every derivative of e^x is e^x, which equals e at $x = 1$. Use (1) at the top of page 307 with $a = 1$.

30. D. Use (2) on page 304 with $a = \dfrac{\pi}{4}$.

31. C. Note that $\ln q$ is defined only if $q > 0$, and that the derivatives must exist at $x = a$ in formula (2) for the Taylor series on page 307.

32. A. Use

$$e^{-x} = 1 - x + \dfrac{x^2}{2!} - \cdots; \qquad e^{-0.1} = 1 - (+0.1) + \dfrac{0.01}{2!} + R_2.$$

$|R_2| < \dfrac{0.001}{3!} < 0.0005$. Or use the series for e^x and let $x = -0.1$.

33. C. $e^u = 1 + u + \dfrac{u^2}{2!} + \cdots$, and $\sin x = x - \dfrac{x^3}{3!} + \cdots$, so

$$e^{\sin x} = 1 + \left(x - \dfrac{x^3}{3!} + \cdots\right) + \dfrac{1}{2!}\left(x - \dfrac{x^3}{3!} + \cdots\right)^2 + \cdots.$$

Or generate the Maclaurin series for $e^{\sin x}$.

34. E. (A), (B), (C), and (D) are all true statements.

35. A. $f(x) = x \ln x,$

$f'(x) = 1 + \ln x,$

$f''(x) = \dfrac{1}{x},$

$f'''(x) = -\dfrac{1}{x^2},$

$f^{(4)}(x) = \dfrac{2}{x^3},$

$f^{(5)}(x) = -\dfrac{3 \cdot 2}{x^4}; \qquad f^{(5)}(1) = -3 \cdot 2.$

So the coefficient of $(x - 1)^5$ is $-\dfrac{3 \cdot 2}{5!} = -\dfrac{1}{20}.$

36. C. $\lim\limits_{n \to \infty} \left| \dfrac{u_{n+1}}{u_n} \right| = \lim\limits_{n \to \infty} \left| \dfrac{x^{n+1} \, (n + 1)^{n+1}}{2^{n+1} \, (n + 1)!} \cdot \dfrac{2^n \cdot n!}{x^n \cdot n^n} \right|$

$= \lim\limits_{n \to \infty} \left| \dfrac{x(n + 1)^{n+1}}{2(n + 1)n^n} \right| = \lim\limits_{n \to \infty} \left| \dfrac{x}{2}\left(\dfrac{n + 1}{n}\right)^n \right|$

$= \lim\limits_{n \to \infty} \left| \dfrac{x}{2}\left(1 + \dfrac{1}{n}\right)^n \right| = \left| \dfrac{x}{2} \cdot e \right|.$

Since the series converges when $\left| \dfrac{x}{2} \cdot e \right| < 1$, that is, when $|x| < \dfrac{2}{e}$, the radius of convergence is $\dfrac{2}{e}$.

37. E.   The error, $|R_4|$, is less than $\dfrac{1}{5!} < 0.009.$

38. A. $\displaystyle\int_0^{0.3} x^2 e^{-x^2} \, dx = \int_0^{0.3} x^2 \left(1 - x^2 + \dfrac{x^4}{2!} - \cdots \right) dx$

$= \displaystyle\int_0^{0.3} \left(x^2 - x^4 + \dfrac{x^6}{2!} - \cdots \right) dx$

$= \left. \dfrac{x^3}{3} - \dfrac{x^5}{5} + \cdots \right|_0^{0.3}$

$= 0.009$ to three decimal places.

39. D. $\displaystyle\int_0^{0.2} \dfrac{e^{-x} - 1}{x} \, dx = \int_0^{0.2} \dfrac{\left(1 - x + \dfrac{x^2}{2!} - \dfrac{x^3}{3!} + \cdots \right) - 1}{x} \, dx$

$= \displaystyle\int_0^{0.2} \left(-1 + \dfrac{x}{2} - \dfrac{x^2}{3!} + \cdots \right) dx$

$= \left. -x + \dfrac{x^2}{2 \cdot 2} - \dfrac{x^3}{3 \cdot 6} + \cdots \right|_0^{0.2}$

$= -0.190; \quad |R_2| < \dfrac{(0.2)^3}{18} < 0.0005.$

40. C. $f(x) = a_0 + a_1x + a_2x^2 + a_3x^3 + \cdots$; if $f(0) = 1$, then $a_0 = 1$.

$f'(x) = a_1 + 2a_2x + 3a_3x^2 + 4a_4x^3 + \cdots$; $f'(0) = -f(0) = -1$.

so $a_1 = -1$. Since $f'(x) = -f(x)$, $f(x) = -f'(x)$:

$$1 - x + a_2x^2 + a_3x^3 + \cdots = -(-1 + 2a_2x + 3a_3x^2 + 4a_4x^3 + \cdots)$$

identically. Thus

$$-2a_2 = -1, \qquad a_2 = \frac{1}{2},$$

$$-3a_3 = a_2, \qquad a_3 = -\frac{1}{3!},$$

$$-4a_4 = a_3, \qquad a_4 = \frac{1}{4!},$$

$$\vdots \qquad\qquad \vdots$$

We see, then, that

$$f(x) = 1 - x + \frac{x^2}{2!} - \frac{x^3}{3!} + \frac{x^4}{4!} - \cdots;$$

$$f(0.2) = 1 - 0.2 + \frac{(0.2)^2}{2!} - \frac{(0.2)^3}{3!} + R_3$$

$$= 0.819; \quad |R_3| < 0.0005.$$

41. D.   Use a calculator to verify that the ratio $\dfrac{\pi^3}{3^\pi}$ (of the given geometric series) equals approximately 0.98. Since the ratio $r < 1$, the sum of the series equals

$$\frac{a}{1-r} \qquad \text{or} \qquad \frac{\pi^3}{3^\pi} \cdot \frac{1}{1 - \dfrac{\pi^3}{3^\pi}}.$$

Simplify to get (D).

42. A.   The error (of the given alternating series) is less in absolute value than the first term dropped, that is, less than $\left|\dfrac{(-1)^{300}}{900-1}\right| \approx 0.0011$. Choice (A) is closest to this approximation.

Practice Section II Problems on Sequences and Series

1. (a) Prove that the series $\sum\limits_{1}^{\infty} \dfrac{1}{n!}$ converges.

    (b) By comparing $\sum\limits_{1}^{\infty} \dfrac{1}{n!}$ with $\sum\limits_{1}^{\infty} \dfrac{1}{2^{n-1}}$ show that, when $\sum\limits_{1}^{\infty} \dfrac{1}{n!}$ is replaced by the sum of its first 11 terms, the error is less than 0.001.

    (c) Does the series $\sum\limits_{k=1}^{\infty} ke^{-k}$ converge? Justify your answer.

2. (a) Find the Maclaurin polynomial of order 4 for $\cos x$.

    (b) Use part (a) to evaluate $\displaystyle\int_0^1 \cos x\, dx$ to five decimal places.

    (c) Estimate the error in (b), justifying your answer.

3. (a) Find the Maclaurin series for $f(x) = \ln(1 + x)$.
    (b) What is the interval of convergence of the series in (a)?
    (c) Use the first five terms in (a) to approximate $\ln(1.2)$.
    (d) Estimate the error in (c), justifying your answer.

4. (a) Show that the series $\sum\limits_{n=1}^{\infty} (-1)^{n+1} \dfrac{1}{\ln(n+1)}$ converges.

    (b) How many terms of the series are needed to get a partial sum within 0.1 of the sum of the whole series?

    (c) Tell whether the series $\sum\limits_{n=2}^{\infty} (-1)^n \dfrac{1}{n \ln n}$ is absolutely convergent, conditionally convergent, or divergent. Justify your answer.

Solutions for Section II Problems: Sequences and Series

1. (a) See Example 24, page 294.

    (b) Since $n! > 2^{n-1}$, $\dfrac{1}{n!} < \dfrac{1}{2^{n-1}}$; also, $\sum\limits_{1}^{\infty} \dfrac{1}{2^{n-1}}$ is a geometric series, with ratio r equal to $\dfrac{1}{2}$. When the first 11 terms are used for $\sum\limits_{1}^{\infty} \dfrac{1}{n!}$, the error is $\sum\limits_{12}^{\infty} \dfrac{1}{n!}$, which is less than $\sum\limits_{12}^{\infty} \dfrac{1}{2^{n-1}}$. So

$$\sum_{12}^{\infty} \frac{1}{n!} < \sum_{12}^{\infty} \frac{1}{2^{n-1}} = \frac{1}{2^{11}} \cdot \frac{1}{1 - \dfrac{1}{2}} = \frac{1}{2^{10}} < 0.001.$$

(c) We use the Integral Test:

$$\int_1^\infty xe^{-x}\,dx = \lim_{b\to\infty} -e^{-x}(1+x)\Big|_1^b = -\lim_{b\to\infty}\left(\frac{1+b}{e^b} - \frac{2}{e}\right)$$

and, by applying L'Hôpital's rule to the first fraction within the parentheses, the integral equals $\frac{2}{e}$. Since the improper integral converges, so does the given series.

2.   (a) For $f(x) = \cos x,\ f'(x) = -\sin x,\ f''(x) = -\cos x,\ f'''(x) = \sin x,$ $f^{(4)}(x) = \cos x, f^{(5)}(x) = -\sin x, f^{(6)}(x) = -\cos x.$ The Maclaurin polynomial of order 4 is

$$\cos x = 1 - \frac{x^2}{2!} + \frac{x^4}{4!}.$$

Note that the next term of the alternating Maclaurin series for $\cos x$ is $-\dfrac{x^6}{6!}$.

(b) $\displaystyle\int_0^1 \cos x\,dx = x - \frac{x^3}{3\cdot 2!} + \frac{x^5}{5\cdot 4!}\Big|_0^1 = 1 - \frac{1}{6} + \frac{1}{120} \approx 0.84167.$

(c) The error in (b) is less than $\displaystyle\int_0^1 \frac{x^6}{6!}\,dx = \frac{x^7}{7!}\Big|_0^1 < 0.00020.$

3.   (a) Let $f(x) = \ln(1+x)$. Then $f'(x) = \dfrac{1}{1+x}, f''(x) = -\dfrac{1}{(1+x)^2}, f'''(x) = \dfrac{2}{(1+x)^3}, f^{(4)}(x) = -\dfrac{3!}{(1+x)^4}, f^{(5)}(x) = \dfrac{4!}{(1+x)^5}.$ At $x = 0, f(0) = 0, f'(0) = 1, f''(0) = -1, f'''(0) = 2, f^{(4)}(0) = -(3!),$ and $f^{(5)}(0) = 4!.$ So

$$\ln(1+x) = x - \frac{x^2}{2} + \frac{x^3}{3} - \frac{x^4}{4} + \frac{x^5}{5} - \cdots.$$

(b) Using the Ratio Test, we know that the series converges when

$\displaystyle\lim_{n\to\infty}\left|\frac{x^{n+1}}{n+1}\cdot\frac{n}{x^n}\right| < 1,$ that is, when $|x| < 1,$ or $-1 < x < 1.$ We test the endpoints: When $x = 1$, we get $1 - \dfrac{1}{2} + \dfrac{1}{3} - \dfrac{1}{4} + \dfrac{1}{5} - \cdots,$ which converges; when $x = -1$, we get $-1 - \dfrac{1}{2} - \dfrac{1}{3} - \dfrac{1}{4} - \dfrac{1}{5} - \cdots,$ which diverges. So the series converges if $-1 < x \le 1.$

(c) $\ln(1.2) = 0.2 - \dfrac{(0.2)^2}{2} + \dfrac{(0.2)^3}{3} - \dfrac{(0.2)^4}{4} + \dfrac{(0.2)^5}{5} = 0.18233.$

(d) Since the series is alternating, the error in the answer for (c) is less than $\dfrac{(0.2)^6}{6} \approx 0.00001.$

4. (a) The given series is alternating. Since $\lim\limits_{n\to\infty} \ln (n + 1)= \infty$, $\lim\limits_{n\to\infty} \dfrac{1}{\ln (n + 1)} = 0$. Since $\ln x$ is an increasing function,

$$\ln (n + 1) > \ln n \quad \text{and} \quad \frac{1}{\ln (n + 1)} < \frac{1}{\ln n}.$$

The series therefore converges.

(b) For an alternating series, the error in using the first n terms for the sum of the whole series is less than the absolute value of the $(n + 1)$st term. Thus the error is less than $\dfrac{1}{\ln (n + 1)}$. We solve for n using $\dfrac{1}{\ln (n + 1)} < 0.1$:

$$\ln (n + 1) > 10,$$

$$(n + 1) > e^{10},$$

$$n > e^{10} - 1 > 22{,}025.$$

The given series converges very slowly!

(c) The series $\displaystyle\sum_{2}^{\infty} (-1)^n \frac{1}{n \ln n}$ is conditionally convergent. The given alternating series converges since the nth term approaches 0 and $\dfrac{1}{(n + 1) \ln (n + 1)} < \dfrac{1}{n \ln n}$. However, the *positive* series diverges by the Integral Test, since

$$\int_2^{\infty} \frac{1}{x \ln x} \, dx = \lim_{b\to\infty} \ln (\ln x) \Big|_2^b = \infty.$$

FURTHER PRACTICE

Miscellaneous Multiple-Choice Questions

1. The line through the point $(2, -1)$ and perpendicular to the line $3x - y = 4$ has y-intercept

   (A) $\dfrac{1}{3}$ (B) $-\dfrac{5}{2}$ (C) 5 (D) $-\dfrac{1}{3}$ (E) $-\dfrac{5}{3}$

2. The equation of the line with x-intercept $-\dfrac{1}{3}$ and slope $\dfrac{1}{2}$ is

   (A) $3x - 6y - 1 = 0$ (B) $3x - 6y + 2 = 0$ (C) $3x - 6y - 2 = 0$
   (D) $6x - 3y + 2 = 0$ (E) $3x - 6y + 1 = 0$

3. If the lines $ax + by = 3$ and $a'x + b'y = 4$ are perpendicular, then it follows that

   (A) $\dfrac{a}{b'} = \dfrac{b}{a'}$ (B) $ab' = -a'b$ (C) $aa' = -bb'$ (D) $\dfrac{a}{a'} = \dfrac{b}{b'}$
   (E) none of these is true

4. The distance between the point $(3, 0)$ and the line $x - 2y + 2 = 0$ equals

   (A) $\sqrt{5}$ (B) $5\sqrt{5}$ (C) 1 (D) $\sqrt{2}$ (E) none of these

5. The equation of the circle with center $(2, -3)$ and radius $\sqrt{13}$ is

   (A) $x^2 + y^2 - 4x + 6y = 0$ (B) $(x + 2)^2 + (y - 3)^2 = 13$
   (C) $x^2 + y^2 + 4x - 6y = 0$ (D) $x^2 + y^2 + 4x - 6y = 26$
   (E) $(x - 2)^2 + (y + 3)^2 = \sqrt{13}$

6. The graph of the equation $x^2 - 2x + 4y - 7 = 0$ is a

   (A) hyperbola (B) pair of straight lines (C) circle
   (D) ellipse (E) parabola

7. The equation of the parabola with vertex at $(2, 1)$ and focus at $(2, 5)$ is

   (A) $(x - 2)^2 = 16(y - 1)$ (B) $(y - 2)^2 = 8(x - 1)$
   (C) $(x - 1)^2 = 2(y - 2)$ (D) $(y - 1)^2 = 16(x - 2)$
   (E) $x^2 - 4x - y = 0$

8. The graph of $y^2 = 1 + 4x + x^2$ is symmetric to

    (A)   the x-axis   **(B)**   the x-axis and the y-axis   **(C)**   the origin
    (D)   the x- and y-axes and the origin   **(E)**   the y-axis

*9. The rectangular equation of the curve given parametrically by $x = 1 - \sin t$ and $y = 4 - 2 \cos t$ is

    (A) $4(x - 1)^2 + (y - 4)^2 = 1$ **(B)** $4(x - 1)^2 + (y - 4)^2 = 4$
    (C) $(x - 1)^2 + 4(y - 4)^2 = 1$ **(D)** $(x - 1)^2 + (y - 4)^2 = 4$
    (E)   none of these

*10. The graph of the pair of parametric equations $x = \sin t - 2$, $y = \cos^2 t$ is

    (A)   part of a circle   **(B)**   part of a parabola   **(C)**   a hyperbola
    (D)   a line   **(E)**   a cycloid

11. The distance between the centers of the two circles $x^2 - 4x + y^2 + 2y = 4$ and $x^2 + y^2 - 6y = 0$ is

    (A) $\sqrt{5}$ **(B)** $2\sqrt{5}$ **(C)** $2\sqrt{2}$ **(D)** $\sqrt{2}$ **(E)**   1

12. The set of x for which $|x - 3| \leq 2$ and for which $|x| > 4$ is

    (A) $4 < x \leq 5$ **(B)** $-4 < x \leq 1$ **(C)** $x < -4$ or $x \geq 1$
    (D) $1 \leq x < 3$ **(E)**   none of these

*13. If $x = 2 \sin u$ and $y = \cos 2u$, then a single equation in x and y is

    (A) $x^2 + y^2 = 1$ **(B)** $x^2 + 4y^2 = 4$ **(C)** $x^2 + 2y = 2$
    (D) $x^2 + y^2 = 4$ **(E)** $x^2 - 2y = 2$

14. If $f(x) = \begin{cases} x^2 & \text{for } x \leq 1 \\ 2x - 1 & \text{for } x > 1 \end{cases}$, then

    (A) $f(x)$ is not continuous at $x = 1$
    (B) $f(x)$ is continuous at $x = 1$ but $f'(1)$ does not exist
    (C) $f'(1)$ exists and equals 1
    (D) $f'(1) = 2$
    (E) $\lim\limits_{x \to 1} f(x)$ does not exist

*15. The curve of the pair of parametric equations $x = 2e^t$, $y = e^{-t}$ is

    (A)   a straight line   **(B)**   a parabola   **(C)**   a hyperbola
    (D)   an ellipse   **(E)**   none of these

16. The number of points in the set for which the inequalities $x^2 + y^2 < 9$ and $x + y \geq 5$ both hold is

    (A)   0   **(B)**   1   **(C)**   2   **(D)**   infinite   **(E)**   none of these

*Questions preceded by an asterisk are likely to appear only on the Calculus BC Examination.

17. The curve of $y = \dfrac{2x^2}{4 - x^2}$ has

    (A)   two horizontal asymptotes
    (B)   two horizontal asymptotes and one vertical asymptote
    (C)   two vertical but no horizontal asymptotes
    (D)   one horizontal and one vertical asymptote
    (E)   one horizontal and two vertical asymptotes

18. The curve of $f(x) = x \sin \dfrac{1}{x}$ is symmetric to

    (A)   the y-axis   **(B)**   the x-axis   **(C)**   the origin
    (D)   the line $y = x$ **(E)**   none of these

19. Which of the following sets does not define a function of x?

    (A) $\{(0, 1), (1, 3), (2, 3)\}$ **(B)** $\{(x, y) \mid -\infty \leqq x \leqq \infty, y = x^2\}$
    (C) $\{(1, 2), (2, 1), (4, 3), (5, 4)\}$ **(D)** $\{(x, y) \mid -1 \leqq x, y^2 = (x + 1)\}$
    (E) $\{(x, y) \mid x \text{ is real and } y = \pi\}$

*20. The locus of the polar equation $r = 2 \sin \theta$ is

    (A)   a circle with center on the x-axis
    (B)   a circle with center on the y-axis   **(C)**   the line $y = 2$
    (D)   the line $x = 2$ **(E)**   a circle with radius 2

21. Which one of the following functions has a derivative at $x = 0$?

    (A) $f(x) = \sin \dfrac{1}{x}$ **(B)** $f(x) = |x|$ **(C)** $f(x) = x|x|$

    (D) $f(x) = x \sin \dfrac{1}{x}$ **(E)** $f(x) = [x]$ (greatest-integer function)

22. The domain of the function $f(x) = \sqrt{x^2 - 4}$ is

    (A) $|x| \geqq 2$ **(B)** $|x| \leqq 2$ **(C)** $x > 2$ or $x < -2$
    (D)   all x except $x = 2$ or -2 **(E)**   all x

*23. The area bounded by the lemniscate with polar equation $r^2 = 2 \cos 2\theta$ is equal to

    (A)   4   **(B)**   1   **(C)** $\dfrac{1}{2}$ **(D)**   2   **(E)**   none of these

*24. The area inside the circle $r = 3 \sin \theta$ and outside the cardioid $r = 1 + \sin \theta$ is given by

    (A) $\displaystyle\int_{\pi/6}^{\pi/2} [9 \sin^2 \theta - (1 + \sin \theta)^2]\, d\theta$ **(B)** $\displaystyle\int_{\pi/6}^{\pi/2} (2 \sin \theta - 1)^2\, d\theta$

    (C) $\dfrac{1}{2}\displaystyle\int_{\pi/6}^{5\pi/6} (8 \sin^2 \theta - 1)\, d\theta$ **(D)** $\dfrac{9\pi}{4} - \dfrac{1}{2}\displaystyle\int_{\pi/6}^{5\pi/6} (1 + \sin \theta)^2\, d\theta$

    (E)   none of these

*25. The graph of the polar equation $r = \theta$, where θ is a real number, is
  (A) a circle
  (B) a hyperbolic spiral asymptotic to the line $y = 1$
  (C) a straight line of slope 1
  (D) a pair of straight lines passing through the origin
  (E) a double spiral which passes through the origin

26. The graphs of $y = x^2 + 1$ and $x^2 - y^2 = 1$ have the following points in common:
  (A) $(1, 2)$ and $(-1, 2)$ (B) $(1, 0)$ (C) $(0, 1)$
  (D) $(1, 0), (-1, 0), (0, 1),$ and $(0, -1)$ (E) none of these

27. Which of the following functions is continuous at $x = 0$?

  (A) $f(x) \begin{cases} = \sin \dfrac{1}{x} & \text{for } x \neq 0 \\[2mm] = 0 & \text{for } x = 0 \end{cases}$

  (B) $f(x) = [x]$ (greatest-integer function)

  (C) $f(x) \begin{cases} = \dfrac{x}{x} & \text{for } x \neq 0 \\[2mm] = 0 & \text{for } x = 0 \end{cases}$

  (D) $f(x) \begin{cases} = x \sin \dfrac{1}{x} & \text{for } x \neq 0 \\[2mm] = 0 & \text{for } x = 0 \end{cases}$

  (E) $f(x) = \dfrac{x + 1}{x}$

28. If the curve of the function $y = f(x)$ is symmetric to the origin, then it follows that

  (A) $f(0) = 0$ (B) $f(-x) = -f(x)$ (C) $f(-x) = f(x)$
  (D) the curve is also symmetric to both the x- and y-axes
  (E) none of the preceding is necessarily true

29. The locus of points whose distance from the line $x = 1$ is twice that from the point $(-1, 0)$ is

  (A) a parabola   (B) a circle   (C) an ellipse   (D) a hyperbola
  (E) a straight line

30. If the graph of a function is as shown, then the function $f(x)$ could be given by which of the following?

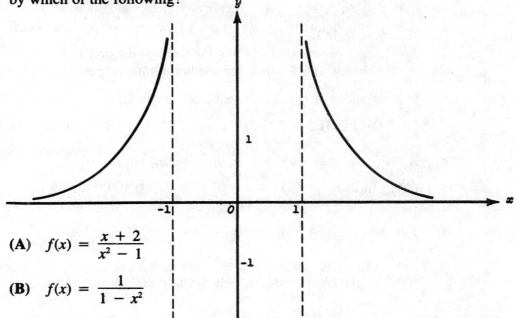

(A) $f(x) = \dfrac{x + 2}{x^2 - 1}$

(B) $f(x) = \dfrac{1}{1 - x^2}$

(C) $f(x) = \dfrac{x^2 - 1}{x^2 + 1}$

(D) $f(x) = \dfrac{2}{x^2 - 1}$

(E) $f(x) = \dfrac{2}{1 - x^2}$

31. If the graph of a function is as shown, then the function $f(x)$ could be given by which of the following?

(A) $f(x) = x^4 - 2x^2$ (B) $f(x) = x^3 - 2x^2$ (C) $f(x) = x^3 - 2x$

(D) $f(x) = x(x - 2)^2$ (E) $f(x) = 2x^2 - x^4$

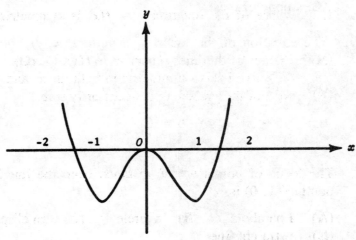

32. If $\lim\limits_{x \to c} f(x) = L$ (L a finite number), then it follows that

(A) $f'(c)$ exists   (B) $f(x)$ is continuous at $x = c$ (C) $f(c) = L$
(D) $f(c)$ is defined   (E) none of the preceding is necessarily true

33. A function $f(x)$ equals $\dfrac{x^2 - x}{x - 1}$ for all x except $x = 1$. In order that the function be continuous at $x = 1$, the value of $f(1)$ must be

(A) 0   (B) 1   (C) 2   (D) ∞ (E) none of these

34. If $f(u) = e^u$ and $g(u) = \ln u$, then $f(g(u))$ equals

(A) $(\ln u)\, e$ (B) 1   (C) u (D) e^{-u} (E) e

*35. The graph of the polar equation $r = \dfrac{1}{\sin \theta - 2 \cos \theta}$ is

(A) a line with slope 2   (B) a line with slope 1   (C) a circle
(D) a parabola   (E) a semicircle

36. If the curve of the equation $ky^2 + xy = 2 - k$ passes through the point $(-2, 1)$, then k equals

(A) 0   (B) -2 (C) 1   (D) -1 (E) 2

37. Which of the following statements about the graph of $y = \dfrac{x^2 + 1}{x^2 - 1}$ is not true?

(A) The graph is symmetric to the y-axis.
(B) The graph has two vertical asymptotes.
(C) There is no y-intercept.
(D) The graph has one horizontal asymptote.
(E) There is no x-intercept.

38. A circle has center at $(2, -1)$ and is tangent to the line $x + 2y + 5 = 0$. Its equation is

(A) $x^2 + y^2 - 4x + 2y = 10$ (B) $x^2 + y^2 - 4x + 2y = 0$
(C) $(x - 2)^2 + (y + 1)^2 = 25$ (D) $x^2 + y^2 + 4x - 2y = 0$
(E) none of these

39. The equation of the locus of a point $P(x, y)$, which moves so that the product of the slopes of the lines joining it to $(2, 0)$ and to $(-3, 1)$ is 1, is

(A) $y^2 - x^2 - x - y + 6 = 0$ (B) $y = x^2 + x - 5$
(C) $x^2 + y^2 + x + y - 6 = 0$ (D) $x^2 - y^2 - x - y + 6 = 0$
(E) $y^2 - x^2 + y - 6 = 0$

40. If $f(x) = 2x - \dfrac{2}{x}$, then $f\left(\dfrac{1}{x}\right)$ equals

(A) $f(x)$ (B) $f\left(-\dfrac{1}{x}\right)$ (C) $-f(-x)$ (D) $-f(x)$
(E) none of these

41.   Suppose

$$f(x) = \begin{cases} x^2 & \text{if } x < -2, \\ 4 & \text{if } -2 < x \leqq 1, \\ 6 - x & \text{if } x > 1. \end{cases}$$

Which statement is true?
(A) f is discontinuous only at $x = -2$.
(B) f is discontinuous only at $x = 1$.
(C) If $f(-2)$ is defined to be 4, then f will be continuous everywhere.
(D) f is continuous everywhere.
(E) f is discontinuous at $x = -2$ and at $x = 1$.

42.   To which of the following lines is $x + by = c$ perpendicular?

I. $y = \frac{1}{b}x$ II. $bx + y = c$ III. $bx - y = 1$

(A) I only   **(B)** II only   **(C)** III only   **(D)** I and II   **(E)** I and III

43.   Suppose $y = \cos(\pi t) + 3t$. Approximately how much change is there in y when t changes from 0.50 to 0.53?

(A) -0.0084 **(B)** -0.0042 **(C)** 0.0042   **(D)** 0.0874   **(E)** 0.0896

44.   With respect to which of the following is the graph of $y = 3^x - 3^{-x}$ symmetric?

I. The x-axis   II. The y-axis   III. The origin

(A) I only   **(B)** II only   **(C)** III only   **(D)** I and II   **(E)** II and III

45.   The solution set of the inequality $\left| \frac{x}{2} + 5 \right| \geqq 2$ is

(A) $(-\infty, -14) \cup (-6, \infty)$ **(B)** $(-\infty, -6) \cup [14, \infty]$ **(C)** $[-14, 6]$
(D) $[-6, 14]$ **(E)** $(-\infty, -14] \cup [-6, \infty]$

46.   The equation of the circle with diameter at endpoints $(10, 2)$ and $(4, 6)$ is

(A) $(x - 7)^2 + (y - 4)^2 = 12$ **(B)** $x^2 + y^2 - 14x + 8y + 52 = 0$
(C) $x^2 + y^2 + 14x + 8y - 52 = 0$ **(D)** $(x - 7)^2 + (y - 4)^2 = 13$
(E) none of these

47. $\lim\limits_{x \to 1^-} ([x] - |x|) =$

(A) -1 **(B)** 0   **(C)** 1   **(D)** 2   **(E)** none of these

48.   Suppose $f(3) = 2$, $f'(3) = 5$, and $f''(3) = -2$. Then $\frac{d^2}{dx^2}(f^2(x))$ at $x = 3$ is equal to

(A) -20 **(B)** 10   **(C)** 20   **(D)** 38   **(E)** 42

49. The distance between the centers of the circles $x^2 + 2x + y^2 - 4y = 6$ and $x^2 - 6x + y^2 + 4y = -1$ is

(A) $\sqrt{20}$ (B) 4   (C) $4\sqrt{2}$ (D) 5   (E) $5\sqrt{2}$

50. $\lim\limits_{x \to 0} \dfrac{\cos x - 1}{x}$ is

(A) -1 (B) 0   (C) 1   (D) ∞ (E) nonexistent

51. The x-coordinate of the point on the curve $y = x^2 - 2x + 3$ at which the tangent is perpendicular to the line $x + 3y + 3 = 0$ is

(A) $-\dfrac{5}{2}$ (B) $-\dfrac{1}{2}$ (C) $\dfrac{7}{6}$ (D) $\dfrac{5}{2}$ (E) none of these

52. $\lim\limits_{x \to 1} \dfrac{\frac{3}{x} - 3}{x - 1}$ is

(A) -3 (B) -1 (C) 1   (D) 3   (E) nonexistent

53. Which statement is false?
(A) The sum of two increasing functions on $[a, b]$ is an increasing function on $[a, b]$.
(B) The product of two increasing functions on $[a, b]$ is an increasing function on $[a, b]$.
(C) If $f'(x) > 0$ for all x in an interval, then f is increasing on the interval.
(D) If f is increasing and differentiable on $[a, b]$, then $f'(x) \geqq 0$ on $[a, b]$.
(E) None of the preceding is false.

54. $\lim\limits_{x \to 3} \dfrac{x + 3}{x^2 - 9}$ is

(A) $+\infty$ (B) 0   (C) $\dfrac{1}{6}$ (D) $-\infty$ (E) nonexistent

55. The only function that does not satisfy the Mean Value Theorem on the interval specified is

(A) $f(x) = x^2 - 2x$ on $[-3, 1]$.   (B) $f(x) = \dfrac{1}{x}$ on $[1, 3]$.
(C) $f(x) = \dfrac{x^3}{3} - \dfrac{x^2}{2} + x$ on $[-1, 2]$.   (D) $f(x) = x + \dfrac{1}{x}$ on $[-1, 1]$.
(E) $f(x) = x^{2/3}$ on $\left[\dfrac{1}{2}, \dfrac{3}{2}\right]$.

56. The equation of the tangent to the graph of $y = 3x^4 - 4x^3 - 2$ at its local minimum point is

(A) $y = -3$ (B) $y = -2$ (C) $y = 0$ (D) $y = 1$ (E) none of these

57. $\lim\limits_{x \to 2^-} \dfrac{|x-2|}{x-2}$ is

(A) $-\infty$ (B) -1 (C) 1   (D) ∞ (E) nonexistent

58. Which statement is true?

(A) If $f(x)$ is continuous at $x = c$, then $f'(c)$ exists.
(B) If $f'(c) = 0$, then f has a local maximum or minimum at $(c, f(c))$.
(C) If $f''(c) = 0$, then f has an inflection point at $(c, f(c))$.
(D) If f is differentiable at $x = c$, then f is continuous at $x = c$.
(E) If f is continuous on (a, b), then f attains a maximum value on (a, b).

59. The solution of the differential equation $\frac{dy}{dx} = 2xy^2$ for which $y = -1$ when $x = 1$ is

(A) $y = -\dfrac{1}{x^2}$ (B) $\ln y^2 = x^2 - 1$ (C) $\dfrac{y^3}{3} = x^2 - \dfrac{4\cdot}{3}$

(D) $y = -\dfrac{1}{x}$ (E) none of these

60. The function $f(x) = 5x^4 - x^5$ is increasing on the following interval(s):

(A) $(-\infty, 0)$ (B) $(4, \infty)$ (C) $\left(0, \dfrac{1}{4}\right)$ (D) $(0, 4)$
(E) $(-\infty, 0)$ and $(4, \infty)$

61. Suppose that a ball, thrown up from the surface of the earth with an initial velocity of 96 ft/sec, attains a height of $(-16t^2 + 96t)$ ft at time t. The ball's velocity, in feet per second, when it hits the ground is

(A) -128 (B) -96 (C) -64 (D) 0 (E) 96

62. If $f(t) = \displaystyle\int_0^{t^2} \dfrac{1}{1 + x^2}\, dx$, then $f'(t)$ equals

(A) $\dfrac{1}{1 + t^2}$ (B) $\dfrac{2t}{1 + t^2}$ (C) $\dfrac{1}{1 + t^4}$ (D) $\dfrac{2t}{1 + t^4}$ (E) $\tan^{-1} t^2$

63. Suppose $f'(x) = x(x - 2)^2(x + 3)^3$. Which of the following is (are) true?
I. f has a local maximum at $x = -3$.
II. f has a local minimum at $x = 0$.
III. f has neither a local maximum nor a local minimum at $x = 2$.

(A) I only   (B) II only   (C) III only   (D) I and II only
(E) I, II, and III

64. If f and g are integrable and if $f(x) \geq g(x)$ for all x in $[a, b]$, then which statement is true?

(A) $\displaystyle\int_a^b |f(x)|\, dx \geqq \int_a^b |g(x)|\, dx$ (B) $\left|\displaystyle\int_a^b f(x)\, dx\right| \geqq \left|\displaystyle\int_a^b g(x)\, dx\right|$

(C) $\displaystyle\int_a^b f(x)\, dx \geqq \int_a^b g(x)\, dx$ (D) $f'(x) \geqq g'(x)$ for all x in $[a, b]$

(E) None of the preceding is true.

65. If $y = uv$, where u and v are differentiable functions of x, then y'' equals

(A) $u''v''$ (B) $u'v'' + v'u''$ (C) $uv'' + vu''$
(D) $uv'' + u'v' + vu''$ (E) $uv'' + 2u'v' + vu''$

66. The number of inflection points of $f(x) = 3x^5 - 10x^3$ is

(A) 4   (B) 3   (C) 2   (D) 1   (E) 0

67. Let V denote the volume and S the surface area of a sphere of radius r. If an error of 0.1 is made in measuring the radius, then the approximate error in computing the volume is

(A) 0.001   (B) 0.1V (C) 0.01S (D) 0.1S

(E) not determinable from the given information

68. The average value of $f(x) = 3 + |x|$ on the interval $[-2, 4]$ is

(A) $2\frac{2}{3}$ (B) $3\frac{1}{3}$ (C) $4\frac{2}{3}$ (D) $5\frac{1}{3}$ (E) 6

69. Suppose $f(x) = \int_0^x \frac{4+t}{t^2+4} \, dt$. It follows that

(A) f increases for all x (B) f increases only if $x > -4$
(C) f has a local min at $x = -4$ (D) f has a local max at $x = -4$
(E) f has no critical points

70. $f(x) = \int_0^{x^2+2} \sqrt{1 + \cos t} \, dt$. Then $f'(x) =$

(A) $2x\sqrt{1 + \cos(x^2 + 2)}$ (B) $2x\sqrt{1 - \sin x}$

(C) $(x^2 + 2)\sqrt{1 - \sin x}$ (D) $\sqrt{1 + \cos(x^2 + 2)}$

(E) $\sqrt{[1 + \cos(x^2 + 2)]} \cdot 2x$

71. $\ln(3e^{x+2}) - \ln(3e^x) =$

(A) 2   (B) e^2 (C) $2\ln 3$ (D) $\ln 3 + 2$ (E) $2\ln 3 + 2$

72. $\int_0^6 |x - 4| \, dx =$

(A) 6   (B) 8   (C) 10   (D) 11   (E) 12

73. The number c satisfying the Mean Value Theorem for $f(x) = \sin x$ on the interval $[1, 1.5]$ is, correct to three decimal places,

(A) 0.995   (B) 1.058   (C) 1.239   (D) 1.253   (E) 1.399

74. An integral for the volume obtained by revolving the region bounded by $y = 2x - x^2$ and the x-axis about the y-axis is

(A) $4\pi \int_0^1 x(2x - x^2) \, dx$ (B) $2\pi \int_0^2 x(2x - x^2) \, dx$

(C) $\pi \int_0^2 (2x - x^2)^2 \, dx$ (D) $8\pi \int_0^1 \left(1 + \sqrt{1 - y}\right) dy$

(E) none of these

75. $\displaystyle\int_e^{e^4} \frac{1}{x}\, dx =$

    (A) 3   **(B)** 4   **(C)** ln 3   **(D)** $e^4 - 1$ **(E)** $e^4 - e$

76. Suppose $f'(x) = x^2(x - 1)$. Then $f''(x) = x(3x - 2)$. Over which interval(s) is the graph of f both increasing and concave up?

    I. $x < 0$ II. $0 < x < \frac{2}{3}$ III. $\frac{2}{3} < x < 1$ IV. $x > 1$

    (A) I only   **(B)** II only   **(C)** II and IV   **(D)** I and IV   **(E)** IV only

77. Which of the following statements is true about the graph of $f(x)$ in Question 76?
    (A) The graph has no relative extrema.
    (B) The graph has one relative extremum and one inflection point.
    (C) The graph has two relative extrema and one inflection point.
    (D) The graph has two relative extrema and two inflection points.
    (E) None of the preceding statements is true.

78. If the half-life of a radioactive substance is 8 years, how long will it take, in years, for two thirds of the substance to decay?

    (A) 4.68   **(B)** 7.69   **(C)** 12   **(D)** 12.21   **(E)** 12.68

79. What is the period of f if $f(x) = \frac{3}{\pi} \sin\left(\frac{\pi}{3} \cdot x\right)$?

    (A) $\frac{3}{\pi}$ **(B)** $\frac{2\pi}{3}$ **(C)** 3   **(D)** 6   **(E)** 6π

80. $\displaystyle\lim_{x\to\infty} \frac{3 + x - 2x^2}{4x^2 + 9}$ is

    (A) $-\frac{1}{2}$ **(B)** $\frac{1}{2}$ **(C)** 1   **(D)** 3   **(E)** nonexistent

81. $\int(\sqrt{x} - 2)x^2\, dx =$

    (A) $\frac{2}{3}x^{3/2} - 2x + C$ **(B)** $\frac{5}{2}x^{3/2} - 4x + C$

    (C) $\frac{2}{3}x^{3/2} - 2x + \frac{x^3}{3} + C$ **(D)** $\frac{2}{5}x^{5/2} - \frac{2}{3}x^3 + C$

    (E) $\frac{2}{7}x^{7/2} - \frac{2}{3}x^3 + C$

82. In which of the following intervals is $\ln(x^2 - 3x)$ defined?

    I. $x \leqq 0$ II. $0 < x < 3$ III. $x \geqq 3$

    (A) I only   **(B)** II only   **(C)** III only   **(D)** I and III
    (E) none of the intervals

83. Let

$$f(x) = \begin{cases} \dfrac{x^2 - 36}{x - 6} & \text{if } x \neq 6, \\[2mm] 12 & \text{if } x = 6. \end{cases}$$

Which of the following statements is (are) true?

    I. f is defined at $x = 6$.   II. $\lim\limits_{x \to 6} f(x)$ exists.

    III. f is continuous at $x = 6$.

    (A) I only   **(B)** II only   **(C)** I and II only   **(D)** I, II, and III

    (E) none of the statements

84. Let $f(x) = x^5 + 3x - 2$, and let f^{-1} denote the inverse of f. Then $(f^{-1})'(2)$ equals

    (A) $\dfrac{1}{83}$ **(B)** $\dfrac{1}{8}$ **(C)** 1 **(D)** 8 **(E)** 83

85. Suppose $f(x) = \dfrac{1}{3} x^3 + x$, $x > 0$, and x increasing. The value of x for which the rate of increase of f is 10 times that of x is

    (A) 1 **(B)** 2 **(C)** $\sqrt[3]{10}$ **(D)** 3 **(E)** $\sqrt{10}$

86. The nth derivative of $\ln (x + 1)$ at $x = 2$ equals

    (A) $\dfrac{(-1)^{n-1}}{3^n}$ **(B)** $\dfrac{(-1)^n \cdot n!}{3^{n+1}}$ **(C)** $\dfrac{(-1)^{n-1}(n - 1)!}{3^n}$

    (D) $\dfrac{(-1)^n \cdot n}{3^n}$ **(E)** $\dfrac{(-1)^{n+1}}{3^{n+1}}$

87. Use the trapezoidal rule with two equal subintervals to approximate, correct to three decimal places, $\displaystyle\int_1^2 \sin \sqrt{x}\, dx$.

    (A) 0.042 **(B)** 0.692 **(C)** 0.915 **(D)** 0.928 **(E)** 1.876

88. The area of the region in the xy-plane bounded by the curves $y = e^x$, $y = e^{-x}$, and $x = 1$ is equal to

    (A) $e + \dfrac{1}{e} - 2$ **(B)** $e - \dfrac{1}{e}$ **(C)** $e + \dfrac{1}{e}$ **(D)** $2e - 2$

    (E) none of the preceding

89. If $y = x^{\ln x}$ $(x > 0)$, then $D_x y =$

    (A) $(\ln x) \cdot x^{\ln x - 1}$ **(B)** $(2 \ln x) \cdot x^{\ln x}$ **(C)** $2x^{\ln x - 1} \cdot \ln x$

    (D) $\dfrac{2 \ln x}{x}$ **(E)** $1 + \ln x$

90. $\displaystyle\int_1^e \dfrac{\ln^3 x}{x}\, dx =$

    (A) $\dfrac{1}{4}$ **(B)** $\dfrac{1}{4} e$ **(C)** $\dfrac{1}{4} (e - 1)$ **(D)** $\dfrac{e^4}{4}$ **(E)** $\dfrac{e^4 - 1}{4}$

91. The area bounded by the curves $y = x^2 - 3x$ and $y = 3 - x$ is given by the integral

    (A) $\displaystyle\int_{-1}^3 (x^2 - 2x - 3)\, dx$ **(B)** $\displaystyle\int_{-1}^3 (3 + 2x - x^2)\, dx$

    (C) $\displaystyle\int_{-3}^1 (3 - 4x - x^2)\, dx$ **(D)** $\displaystyle\int_0^4 (3 + 2x - x^2)\, dx$

    (E) none of these

*92. $\displaystyle\int_{-\infty}^{\infty} \frac{dx}{x^2 + 1} =$

(A) 0   (B) $\frac{\pi}{2}$ (C) π (D) 2π (E) none of these

*93. The base of a solid is the region bounded by the parabola $y^2 = 4x$ and the line $x = 2$. Each plane section perpendicular to the x-axis is a square. The volume of the solid is

(A) 6   (B) 8   (C) 10   (D) 16   (E) 32

*94. The natural length of a spring is 1 ft. If takes a force of 4 lb. to keep the spring extended 3 in. The work, in inch-pounds, done in stretching the spring to a length of 16 in. is

(A) $2\frac{2}{3}$ (B) $5\frac{1}{3}$ (C) $10\frac{2}{3}$ (D) $21\frac{1}{3}$ (E) $74\frac{2}{3}$

*95. $\int \cos^3 x\, dx =$

(A) $\frac{\cos^4 x}{4} + C$ (B) $\frac{\sin^4 x}{4} + C$

(C) $\sin x - \frac{\sin^3 x}{3} + C$ (D) $\sin x + \frac{\sin^3 x}{3} + C$

(E) $\cos x - \frac{\cos^3 x}{3} + C$

*96. $\displaystyle\int_{1}^{2} \frac{dx}{\sqrt{4 - x^2}}$ is

(A) $-\frac{\pi}{3}$ (B) $\frac{\pi}{6}$ (C) $\frac{\pi}{4}$ (D) $\frac{\pi}{3}$ (E) nonexistent

*97. The first four terms of the Maclaurin series (the Taylor series about $x = 0$) for $f(x) = \dfrac{1}{1 - 2x}$ are

(A) $1 + 2x + 4x^2 + 8x^3$ (B) $1 - 2x + 4x^2 - 8x^3$
(C) $-1 - 2x - 4x^2 - 8x^3$ (D) $1 - x + x^2 - x^3$
(E) $1 + x + x^2 + x^3$

*98. $\int x^2 e^{-x}\, dx =$

(A) $\frac{1}{3} x^3 e^{-x} + C$ (B) $-\frac{1}{3} x^3 e^{-x} + C$ (C) $-x^2 e^{-x} + 2xe^{-x} + C$

(D) $-x^2 e^{-x} - 2xe^{-x} - 2e^{-x} + C$ (E) $-x^2 e^{-x} + 2xe^{-x} - 2e^{-x} + C$

*99. If for any positive number ϵ, however small, there exists a positive number N and a number L such that, if $x > N$, then $|f(x) - L| < \epsilon$, then

(A) $\lim\limits_{x \to 0} f(x) = L$ (B) $\lim\limits_{x \to \infty} f(x) = L.$

(C) $\lim\limits_{x \to L} f(x) = \infty$ (D) $\lim\limits_{x \to N} f(x) = L$ (E) $\lim\limits_{x \to \infty} f(x) = \infty$

*100. When partial fractions are used, the decomposition of $\dfrac{x-1}{x^2+3x+2}$ is equal to

(A) $\dfrac{2}{x+1} - \dfrac{3}{x+2}$ (B) $-\dfrac{2}{x+1} + \dfrac{3}{x+2}$ (C) $\dfrac{3}{x+1} - \dfrac{2}{x+2}$

(D) $\dfrac{2}{x+1} + \dfrac{3}{x+2}$ (E) $-\dfrac{2}{x+1} - \dfrac{3}{x+2}$

Answers: Miscellaneous Multiple-Choice Questions

1.	D	**21.**	C	**41.**	E	**61.**	B	**81.**	E
2.	E	**22.**	A	**42.**	C	**62.**	D	**82.**	E
3.	C	**23.**	D	**43.**	B	**63.**	E	**83.**	D
4.	A	**24.**	A	**44.**	C	**64.**	C	**84.**	B
5.	A	**25.**	E	**45.**	E	**65.**	E	**85.**	D
6.	E	**26.**	E	**46.**	D	**66.**	B	**86.**	C
7.	A	**27.**	D	**47.**	A	**67.**	D	**87.**	D
8.	A	**28.**	B	**48.**	E	**68.**	C	**88.**	A
9.	B	**29.**	C	**49.**	C	**69.**	C	**89.**	C
10.	B	**30.**	D	**50.**	B	**70.**	A	**90.**	A
11.	B	**31.**	A	**51.**	D	**71.**	A	**91.**	B
12.	A	**32.**	E	**52.**	A	**72.**	C	**92.**	C
13.	C	**33.**	B	**53.**	B	**73.**	D	**93.**	E
14.	D	**34.**	C	**54.**	E	**74.**	B	**94.**	C
15.	C	**35.**	A	**55.**	D	**75.**	A	**95.**	C
16.	A	**36.**	E	**56.**	A	**76.**	C	**96.**	D
17.	E	**37.**	C	**57.**	B	**77.**	E	**97.**	A
18.	A	**38.**	B	**58.**	D	**78.**	E	**98.**	D
19.	D	**39.**	A	**59.**	A	**79.**	D	**99.**	B
20.	B	**40.**	D	**60.**	D	**80.**	A	**100.**	B

1. D.   The equation of the line is $x + 3y = -1$.

2. E.   Use the point-slope form: $y - 0 = \frac{1}{2}\left(x + \frac{1}{3}\right)$.

3. C.   Since the slopes are negative reciprocals of each other, $-\dfrac{a}{b} = \dfrac{b'}{a'}$.

4. A.   Recall that the distance from (x_1, y_1) to the line with equation $ax + by + c = 0$ is $\dfrac{|ax_1 + by_1 + c|}{\sqrt{a^2 + b^2}}$.

5. A.   Using Equation (40) on page 485 yields $(x - 2)^2 + (y + 3)^2 = 13$.

6. E.   The equation can be written in the form $(x - 1)^2 = -4(y - 2)$. See Equation (44) on page 485.

7. A. Hint: If new axes, x' and y', are chosen with origin at $(2, 1)$, the equation of the parabola referred to these is $x'^2 = 16y'$.

8. A. A graph is symmetric to the x-axis if, whenever (x_1, y_1) is on the graph, so is $(x_1, -y_1)$. In the equation of a polynomial or rational function y can occur only to even powers.

9. B. Rewrite the equations as

$$x - 1 = -\sin t, \qquad \frac{y - 4}{2} = -\cos t;$$

square and add. The locus is an ellipse with center at $(1, 4)$.

10. B. Since $x + 2 = \sin t$ and $y = \cos^2 t$, we get

$$(x + 2)^2 + y = 1,$$

where $-3 \leqq x \leqq -1$ and $0 \leqq y \leqq 1$.

11. B. The circles have centers at $(2, -1)$ and $(0, 3)$, respectively.

12. A. $|x - 3| \leqq 2$ when $1 \leqq x \leqq 5$; $|x| > 4$ when $x < -4$ or $x > 4$.

13. C. The equations may be rewritten as

$$\frac{x}{2} = \sin u \qquad \text{and} \qquad y = 1 - 2\sin^2 u,$$

giving $y = 1 - 2 \cdot \dfrac{x^2}{4}$.

14. D. The graph of f is shown in Figure 11–1; f is defined and

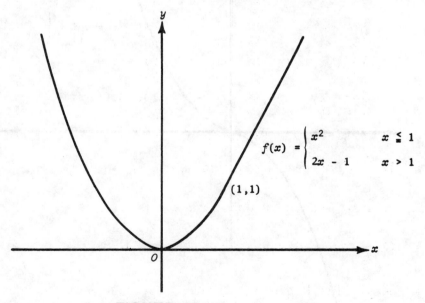

$$f(x) = \begin{cases} x^2 & x \leqq 1 \\ 2x - 1 & x > 1 \end{cases}$$

$(1,1)$

FIGURE 11–1

continuous at all x, including $x = 1$. Since

$$\lim_{x \to 1^-} f'(x) = 2 = \lim_{x \to 1^+} f'(x),$$

$f'(1)$ exists and is equal to 2.

15. C. The Cartesian equation is $xy = 2$.

16. A. The set of points such that $x^2 + y^2 < 9$ is the interior of the circle with center at $(0, 0)$ and radius 3. The inequality $x + y \geq 5$ is satisfied by points on or above the line $x + y = 5$. The intersection of these sets is empty.

17. E. The curve has vertical asymptotes at $x = 2$ and $x = -2$ and a horizontal asymptote at $y = -2$.

18. A. Since $\sin(-\theta) = -\sin\theta$, $f(-x)$ here is equal to $f(x)$ and the function is even.

19. D. The relation $y^2 = x + 1$ yields two y's for each x in the domain.

20. B. The rectangular equation is $x^2 + y^2 - 2y = 0$, that of a circle with center at $(0, 1)$ and radius 1.

21. C. The graph of f is shown in Figure 11–2. Here, $f(x) = x|x|$ is equivalent to

$$f(x) = \begin{cases} x^2 & \text{for } x \geq 0 \\ -x^2 & \text{for } x < 0 \end{cases}$$

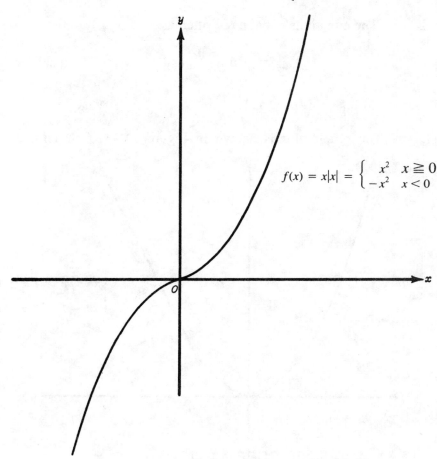

$$f(x) = x|x| = \begin{cases} x^2 & x \geq 0 \\ -x^2 & x < 0 \end{cases}$$

FIGURE 11–2

Here

$$\lim_{x \to 0^-} f'(x) = 0 = \lim_{x \to 0^+} f'(x),$$

so the derivative exists at $x = 0$.

22. A.   We seek x such that $x^2 - 4 \geqq 0$, that is, $x \leqq -2$ or $x \geqq 2$.

23. D.   **Using the formula for area in polar coordinates,**

$$A = \frac{1}{2} \int_{\alpha}^{\beta} r^2 \, d\theta,$$

we see that the required area is given by

$$4 \cdot \frac{1}{2} \int_{0}^{\pi/4} 2 \cos 2\theta \, d\theta.$$

24. A.   The required area is lined in Figure 11–3.

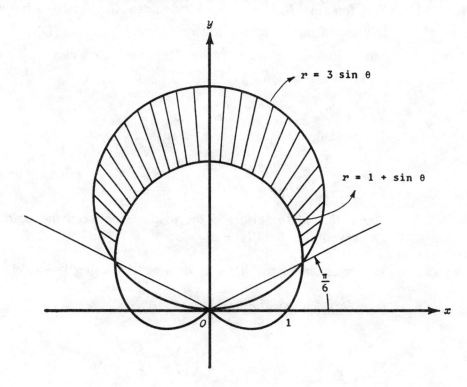

FIGURE 11–3

25. E.   Remember that θ can also be negative.

26. E.   **Any points of intersection of the two graphs would have ordinates that satisfy the equation $y^2 - y + 2 = 0$. Since this equation has no real roots, the graphs do not intersect.**

27. D.   If $f(x) = x \sin \dfrac{1}{x}$ for $x \neq 0$ and $f(0) = 0$, then

$$\lim_{x \to 0} f(x) = 0 = f(0);$$

thus this function is continuous at 0. Explain why each of the other functions is not continuous at $x = 0$.

28. B.   The statement in (B) defines symmetry of a curve to the origin.

29. C.   The locus leads to the equation

$$(x - 1)^2 = 4[(x + 1)^2 + y^2].$$

30. D.   The graph shown has the following characteristics: it has no x-intercepts; the y-intercept is -2; it has vertical asymptotes $x = 1$ and $x = -1$; and it has the x-axis as horizontal asymptote.

31. A.   For the graph shown, since the tangent is horizontal at three points, the function must be of fourth degree. Since $f(x) < 0$ when $|x| < \sqrt{2}$, only the function in (A) can be correct.

32. E.   The function $f(x) = \dfrac{x^2 - 1}{x - 1}$ is such that $\lim\limits_{x \to 1} f(x) = 2$ but it does not satisfy any of the conditions given in (A), (B), (C), or (D).

33. B.   Since $\lim\limits_{x \to 1} f(x) = 1$, to render $f(x)$ continuous at $x = 1$ we must define $f(1)$ to be 1.

34. C.   Here, $f(g(u)) = e^{\ln u} = u$.

35. A.   The rectangular equation is $y - 2x = 1$.

36. E.   Replace (x, y) by $(-2, 1)$ and solve for k.

37. C.   To find the y-intercept, let $x = 0$; $y = 1$.

38. B.   The radius of the circle is the distance between the point $(2, -1)$ and the tangent line $x + 2y + 5 = 0$.

39. A.   The equation of the locus is obtained by using $\left(\dfrac{y}{x - 2}\right)\left(\dfrac{y - 1}{x + 3}\right) = 1$.

40. D.   Here, since $f(x) = 2x - \dfrac{2}{x}$,

$$f\left(\frac{1}{x}\right) = \frac{2}{x} - 2x,$$

which is equal to $-f(x)$.

41. E.   The function is not defined at $x = -2$; $\lim\limits_{x \to 1^-} f(x) \neq \lim\limits_{x \to 1^+} f(x)$. Defining $f(-2) = 4$ will make f continuous at $x = -2$, but f will still be discontinuous at $x = 1$.

42. C.   The slope of $x + by = c$ is $-\dfrac{1}{b}$. A line perpendicular to this line must have slope b.

43. B.   Using $dy = [-\pi \sin(\pi t) + 3]\, dt$, with a calculator we compute dy for $t = 0.50$ and $dt = 0.03$; $dy = (-\pi + 3)(0.03)$.

44. C.   Since $3^{-x} - 3^x = -y$, this is equivalent to $3^x - 3^{-x} = y$. So $(-x, -y)$ is on the graph whenever (x, y) is.

45. E. $\frac{x}{2} + 5 \geqq 2$ implies $\frac{x}{2} \geqq -3$ implies $x \geqq -6$

$\frac{x}{2} + 5 \leqq -2$ implies $\frac{x}{2} \leqq -7$ implies $x \leqq -14$

46. D.   The midpoint of the diameter, $(7, 4)$, is the center of the circle. Its radius is $\sqrt{(7 - 10)^2 + (4 - 2)^2} = \sqrt{13}$.

47. A. $\lim\limits_{x \to 1^-} [x] - \lim\limits_{x \to 1^-} |x| = 0 - 1 = -1.$

48. E. $\frac{d}{dx}(f^2(x)) = 2f(x)f'(x),$

$\frac{d^2}{dx^2}(f^2(x)) = 2[f(x)f''(x) + f'(x)f'(x)]$
$= 2[ff'' + (f')^2].$
At $x = 3$; the answer is $2[2(-2) + 5^2] = 42.$

49. C.   The center of $x^2 + 2x + y^2 - 4y = 6$ is $(-1, 2)$; the center of $x^2 - 6x + y^2 + 4y = -1$ is $(3, -2)$.

50. B. $\lim\limits_{x \to 0} \frac{\cos x - 1}{x} = \lim\limits_{x \to 0} \frac{\cos x - 1}{x} \cdot \frac{\cos x + 1}{\cos x + 1} = \lim\limits_{x \to 0} \frac{\cos^2 x - 1}{x(\cos x + 1)}$

$= \lim\limits_{x \to 0} \frac{-\sin^2 x}{x(\cos x + 1)} = \lim\limits_{x \to 0} - \frac{\sin x}{x} \cdot \lim\limits_{x \to 0} \frac{\sin x}{\cos x + 1}$

$= -1 \cdot \frac{0}{2} = 0.$

Or, since the given limit is of the form $\frac{0}{0}$, use L'Hôpital's rule to get

$$\lim\limits_{x \to 0} \frac{-\sin x}{1} = -\frac{0}{1} = 0.$$

51. D.   The line $x + 3y + 3 = 0$ has slope $-\frac{1}{3}$; a line perpendicular to it has slope 3. The slope of the tangent to $y = x^2 - 2x + 3$ at any point is the derivative $2x - 2$. Set $2x - 2$ equal to 3.

52. A. $\lim\limits_{x \to 1} \frac{\frac{3}{x} - 3}{x - 1}$ is $f'(1)$, where $f(x) = \frac{3}{x}$, $f'(x) = -\frac{3}{x^2}$. Or, simplify the given fraction to

$$\frac{3 - 3x}{x(x - 1)} = \frac{3(1 - x)}{x(x - 1)} = \frac{-3}{x} \quad (x \neq 1).$$

53. B.   Only (B) is false. A counterexample is the pair of functions x and x^3. Each is an increasing function on the reals, but their product, x^4, is not. Verify that (A), (C), and (D) are all true.

54. E. $\frac{x + 3}{x^2 - 9} = \frac{x + 3}{(x + 3)(x - 3)} = \frac{1}{x - 3}$; $\lim\limits_{x \to 3^+} \frac{1}{x - 3} = +\infty$; $\lim\limits_{x \to 3^-} \frac{1}{x - 3} = -\infty.$

55. D.   In (D), $f(x)$ is not defined at $x = 0$. Verify that each of the other functions satisfies both conditions of the Mean Value Theorem.

56. A. $y' = 12x^3 - 12x^2 = 12x^2(x - 1)$. Since y' does not change sign as x passes through 0, but does change from $-$ to $+$ as x passes through 1, the local minimum of y occurs at $x = 1$; the corresponding y-value is -3. Since the slope at $x = 1$ is 0, the tangent there has equation $y = -3$.

57. B.   Since $|x - 2| = 2 - x$ if $x < 2$, the limit as $x \to 2^-$ is $\dfrac{2 - x}{x - 2} = -1$.

58. D.   Counterexamples are, respectively: for (A), $f(x) = |x|$, $c = 0$; for (B), $f(x) = x^3$, $c = 0$; for (C), $f(x) = x^4$, $c = 0$; for (E), $f(x) = x^2$ on $(-1, 1)$.

59. A.   Separate to get $\dfrac{dy}{y^2} = 2x \, dx$, $-\dfrac{1}{y} = x^2 + C$. Since $-(-1) = 1 + C$ implies that $C = 0$, the solution is

$$-\frac{1}{y} = x^2 \qquad \text{or} \qquad y = -\frac{1}{x^2}.$$

60. D. $f'(x) = 20x^3 - 5x^4 = 5x^3(4 - x)$. The signs are as follows:

$$\begin{array}{ccccc} & - & & + & & - \\ \hline & & 0 & & 4 & \end{array} \quad \to x \quad \leftarrow \text{signs of } f'$$

Since $f'(x) > 0$ on $(0, 4)$, f is increasing there.

61. B. $s = -16t^2 + 96t$ implies $v = -32t + 96$. Since $s = 0$ when $t = 0$ or $t = 6$, the velocity when the ball hits the ground (when $t = 6$) is -96 ft/sec.

62. D. $\dfrac{d}{du} \displaystyle\int_0^u \dfrac{1}{1 + x^2} \, dx = \dfrac{1}{1 + u^2}$. When $u = t^2$,

$$\frac{d}{dt} \int_0^u \frac{1}{1 + x^2} \, dx = \frac{1}{1 + u^2} \frac{du}{dt} = \frac{1}{1 + t^4} (2t).$$

63. E.   The signs within the intervals bounded by the critical points are given by

$$\begin{array}{ccccccc} + & & - & & + & & + \\ \hline & -3 & & 0 & & 2 & \end{array} \quad \to x \quad \leftarrow \text{signs of } f'$$

f has a local maximum at -3 and a local minimum at 0.

64. C.   Draw graphs to show that (A), (B), and (D) may be false.

65. E. $y = uv$ implies $y' = uv' + vu'$ implies $y'' = uv'' + v'u' + vu'' + v'u'$
$\qquad\qquad\qquad\qquad\qquad\qquad\qquad\qquad = uv'' + 2u'v' + vu''.$

66. B. $f'(x) = 15x^4 - 30x^2$; $f''(x) = 60x^3 - 60x = 60x(x + 1)(x - 1)$; this equals 0 when $x = -1$, 0, or 1. Here are the signs within the intervals:

$$\underset{\overset{|}{-1}\quad\overset{|}{0}\quad\overset{|}{1}}{\overset{-\qquad +\qquad -\qquad +}{\underline{\qquad\qquad\qquad\qquad\qquad\qquad}}} \rightarrow x \leftarrow\text{signs of } f''$$

67. D. Since $V = \frac{4}{3}\pi r^3$, $dV = 4\pi r^2\, dr$. When $dr = 0.1$, $dV = 4\pi r^2(0.1)$. But $S = 4\pi r^2$, so $dV = 0.1S$.

68. C. $(f_{av})_x = \dfrac{1}{4 - (-2)} \displaystyle\int_{-2}^{4} (3 + |x|)\, dx = \dfrac{1}{6}\left[\displaystyle\int_{-2}^{0}(3 - x)\,dx + \displaystyle\int_{0}^{4}(3 + x)dx\right]$

$= \dfrac{1}{6}\left[\left(3x - \dfrac{x^2}{2}\right)\Big|_{-2}^{0} + \left(3x + \dfrac{x^2}{2}\right)\Big|_{0}^{4}\right] = \dfrac{1}{6}[0 - (-6 - 2) + (12 + 8)] = \dfrac{14}{3}$.

69. C. Note that $f'(x) = \dfrac{4 + x}{x^2 + 4}$, so f has a critical value at $x = -4$. As x passes through -4, the sign of f' changes from $-$ to $+$, so f has a local minimum at $x = -4$.

70. A. Let $u = x^2 + 2$. Then

$$\frac{d}{du}\int_{0}^{u}\sqrt{1 + \cos t}\,dt = \sqrt{1 + \cos u}$$

and

$$\frac{d}{dx}\int_{0}^{u}\sqrt{1 + \cos t}\,dt = \sqrt{1 + \cos u}\,\frac{du}{dx} = \sqrt{1 + \cos(x^2 + 2)}\cdot(2x).$$

71. A. $\ln(3e^{x+2}) - \ln(3e^x) = \ln\dfrac{3e^{x+2}}{3e^x} = \ln e^2 = 2\ln e = 2$.

72. C. $\displaystyle\int_{0}^{6}|x - 4|\,dx = \int_{0}^{4}(4 - x)\,dx + \int_{4}^{6}(x - 4)dx = \left(4x - \dfrac{x^2}{2}\right)\Big|_{0}^{4} + \left(\dfrac{x^2}{2} - 4x\right)\Big|_{4}^{6}$

$= 8 + [(18 - 24) - (8 - 16)] = 8 + (-6 + 8) = 10.$

Save time by finding the area under $y = |x - 4|$ from a sketch! See Figure 11–4.

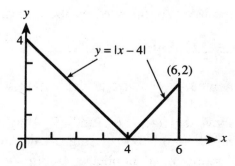

Area $= \dfrac{1}{2}(4 \cdot 4) + \dfrac{1}{2}(2 \cdot 2) = 10.$

FIGURE 11–4

73. D.   According to the Mean Value Theorem, the number c satisfies $f'(c) = \dfrac{f(1.5) - f(1)}{1.5 - 1}$. We seek c such that

$$\cos c = \frac{\sin 1.5 - \sin 1}{0.5}$$

Use a calculator in the radian mode.

74. B.

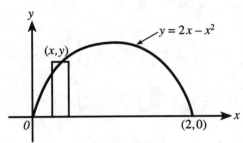

FIGURE 11–5

SHELL.

$$\Delta V = 2\pi RHT$$

$$= 2\pi x(2x - x^2)\,\Delta x;$$

$$V = 2\pi \int_0^2 x(2x - x^2)\,dx.$$

75. A. $\displaystyle\int_e^{e^4} \frac{1}{x}\,dx = \ln x \,\Big|_e^{e^4} = \ln e^4 - \ln e = 4 - 1 = 3.$

76. C.   Here are the relevant sign lines:

INTERVAL:	$x < 0$	$0 < x < \frac{2}{3}$	$\frac{2}{3} < x < 1$	$x > 1$
SIGN OF f':	$-$	$-$	$-$	$+$
SIGN OF f'':	$+$	$-$	$+$	$+$

We see that f' and f'' are both positive only if $x > 1$.

77. E.   We see from the sign lines in Question 76 that f has a local minimum (at $x = 1$) and two points of inflection (at $x = 0$ and at $x = \frac{2}{3}$).

78. E.   If Q_0 is the initial amount of the substance and Q is the amount at time t, then

$$Q = Q_0 e^{-kt}.$$

When $t = 8$, $Q = \frac{1}{2} Q_0$, so

$$\frac{1}{2} Q_0 = Q_0 e^{-8k}$$

and $\frac{1}{2} = e^{-8k}$. Thus $k = 0.08664$. Using a calculator, we seek t when $Q = \frac{1}{3} Q_0$. $\frac{1}{3} = e^{-0.08664t}$, so $t \approx 12.68$. Don't round off k too quickly.

79. D.   The period of $\frac{3}{\pi} \sin\left(\frac{\pi}{3} \cdot x\right)$ is $\dfrac{2\pi}{\frac{\pi}{3}} = 6$.

80. A.   Since the degrees of numerator and denominator are the same, the limit as $x \to \infty$ is the ratio of the coefficients of the terms of highest degree: $\frac{-2}{4}$.

81. E. $\int(\sqrt{x} - 2)\, x^2 \, dx = \int(x^{5/2} - 2x^2)\, dx = \frac{2}{7}x^{7/2} - \frac{2}{3}x^3 + C$.

82. E.   The domain of $\ln u$ is $u > 0$. Since $x^2 - 3x = x(x - 3)$, we note that $x(x - 3) > 0$ if $x < 0$ or $x > 3$, but that $x(x - 3) = 0$ if $x = 0$ or $x = 3$.

83. D.   Note that $f(x) = x + 6$ if $x \neq 6$, that $f(6) = 12$, and that $\lim\limits_{x \to 6} f(x) = 12$. So f is continuous at $x = 6$, which implies the truth of I and II.

84. B. $(f^{-1})'(y_0)$ denotes the derivative of the inverse of $f(x)$ when $y = y_0$. Setting $2 = x^5 + 3x - 2$, we get $4 = x^5 + 3x$, so $x = 1$ (when $y = 2$) by observation. Then

$$(f^{-1})'(y) = \frac{1}{5x^4 + 3} \qquad \text{and} \qquad (f^{-1})'(2) = \frac{1}{5 \cdot 1 + 3} = \frac{1}{8}.$$

85. D. $\frac{df}{dt} = (x^2 + 1)\frac{dx}{dt}$. Find x when $\frac{df}{dt} = 10\frac{dx}{dt}$.

$$10\frac{dx}{dt} = (x^2 + 1)\frac{dx}{dt}$$

implies that $x = 3$.

86. C.   The derivatives of $\ln(x + 1)$ are $\dfrac{1}{x + 1}, \dfrac{-1}{(x + 1)^2}, \dfrac{+2!}{(x + 1)^3}, \dfrac{-(3!)}{(x + 1)^4}, \ldots$

The nth derivative at $x = 2$ is $\dfrac{(-1)^{n-1}(n - 1)!}{3^n}$.

87. D.   With two subdivisions, the interval from 1 to 2 is divided at 1.5 and $h = \frac{1}{2}$. Then

$$\int_1^2 \sin \sqrt{x}\, dx \approx \frac{1}{2} \cdot \frac{1}{2}\left(\sin \sqrt{1} + 2 \sin \sqrt{1.5} + \sin \sqrt{2}\right).$$

Use the radian mode on your calculator to get 0.928.

88. A.

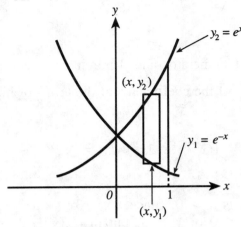

$$\Delta A = (y_2 - y_1)\, \Delta x;$$

$$A = \int_0^1 (e^x - e^{-x})\, dx$$

$$= (e^x + e^{-x})\Big|_0^1 = e + \frac{1}{e} - 2.$$

FIGURE 11–6

89. C. $\ln y = (\ln x)(\ln x) = \ln^2 x,$

$\dfrac{1}{y} y' = \dfrac{2 \ln x}{x}$ implies $y' = y\left(\dfrac{2 \ln x}{x}\right),$

$y' = x^{\ln x}\left(\dfrac{2 \ln x}{x}\right) = 2x^{\ln x - 1}\, \ln x.$

90. A. $\displaystyle\int_1^e \frac{\ln^3 x}{x}\, dx = \frac{1}{4}\, \ln^4 x\,\Big|_1^e = \frac{1}{4}(\ln^4 e - 0) = \frac{1}{4}.$

91. B

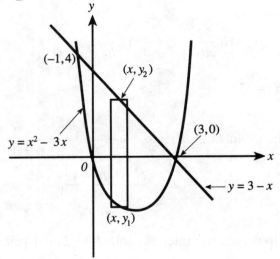

Solve for the points of intersection.

$$\Delta A = (y_2 - y_1)\, \Delta x;$$

$$A = \int_{-1}^3 [3 - x - (x^2 - 3x]\, dx$$

$$= \int_{-1}^3 (3 + 2x - x^2)\, dx.$$

FIGURE 11–7

92. C. $\displaystyle\int_{-\infty}^{\infty} \frac{dx}{x^2 + 1} = \lim_{b \to \infty} \tan^{-1} x\,\Big|_{-b}^{b} = \frac{\pi}{2} - \left(-\frac{\pi}{2}\right) = \pi.$

93. E.

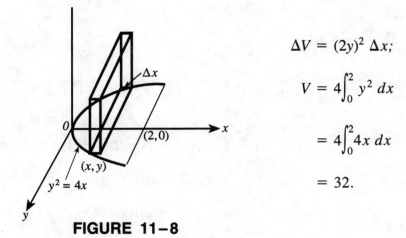

$$\Delta V = (2y)^2 \, \Delta x;$$

$$V = 4 \int_0^2 y^2 \, dx$$

$$= 4 \int_0^2 4x \, dx$$

$$= 32.$$

FIGURE 11–8

94. C.   According to Hooke's law, $F = kx$ (page 261), so $4 = 3k$ and $k = \frac{4}{3}$. The work

$$W = \int_0^4 \frac{4}{3} x \, dx = \frac{4}{3} \cdot \frac{x^2}{2} \Big|_0^4 = \frac{32}{3} \text{ in.-lb.}$$

95. C. $\int \cos^3 x \, dx = \int \cos^2 x \cdot \cos x \, dx = \int (1 - \sin^2 x) \cos x \, dx$

$$= \int \cos x \, dx - \int \sin^2 x \cdot \cos x \, dx = \sin x - \frac{\sin^3 x}{3} + C.$$

96. D. $\displaystyle\int_1^2 \frac{1}{\sqrt{4 - x^2}} \, dx = \lim_{h \to 2^-} \int_1^h \frac{1}{\sqrt{4 - x^2}} \, dx = \lim_{h \to 2^-} \sin^{-1} \frac{x}{2} \Big|_1^h$

$$= \lim_{h \to 2^-} \left(\sin^{-1} \frac{h}{2} - \sin^{-1} \frac{1}{2} \right) = \frac{\pi}{2} = \frac{\pi}{6} = \frac{\pi}{3}.$$

97. A.   The first three derivatives of $\dfrac{1}{1 - 2x}$ are $\dfrac{2}{(1-2x)^2}$, $\dfrac{8}{(1 - 2x)^3}$, and $\dfrac{48}{(1 - 2x)^4}$. The first four terms of the Maclaurin series (about $x = 0$) are 1, $+ 2x$, $+ \dfrac{8x^2}{2!}$, and $+ \dfrac{48x^3}{3!}$.

98. D.   Use parts (and tic-tac-toe), with $u = x$ and $dv = e^{-x} \, dx$.

$$
\begin{array}{ll}
x^2 & +\ e^{-x} \\
2x & -\ -e^{-x} \\
2 & +\ e^{-x} \\
0 & -e^{-x}
\end{array}
$$

The antiderivative is $-x^2 e^{-x} - 2x e^{-x} - 2e^{-x} + C.$

99. B.   The definition given is for $\lim\limits_{x \to \infty} f(x) = L$.

100. B.   We set

$$\frac{x - 1}{x^2 + 3x + 2} = \frac{x - 1}{(x + 1)(x + 2)} = \frac{A}{x + 1} + \frac{B}{x + 2}.$$

$$x - 1 = A(x + 2) + B(x + 1);$$

$$x = -2 \text{ implies } -3 = -B, \text{ or } B = 3;$$

$$x = -1 \text{ implies } -2 = A, \text{ or } A = -2.$$

AB Practice Examination 1

Section I

1. $\lim\limits_{x \to \infty} \dfrac{3x^2 - 4}{2 - 7x - x^2}$ is

   (A) 3   **(B)** 1   **(C)** -3 **(D)** ∞ **(E)** 0

2. The value of k for which the line $3x + ky - 5 = 0$ is perpendicular to the line $x - 2y + 7 = 0$ is

   (A) $\dfrac{3}{2}$ **(B)** $-\dfrac{3}{2}$ **(C)** 6   **(D)** -6 **(E)** $\dfrac{2}{3}$

3. $\lim\limits_{h \to 0} \dfrac{\cos\left(\dfrac{\pi}{2} + h\right)}{h}$ is

   (A) 1   **(B)** nonexistent   **(C)** 0   **(D)** -1 **(E)** none of these

4. The distance from the point $(6, -4)$ to the circle $x^2 + y^2 - 8y = 0$ is

   (A) $\sqrt{10}$ **(B)** 4   **(C)** 6   **(D)** 10   **(E)** none of these

5. The equation of the parabola with vertex at $(2, 1)$ and focus at $(2, 0)$ is

   (A) $x^2 - 4x - 4y + 8 = 0$ **(B)** $4y^2 - 8y + x + 2 = 0$
   (C) $y^2 - 4x - 2y + 9 = 0$ **(D)** $x^2 - 2x + 4y - 7 = 0$
   (E) $x^2 - 4x + 4y = 0$

6. The graph of $(x + 1)^2 - (y - 3)^2 = 0$ consists of

   (A) a hyperbola   **(B)** a point   **(C)** an ellipse
   (D) two intersecting lines   **(E)** two parallel lines

7. The maximum value of the function $f(x) = x^4 - 4x^3 + 6$ on the closed interval $[1, 4]$ is

   (A) 1   **(B)** 0   **(C)** 3   **(D)** 6   **(E)** none of these

8. The equation of the cubic with a relative maximum at $(0, 0)$ and a point of inflection at $(1, -2)$ is

   (A) $y = x^3 - 3x^2$ **(B)** $y = -x^3 + 3x^2$ **(C)** $y = 4x^3 - 6x^2$
   (D) $y = x^3 - 3x$ **(E)** $y = 3x - x^3$

9. If $f(x)$ is continuous at the point where $x = a$, which of the following statements may be false?

   (A) $\lim\limits_{x \to a} f(x)$ exists.   **(B)** $\lim\limits_{x \to a} f(x) = f(a)$.   **(C)** $f'(a)$ exists.

   (D) $f(a)$ is defined.   **(E)** $\lim\limits_{x \to a^-} f(x) = \lim\limits_{x \to a^+} f(x)$.

10. Which of the following functions is not everywhere continuous?

    (A) $y = |x|$ **(B)** $y = \dfrac{x}{x^2 + 1}$ **(C)** $y = \sqrt{x^2 + 8}$ **(D)** $y = x^{2/3}$

    (E) $y = \dfrac{4}{(x + 1)^2}$

11. The equation of the tangent to the curve of $y = x^2 - 4x$ at the point where the curve crosses the y-axis is

    (A) $y = 8x - 4$ **(B)** $y = -4x$ **(C)** $y = -4$ **(D)** $y = 4x$
    (E) $y = 4x - 8$

12. If y is a differentiable function of x, then the slope of the curve of $xy^2 - 2y + 4y^3 = 6$ at the point where $y = 1$ is

    (A) $-\dfrac{1}{18}$ **(B)** $-\dfrac{1}{26}$ **(C)** $\dfrac{5}{18}$ **(D)** $-\dfrac{11}{18}$ **(E)** 2

13. On the graph of $y = f(x)$, $f'(x)$ and $f''(x)$ are both positive on which interval?

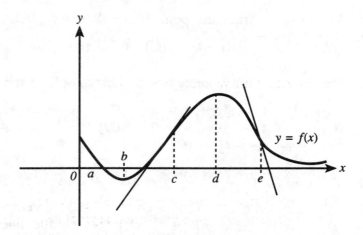

    (A) $0 < x < a$ **(B)** $b < x < c$ **(C)** $c < x < d$ **(D)** $d < x < e$
    (E) $x > e$

14. Suppose $\int_0^3 f(x + k)\,dx = 4$, where k is a constant. Then $\int_k^{3+k} f(x)\,dx$ equals

(A) 3   (B) $4 - k$ (C) 4   (D) $4 + k$ (E) none of these

15. If, for all x, $f'(x) = (x - 2)^4(x - 1)^3$, it follows that the function f has

(A) a relative minimum at $x = 1$
(B) a relative maximum at $x = 1$
(C) both a relative minimum at $x = 1$ and a relative maximum at $x = 2$
(D) neither a relative maximum nor a relative minimum
(E) relative minima at $x = 1$ and at $x = 2$

16. If a particle's motion along a straight line is given by

$$s = t^3 - 6t^2 + 9t + 2,$$

then s is increasing

(A) when $1 < t < 3$ (B) when $-1 < t < 3$ (C) for all t
(D) when $|t| > 3$ (E) when $t < 1$ or $t > 3$

17. Which of the following statements is *not* true of the graph of $y^2 = x^3 - x$?

(A) It is symmetric to the x-axis.
(B) It intersects the x-axis at 0, 1, and -1.
(C) It exists only for $|x| \leq 1$.
(D) It has no horizontal asymptotes.
(E) It has no vertical asymptotes.

18. The area in the first quadrant bounded by the curve $y = x^2$ and the line $y - x - 2 = 0$ is equal to

(A) $\dfrac{3}{2}$ (B) $\dfrac{2}{3}$ (C) $\dfrac{7}{6}$ (D) $\dfrac{10}{3}$ (E) $\dfrac{9}{2}$

19. Let $\begin{cases} f(x) = \dfrac{\sqrt{x + 4} - 3}{x - 5} & \text{if } x \neq 5 \\ f(5) = c, \end{cases}$
and let f be continuous at $x = 5$. Then $c =$

(A) $-\dfrac{1}{6}$ (B) 0   (C) $\dfrac{1}{6}$ (D) 1   (E) 6

20. The curve of the equation $(x^2 - 1)y = x^2 - 4$ has

(A) one horizontal and one vertical asymptote
(B) two vertical but no horizontal asymptotes
(C) one horizontal and two vertical asymptotes
(D) two horizontal and two vertical asymptotes
(E) neither a horizontal nor a vertical asymptote

21. If $f(x) = x^5 + 2$, then its inverse function $f^{-1}(x)$ is

    (A) $\dfrac{1}{x^5 + 2}$ (B) $\sqrt[5]{x + 2}$ (C) $\sqrt[5]{x} + 2$ (D) $\dfrac{1}{\sqrt[5]{x} - 2}$

    (E) $\sqrt[5]{x - 2}$

22. If $f(x) = \cos x \sin 3x$, then $f'\left(\dfrac{\pi}{6}\right)$ is equal to

    (A) $\dfrac{1}{2}$ (B) $-\dfrac{\sqrt{3}}{2}$ (C) 0 (D) 1 (E) $-\dfrac{1}{2}$

23. The region in the first quadrant under the curve of $y = e^{-x}$ and bounded at the right by $x = k$ is rotated about the x-axis. The volume of the solid obtained, as $k \to \infty$, equals

    (A) π (B) 2π (C) $\dfrac{1}{2}$ (D) $\dfrac{\pi}{2}$ (E) none of these

24. If $y = x^2 \ln x \ (x > 0)$, then y'' is equal to

    (A) $3 + \ln x$ (B) $3 + 2 \ln x$ (C) $3 \ln x$ (D) $3 + 3 \ln x$
    (E) $2 + x + \ln x$

25. $\displaystyle\int_0^1 \dfrac{x \, dx}{x^2 + 1}$ is equal to

    (A) $\dfrac{\pi}{4}$ (B) $\ln \sqrt{2}$ (C) $\dfrac{1}{2}(\ln 2 - 1)$ (D) $\dfrac{3}{2}$ (E) $\ln 2$

26. $\displaystyle\int_0^{\pi/2} \cos^2 x \sin x \, dx =$

    (A) -1 (B) $-\dfrac{1}{3}$ (C) 0 (D) $\dfrac{1}{3}$ (E) 1

27. The acceleration of a particle moving along a straight line is given by $a = 6t$. If, when $t = 0$, its velocity, v, is 1 and its distance, s, is 3, then at any time t

    (A) $s = t^3 + 3$ (B) $s = t^3 + 3t + 1$ (C) $s = t^3 + t + 3$

    (D) $s = \dfrac{t^3}{3} + t + 3$ (E) $s = \dfrac{t^3}{3} + \dfrac{t^2}{2} + 3$

28. If $y = f(x^2)$ and $f'(x) = \sqrt{5x - 1}$, then $\dfrac{dy}{dx}$ is equal to

    (A) $2x\sqrt{5x^2 - 1}$ (B) $\sqrt{5x - 1}$ (C) $2x\sqrt{5x - 1}$

    (D) $\dfrac{\sqrt{5x - 1}}{2x}$ (E) none of these

29. The equation of the ellipse with center at the origin and with a focus and a vertex respectively at $(2, 0)$ and $(3, 0)$ is

    (A) $13x^2 + 9y^2 = 117$ (B) $9x^2 + 5y^2 = 45$ (C) $5x^2 - 9y^2 = 45$
    (D) $5x^2 + 9y^2 = 45$ (E) $5x^2 + 4y^2 = 20$

30. $\int_0^1 xe^x \, dx$ equals

    (A) 1   (B) -1 (C) $2 - e$ (D) $\dfrac{e^2}{2} - e$ (E) $e - 1$

31. A 26-ft ladder leans against a building so that its foot moves away from the building at the rate of 3 ft/sec. When the foot of the ladder is 10 ft from the building, the top is moving down at the rate of r ft/sec, where r is

    (A) $\dfrac{46}{3}$ (B) $\dfrac{3}{4}$ (C) $\dfrac{5}{4}$ (D) $\dfrac{5}{2}$ (E) $\dfrac{4}{5}$

32. If $\sin x = \ln y$ and $0 < x < \pi$, then, in terms of x, $\dfrac{dy}{dx}$ equals

    (A) $e^{\sin x} \cos x$ (B) $e^{-\sin x} \cos x$ (C) $\dfrac{e^{\sin x}}{\cos x}$ (D) $e^{\cos x}$

    (E) $e^{\sin x}$

33. If $\dfrac{dx}{dt} = kx$, and if $x = 2$ when $t = 0$ and $x = 6$ when $t = 1$, then k equals

    (A) $\ln 4$ (B) 8   (C) e^3 (D) 3   (E) none of these

34. Let $f(x) = 2 - \sin^2 \dfrac{\pi x}{3}$. The period of f is

    (A) 1   (B) 2   (C) 3   (D) 4   (E) 6

35. If $f(x) = \sqrt{1 - \dfrac{2}{x}}$ and $g(x) = \dfrac{1}{x}$ $(x \neq 0)$, then the derivative of $f(g(x))$

    (A) $= 0$ (B) $= \dfrac{1}{x^2 \sqrt{1 - \dfrac{2}{x}}}$ (C) does not exist

    (D) $= -\dfrac{1}{\sqrt{1 - 2x}}$ (E) $= \dfrac{-2}{(x^2 - 2)^{3/2}}$

36. If $F(x) = \int_1^{2x} \dfrac{1}{1 - t^3} \, dt$, then $F'(x) =$

    (A) $\dfrac{1}{1 - x^3}$ (B) $\dfrac{1}{1 - 2x^3}$ (C) $\dfrac{2}{1 - 2x^3}$ (D) $\dfrac{1}{1 - 8x^3}$

    (E) $\dfrac{2}{1 - 8x^3}$

37. Which of the following functions does not satisfy the hypotheses of Rolle's theorem on the interval $[0, 1]$?

   (A) $f(x) = \dfrac{x^2 - x}{2x - 1}$ (B) $f(x) = x^2 - x$ (C) $f(x) = x^3 - x^2$

   (D) $f(x) = (x - 1)(e^x - 1)$ (E) $f(x) = \dfrac{x^2 - x}{x + 1}$

38. The number of real zeros of the function $f(x) = 2x^5 + 3x^3 + 4x + 1$ is

   (A) 0   (B) 1   (C) 2   (D) 3   (E) 5

39. The total area of the region bounded by the graph of $y = x\sqrt{1 - x^2}$ and the x-axis is

   (A) $\dfrac{1}{3}$ (B) $\dfrac{1}{3}\sqrt{2}$ (C) $\dfrac{1}{2}$ (D) $\dfrac{2}{3}$ (E) 1

40. The curve of $y = \dfrac{1 - x}{x - 3}$ is concave up when

   (A) $x > 3$ (B) $1 < x < 3$ (C) $x > 1$ (D) $x < 1$ (E) $x < 3$

41. The area of the largest isosceles triangle that can be drawn with one vertex at the origin and with the others on a line parallel to and above the x-axis and on the curve $y = 27 - x^2$ is

   (A) 108   (B) 27   (C) $12\sqrt{3}$ (D) 54   (E) $24\sqrt{3}$

42. The average (mean) value of $\tan x$ on the interval from $x = 0$ to $x = \dfrac{\pi}{3}$ is

   (A) $\ln 2$ (B) $\dfrac{3}{\pi} \ln 2$ (C) $\ln \dfrac{1}{2}$ (D) $\dfrac{9}{\pi}$ (E) $\dfrac{\sqrt{3}}{2}$

43. Which of the following intervals is in the domain of the function $f(x) = \ln (\cos x)$?

   (A) $-\dfrac{\pi}{2} \leqq x \leqq \dfrac{\pi}{2}$ (B) $-\dfrac{\pi}{2} < x < \dfrac{\pi}{2}$ (C) $0 \leqq x \leqq \pi$

   (D) $0 < x < \pi$ (E) $0 \leqq x \leqq \dfrac{\pi}{2}$

44. If $y = \sin (x^2 - 1)$ and $x = \sqrt{u^2 + 1}$, then $\dfrac{dy}{du}$ equals

   (A) $\dfrac{u \cos u^2}{\sqrt{1 + u^2}}$ (B) $\cos u^2$ (C) $\dfrac{\cos u^2}{2\sqrt{1 + u^2}}$ (D) $2u \cos u^2$

   (E) $\dfrac{u \cos (x^2 - 1)}{\sqrt{1 + u^2}}$

45. If $f(x) = x^n$, n a positive integer, the smallest integer q for which the qth derivative equals zero is

(A) 1   **(B)** $n - 1$ **(C)** n **(D)** $n + 1$ **(E)** $n + 2$

AB Practice Examination 1, Section II

1. (a) Sketch the region bounded by the curve of $y^2 = x^3$ and the line segment joining the points $(1, 1)$ and $(4, -8)$.

(b) Find the area of this region.

2. If the parabola $2y = x^2 - 6$ is rotated about the y-axis, it generates a paraboloid of revolution. Suppose a vessel having the shape of this paraboloid contains water to a depth of 4 ft. Determine the amount of water which must be removed to lower the surface by 2 ft.

3. A particle moves along a line so that its position s at time t is given by

$$s(t) = \int_0^t (3w^2 - 6w)\, dw.$$

(a) Find its acceleration at any time t.

(b) Determine the values of t for which the particle is moving in a positive direction.

(c) Find the values of t for which the particle is slowing down.

4. The vertices of a triangle are $(0, 0)$, $(x, \sin x)$, and $(\cos^3 x, 0)$, where $0 \leqq x \leqq \frac{\pi}{2}$. Find the value of x for which the area of the triangle is a maximum. Justify your answer.

5. Consider the family of straight lines with a given slope m, and the chords whose endpoints are the intersections of these lines with the parabola $y = x^2$. Find an equation of the locus of the midpoints of these chords.

6. Let u and v be functions defined for all real numbers and satisfying the conditions:

(a) $u(a + b) = u(a) \cdot u(b)$ for all reals a and b;

(b) $u(x) = 1 + xv(x)$;

(c) $\lim\limits_{x \to 0} v(x) = 1$.

Prove that u has a derivative for each x and that $u'(x) = u(x)$.

Answers to AB Practice Examination 1

SECTION I

1. C	**9.** C	**17.** C	**25.** B	**33.** E	**41.** D
2. A	**10.** E	**18.** D	**26.** D	**34.** C	**42.** B
3. D	**11.** B	**19.** C	**27.** C	**35.** D	**43.** B
4. C	**12.** A	**20.** C	**28.** A	**36.** E	**44.** D
5. E	**13.** B	**21.** E	**29.** D	**37.** A	**45.** D
6. D	**14.** C	**22.** E	**30.** A	**38.** B	
7. D	**15.** A	**23.** D	**31.** C	**39.** D	
8. A	**16.** E	**24.** B	**32.** A	**40.** E	

1. C. Use the Rational Function Theorem on page 23.

2. A. Find k such that $-\dfrac{3}{k} = -2$.

3. D. $\lim\limits_{h \to 0} \dfrac{\cos\left(\dfrac{\pi}{2} + h\right)}{h} = \lim\limits_{h \to 0} \dfrac{-\sin h}{h} = -1$.

4. C. The center of the circle is $(0, 4)$; the radius is 4. The distance from $(6, -4)$ to $(0, 4)$ is 10; the distance to the circle is therefore 6.

5. E. See Equation 44 on page 485. Since $p = -1$, the answer is
$$(x - 2)^2 = -4(y - 1).$$

6. D. The two lines have equations $(x + 1) \pm (y - 3) = 0$.

7. D. On the interval $[1, 4]$, $f'(x) = 0$ only for $x = 3$. Since $f(3)$ is a relative minimum, we check the endpoints and find that $f(4) = 6$ is the absolute maximum of the function.

8. A. Let $y = ax^3 + bx^2 + cx + d$; use the equations

$$y(0) = 0, \quad y'(0) = 0, \quad y''(1) = 0, \quad \text{and } y(1) = -2$$

to determine the constants and use the following derivatives:

$$y' = 3ax^2 + 2bx + c, \quad y'' = 6ax + 2b.$$

Then $0 = d$, $0 = c$, and $0 = 6a + 2b$. So $b = -3a$. From $y(1) = -2$, we have $-2 = a - 3a$ and $a = 1$. Then $b = -3$. The given function is $x^3 - 3x^2$.

9. C.   The absolute value function $f(x) = |x|$ is continuous at $x = 0$, but $f'(0)$ does not exist.

10. E.   Since the function in (E) is not defined at $x = -1$, it is not continuous at that point.

11. B.   The curve crosses the y-axis only at the origin. Its slope there is -4.

12. A.   Differentiating implicitly yields $2xyy' + y^2 - 2y' + 12y^2y' = 0$. When $y = 1$, $x = 4$. Substitute to find y'.

13. B.   The curve both rises and is concave up only on the interval $b < x < c$.

14. C.   Let $F'(x) = f(x)$; then $F'(x + k) = f(x + k)$;

$$\int_0^3 f(x + k)\, dx = F(3 + k) - F(k);$$

$$\int_k^{3+k} f(x)\, dx = F(3 + k) - F(k).$$

Or let $u = x + k$. Then $dx = du$; when $x = 0$, $u = k$; when $x = 3$, $u = 3 + k$.

15. A.   Although $f'(2) = f'(1) = 0$, $f'(x)$ changes sign only as x increases through 1, and in this case $f'(x)$ changes from negative to positive.

16. E.   The distance s is increasing when $v > 0$. Here

$$v = 3t^2 - 12t + 9 = 3(t - 1)(t - 3).$$

So $v > 0$ when $t < 1$ or $t > 3$.

17. C.   The domain of this relation is $\{x \,|\, -1 \leq x \leq 0 \text{ or } x \geq 1\}$.

18. D.   The area is given, here, by $\int_0^2 (2 + x - x^2)\, dx$.

19. C.   To find $\lim f$ as $x \to 5$ (if it exists), multiply f by $\dfrac{\sqrt{x + 4} + 3}{\sqrt{x + 4} + 3}$.
Then

$$f(x) = \frac{x - 5}{(x - 5)(\sqrt{x + 4} + 3)},$$

and if $x \neq 5$ this equals $\dfrac{1}{\sqrt{x + 4} + 3}$. So $\lim f(x)$ as $x \to 5$ is $\dfrac{1}{6}$. For f to be continuous at $x = 5$, $f(5)$ or c must also equal $\dfrac{1}{6}$.

20. C. The graph of $y = \frac{x^2 - 4}{x^2 - 1}$ has a horizontal asymptote at $y = 1$ and two vertical asymptotes at $x = 1$ and $x = -1$.

21. E. Let $y = x^5 + 2$. Then $x = \sqrt[5]{y - 2}$. Interchange x and y.

22. E. $f'(x) = 3 \cos x \cos 3x - \sin x \sin 3x$.

23. D. The volume is given by $\lim\limits_{k \to \infty} \pi \int_0^k e^{-2x} \, dx = \frac{\pi}{2}$.

24. B. $y' = x + 2x \ln x$ and $y'' = 3 + 2 \ln x$.

25. B. Evaluate $\frac{1}{2} \ln (x^2 + 1) \Big|_0^1$.

26. D. Evaluate $-\frac{1}{3} \cos^3 x \Big|_0^{\pi/2}$.

27. C. $v = 3t^2 + 1$ and $s = t^3 + t + 3$.

28. A. Let $u = x^2$. Then

$$\frac{dy}{dx} = \frac{dy}{du} \cdot \frac{du}{dx} = \frac{df}{du} \cdot \frac{du}{dx} = f'(u) \frac{du}{dx} = \sqrt{5u - 1} \cdot 2x = 2x\sqrt{5x^2 - 1}.$$

29. D. See Equation 45 on page 485. Find b by using the fact that $a^2 = b^2 + c^2$.

30. A. See page 195, Answer 19.

31. C. See Figure AB1–1. Since $x^2 + y^2 = 26^2$ and since it is given that $\frac{dx}{dt} = 3$, it follows that

$$2x\frac{dx}{dt} + 2y\frac{dy}{dt} = 0 \qquad \text{and} \qquad \frac{dy}{dt} = -\frac{x}{y}(3)$$

at any time t. When $x = 10$, then $y = 24$ and $\frac{dy}{dt} = \frac{-5}{4}$.

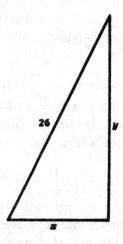

26

y

x

FIGURE AB1–1

32. A. $\cos x = \dfrac{1}{y}\,\dfrac{dy}{dx}$ and $\dfrac{dy}{dx} = y \cos x$. From the equation given, $y = e^{\sin x}$.

33. E.   Separating variables yields $\dfrac{dx}{x} = k\,dt$. Integrating, we get $\ln x = kt + C$. Since $x = 2$ when $t = 0$, $\ln 2 = C$. Then $\ln \frac{x}{2} = kt$. Using $x = 6$ when $t = 1$, it follows that $\ln 3 = k$.

34. C.   Since the period of $\sin bx$ is $\frac{2\pi}{b}$, the period of $\sin^2 bx$ is $\frac{\pi}{b}$.

35. D.   Note that $f(g(x)) = \sqrt{1 - \dfrac{2}{1/x}} = \sqrt{1 - 2x}\ (x \neq 0)$.

36. E.   Let $u = 2x$ and note that $F'(u) = \dfrac{1}{1 - u^3}$. Then

$$F'(x) = F'(u)u'(x) = 2F'(u) = 2 \cdot \dfrac{1}{1 - (2x)^3}.$$

37. A.   The function in (A) is not continuous on $[0, 1]$ since $f\left(\frac{1}{2}\right)$ does not exist.

38. B. $f'(x) = 10x^4 + 9x + 4$, which is positive for all x. So f is an increasing function. Since $f(-1) < 0$ while $f(0) > 0$, f has exactly one real zero (between $x = -1$ and $x = 0$).

39. D.   Note that the domain of y is all x such that $|x| \leqq 1$ and that the graph is symmetric to the origin. The area is given by

$$2 \int_0^1 x\sqrt{1 - x^2}\, dx.$$

40. E.   Since

$$y' = 2(x - 3)^{-2} \quad \text{and} \quad y'' = -4(x - 3)^{-3} = \dfrac{-4}{(x - 3)^3},$$

we see that y is positive when $x < 3$.

41. D.   Draw a figure and let (x, y) be the point in the first quadrant where the line parallel to the x-axis meets the parabola. The area of the triangle is given by $A = xy = x(27 - x^2)$. Show that A is a maximum for $x = 3$.

42. B. $\dfrac{1}{\pi/3} \displaystyle\int_0^{\pi/3} \tan x\, dx = \dfrac{3}{\pi}\left[-\ln \cos x\right]_0^{\pi/3} = \dfrac{3}{\pi}\left(-\ln \dfrac{1}{2}\right)$.

43. B.   Since $\ln u$ is defined only if $u > 0$, the domain of $\ln (\cos x)$ can contain only x for which $\cos x > 0$.

44. D. $\dfrac{dy}{du} = \dfrac{2x \cos (x^2 - 1) \cdot u}{\sqrt{u^2 + 1}}$.

45. D.   The first derivative is of degree $(n - 1)$, the second of degree $(n - 2)$, . . . , the nth of degree $(n - n)$ or 0. Since the derivative of the nth degree is a constant, the $(n + 1)$st is equal to 0.

SECTION II

1. (a) See Figure AB1 – 2 below.

(b) The required area is labeled ROQ. The equation of the line RQ through $(1, 1)$ and $(4, -8)$ is

$$y - 1 = -3(x - 1) \quad \text{or} \quad x = \frac{4 - y}{3}.$$

We describe two methods for finding the area.

METHOD 1. If the area is found by summing horizontal elements, then

$$A = \int_{-8}^{1} (x_2 - x_1)\, dy,$$

where x_2 is an abscissa on the line and x_1 on the curve. So

$$A = \int_{-8}^{1} \left(\frac{4 - y}{3} - y^{2/3} \right) dy$$

$$= \frac{4y}{3} - \frac{y^2}{6} - \frac{3}{5} y^{5/3} \Big|_{-8}^{1} = \frac{27}{10}.$$

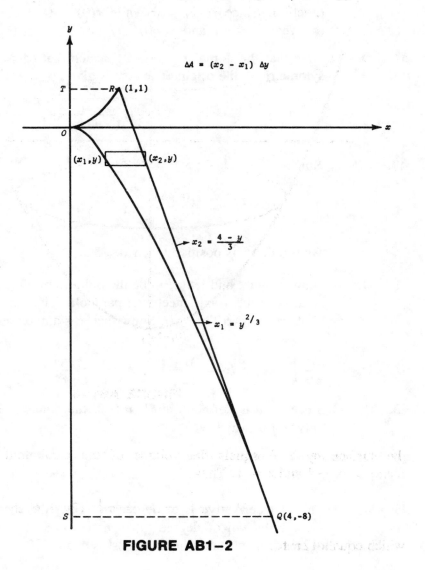

FIGURE AB1–2

METHOD 2. *A* can also be found by subtracting from trapezoid *RQST* the area bounded by the curve, the *y*-axis, and the horizontal lines $y = -8$ and $y = 1$. Then

$$A = \frac{1}{2}(1 + 4)\cdot 9 - \int_{-8}^{1} y^{2/3}\, dy = \frac{45}{2} - \frac{99}{5} = \frac{27}{10}.$$

2. See Figure AB1–3. The amount of water that must be removed to lower

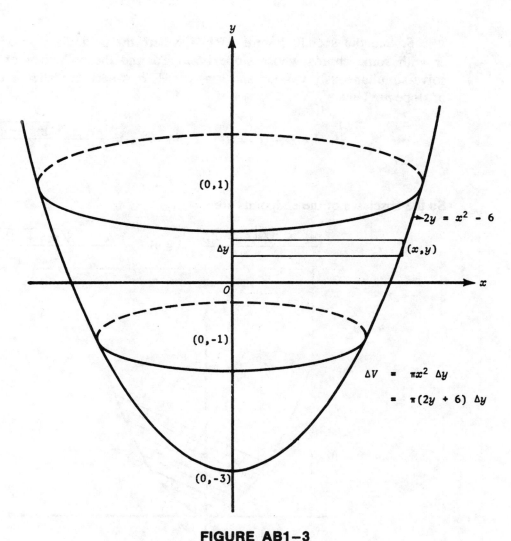

FIGURE AB1–3

the surface by 2 ft equals the volume of the paraboloid contained between the planes $x = -1$ and $x = 1$. Thus

$$V = \pi \int_{-1}^{1} x^2\, dy = \pi \int_{-1}^{1} (2y + 6)\, dy,$$

which equals 12π ft³.

3. (a) Since $v = \dfrac{ds}{dt} = 3t^2 - 6t$, $a = \dfrac{dv}{dt} = 6t - 6$.

   (b) $v > 0$ if $t < 0$ or $t > 2$.

   (c) The particle is slowing down whenever v and a have opposite signs. Since a is positive when $t > 1$ and negative otherwise, the particle is slowing down if $t < 0$ or if $1 < t < 2$.

4. The area A of the triangle equals $\dfrac{1}{2} \cos^3 x \sin x$, and this is a maximum when $x = \dfrac{\pi}{6}$. Note that $\dfrac{d^2A}{dx^2}$ is negative for this value of x.

5. See the sketch, Figure AB1–4, where the parabola $y = x^2$ is shown together with some chords whose slope is m. To find the endpoints of these chords, we solve simultaneously $y = x^2$ and $y = mx + b$, where the latter is the family of lines of slope m. Thus

$$x^2 - mx - b = 0 \quad \text{and} \quad x = \frac{m \pm \sqrt{m^2 + 4b}}{2}.$$

So the abscissas of the endpoints are

$$\frac{m + \sqrt{m^2 + 4b}}{2} \quad \text{and} \quad \frac{m - \sqrt{m^2 + 4b}}{2}.$$

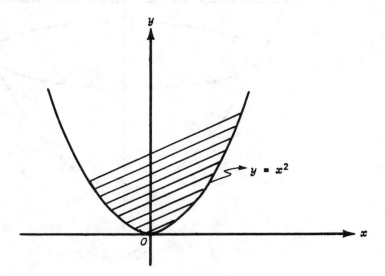

FIGURE AB1–4

The x value of the midpoint of any of these chords is $\dfrac{1}{2}m$. The locus of the centers of the chords, then, is the line $x = \dfrac{1}{2}m$.

6. The definition of $u'(x)$ is $\lim\limits_{h \to 0} \dfrac{u(x + h) - u(x)}{h}$ if this limit exists. We know that (1) $u(a + b) = u(a) \cdot u(b)$ for all real a and b; (2) $u(x) = 1 + xv(x)$; and (3) $\lim\limits_{x \to 0} v(x) = 1$. So

$$\frac{u(x + h) - u(x)}{h} = \frac{u(x) \cdot u(h) - u(x)}{h} \qquad \text{by (1);}$$

$$= u(x) \frac{[u(h) - 1]}{h}$$

$$= u(x) \frac{[1 + hv(h) - 1]}{h} \qquad \text{by (2);}$$

$$= u(x) \frac{[hv(h)]}{h} = u(x)v(h) \qquad (h \neq 0).$$

Now $u'(x) = \lim\limits_{h \to 0} [u(x)v(h)]$ if it exists. But

$$\lim\limits_{h \to 0} [u(x)v(h)] = \lim\limits_{h \to 0} u(x) \lim\limits_{h \to 0} v(h),$$

since the limit of a product equals the product of the limits, if they exist. Thus $u'(x) = u(x) \cdot 1 = u(x)$, since we know that $u(x)$ is defined for all x and that $\lim\limits_{h \to 0} v(h) = 1$ by (3).

AB Practice Examination 2

Section I

1. $\lim\limits_{x \to \infty} \dfrac{20x^2 - 13x + 5}{5 - 4x^3}$ is

   (A) -5 **(B)** ∞ **(C)** 0 **(D)** 5 **(E)** 1

2. $\lim\limits_{x \to \frac{\pi}{2}} \dfrac{\cos x}{x - \dfrac{\pi}{2}}$ is

   (A) -1 **(B)** 1 **(C)** 0 **(D)** ∞ **(E)** none of these

3. $\lim\limits_{x \to 0} x \sin \dfrac{1}{x}$ is

   (A) 1 **(B)** 0 **(C)** ∞ **(D)** -1 **(E)** none of these

4. $\lim\limits_{h \to 0} \dfrac{\ln(2 + h) - \ln 2}{h}$ is

   (A) 0 **(B)** $\ln 2$ **(C)** $\dfrac{1}{2}$ **(D)** $\dfrac{1}{\ln 2}$ **(E)** ∞

5. If $y = \dfrac{x - 3}{2 - 5x}$, then $\dfrac{dy}{dx}$ equals

   (A) $\dfrac{17 - 10x}{(2 - 5x)^2}$ **(B)** $\dfrac{13}{(2 - 5x)^2}$ **(C)** $\dfrac{x - 3}{(2 - 5x)^2}$ **(D)** $\dfrac{17}{(2 - 5x)^2}$

   (E) $\dfrac{-13}{(2 - 5x)^2}$

6. If $y = e^{-x^2}$, then $y''(0)$ equals

   (A) 2   (B) -2 (C) $\dfrac{2}{e}$ (D) 0   (E) -4

7. If $f(x) = x \cos \dfrac{1}{x}$, then $f'\left(\dfrac{2}{\pi}\right)$ equals

   (A) $\dfrac{\pi}{2}$ (B) $-\dfrac{2}{\pi}$ (C) -1 (D) $-\dfrac{\pi}{2}$ (E) 1

8. If $xy^2 - 3x + 4y - 2 = 0$ and y is a differentiable function of x, then $\dfrac{dy}{dx}$ equals

   (A) $-\dfrac{1 + y^2}{2xy}$ (B) $\dfrac{3}{2y + 4}$ (C) $\dfrac{3}{2xy + 4}$ (D) $\dfrac{3 - y^2}{2xy + 4}$

   (E) $\dfrac{5 - y^2}{2xv + 4}$

9. An equation of the reflection of $y = 3 - e^{-x}$ in the x-axis is

   (A) $y = e^{-x} - 3$ (B) $y = e^x - 3$ (C) $y = 3 - e^x$
   (D) $y = 3 + e^x$ (E) $y = -\ln(3 - x)$

10. If y is a differentiable function of x, then the derivative of $\sin^2(x + y)$ with respect to x is

    (A) $2[\sin(x + y)]\dfrac{dy}{dx}$ (B) $[\cos^2(x + y)]\left(1 + \dfrac{dy}{dx}\right)$

    (C) $[\sin 2(x + y)]\left(1 + \dfrac{dy}{dx}\right)$ (D) $\cos^2(x + y)$ (E) $2\sin(x + y)$

11. The equation of the tangent to the curve $y = e^x \ln x$, where $x = 1$, is

    (A) $y = ex$ (B) $y = e^x + 1$ (C) $y = e(x - 1)$ (D) $y = ex + 1$
    (E) $y = x - 1$

12. Let $f(x) = x^5 + 1$ and let g be the inverse function of f. What is the value of $g'(0)$?

    (A) -1 (B) $\dfrac{1}{5}$ (C) 1   (D) $g'(0)$ does not exist.

    (E) It cannot be determined from the given information.

13. The hypotenuse AB of a right triangle ABC is 5 ft, and one leg, AC, is decreasing at the rate of 2 ft/sec. The rate, in square feet per second, at which the area is changing when $AC = 3$ is

    (A) $\dfrac{25}{4}$ (B) $\dfrac{7}{4}$ (C) $-\dfrac{3}{2}$ (D) $-\dfrac{7}{4}$ (E) $-\dfrac{7}{2}$

14. The derivative of a function f is given for all x by

$$f'(x) = x^2(x + 1)^3(x - 4)^2.$$

The set of x for which f is a relative maximum is

(A) $\{0, -1, 4\}$ (B) $\{-1\}$ (C) $\{0, 4\}$ (D) $\{1\}$
(E)   none of these

15. The position of a point P on a line at time t is given by

$$s = t^3 + t^2 - t - 3.$$

P is moving to the right for

(A) $t > -1$ (B) $t < -\dfrac{1}{3}$ or $t > 1$ (C) $t < -1$ or $t > \dfrac{1}{3}$

(D) $-1 < t < \dfrac{1}{3}$ (E) $t < \dfrac{1}{3}$

16. If the displacement from the origin of a particle on a line is given by $s = 3 + (t - 2)^4$, then the number of times the particle reverses direction is

(A) 0   (B) 1   (C) 2   (D) 3   (E)   none of these

17. At which point on the graph of $y = f(x)$ below is $f'(x) < 0$ and $f''(x) > 0$?

(A) A (B) B (C) C (D) D (E) E

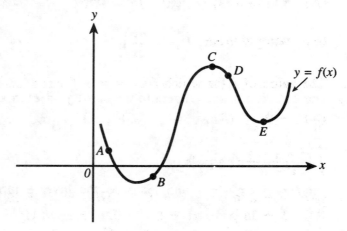

18. The maximum value of the function $f(x) = xe^{-x}$ is

(A) $\dfrac{1}{e}$ (B) e (C) 1   (D) -1 (E)   none of these

19. A rectangle of perimeter 18 in. is rotated about one of its sides to generate a right circular cylinder. The rectangle which generates the cylinder of largest volume has area, in square inches, of

(A) 14   (B) 20   (C) $\dfrac{81}{4}$ (D) 18   (E) $\dfrac{77}{4}$

20. If $f'(x)$ exists on the closed interval $[a, b]$, then it follows that

  (A) $f(x)$ is constant on $[a, b]$
  (B)   there exists a number c, $a < c < b$, such that $f'(c) = 0$
  (C)   the function has a maximum value on the open interval (a, b)
  (D)   the function has a minimum value on the open interval (a, b)
  (E)   the mean-value theorem applies

21. $\int_1^2 (3x - 2)^3 \, dx$ is equal to

  (A) $\dfrac{16}{3}$ (B) $\dfrac{63}{4}$ (C) $\dfrac{13}{3}$ (D) $\dfrac{85}{4}$ (E)   none of these

22. $\int_{\pi/4}^{\pi/2} \sin^3 \alpha \cos \alpha \, d\alpha$ is equal to

  (A) $\dfrac{3}{16}$ (B) $\dfrac{1}{8}$ (C) $-\dfrac{1}{8}$ (D) $-\dfrac{3}{16}$ (E) $\dfrac{3}{4}$

23. $\int x \cos x^2 \, dx$ equals

  (A) $\sin x^2 + C$ (B) $2 \sin x^2 + C$ (C) $-\dfrac{1}{2} \sin x^2 + C$

  (D) $\dfrac{1}{4} \cos^2 x^2 + C$ (E) $\dfrac{1}{2} \sin x^2 + C$

24. $\int_0^1 \dfrac{e^x}{(3 - e^x)^2} \, dx$ equals

  (A) $3 \ln (e - 3)$ (B) 1 (C) $\dfrac{1}{3 - e}$ (D) $\dfrac{e - 2}{3 - e}$

  (E)   none of these

25. The value of c for which $f(x) = x + \dfrac{c}{x}$ has a local minimum at $x = 3$ is

  (A) -9 (B) -6 (C) -3 (D) 6 (E) 9

26. $\int \dfrac{x^2 + x - 1}{x^2 - x} \, dx$ equals

  (A) $x + 2 \ln |x^2 - x| + C$ (B) $\ln |x| + \ln |x - 1| + C$
  (C) $1 + \ln |x^2 - x| + C$ (D) $x + \ln |x^2 - x| + C$
  (E) $x - \ln |x| - \ln |x - 1| + C$

27. $\int_{-1}^1 (1 - |x|) \, dx$ equals

  (A) 0 (B) $\dfrac{1}{2}$ (C) 1 (D) 2 (E)   none of these

28. $\int_{-1}^0 e^{-x} \, dx$ equals

  (A) $1 - e$ (B) $\dfrac{1 - e}{e}$ (C) $e - 1$ (D) $1 - \dfrac{1}{e}$ (E) $e + 1$

29. The set of real x for which $f(g(x)) = g(f(x))$, where $f(x) = x + 2$ and $g(x) = \frac{3}{x}$, is

(A) $\{0, 2\}$ (B) $\{0, -2\}$ (C) $\{-1, 3\}$ (D) $\{-3, +1\}$
(E) the empty set

30. If $F'(x) = G'(x)$ for all x and k is a constant, then it is necessary that

(A) $F(x) = G(x) + k$ (B) $F(x) = G(x)$ (C) $F(k) = G(k)$
(D) $F(x) = kG(x)$ (E) $F(x) = G(x + k)$

31. The area enclosed by the curve of $y^2 = x^3$ and the line segment joining $(1, 1)$ and $(4, -8)$ is given by

(A) $2\int_0^1 x^{3/2}\,dx + \int_1^4 (-3x + 4 - x^{3/2})\,dx$

(B) $2\int_0^1 x^{3/2}\,dx + \int_1^4 (4 - 3x + x^{3/2})\,dx$ (C) $\int_0^4 (4 - 3x - x^{3/2})\,dx$

(D) $2\int_0^1 x^{3/2}\,dx + \int_{-8}^1 \left(\frac{4 - y}{3} - y^{2/3}\right) dy$ (E) $\int_1^4 (x^{3/2} + 3x - 4)\,dx$

32. The first-quadrant area bounded below by the x-axis and laterally by the curves of $y = x^2$ and $y = 4 - x^2$ equals

(A) $6 - 2\sqrt{3}$ (B) $\frac{10}{3}$ (C) $\frac{8}{3}(2 - \sqrt{2})$ (D) $\frac{4\sqrt{2}}{3}$

(E) none of these

33. The equation of the curve shown below is $y = \frac{4}{1 + x^2}$. What does the area of the shaded region of the rectangle equal?

(A) $4 - \frac{\pi}{4}$ (B) $8 - 2\pi$ (C) $8 - \pi$ (D) $8 - \frac{\pi}{2}$

(E) $2\pi - 4$

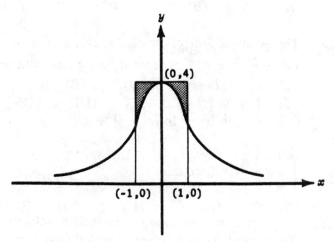

34. The amplitude of the function $f(x) = \sqrt{3}\sin x + 3\cos x$ is

(A) $3 + \sqrt{3}$ (B) $2\sqrt{3}$ (C) 3 (D) $\sqrt{3}$ (E) $3 - \sqrt{3}$

35. The volume of an "inner tube" with inner diameter 4 ft and outer diameter 8 ft is, in cubic feet,

   (A) $4\pi^2$ (B) $12\pi^2$ (C) $8\pi^2$ (D) $24\pi^2$ (E) $6\pi^2$

36. The volume generated by rotating the region bounded by $y = \dfrac{1}{\sqrt{x}}$, $x = 1$, $x = 4$, and $y = 0$ about the y-axis is

   (A) $\dfrac{28\pi}{3}$ (B) $\pi \ln 4$ (C) $\dfrac{64\pi}{3}$ (D) $\dfrac{14\pi}{3}$ (E) $\dfrac{16\pi}{3}$

37. If $f(x)$ and $g(x)$ are both continuous functions on the closed interval $[a, b]$, if $f(a) = g(a)$ and $f(b) = g(b)$, and if, further, $f(x) > g(x)$ for all x, $a < x < b$, then it follows that

   (A) $\displaystyle\int_a^b f(x)\,dx \geqq 0$

   (B) $\displaystyle\int_a^b |g(x)|\,dx > \int_a^b |f(x)|\,dx$

   (C) $\displaystyle\int_a^b |f(x)|\,dx > \int_a^b |g(x)|\,dx$

   (D) $\displaystyle\int_a^b f(x)\,dx > \int_a^b g(x)\,dx$

   (E) none of the preceding is necessarily true

38. If $\dfrac{dy}{dx} = -y$ and $y = 1$ when $x = 2$, then y equals

   (A) $2e^{-x}$ (B) $-2e^{-x}$ (C) e^{2-x} (D) e^{x-2} (E) $\dfrac{1}{2}e^{-x}$

39. The point of inflection of the graph of $y = \sqrt{3}\sin x - 3\cos x$ on the interval $-\dfrac{\pi}{2} < x < \dfrac{\pi}{2}$ occurs at the point whose coordinates are

   (A) $\left(-\dfrac{\pi}{3}, -3\right)$ (B) $\left(-\dfrac{\pi}{6}, -2\sqrt{3}\right)$ (C) $(0, -3)$

   (D) $\left(\dfrac{\pi}{6}, -\sqrt{3}\right)$ (E) $\left(\dfrac{\pi}{3}, 0\right)$

40. A particle moves along a line with velocity, in feet per second, $v = t^2 - t$. The total distance, in feet, traveled from $t = 0$ to $t = 2$ equals

   (A) $\dfrac{1}{3}$ (B) $\dfrac{2}{3}$ (C) 2 (D) 1 (E) $\dfrac{4}{3}$

41. The area of the shaded region in the figure is exactly equal to

(A) $\int_1^3 \frac{1}{x}\, dx$ (B) $\int_1^4 \frac{1}{x}\, dx$ (C) $(\ln 4) - 2$ (D) $\ln 2$ (E) $\ln 4$

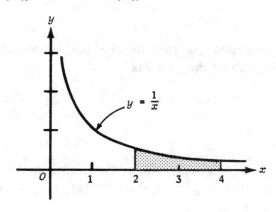

42. The average (or mean) value of $\frac{1}{2} t^2 - \frac{1}{3} t^3$ over the interval $-2 \leq t \leq 1$ is

(A) $\frac{1}{36}$ (B) $\frac{1}{12}$ (C) $\frac{11}{12}$ (D) 2 (E) $\frac{33}{12}$

43. The lines $x - ky = 3$ and $kx + y = 4$ are perpendicular for

(A) no real k (B) $k = 0$ only   (C) $k = 1$ only
(D) $k = -1$ only   (E) all real k

44. If

$$f(x) = \begin{cases} x^2 & \text{for } x \leq 2 \\ 4x - x^2 & \text{for } x > 2 \end{cases},$$

then $\int_{-1}^4 f(x)\, dx$ equals

(A) 7 (B) $\frac{23}{3}$ (C) $\frac{25}{3}$ (D) 9 (E) $\frac{65}{3}$

45. If x and y are real numbers, the domain of the function $f(x) = \frac{\sqrt{x^2 + 1}}{4 - x}$ is

(A) all x except $x = 2$ or $x = -2$ (B) all x except $x = 4$
(C) $|x| \leq 1$ (D) $x > 1$ or $x < -1$ (E) all x

AB Practice Examination 2, Section II

1. (a) Sketch the region bounded by the curves of $\sqrt{x} + \sqrt{y} = 3$ and $x + y = 5$.

(b) Find the area of this region.

2. Let the function f be defined as follows:

$$f(x) = x^3 \qquad \text{for } x > 1,$$
$$f(x) = 3x - 2 \quad \text{for } x \leqq 1.$$

(a) Show that f is continuous at $x = 1$.
(b) Use the definition of derivative to show that $f'(1)$ exists.
(c) Is f' continuous at $x = 1$? Why?
(d) Evaluate $\displaystyle\int_0^2 f(x)\, dx$.

3. Consider, in the first quadrant, the graphs of

$$y^2 = x \qquad \text{and} \qquad y^2 = 4 - x.$$

A line segment parallel to the x-axis and with endpoints on these curves is rotated about the x-axis to generate the surface of a cylinder. Show that the cylinder of greatest volume has its height equal to its diameter.

4. Sketch the graph of $y = x \ln x$, after finding
(a) the domain;
(b) intercepts;
(c) the coordinates of any relative maximum or minimum points;
(d) the coordinates of any points of inflection;
(e) the behavior of y for large x;
(f) $\lim y$ as $x \to 0^+$.

5. A particle moves along a line so that its displacement from the origin at any time t is given by $x = \frac{7}{2} e^{-4t} \sin 2t$.

(a) Prove: $a + 8v + 20x = 0$, where v is the velocity and a the acceleration of the particle.

(b) Show that relative maximum and minimum values of x occur when $\tan 2t = \frac{1}{2}$.

6. A certain chemical substance decomposes at a rate proportional to the amount present. If initially there are 6 gm of the substance and 2 gm decompose in 1 min, how many grams are left after 10 min?

Answers to AB Practice Examination 2

SECTION I

1. C	**9.** A	**17.** A	**25.** E	**33.** B	**41.** D
2. A	**10.** C	**18.** A	**26.** D	**34.** B	**42.** C
3. B	**11.** C	**19.** D	**27.** C	**35.** E	**43.** E
4. C	**12.** B	**20.** E	**28.** C	**36.** A	**44.** C
5. E	**13.** D	**21.** D	**29.** E.	**37.** D	**45.** B
6. B	**14.** E	**22.** A	**30.** A.	**38.** C	
7. A	**15.** C	**23.** E	**31.** B	**39.** E	
8. D	**16.** B	**24.** E	**32.** C	**40.** D	

1. **C.** Use the Rational Function Theorem on page 23.

2. **A.** Since

$$\frac{\cos x}{x - \dfrac{\pi}{2}} = \frac{\sin\left(\dfrac{\pi}{2} - x\right)}{x - \dfrac{\pi}{2}} = \frac{-\sin\left(x - \dfrac{\pi}{2}\right)}{x - \dfrac{\pi}{2}},$$

we can let $\alpha = \left(x - \dfrac{\pi}{2}\right)$ and then find $\lim\limits_{\alpha \to 0} \dfrac{-\sin \alpha}{\alpha} = -1$.

3. **B.** As $x \to 0$, $\sin \dfrac{1}{x}$ oscillates between -1 and 1, remaining finite. Since $-1 \leq \sin \dfrac{1}{x} \leq 1$, we get

$$-x \leq x \sin \frac{1}{x} \leq x \quad \text{when } x > 0,$$

and

$$-x \geq x \sin \frac{1}{x} \geq x \quad \text{when } x < 0.$$

In either case $x \sin \dfrac{1}{x}$ is "squeezed" to 0 as $x \to 0$.

4. **C.** Note that $\lim\limits_{h \to 0} \dfrac{\ln (2 + h) - \ln 2}{h} = f'(2)$, where $f(x) = \ln x$.

5. **E.** Use the quotient rule (formula (6) on page 33.)

6. **B.** Since $y' = -2xe^{-x^2}$, therefore

$$y'' = -2(x \cdot e^{-x^2} \cdot (-2x) + e^{-x^2}).$$

Replace x by 0.

7. **A.** If $f(x) = x \cos \dfrac{1}{x}$, then

$$f'(x) = -x \sin \frac{1}{x}\left(-\frac{1}{x^2}\right) + \cos \frac{1}{x} = \frac{1}{x} \sin \frac{1}{x} + \cos \frac{1}{x},$$

and

$$f'\left(\frac{2}{\pi}\right) = \frac{\pi}{2} \cdot 1 + 0.$$

8. D.   Implicit differentiation yields $2xyy' + y^2 - 3 + 4y' = 0$. Solve for y'.

9. A.   The equation of the reflection of $y = f(x)$ in the x-axis is $y = -f(x)$.

10. C. $\dfrac{d}{dx}\sin^2(x + y) = [\sin 2(x + y)]\left(1 + \dfrac{dy}{dx}\right)$, using formula (19), page 483.

11. C.   If $y = e^x \ln x$, then $\dfrac{dy}{dx} = \dfrac{e^x}{x} + e^x \ln x$, which equals e when $x = 1$. Since also $y = 0$ when $x = 1$, the equation of the tangent is $y = e(x - 1)$.

12. B. $g'(y) = \dfrac{1}{f'(x)} = \dfrac{1}{5x^4}$ To find $g'(0)$, we seek x such that $f(x) = \dfrac{1}{5}$. By inspection, $x = -1$, so $g'(0) = \dfrac{1}{5(-1)^4} = \dfrac{1}{5}$

13. D.   See Figure AB2–1. It is given that $\dfrac{dx}{dt} = -2$; we want $\dfrac{dA}{dt}$, where $A = \dfrac{1}{2}xy$.

$$\frac{dA}{dt} = \frac{1}{2}\left(x\frac{dy}{dt} + y\frac{dx}{dt}\right) = \frac{1}{2}\left[3\cdot\frac{dy}{dt} + y\cdot(-2)\right].$$

Since $y^2 = 25 - x^2$, it follows that $2y\dfrac{dy}{dt} = -2x\dfrac{dx}{dt}$ and when $x = 3$,

$$y = 4 \quad \text{and} \quad \frac{dy}{dt} = \frac{3}{2}.$$

Then $\dfrac{dA}{dt} = -\dfrac{7}{4}$.

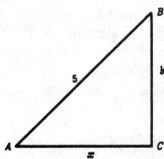

FIGURE AB2–1

14. E.   Since $f'(x)$ exists for all x, it must equal 0 for any x_0 for which it is a relative maximum, and it must also change sign from positive to negative as x increases through x_0. For the given derivative, no x satisfies both of these conditions.

15. C. Here, $v = 3t^2 + 2t - 1 = (3t - 1)(t + 1)$; when v is positive, the particle is moving to the right.

16. B. $v = 4(t - 2)^3$ and changes sign exactly once, when $t = 2$.

17. A. The curve falls when $f'(x) < 0$ and is concave up when $f''(x) > 0$.

18. A. Here, $f'(x)$ is $e^{-x}(1 - x)$; f has maximum value when $x = 1$.

19. D. See Figure AB2–2. V, the volume of the cylinder, equals $\pi x^2 y$, where $2x + 2y = 18$. So

$$V = \pi x^2(9 - x) \qquad \text{and} \qquad V' = \pi(18x - 3x^2).$$

Since $x = 6$ yields maximum volume, the area of the rectangle, xy, equals 18.

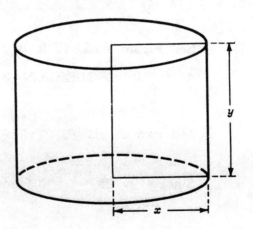

FIGURE AB2–2

20. E. The Mean Value Theorem holds if (1) $f(x)$ is continuous on $[a, b]$ and (2) $f'(x)$ exists on (a, b). The condition given in the question assures both of these. Find counterexamples for (A) through (D).

21. D. Evaluate $\dfrac{1}{12}(3x - 2)^4 \Big|_1^2$.

22. A. Evaluate $\dfrac{1}{4}\sin^4 \alpha \Big|_{\pi/4}^{\pi/2}$.

23. E. Use formula (5) on page 125 with $u = x^2$; $du = 2x\,dx$.

24. E. Evaluating $\dfrac{1}{3 - e^x}\Big|_0^1$ yields $\dfrac{e - 1}{2(3 - e)}$.

25. E. Since $f'(x) = 1 - \dfrac{c}{x^2}$, it equals 0 for $x = \pm\sqrt{c}$. When $x = 3$, $c = 9$; this yields a maximum since $f''(9) < 0$.

26. D.   The given integral is equivalent to $\int \left[1 + \frac{2x-1}{x^2-x} \right] dx$, which integrates into $x + \ln |x^2 - x| + C$.

27. C.   The given integral is equivalent to $\int_{-1}^{0} (1+x)\, dx + \int_{0}^{1} (1-x)\, dx$.

See Figure AB2–3, which shows the graph of $f(x) = 1 - |x|$ on $[-1, 1]$.

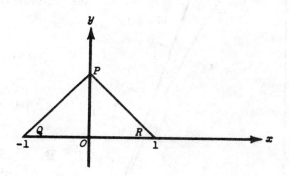

FIGURE AB2–3

The area of triangle PQR is equal to $\int_{-1}^{1} (1 - |x|)\, dx$.

28. C.   Evaluate $-e^{-x} \Big|_{-1}^{0}$.

29. E.   Since $f(g(x)) = \frac{3}{x} + 2$ and $g(f(x)) = \frac{3}{x+2}$, we seek, after simplifying, the solution, if any, of $x^2 + 2x + 3 = 0$.

30. A.   If two functions have the same derivative, then they differ at most by a constant. No one of (B) through (E) is necessarily true.

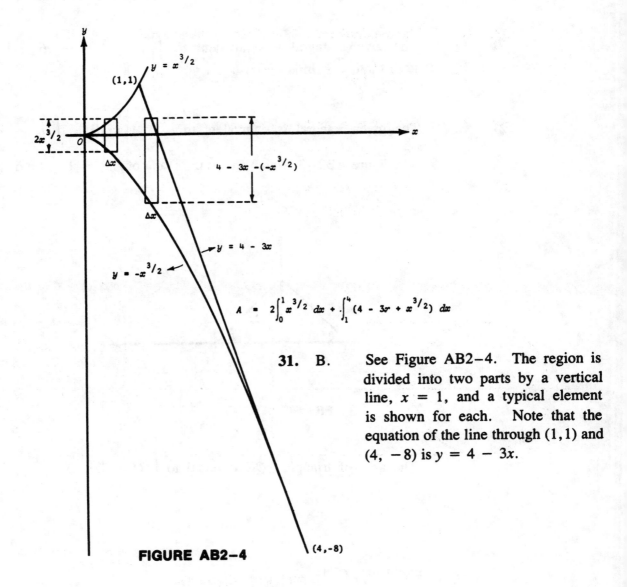

$$A = 2\int_0^1 x^{3/2}\, dx + \int_1^4 (4 - 3x + x^{3/2})\, dx$$

FIGURE AB2-4

31. B. See Figure AB2-4. The region is divided into two parts by a vertical line, $x = 1$, and a typical element is shown for each. Note that the equation of the line through $(1, 1)$ and $(4, -8)$ is $y = 4 - 3x$.

32. C. See Figure AB2-5. A represents the required area.

FIGURE AB2-5

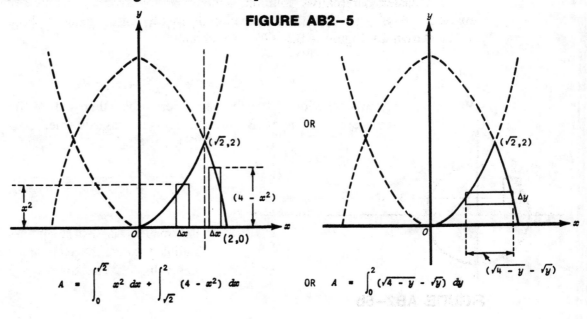

$$A = \int_0^{\sqrt{2}} x^2\, dx + \int_{\sqrt{2}}^2 (4 - x^2)\, dx \qquad \text{OR} \qquad A = \int_0^2 (\sqrt{4 - y} - \sqrt{y})\, dy$$

33. B.   The required area, A, is given by the integral

$$2\int_0^1 \left(4 - \frac{4}{1 + x^2}\right) dx = 2(4x - 4\tan^{-1} x)\Big|_0^1 = 2\left(4 - 4\cdot\frac{\pi}{4}\right).$$

34. B.   Here, f is a maximum at $x = \frac{\pi}{6}$, a minimum at $x = \frac{7\pi}{6}$. The amplitude of the function is $f\left(\frac{\pi}{6}\right)$.

35. E.   See Figure AB2–6a. The equation of the generating circle is $(x - 3)^2 + y^2 = 1$, which yields $x = 3 \pm \sqrt{1 - y^2}$.

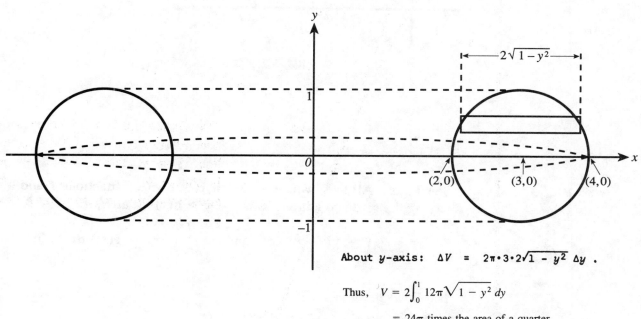

About y-axis: $\Delta V = 2\pi \cdot 3 \cdot 2\sqrt{1 - y^2}\ \Delta y$.

Thus, $V = 2\int_0^1 12\pi\sqrt{1 - y^2}\ dy$

= 24π times the area of a quarter
of a unit circle = $6\pi^2$.

FIGURE AB2–6a

An easier, alternative solution is obtained by rotating the area bounded by the circle $x^2 + y^2 = 1$ about the line $x = 3$. In this case, using shells parallel to $x = 3$ (as shown in Figure AB2–6b), we obtain

$$\Delta V = 2\pi\int_{-1}^1 (3 - x)2y\ dx$$

$$= 4\pi\int_{-1}^1 (3 - x)\sqrt{1 - x^2}\ dx$$

$$= 12\pi\int_{-1}^1 \sqrt{1 - x^2}\ dx - 4\pi\int_{-1}^1 x\sqrt{1 - x^2}\ dx$$

= 12 times the area of a semicircle of
radius 1 (since the second integral
has value 0). The answer is $6\pi^2$.

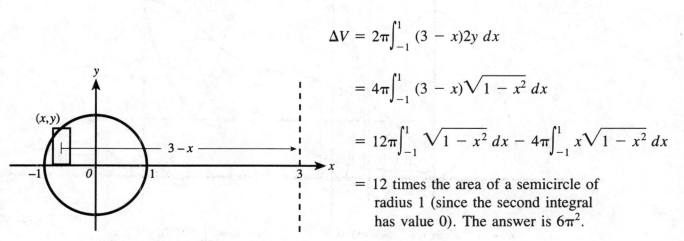

FIGURE AB2–6b

36. A. See Figure AB2–7.

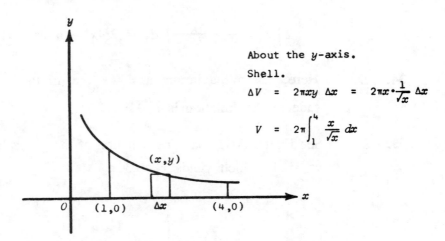

About the y-axis.
Shell.
$$\Delta V = 2\pi xy\ \Delta x = 2\pi x\cdot\frac{1}{\sqrt{x}}\ \Delta x$$

$$V = 2\pi\int_{1}^{4}\frac{x}{\sqrt{x}}\ dx$$

FIGURE AB2–7

37. D. See Figure AB2–8, where a sketch is shown of functions f and g satisfying the given conditions. Since $f(x) > g(x)$ for all x, $a < x < b$,

$$f(x) - g(x) > 0 \quad \text{and} \quad \int_{a}^{b}(f(x) - g(x))\ dx > 0.$$

It follows that

$$\left[\int_{a}^{b}f(x)\ dx - \int_{a}^{b}g(x)\ dx\right] > 0.$$

FIGURE AB2–8

38. C.   The particular solution is $-\ln y = x - 2$, or $\ln y = 2 - x$.

39. E.   Since $y'' = -\sqrt{3}\sin x + 3\cos x$, it is equal to 0 on $\left(-\frac{\pi}{2}, \frac{\pi}{2}\right)$ when $x = \frac{\pi}{3}$. Since y'' changes sign at this value of x, it yields a point of inflection.

40. D.   Note that v is negative from $t = 0$ to $t = 1$, but positive from $t = 1$ to $t = 2$. Thus the distance traveled is given by

$$-\int_0^1 (t^2 - t)\,dt + \int_1^2 (t^2 - t)\,dt.$$

41. D.   The shaded area equals exactly $\int_2^4 \frac{1}{x}\,dx = \ln 4 - \ln 2 = \ln 2$.

42. C.   The average value equals $\frac{1}{3}\left(\frac{t^3}{6} - \frac{t^4}{12}\right)\Big|_{-2}^1$.

43. E.   The product of the slopes, $\frac{1}{k} \cdot -\frac{k}{1}$, is equal to -1 for all k.

44. C. $\int_{-1}^4 f(x)\,dx = \int_{-1}^2 x^2\,dx + \int_2^4 (4x - x^2)\,dx$.

45. B.   Since the numerator is defined for all x, we need eliminate only those values of x for which the denominator equals 0.

SECTION II

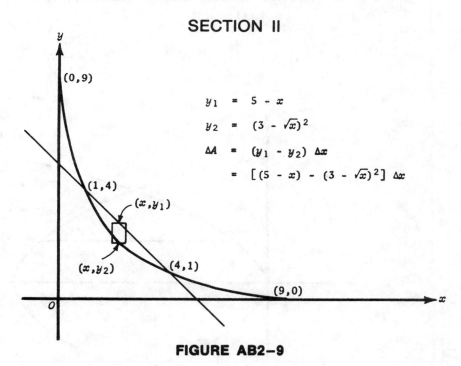

FIGURE AB2–9

  1. (a) The region is sketched in Figure AB2–9. To determine where the graphs intersect, we solve simultaneously $y = 5 - x$ and $y = (3 - \sqrt{x})^2$. Then

$$5 - x = 9 - 6\sqrt{x} + x, \qquad 6\sqrt{x} = 2x + 4$$

or $3\sqrt{x} = x + 2$, $9x = x^2 + 4x + 4$,   and $0 = x^2 - 5x + 4$,

with roots $x = 1$ and $x = 4$. The points are $(1, 4)$ and $(4, 1)$, as shown in the figure.

(b) For the area we thus have

$$A = \int_1^4 [5 - x - (3 - \sqrt{x})^2]\, dx = \int_1^4 [5 - x - (9 - 6\sqrt{x} + x)]\, dx$$

$$= \int_1^4 (6\sqrt{x} - 2x - 4)\, dx = 4x^{3/2} - x^2 - 4x \Big|_1^4$$

$$= 4(8 - 1) - (16 - 1) - 4(4 - 1) = 1.$$

2. **(a)** $f(1) = 1;\ \lim\limits_{x \to 1^-} f(x) = \lim\limits_{x \to 1^+} f(x) = 1$. Since $\lim\limits_{x \to 1} f(x)$ exists and equals $f(1)$, f is continuous at $x = 1$.

(b) $f'(1) = \lim\limits_{h \to 0} \dfrac{f(1 + h) - f(1)}{h}$ if the latter exists. When $h \to 0^+$, we have

$$\lim\limits_{h \to 0^+} \frac{f(1 + h) - f(1)}{h} = \lim\limits_{h \to 0^+} \frac{3(1 + h) - 2 - 1}{h} = 3.$$

When $h \to 0^-$, then

$$\lim\limits_{h \to 0^-} \frac{f(1 + h) - f(1)}{h} = \lim\limits_{h \to 0^-} \frac{(1 + h)^3 - 1}{h} = \lim\limits_{h \to 0^-} 3 + h = 3.$$

Since the left- and right-hand limits are the same, $f'(1)$ exists.

(c) Yes. $\lim\limits_{x \to 1^+} f'(x) = \lim\limits_{x \to 1^-} f'(x) = 3 = f'(1)$.

(d) $\displaystyle\int_0^2 f(x)\, dx = \int_0^1 x^3\, dx + \int_1^2 (3x - 2)\, dx = \frac{13}{4}.$

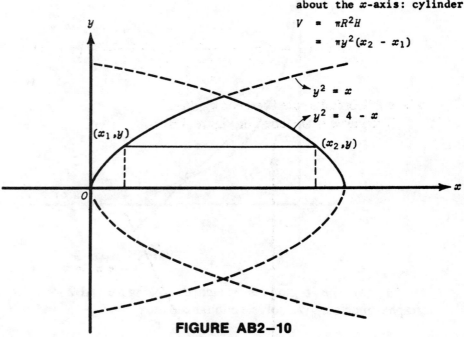

about the x-axis: cylinder
$$V = \pi R^2 H$$
$$= \pi y^2 (x_2 - x_1)$$

$y^2 = x$

$y^2 = 4 - x$

(x_1, y)

(x_2, y)

FIGURE AB2–10

3. See Figure AB2–10. Since the volume V of the cylinder is $\pi R^2 H$ and the radius R, here, equals y, while the height H equals $x_2 - x_1$, we have

$$V = \pi y^2 (x_2 - x_1) = \pi y^2 (4 - y^2 - y^2) = \pi y^2 (4 - 2y^2) = \pi(4y^2 - 2y^4).$$

Then

$$\frac{dV}{dy} = \pi(8y - 8y^3) = 8\pi y(1 - y^2),$$

and the only possible y here for which $\frac{dV}{dy} = 0$ is $y = 1$. Since $\frac{d^2V}{dy^2} = 8\pi(1 - 3y^2)$ and this is negative when $y = 1$, the volume of the cylinder is a maximum for this radius. Thus the diameter is 2 and the height, $4 - 2y^2$, is also 2.

4. Here $y = x \ln x$, $y' = 1 + \ln x$, and $y'' = \frac{1}{x}$.

(a) The domain consists of all positive x.

(b) x cannot be 0, but $y = 0$ if $x = 1$.

(c) Since $y' = 0$ for $\ln x = -1$, $x = \frac{1}{e}$ is a critical value; $y''\left(\frac{1}{e}\right) > 0$, assuring that $\left(\frac{1}{e}, -\frac{1}{e}\right)$ is a relative minimum. Since y is continuous if $x > 0$, the curve has no other maximum or minimum points.

(d) $y'' > 0$ if $x > 0$; the curve is thus concave up and has no inflection points.

(e) $\lim\limits_{x\to\infty} y = \infty$.

(f) $\lim\limits_{x\to 0^+} x \ln x = 0$. This can best be shown by noting that

$$x \ln x = -x \ln x^{-1} = -\frac{\ln \dfrac{1}{x}}{\dfrac{1}{x}}.$$

Then

$$\lim_{x\to 0^+} x \ln x = \lim_{x\to 0^+} -\frac{\ln \dfrac{1}{x}}{\dfrac{1}{x}} = \lim_{x\to 0^+} \frac{\dfrac{1}{x}}{-\dfrac{1}{x^2}} = \lim_{x\to 0^+} (-x) = 0,$$

where L'Hôpital's rule has been applied.

The curve is sketched in Figure AB2−11.

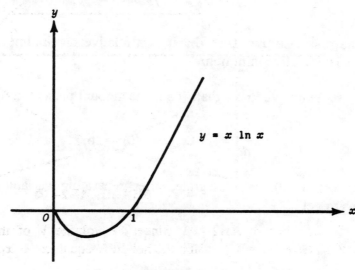

$y = x \ln x$

FIGURE AB2−11

5. Since $x = \frac{7}{2}e^{-4t} \sin 2t$,

$$v = 7e^{-4t}(\cos 2t - 2 \sin 2t) \qquad \text{and} \qquad a = 14e^{-4t}(3 \sin 2t - 4 \cos 2t).$$

(a) Then it can be verified directly that $a + 8v + 20x = 0$.

(b) Since $e^{-4t} > 0$ for all t, $v = 0$ when $\cos 2t = 2 \sin 2t$, or when $\tan 2t = \frac{1}{2}$.

The triangles drawn in Figure AB2–12 show the cases possible when $\tan 2t = \frac{1}{2}$.

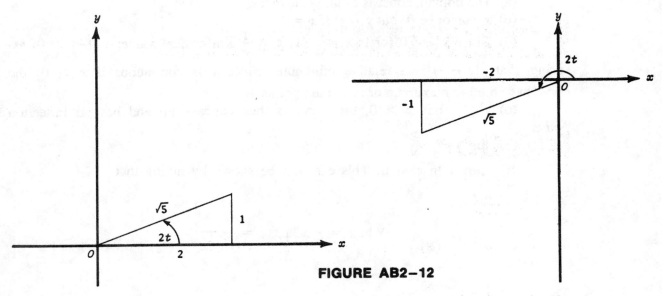

FIGURE AB2–12

If $0 < 2t < \frac{\pi}{2}$,

$$\sin 2t = \frac{1}{\sqrt{5}} \qquad \text{and} \qquad \cos 2t = \frac{2}{\sqrt{5}};$$

if $\pi < 2t < \frac{3\pi}{2}$,

$$\sin 2t = -\frac{1}{\sqrt{5}} \qquad \text{and} \qquad \cos 2t = -\frac{2}{\sqrt{5}}.$$

In the first case $a < 0$ so that x is a relative maximum; in the second case $a > 0$ so that x is a relative minimum.

6. Hint: We know that, if s is the amount present at time t, then

$$\frac{ds}{dt} = -ks, \qquad s(0) = 6, \qquad \text{and} \qquad s(1) = 4.$$

We seek $s(10)$. We integrate, after separating variables, and use the given conditions to get

$$\ln\left(\frac{s}{6}\right) = \ln\left(\frac{2}{3}\right)^t \qquad \text{or} \qquad \frac{s}{6} = \left(\frac{2}{3}\right)^t.$$

Thus, $s(10) = 6\left(\frac{2}{3}\right)^{10}$.

AB Practice
Examination 3

Section I

1. $\lim\limits_{x\to 2} \dfrac{x^2 - 2}{4 - x^2}$ is

   (A) -2 (B) -1 (C) $-\dfrac{1}{2}$ (D) 0   (E) nonexistent

2. $\lim\limits_{x\to\infty} \dfrac{\sqrt{x} - 4}{4 - 3\sqrt{x}}$ is

   (A) $-\dfrac{1}{3}$ (B) -1 (C) ∞ (D) 0   (E) $\dfrac{1}{3}$

3. $\lim\limits_{x\to 0} \dfrac{\sin^2 \frac{x}{2}}{x^2}$ is

   (A) 4   (B) 0   (C) $\dfrac{1}{4}$ (D) 2   (E) nonexistent

4. $\lim\limits_{x\to 0} \dfrac{e^x - 1}{x}$ is

   (A) 1   (B) e (C) $\dfrac{1}{e}$ (D) 0   (E) nonexistent

5. The slope of $y = |x|$ at the point where $x = \dfrac{1}{2}$ is

   (A) -1 (B) 0   (C) $\dfrac{1}{2}$ (D) 1   (E) nonexistent

6. If $y = \dfrac{e^{\ln u}}{u}$, then $\dfrac{dy}{du}$ equals

   (A) $\dfrac{e^{\ln u}}{u^2}$ (B) $e^{\ln u}$ (C) $\dfrac{2e^{\ln u}}{u^2}$ (D) 1   (E) 0

7. If $y = \sin^3 (1 - 2x)$, then $\dfrac{dy}{dx}$ is

   (A) $3 \sin^2 (1 - 2x)$ (B) $-2 \cos^3 (1 - 2x)$ (C) $-6 \sin^2 (1 - 2x)$

   (D) $-6 \sin^2 (1 - 2x) \cos (1 - 2x)$ (E) $-6 \cos^2 (1 - 2x)$

8. If $f(u) = \tan^{-1} u^2$ and $g(u) = e^u$, then the derivative of $f(g(u))$ is

   (A) $\dfrac{2ue^u}{1 + u^4}$ (B) $\dfrac{2ue^{u^2}}{1 + u^4}$ (C) $\dfrac{2e^u}{1 + 4e^{2u}}$ (D) $\dfrac{2e^{2u}}{1 + e^{4u}}$

   (E) $\dfrac{2e^{2u}}{\sqrt{1 - e^{4u}}}$

9. The equation of the tangent to the curve of $xy - x + y = 2$ at the point where $x = 0$ is

   (A) $y = -x$ (B) $y = \dfrac{1}{2}x + 2$ (C) $y = x + 2$ (D) $y = 2$

   (E) $y = 2 - x$

10. If $y = \dfrac{1}{x}$, then $y^{(4)}(1)$ equals

    (A) $-4!$ (B) $-3!$ (C) $4!$ (D) $5!$ (E) $3!$

11. If $y = x^2 e^{1/x}$ $(x \neq 0)$, then $\dfrac{dy}{dx}$ is

    (A) $xe^{1/x}(x + 2)$ (B) $e^{1/x}(2x - 1)$ (C) $\dfrac{-2e^{1/x}}{x}$

    (D) $e^{-x}(2x - x^2)$ (E) none of these

12. The period of the function $f(x) = k \sin \dfrac{1}{k}x$ is 2. Its amplitude is

    (A) $\dfrac{1}{2\pi}$ (B) $\dfrac{1}{\pi}$ (C) $\dfrac{1}{2}$ (D) 1   (E) 2

13. A point moves along the curve $y = x^2 + 1$ so that the x-coordinate is increasing at the constant rate of $\dfrac{3}{2}$ units per second. The rate, in units per second, at which the distance from the origin is changing when the point has coordinates $(1, 2)$ is equal to

    (A) $\dfrac{7\sqrt{5}}{10}$ (B) $\dfrac{3\sqrt{5}}{2}$ (C) $3\sqrt{5}$ (D) $\dfrac{15}{2}$ (E) $\sqrt{5}$

14. The set of x for which the curve of $y = 1 - 6x^2 - x^4$ has inflection points is

(A) {0}   (B) $\{\pm\sqrt{3}\}$ (C) {1}   (D) $\{\pm 1\}$
(E)   none of these

15. If the position of a particle on a line at time t is given by $s = t^3 + 3t$, then the speed of the particle is decreasing when

(A) $-1 < t < 1$ (B) $-1 < t < 0$ (C) $t < 0$ (D) $t > 0$
(E) $|t| > 1$

16. If $\sin(xy) = y$, then $\dfrac{dy}{dx}$ equals

(A) $\sec(xy)$ (B) $y\cos(xy) - 1$ (C) $\dfrac{1 - y\cos(xy)}{x\cos(xy)}$

(D) $\dfrac{y\cos(xy)}{1 - x\cos(xy)}$ (E) $\cos(xy)$

17. If the graph of $f(x) = x^3 + x + c$ has exactly one x-intercept, then it follows that

(A) c must be negative   (B) c must equal zero
(C) c must equal -2 (D) c must be positive
(E) c can be any real number

18. A rectangle with one side on the x-axis is inscribed in the triangle formed by the lines $y = x$, $y = 0$, and $2x + y = 12$. The area of the largest such rectangle is

(A) 6   (B) 3   (C) $\dfrac{5}{2}$ (D) 5   (E) 7

19. The abscissa of the first-quadrant point which is on the curve of $x^2 - y^2 = 1$ and closest to the point $(3, 0)$ is

(A) 1   (B) $\dfrac{3}{2}$ (C) 2   (D) 3   (E)   none of these

20. If c represents the number defined by Rolle's theorem, then, for the function $f(x) = x^3 - 3x^2$ on the interval $0 \leq x \leq 3$, c is equal to

(A) 2   (B) 1   (C) 0   (D) $\sqrt{2}$ (E)   none of these

21. $\displaystyle\int \dfrac{x\,dx}{\sqrt{9 - x^2}}$ equals

(A) $-\dfrac{1}{2}\ln\sqrt{9 - x^2} + C$ (B) $\sin^{-1}\dfrac{x}{3} + C$ (C) $-\sqrt{9 - x^2} + C$

(D) $-\dfrac{1}{4}\sqrt{9 - x^2} + C$ (E) $2\sqrt{9 - x^2} + C$

22. Let $x > 0$. Suppose

$$\frac{d}{dx}f(x) = g(x) \quad \text{and} \quad \frac{d}{dx}g(x) = f(\sqrt{x});$$

then $\dfrac{d^2}{dx^2}f(x^2) =$

(A) $f(x^4)$ (B) $f(x^2)$ (C) $2xg(x^2)$ (D) $\dfrac{1}{2x}f(x)$

(E) $2g(x^2) + 4x^2f(x)$

23. $\displaystyle\int \frac{(y - 1)^2}{2y}\,dy$ equals

(A) $\dfrac{y^2}{4} - y + \dfrac{1}{2}\ln|y| + C$ (B) $y^2 - y + \ln|2y| + C$

(C) $y^2 - 4y + \dfrac{1}{2}\ln|2y| + C$ (D) $\dfrac{(y - 1)^3}{3y^2} + C$

(E) $\dfrac{1}{2} - \dfrac{1}{2y^2} + C$

24. $\displaystyle\int_{\pi/6}^{\pi/2} \cot x\,dx$ equals

(A) $\ln\dfrac{1}{2}$ (B) $\ln 2$ (C) $-\ln(2 - \sqrt{3})$ (D) $\ln(\sqrt{3} - 1)$

(E) none of these

25. $\displaystyle\int_1^e \ln x\,dx$ equals

(A) $\dfrac{1}{2}$ (B) $e - 1$ (C) $e + 1$ (D) 1 (E) -1

26. If $F(x) = \displaystyle\int_1^x \sqrt{t^2 + 3t}\,dt$, then $F'(x)$ equals

(A) $\dfrac{2}{3}[(x^2 + 3x)^{3/2} - 8]$ (B) $\sqrt{x^2 + 3x}$ (C) $\sqrt{x^2 + 3x} - 2$

(D) $\dfrac{1}{2}\dfrac{(2x + 3)}{\sqrt{x^2 + 3x}}$ (E) none of these

27. Suppose $y = \dfrac{P(x)}{Q(x)}$ is a rational function and that, as $x \to c^+$, $\lim y = +\infty$. Then it follows that

(A) the function is not defined if $x \leqq c$
(B) as $x \to c^-$, $y \to +\infty$
(C) as $x \to c^-$, $y \to -\infty$
(D) the line $x = c$ is a vertical asymptote
(E) the line $y = c$ is a horizontal asymptote

28. If $f(x)$ is continuous on the interval $-a \leqq x \leqq a$, then it follows that
$$\int_{-a}^{a} f(x)\, dx$$

(A) equals $2\int_{0}^{a} f(x)\, dx$ (B) equals 0   (C) equals $f(a) - f(-a)$

(D) represents the total area bounded by the curve $y = f(x)$, the x-axis, and the vertical lines $x = -a$ and $x = a$

(E) is not necessarily any of these

29. If $f(x) = 2 \ln (x - 1)$, where $x > 1$, then its inverse function $f^{-1}(x) =$

(A) $1 + \dfrac{x}{2}$ (B) $2 + e^x$ (C) $2 - e^{-x}$ (D) $1 + e^{x/2}$

(E) $1 - e^{x/2}$

30. The area bounded by the cubic $y = x^3 - 3x^2$ and the line $y = -4$ is given by the integral

(A) $\displaystyle\int_{-1}^{2} (x^3 - 3x^2 + 4)\, dx$ (B) $\displaystyle\int_{-1}^{2} (x^3 - 3x^2 - 4)\, dx$

(C) $2\displaystyle\int_{0}^{2} (x^3 - 3x^2 - 4)\, dx$ (D) $\displaystyle\int_{-1}^{2} (4 - x^3 + 3x^2)\, dx$

(E) $\displaystyle\int_{0}^{3} (3x^2 + x^3)\, dx$

31. The area bounded by the curve $y = \dfrac{1}{x + 1}$, the axes, and the line $x = e - 1$ equals

(A) $1 - \dfrac{1}{e^2}$ (B) $\ln (e - 1)$ (C) 1   (D) 2

(E) $2\sqrt{e + 1} - 2$

32. The area bounded by the curve $x = 3y - y^2$ and the line $x = -y$ is represented by

(A) $\displaystyle\int_{0}^{4} (2y - y^2)\, dy$ (B) $\displaystyle\int_{0}^{4} (4y - y^2)\, dy$

(C) $\displaystyle\int_{0}^{3} (3y - y^2)\, dy + \int_{0}^{4} y\, dy$ (D) $\displaystyle\int_{0}^{3} (y^2 - 4y)\, dy$

(E) $\displaystyle\int_{0}^{3} (2y - y^2)\, dy$

33. The minimum value of $f(x) = x^2 + \dfrac{2}{x}$ on the interval $\dfrac{1}{2} \leqq x \leqq 2$ is

(A) $\dfrac{1}{2}$ (B) 1   (C) 3   (D) $4\dfrac{1}{2}$ (E) 5

34. A solid is cut out of a sphere of radius 2 by two parallel planes each 1 unit from the center. The volume of this solid is

(A) 8π (B) $\dfrac{32\pi}{3}$ (C) $\dfrac{25\pi}{3}$ (D) $\dfrac{22\pi}{3}$ (E) $\dfrac{20\pi}{3}$

35. The volume generated by rotating the region bounded by $y = \dfrac{1}{\sqrt{1 + x^2}}$, $y = 0$, $x = 0$, and $x = 1$ about the y-axis is

(A) $\dfrac{\pi^2}{4}$ (B) $\pi\sqrt{2}$ (C) $2\pi(\sqrt{2} - 1)$ (D) $\dfrac{2\pi}{3}(2\sqrt{2} - 1)$

(E) $1 + \dfrac{\pi^2}{4}$

36. The region bounded by $y = e^x$, $y = 1$, and $x = 2$ is rotated about the x-axis. The volume of the solid generated is given by the integral:

(A) $\pi \displaystyle\int_0^2 e^{2x}\, dx$ (B) $2\pi \displaystyle\int_1^{e^2} (2 - \ln y)(y - 1)\, dy$

(C) $\pi \displaystyle\int_0^2 (e^{2x} - 1)\, dx$ (D) $2\pi \displaystyle\int_0^{e^2} y(2 - \ln y)\, dy$

(E) $\pi \displaystyle\int_0^2 (e^x - 1)^2\, dx$

37. A particle moves on a straight line so that its velocity at time t is given by $v = 4s$, where s is its distance from the origin. If $s = 3$ when $t = 0$, then, when $t = \dfrac{1}{2}$, s equals

(A) $1 + e^2$ (B) $2e^3$ (C) e^2 (D) $2 + e^2$ (E) $3e^2$

38. If $f(x) = xe^{-x^2}$, then $f'(1) =$

(A) $-\dfrac{1}{e}$ (B) $\dfrac{1}{e}$ (C) $-2 + \dfrac{1}{e}$ (D) $-\dfrac{2}{e} + e$

(E)   none of the preceding

39. The average (mean) value of $y = (x - 3)^2$ over the interval from $x = 1$ to $x = 3$ equals

(A) 2 (B) $\dfrac{2}{3}$ (C) $\dfrac{4}{3}$ (D) $\dfrac{8}{3}$ (E)   none of these

40. The area bounded by the curves $y = f(x)$ and $y = g(x)$ in the figure equals which of the following?

(A) $\displaystyle\int_a^d [f(x) + g(x)]\, dx$

(B) $\displaystyle\int_a^d [f(x) - g(x)]\, dx$

(C) $\displaystyle\int_a^d [g(x) - f(x)]\, dx$

(D) $\displaystyle\int_a^b [g(x) - f(x)]\, dx +$ $\displaystyle\int_b^d [f(x) - g(x)]\, dx$

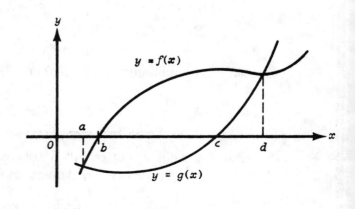

(E)   none of the preceding

41. A particle moves along a line so that its acceleration, a, at time t is $a = -t^2$. If the particle is at the origin when $t = 0$ and 3 units to the right of the origin when $t = 1$, then its velocity at $t = 0$ is

(A) 0   (B) $\dfrac{1}{12}$ (C) $2\dfrac{11}{12}$ (D) $\dfrac{37}{12}$ (E)   none of these

42. Which of the following statements is true of the graph of $y = x^4 - 4x^3$?

(A)   The graph has no relative extreme point.
(B)   The graph has one point of inflection, one relative maximum, and one relative minimum.
(C)   The graph has two points of inflection and one relative minimum.
(D)   The graph has two points of inflection, one relative maximum and one relative minimum.
(E)   None of the preceding.

43. The distance from $(-4, 5)$ to the line $3y + 2x - 7 = 0$ is

(A) $\dfrac{9}{5}$ (B) $\dfrac{9}{\sqrt{13}}$ (C) 1   (D) $\sqrt{13}$ (E)   none of these

44. The set of x for which $|x + 1| > 2$ and $|x - 1| \leqq 2$ is

(A)   the null set   (B) $x < -3$ or $x > -1$ (C) $x < -3$ or $x \geqq -1$

(D) $1 < x \leqq 3$ (E)   none of these

45. Which of the following functions could have the graph sketched below?

(A) $f(x) = xe^x$ (B) $f(x) = xe^{-x}$ (C) $f(x) = \dfrac{e^x}{x}$

(D) $f(x) = \dfrac{x}{x^2 + 1}$ (E) $f(x) = \dfrac{x^2}{x^3 + 1}$

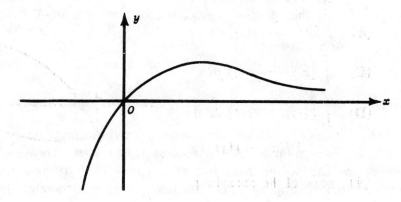

AB Practice Examination 3, Section II

1. An open box with a capacity of 4 ft³ and with a square base is to be made from sheet metal costing 50¢ per square foot. If welding costs 10¢ per foot and all the edges will need to be welded,

   (a) find the dimensions for the box that will minimize the cost;

   (b) show that the dimensions that yield minimum cost are independent of the costs of the sheet metal and of the welding.

2. (a) Integrate $\int_1^x \frac{dt}{t}$, where $x > 1$.

   (b) Interpret the integral in (a) as an area.

   (c) Show that $x - 1 > \ln x > \frac{x-1}{x}$ if $x > 1$.

3. The region bounded by the curve of the function $y = f(t)$, the t-axis, $t = 0$, and $t = x$ is rotated about the t-axis. The volume generated for all x is $2\pi(x^2 + 2x)$. Find $f(t)$.

4. A particle moving along the x-axis starts at the origin when $t = 0$. Its acceleration at any time t is $a = -6t$. Find its velocity when $t = 0$ if the maximum displacement of the particle in the positive direction is 16 units.

5. (a) If $f(x) = |x|^3$, find $f'(0)$ by using the definition of derivative.

   (b) Is $f'(x)$ continuous at $x = 0$? Why or why not?

   (c) Is the function $x|x|$ continuous at $x = 0$? Why or why not?

   (d) Evaluate $\int_{-1}^1 (1 - |x|)\ dx$.

6. At time $t = 0$ an insect starts crawling along a straight line at the rate of 3 ft/min. Two minutes later a second insect starts crawling in a direction perpendicular to that of the first and at a speed of 5 ft/min. How fast is the distance between them changing when the first insect has traveled 12 ft?

Answers to AB Practice Examination 3

SECTION I

1. E	**8.** D	**15.** C	**22.** E	**29.** D	**36.** C
2. A	**9.** E	**16.** D	**23.** A	**30.** A	**37.** E
3. C	**10.** C	**17.** E	**24.** B	**31.** C	**38.** A
4. A	**11.** B	**18.** A	**25.** D	**32.** B	**39.** C
5. D	**12.** B	**19.** B	**26.** B	**33.** C	**40.** B
6. E	**13.** B	**20.** A	**27.** D	**34.** D	**41.** D
7. D	**14.** E	**21.** C	**28.** E	**35.** C	**42.** C

43. E   **44.** D   **45.** B

1. E. $\dfrac{x^2 - 2}{4 - x^2} \to \infty$ as $x \to 2$.

2. A.   Divide both numerator and denominator by $\sqrt{x}$.

3. C.   Note that

$$\frac{\sin^2 \dfrac{x}{2}}{x^2} = \frac{\sin^2 \dfrac{x}{2}}{4\dfrac{x^2}{4}} = \frac{1}{4} \lim_{\theta \to 0} \left[\frac{\sin \theta}{\theta} \right]^2,$$

where we let $\dfrac{x}{2} = \theta$.

4. A.   This is $f'(0)$, where $f(x) = e^x$; or one can use L'Hôpital's rule here, getting

$$\lim_{x \to 0} \frac{e^x - 1}{x} = \lim_{x \to 0} \frac{e^x}{1} = 1.$$

5. D.   The slope of $y = |x|$ equals 1 for $x > 0$ and -1 for $x < 0$.

6. E.   Since $e^{\ln u} = u, y = 1$.

7. D.   Here $y' = 3 \sin^2 (1 - 2x) \cos (1 - 2x) \cdot (-2)$.

8.   **D.**   We want $\dfrac{d}{du} \tan^{-1} e^{2u}$,   where

$$\frac{d}{dx} \tan^{-1} v = \frac{\dfrac{dv}{dx}}{1 + v^2}.$$

9.   **E.**   Differentiating the relation $xy - x + y = 2$ implicitly yields $\dfrac{dy}{dx} = \dfrac{1 - y}{x + 1}$. Since when $x = 0$, $y = 2$, the slope at this point is -1 and the equation of the tangent is thus $y = -x + 2$.

10.   **C.**   Since $y = x^{-1}$, $y' = -x^{-2}$; $y'' = 2x^{-3}$; $y''' = -6x^{-4}$; $y^{(4)} = 24x^{-5}$; and $y^{(4)}(1) = 24$.

11.   **B.** $\dfrac{d}{dx}(x^2 e^{x^{-1}}) = x^2 e^{x^{-1}}\left(-\dfrac{1}{x^2}\right) + 2x e^{x^{-1}}$.

12.   **B.**   Since the period of $a \sin bx$ is $\dfrac{2\pi}{b}$, we know that

$$\frac{2\pi}{1/k} = 2.$$

13.   **B.**   Let s be the distance from the origin: then

$$s = \sqrt{x^2 + y^2} \qquad \text{and} \qquad \frac{ds}{dt} = \frac{x\dfrac{dx}{dt} + y\dfrac{dy}{dt}}{\sqrt{x^2 + y^2}}.$$

Since

$$\frac{dy}{dt} = 2x\frac{dx}{dt} \qquad \text{and} \qquad \frac{dx}{dt} = \frac{3}{2},$$

$\dfrac{dy}{dt} = 3x$.   Substituting yields $\dfrac{ds}{dt} = \dfrac{3\sqrt{5}}{2}$.

14.   **E.**   Here,

$$\frac{dy}{dx} = -12x - 4x^3 \qquad \text{and} \qquad \frac{d^2y}{dx^2} = -12 - 12x^2 = -12(1 + x^2).$$

Since the second derivative never changes sign (the curve is everywhere concave down), there are no inflection points.

15.   **C.**   Since $v = 3t^2 + 3$, it is always positive, while $a = 6t$ and is positive for $t > 0$ but negative for $t < 0$. The speed therefore increases for $t > 0$ but decreases for $t < 0$.

16.   **D.**   Let $y' = \dfrac{dy}{dx}$. Then $\cos(xy)[xy' + y] = y'$. Solve for y'.

17. **E.** Since $f'(x) = 3x^2 + 1$, the function increases for all x; further, no matter what c is, $f(x) \to \infty$ as $x \to \infty$ and $f(x) \to -\infty$ as $x \to -\infty$.

18. **A.** See Figure AB3–1. The area, A, of a typical rectangle is

$$A = (x_2 - x_1) \cdot y = \left(\frac{12 - y}{2} - y \right) \cdot y = 6y - \frac{3y^2}{2}.$$

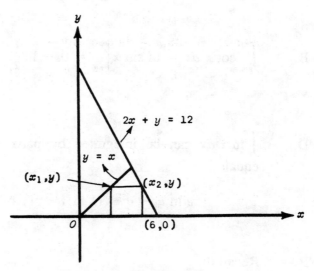

FIGURE AB3–1

For $y = 2$, $\frac{dA}{dy} = 0$. Note that $\frac{d^2A}{dy^2}$ is always negative.

19. **B.** If S represents the square of the distance from $(3, 0)$ to a point (x, y) on the curve, then $S = (3 - x)^2 + y^2$. Setting $\frac{dS}{dx} = 0$ yields $\frac{dy}{dx} = \frac{3 - x}{y}$, while differentiating the relation $x^2 - y^2 = 1$ with respect to x yields $\frac{dy}{dx} = \frac{x}{y}$. These derivatives are equal for $x = \frac{3}{2}$.

20. **A.** Note that $f'(c) = 3c^2 - 6c = 3c(c - 2)$.

21. **C.** The integral is equivalent to $-\frac{1}{2} \int (9 - x^2)^{-1/2}(-2x \; dx)$. Use (3) on page 125 with $u = 9 - x^2$ and $n = \frac{1}{2}$.

22. **E.**
$$\frac{d^2}{dx^2} f(x^2) = \frac{d}{dx}\left[\frac{d}{dx} f(x^2) \right] = \frac{d}{dx}\left[\frac{d}{dx^2} f(x^2) \cdot \frac{dx^2}{dx} \right] = \frac{d}{dx}\left[g(x^2) \cdot 2x \right]$$
$$= g(x^2) \frac{d}{dx}(2x) + 2x \frac{d}{dx} g(x^2) = g(x^2) \cdot 2 + 2x \frac{d}{dx^2} g(x^2) \frac{dx^2}{dx}$$
$$= 2g(x^2) + 2x \cdot f(\sqrt{x^2}) \cdot 2x = 2g(x^2) + 4x^2 \cdot f(x).$$

23. A. The integral is rewritten as

$$\frac{1}{2}\int \frac{y^2 - 2y + 1}{y}\, dy,$$

$$= \frac{1}{2}\int \left(y - 2 + \frac{1}{y}\right) dy,$$

$$= \frac{1}{2}\left(\frac{y^2}{2} - 2y + \ln y\right) + C.$$

24. B. $\displaystyle\int_{\pi/6}^{\pi/2} \cot x\, dx = \ln \sin x \Big|_{\pi/6}^{\pi/2} = 0 - \ln \frac{1}{2}.$

25. D. $\displaystyle\int_{1}^{e} \ln x\, dx$ can be integrated by parts to yield $(x \ln x - x)\Big|_{1}^{e}$, which equals

$$e \ln e - e - (1 \ln 1 - 1) = e - e - (0 - 1) = 1.$$

26. B. Recall that

$$\frac{d}{dx}\int_{a}^{x} f(t)\, dt = f(x).$$

27. D. Draw a sketch of a function where $y \to \infty$ as $x \to c^+$. Note that none of the other answers need follow.

28. E. $\displaystyle\int_{-a}^{a} f(x)\, dx = 2\int_{0}^{a} f(x)\, dx$ only if $f(x)$ is even;

$$= 0 \qquad \text{only if } f(x) \text{ is odd;}$$

$$= F(a) - F(-a), \quad \text{where } \frac{dF(x)}{dx} = f(x);$$

$\displaystyle\int_{-a}^{a} f(x)\, dx$ represents the total area bounded by $y = f(x)$, the x-axis, and the vertical lines $x = -a$ and $x = a$, only if $f(x) \geqq 0$ on $[-a, a]$.

29. D. Let $y = 2 \ln (x - 1)$; then

$$\frac{y}{2} = \ln (x - 1), \qquad e^{y/2} = x - 1. \text{ Interchange } x \text{ and } y.$$

30. A.   See Figure AB3–2.

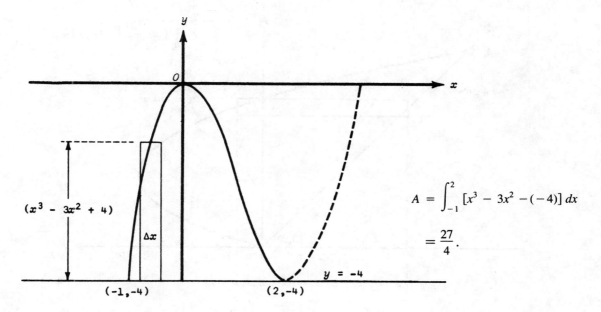

$$A = \int_{-1}^{2} [x^3 - 3x^2 - (-4)]\, dx$$

$$= \frac{27}{4}.$$

FIGURE AB3–2

31. C.   See Figure AB3–3.

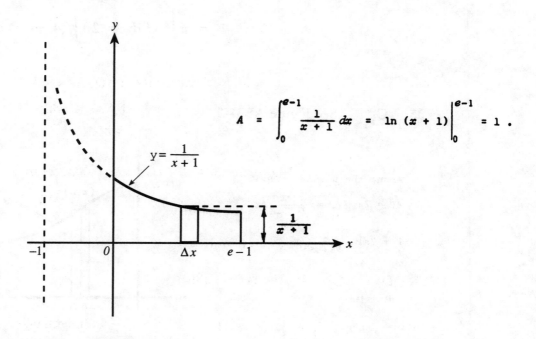

$$A = \int_{0}^{e-1} \frac{1}{x+1}\, dx = \ln(x+1)\Big|_{0}^{e-1} = 1.$$

FIGURE AB3–3

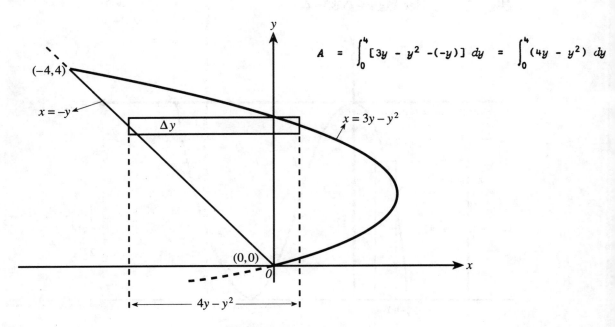

$$A = \int_0^4 [3y - y^2 - (-y)] \, dy = \int_0^4 (4y - y^2) \, dy$$

FIGURE AB3-4

33. C. $f'(x) = 0$ for $x = 1$ and $f''(1) < 0$.

34. D. The generating circle has equation $x^2 + y^2 = 4$. The volume, V, is given by

$$V = \pi \int_{-1}^{1} x^2 \, dy = 2\pi \int_0^1 (4 - y^2) \, dy.$$

35. C. See Figure AB3-5.

About the y-axis. Shell.

$$\Delta V = 2\pi xy \, \Delta x = \frac{2\pi x}{\sqrt{1 + x^2}} \, \Delta x$$

$$V = 2\pi \int_0^1 \frac{x}{\sqrt{1 + x^2}} \, dx$$

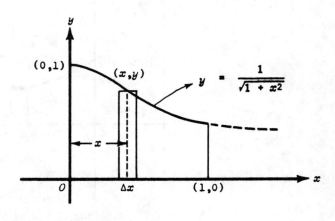

FIGURE AB3-5

36. C.   See Figure AB3–6.

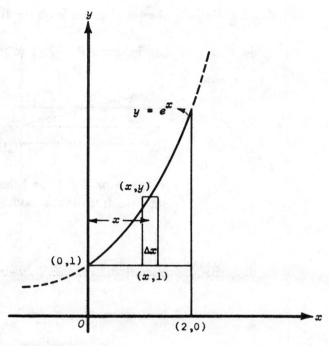

About the x-axis. Washer.

$\Delta V = \pi(y^2 - 1^2) \, \Delta x.$

$V = \pi \displaystyle\int_0^2 (e^{2x} - 1) \, dx.$

FIGURE AB3–6

37. E.   Since $v = \dfrac{ds}{dt} = 4s$, we have $\dfrac{ds}{s} = 4dt$. So $\ln s = 4t + C$, where, using $s(0) = 3$, $C = \ln 3$. Then, since $s = 3e^{4t}$, $s\left(\dfrac{1}{2}\right) = 3e^2$.

38. A.   Note that $f'(x) = -2x^2 e^{-x^2} + e^{-x^2}$.

39. C. $y_{av} = \dfrac{1}{3 - 1} \displaystyle\int_1^3 (x - 3)^2 \, dx.$

40. B.   A typical rectangle (element of area) perpendicular to the x-axis, on $[a, d]$, has height $f(x) - g(x)$.

41. D. $a = \dfrac{dv}{dt} = -t^2$ yields $v = -\dfrac{t^3}{3} + C_1$, and

$$s = -\frac{t^4}{12} + C_1 t + C_2.$$

Since $s(0) = 0$, $C_2 = 0$; and since $s(1) = 3$, $C_1 = \dfrac{37}{12}$. Thus $v(0) = \dfrac{37}{12}$.

42. C.   We have

$$y' = 4x^3 - 12x^2 = 4x^2(x - 3);$$
$$y'' = 12x^2 - 24x = 12x(x - 2).$$

Since $y'(3) = 0$ and $y''(3) > 0$, $x = 3$ yields a minimum. Since $y''(0)$ and $y''(2)$ both equal 0 and y'' changes sign at both $x = 0$ and $x = 2$, these yield points of inflection.

43. E. The point given is *on* the line. The distance is therefore zero.

44. D. See Figure AB3−7. The set of x for which $|x + 1| > 2$ is

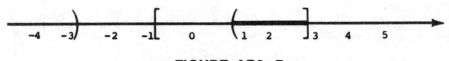

FIGURE AB3−7

$x < -3$ or $x > 1$, and the set for which $|x - 1| \leqq 2$ is $-1 \leqq x \leqq 3$. The set which satisfies both inequalities is shown by a heavy line in the figure.

45. B. Note that

$$\lim_{x \to \infty} xe^x = \infty, \qquad \lim_{x \to \infty} \frac{e^x}{x} = \infty, \qquad \lim_{x \to -\infty} \frac{x}{x^2 + 1} = 0,$$

$$\text{and} \qquad \frac{x^2}{x^3 + 1} \geqq 0 \quad \text{for } x > -1.$$

SECTION II

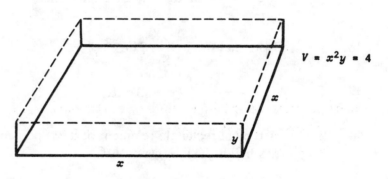

FIGURE AB3−8

1. (a) In Figure AB3−8 we let the dimensions of the box be $x \times x \times y$. The volume $x^2y = 4$. The cost C, in cents, is given by

$$C = 50(x^2 + 4xy) + 10(4x + 4y).$$

Using the expression for the volume, we get

$$C = 50\left(x^2 + \frac{16}{x}\right) + 10\left(4x + \frac{16}{x^2}\right), \tag{1}$$

whence

$$\frac{dC}{dx} = 50\left(2x - \frac{16}{x^2}\right) + 10\left(4 - \frac{32}{x^3}\right).$$

Setting this equal to zero yields

$$10\left(10\frac{x^3 - 8}{x^2} + 4\frac{x^3 - 8}{x^3}\right) = 0,$$

$$\frac{20(x^3 - 8)(5x + 2)}{x^3} = 0.$$

If $x > 0$, note that $x = 2$ is the only critical value, that C is a continuous function of x, and that $\frac{dC}{dx}$ changes from negative to positive as x increases through 2. Thus $x = 2$ yields minimum cost, and the dimensions of the box are $2 \times 2 \times 1$.

(b) To show that the specific costs of the sheet metal and welding do not affect the dimensions that yield a minimum total cost, let p be the cost per square foot of the sheet metal and q the cost per foot of the welding. Then (1) above becomes

$$C = p\left(x^2 + \frac{16}{x}\right) + q\left(4x + \frac{16}{x^2}\right)$$

and

$$\frac{dC}{dx} = p\left(2x - \frac{16}{x^2}\right) + q\left(4 - \frac{32}{x^3}\right)$$

$$= 2p\frac{x^3 - 8}{x^2} + 4q\frac{x^3 - 8}{x^3}$$

$$= \frac{2(x^3 - 8)(px + 2q)}{x^3}.$$

Since x must be positive, note that the only x for which $\frac{dC}{dx} = 0$ is, as before, $x = 2$; furthermore, this does not depend on p or q.

2. (a) $\ln x$.

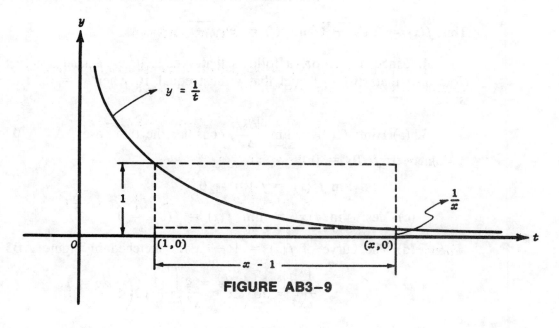

FIGURE AB3–9

(b) See Figure AB3–9: $\ln x$ is the area above the t-axis, below the curve $y = \dfrac{1}{t}$, and bounded by $t = 1$ and $t = x$.

(c) The area A described in (b) is less than that of the circumscribed rectangle of height 1 and length $(x - 1)$, and greater than that of the inscribed rectangle of height $\dfrac{1}{x}$ and length $(x - 1)$. Thus, if $x > 1$, then $x - 1 > \ln x > (x - 1) \cdot \dfrac{1}{x}$.

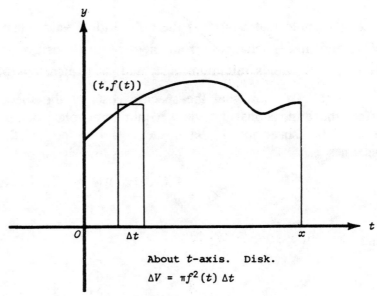

About t-axis.   Disk.

$$\Delta V = \pi f^2(t)\,\Delta t$$

FIGURE AB3–10

3. See Figure AB3–10. The volume is given by the integral

$$V = \pi \int_{t=0}^{x} y^2\, dt = \pi \int_{0}^{x} f^2(t)\, dt.$$

Since $\dfrac{dV}{dx} = \pi f^2(x)$ and since $v = 2\pi(x^2 + 2x)$, then

$$\pi f^2(x) = 2\pi(2x + 2) \qquad \text{and} \qquad f^2(x) = 4x + 4.$$

Thus $f(x) = 2\sqrt{x + 1}$ and $f(t) = 2\sqrt{t + 1}$.

4. Since $a = -6t$, it follows that $v = -3t^2 + C$ and $s = -t^3 + Ct + C'$. Note also that $s(0) = 0$ and that s must equal 16 when $v = 0$. The initial velocity is 12.

5. (a) Hint: $f'(0) = \lim\limits_{\Delta x \to 0} \dfrac{|\Delta x|^3}{\Delta x}$; consider the two cases, $\Delta x \to 0^+$ and $\Delta x \to 0^-$.
Answer: $f'(0) = 0$.

(b) Yes. $\lim\limits_{x \to 0} f'(x) = f'(0) = 0$.

(c) Yes: $\lim\limits_{x \to 0^-} f(x) = \lim\limits_{x \to 0^+} f(x) = f(0) = 0$.

(d) The curve of $f(x) = 1 - |x|$ is sketched in Figure AB3–11.

$$\int_{-1}^{1} (1 - |x|)\, dx = 2 \int_{0}^{1} (1 - x)\, dx = 1.$$

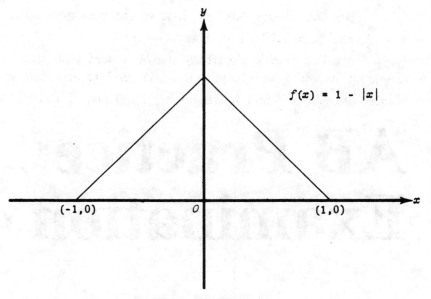

FIGURE AB3–11

6. See Figure AB3–12:

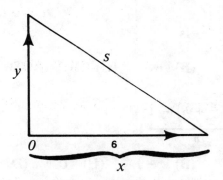

FIGURE AB3–12

$$s^2 = x^2 + y^2;$$

$$2s \frac{ds}{dt} = 2x \frac{dx}{dt} + 2y \frac{dy}{dt}.$$

We are given that $\frac{dx}{dt} = 3$, $\frac{dy}{dt} = 5$, and are asked to find $\frac{ds}{dt}$ when $x = 12$. Note that $y = 10$ and $s = \sqrt{244}$.

$$\frac{ds}{dt} = \frac{12(3) + 10(5)}{\sqrt{244}} = \frac{43\sqrt{61}}{61} \text{ ft/min.}$$

AB Practice Examination 4

Section I

1. $\lim\limits_{x \to 0} \dfrac{x^3 - 3x^2}{x}$ is

   (A) 0   **(B)** ∞   **(C)** −3   **(D)** 1   **(E)** 3

2. $\lim\limits_{h \to 0} \dfrac{\sin\left(\dfrac{\pi}{2} + h\right) - 1}{h}$ is

   (A) 1   **(B)** −1   **(C)** 0   **(D)** ∞   **(E)** none of these

3. $\lim\limits_{x \to 2} [x]$ (where $[x]$ is the greatest integer in x) is

   (A) 1   **(B)** 2   **(C)** 3   **(D)** ∞   **(E)** nonexistent

4. $\lim\limits_{x \to \infty} x \tan \dfrac{\pi}{x}$ is

   (A) 0   **(B)** 1   **(C)** $\dfrac{1}{\pi}$ **(D)** π **(E)** ∞

5. If $y = \ln \dfrac{x}{\sqrt{x^2 + 1}}$, then $\dfrac{dy}{dx}$ is

   (A) $\dfrac{1}{x^2 + 1}$ **(B)** $\dfrac{1}{x(x^2 + 1)}$ **(C)** $\dfrac{2x^2 + 1}{x(x^2 + 1)}$ **(D)** $\dfrac{1}{x\sqrt{x^2 + 1}}$

   (E) $\dfrac{1 - x^2}{x(x^2 + 1)}$

412

6. If $y = \sqrt{x^2 + 16}$, then $\dfrac{d^2y}{dx^2}$ is

(A) $-\dfrac{1}{4(x^2 + 16)^{3/2}}$ (B) $4(3x^2 + 16)$ (C) $\dfrac{16}{\sqrt{x^2 + 16}}$

(D) $\dfrac{2x^2 + 16}{(x^2 + 16)^{3/2}}$ (E) $\dfrac{16}{(x^2 + 16)^{3/2}}$

7. The graph of $y = e^{x/2}$ is reflected in the line $y = x$. An equation of the reflection is $y =$

(A) $2 \ln x$ (B) $\ln \dfrac{x}{2}$ (C) e^{2x} (D) $e^{-x/2}$ (E) $\dfrac{1}{2} \ln x$

8. The equation of the tangent to the curve $2x^2 - y^4 = 1$ at the point $(-1, 1)$ is

(A) $y = -x$ (B) $y = 2 - x$ (C) $4y + 5x + 1 = 0$

(D) $x - 2y + 3 = 0$ (E) $x - 4y + 5 = 0$

9. If $f(x) = x \ln x$, then $f'''(e)$ equals

(A) $\dfrac{1}{e}$ (B) 0 (C) $-\dfrac{1}{e^2}$ (D) $\dfrac{1}{e^2}$ (E) $\dfrac{2}{e^3}$

10. If $y = \sqrt{\ln (x^2 + 1)}$ $(x > 0)$, then the derivative of y^2 with respect to $\ln x$ is equal to

(A) $\dfrac{2x}{x^2 + 1}$ (B) $\dfrac{2}{x^2 + 1}$ (C) $\dfrac{2x}{\ln x(x^2 + 1)}$ (D) $\dfrac{2x^2}{x^2 + 1}$

(E) none of these

11. If $f(t) = \dfrac{1}{t^2} - 4$ and $g(t) = \cos t$, then the derivative of $f(g(t))$ is

(A) $2 \sec^2 t \tan t$ (B) $\tan t$ (C) $2 \sec t \tan t$ (D) $\dfrac{2}{t^3 \sin t}$

(E) $-\dfrac{2}{\cos^3 t}$

12. If $f(x) = \dfrac{e^{2 \ln x}}{x}$, then $f(1)$ is

(A) 0 (B) 1 (C) e (D) $\dfrac{e}{2}$ (E) not defined

13. A particle moves on a line according to the law $s = f(t)$ so that its velocity $v = ks$, where k is a nonzero constant. Its acceleration is

(A) k^2v (B) k^2s (C) k (D) 0 (E) none of these

14. The intervals on which the function $f(x) = x^4 - 4x^3 + 4x^2 + 6$ increases are

(A) $x < 0$ and $1 < x < 2$ (B) only $x > 2$ (C) $0 < x < 1$ and $x > 2$
(D) only $0 < x < 1$ (E) only $1 < x < 2$

15. A relative maximum value of the function $y = \dfrac{\ln x}{x}$ is

(A) 1 (B) e (C) $\dfrac{2}{e}$ (D) $\dfrac{1}{e}$ (E) none of these

16. If a particle moves on a line according to the law $s = t^5 + 2t^3$, then the number of times it reverses direction is

(A) 4 (B) 3 (C) 2 (D) 1 (E) 0

17. If $f(x) = \dfrac{1}{2x - 1}$, then $f^{-1}(x)$, the inverse function, equals

(A) $2x - 1$ (B) $\dfrac{2(x + 1)}{x}$ (C) $\dfrac{x - 1}{x}$ (D) $\dfrac{x + 1}{x}$ (E) $\dfrac{x + 1}{2x}$

18. A rectangular pigpen is to be built against a wall so that only three sides will require fencing. If p feet of fencing are to be used, the area of the largest possible pen is

(A) $\dfrac{p^2}{2}$ (B) $\dfrac{p^2}{4}$ (C) $\dfrac{p^2}{8}$ (D) $\dfrac{p^2}{9}$ (E) $\dfrac{p^2}{16}$

19. A smooth curve with equation $y = f(x)$ is such that its slope at each x equals x^2. If the curve goes through the point $(-1, 2)$, then its equation is

(A) $y = \dfrac{x^3}{3} + 7$ (B) $x^3 - 3y + 7 = 0$ (C) $y = x^3 + 3$

(D) $y - 3x^3 - 5 = 0$ (E) none of these

20. The region bounded by the parabolas $y = x^2$ and $y = 6x - x^2$ is rotated about the x-axis so that a vertical line segment cut off by the curves generates a ring. The value of x for which the ring of largest area is obtained is

(A) 4 (B) 3 (C) $\dfrac{5}{2}$ (D) 2 (E) $\dfrac{3}{2}$

21. $\displaystyle\int \dfrac{\cos x}{4 + 2 \sin x}\, dx$ equals

(A) $\sqrt{4 + 2 \sin x} + C$ (B) $-\dfrac{1}{2(4 + \sin x)} + C$

(C) $\ln \sqrt{4 + 2 \sin x} + C$ (D) $2 \ln |4 + 2 \sin x| + C$

(E) $\dfrac{1}{4} \sin x - \dfrac{1}{2} \csc^2 x + C$

22. $\int \dfrac{e^u}{4 + e^{2u}}\, du$ is equal to

   (A) $\ln (4 + e^{2u}) + C$ (B) $\dfrac{1}{2} \ln |4 + e^{2u}| + C$ (C) $\dfrac{1}{2} \tan^{-1} \dfrac{e^u}{2} + C$

   (D) $\tan^{-1} \dfrac{e^u}{2} + C$ (E) $\dfrac{1}{2} \tan^{-1} \dfrac{e^{2u}}{2} + C$

23. $\int_{-2}^{3} |x - 1|\, dx =$

   (A) $\dfrac{5}{2}$ (B) $\dfrac{7}{2}$ (C) $\dfrac{9}{2}$ (D) $\dfrac{11}{2}$ (E) $\dfrac{13}{2}$

24. $\int \dfrac{dx}{x \ln x}$ equals

   (A) $\ln (\ln x) + C$ (B) $-\dfrac{1}{\ln^2 x} + C$ (C) $\dfrac{(\ln x)^2}{2} + C$

   (D) $\ln x + C$ (E)   none of these

25. If we replace $\sqrt{x - 2}$ by u, then $\int_{3}^{6} \dfrac{\sqrt{x - 2}}{x}\, dx$ is equivalent to

   (A) $\int_{1}^{2} \dfrac{u\, du}{u^2 + 2}$ (B) $2 \int_{1}^{2} \dfrac{u^2\, du}{u^2 + 2}$ (C) $\int_{3}^{6} \dfrac{2\, u^2}{u^2 + 2} du$

   (D) $\int_{3}^{6} \dfrac{u\, du}{u^2 + 2}$ (E) $\dfrac{1}{2} \int_{1}^{2} \dfrac{u^2}{u^2 + 2} du$

26. What is $\displaystyle\lim_{x \to \infty} \dfrac{x^2 + 9}{(3 + x)(3 - x)}$?

   (A) -1 (B) 0 (C) 1 (D) 9 (E)   Limit does not exist.

27. The graph of $y^2 - 4y - x^2 = 1$ is symmetric to

   (A)   the origin   (B)   the x-axis   (C)   the y-axis

   (D)   the line $y = x$ (E)   none of the preceding

28. If $\dfrac{dQ}{dt} = \dfrac{Q}{10}$ and $Q = Q_0$ when $t = 0$, then Q, at any time t, equals

   (A) $Q_0 10^t$ (B) $Q_0 \cdot e^{10t}$ (C) $Q_0 \cdot t^{1/10}$ (D) $Q_0 \cdot e^{t/10}$

   (E) $\dfrac{10 Q_0}{10 - t}$

29. Suppose

$$f(x) = \begin{cases} 2x & \text{for } x \leq 1 \\ 3x^2 - 2 & \text{for } x > 1 \end{cases}$$

Then $\int_0^2 f(x)\, dx$ equals

(A) 6   (B) 5   (C) 4   (D) 3   (E) none of these

30. The area bounded by the parabola $y = x^2$ and the lines $y = 1$ and $y = 9$ equals

(A) 8   (B) $\dfrac{84}{3}$ (C) $\dfrac{64}{3}\sqrt{2}$ (D) 32   (E) $\dfrac{104}{3}$

31. The total area bounded by the curves $y = x^3$ and $y = x^{1/3}$ is equal to

(A) $\dfrac{1}{2}$ (B) $\dfrac{5}{6}$ (C) 1   (D) 2   (E) none of these

32. The area under the curve $y = \dfrac{1}{\sqrt{x-1}}$, above the x-axis, and bounded vertically by $x = 2$ and $x = 5$ equals

(A) 1   (B) 2   (C) 4   (D) ln 2   (E) 2 ln 2

33. The total area enclosed by the curves $y = 4x - x^3$ and $y = 4 - x^2$ is represented by

(A) $\displaystyle \int_{-2}^{2} (4x - x^3 + x^2 - 4)\, dx$

(B) $\displaystyle \int_{-2}^{2} (4 - x^2 - 4x + x^3)\, dx$

(C) $\displaystyle \int_{-2}^{1} (4x - x^3 + x^2 - 4)\, dx + \int_{1}^{2} (4 - x^2 - 4x + x^3)\, dx$

(D) $\displaystyle \int_{-2}^{1} (4 - x^2 - 4x + x^3)\, dx + \int_{1}^{2} (4x - x^3 + x^2 - 4)\, dx$

(E) $\displaystyle \int_{-2}^{1} (4x - x^3 + x^2 - 4)\, dx$

34. The graph of $y = \dfrac{3x^2}{x^2 - 9}$ has

(A) one horizontal and one vertical asymptote
(B) two horizontal asymptotes, no vertical ones
(C) two horizontal and one vertical asymptotes
(D) one horizontal and two vertical asymptotes
(E) two vertical asymptotes, no horizontal ones

35. The volume obtained by rotating the region bounded by $x = y^2$ and $x = 2 - y^2$ about the y-axis is equal to

(A) $\dfrac{16\pi}{3}$ (B) $\dfrac{32\pi}{3}$ (C) $\dfrac{32\pi}{15}$ (D) $\dfrac{64\pi}{15}$ (E) $\dfrac{8\pi}{3}$

36. Suppose $f(x) = \dfrac{x^2 + x}{x}$ if $x \neq 0$ and $f(0) = 1$. Which of the following statements are true of f? I. f is defined at $x = 0$. II. $\lim\limits_{x \to 0} f(x)$ exists. III. f is continuous at $x = 0$.

(A) I only   (B) II only   (C) I and II only
(D) None of the statements is true.   (E) All are true.

37. Which function could have the graph shown below?

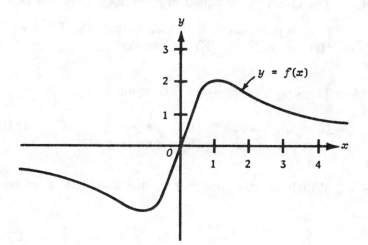

(A) $y = \dfrac{x}{x^2 + 1}$ (B) $y = \dfrac{4x}{x^2 + 1}$ (C) $y = \dfrac{2x}{x^2 - 1}$

(D) $y = \dfrac{x^2 + 3}{x^2 + 1}$ (E) $y = \dfrac{4x}{x + 1}$

38. If $f(x) = 2 \ln x$ and $g(x) = e^{x/2}$, then $f(g(x)) =$
(A) x (B) e^x (C) 1 (D) 2 (E) none of the preceding

39. The average (mean) value of $y = \ln x$ over the interval from $x = 1$ to $x = e$ equals

(A) $\dfrac{1}{2}$ (B) $\dfrac{1}{2(e - 1)}$ (C) $\dfrac{2}{e - 1}$ (D) $\dfrac{1}{e - 1}$

(E) none of these

40. A particle moves along a line with acceleration $a = 6t$. If, when $t = 0$, $v = 1$, then the total distance traveled between $t = 0$ and $t = 3$ equals

(A) 30   (B) 28   (C) 27   (D) 26   (E) none of these

41. The area in the first quadrant bounded above by $y = \frac{1}{x}$, at the left by $x = 1$, and at the right by $x = e$ equals

(A) $\frac{1}{e}$ (B) 1   (C) $e - 1$ (D) e (E) $\ln (e - 1)$

42. If $x = f(t)$ is the law of motion for a particle on a line, the particle is said to be in *simple harmonic motion* if the acceleration is given by $\frac{d^2x}{dt^2} = -k^2x$, k a constant. Which of the following equations does *not* define a simple harmonic motion?

(A) $x = e^t + e^{-t}$ (B) $x = 4 \cos 2t$ (C) $x = 2 \sin (\pi t + 2)$
(D) $x = \sin 2t + \cos 2t$ (E) $x = \cos (\omega t + \phi)$, ω and ϕ constants

43. The graph of the equation $y^2 = 4(x + 1)^2 + 4$ is

(A) a parabola   (B) an ellipse   (C) two intersecting straight lines
(D) a circle   (E) a hyperbola

44. The graph of $y = \sin \frac{1}{x}$ is symmetric to

(A) the x-axis   (B) the y-axis   (C) the line $y = x$
(D) the origin   (E) none of these

45. Which of the following functions has a graph which is asymptotic to the y-axis?

(A) $y = \frac{x}{x^2 - 1}$ (B) $y = \frac{x}{x^2 + 1}$ (C) $y = x - \frac{2}{x}$

(D) $y = \ln (x - 1)$ (E) $y = \frac{\sqrt{x - 1}}{x}$

AB Practice Examination 4, Section II

1. Sketch the curve of $y = \ln x$, and find the volume of the solid obtained if the region in the first quadrant bounded by the curve and the line $x = e$ is rotated about the y-axis.

2. Sketch the curve of $y = \sin x (1 + \cos x)$, after having found any:
(a) intercepts;
(b) relative maximum or minimum points;
(c) inflection points.

3. Assume that a falling balloon loses volume at a rate proportional to its volume. If it loses 10% of its initial volume during the first minute of fall, find how long it will take to lose 50% of its initial volume.

4. A point $P(x, y)$ moves along the curve $y = \ln x$ so that its x-coordinate increases at the rate of $\sqrt{3}$ units per second. If P is one vertex of an equilateral triangle whose other two vertices are on the x-axis, find how fast the area of the triangle is changing when the y-coordinate of P is 1.

5. If a tangent is drawn to the parabola $y = 3 - x^2$ at any point other than the vertex, then it forms a right angle with the axes. At what point on the curve should the tangent be drawn to form the triangle of least area?

6. Suppose $F(x) = \displaystyle\int_1^x \frac{\sin t}{t}\, dt$.

    (a) Determine $\displaystyle\lim_{x \to 0} F'(x)$, if it exists.

    (b) Prove that $\displaystyle\lim_{x \to 0} F'\left(\frac{1}{x}\right)$ exists and equals zero. State any theorems you use.

Answers to AB Practice Examination 4

SECTION I

1. A	9. C	17. E	25. B	33. D	41. B
2. C	10. D	18. C	26. A	34. D	42. A
3. E	11. A	19. B	27. C	35. A	43. E
4. D	12. B	20. D	28. D	36. E	44. D
5. B	13. B	21. C	29. A	37. B	45. C
6. E	14. C	22. C	30. E	38. A	
7. A	15. D	23. E	31. C	39. D	
8. A	16. E	24. A	32. B	40. A	

1. A. $\lim\limits_{x\to 0} \dfrac{x^3 - 3x^2}{x} = \lim\limits_{x\to 0}(x^2 - 3x) = 0.$

2. C. The given limit equals $f'\left(\dfrac{\pi}{2}\right)$, where $f(x) = \sin x.$

3. E. Here, $\lim\limits_{x\to 2^-}[x] = 1$, while $\lim\limits_{x\to 2^+}[x] = 2.$

4. D. $\lim\limits_{x\to\infty} x \tan\dfrac{\pi}{x} = \lim\limits_{x\to\infty}\dfrac{\tan\dfrac{\pi}{x}}{\dfrac{1}{x}} = \lim\limits_{x\to\infty}\dfrac{\pi\tan\dfrac{\pi}{x}}{\dfrac{\pi}{x}}.$ If we set $y = \dfrac{\pi}{x}$, this yields

$$\pi\lim\limits_{y\to 0}\dfrac{\tan y}{y} = \pi\lim\limits_{y\to 0}\left[\dfrac{\sin y}{y}\cdot\dfrac{1}{\cos y}\right] = \pi\cdot 1\cdot 1 = \pi.$$

5. B. Since $\ln\dfrac{x}{\sqrt{x^2 + 1}} = \ln x - \dfrac{1}{2}\ln(x^2 + 1)$, then

$$\dfrac{dy}{dx} = \dfrac{1}{x} - \dfrac{1}{2}\cdot\dfrac{2x}{x^2 + 1} = \dfrac{1}{x(x^2 + 1)}.$$

6. E. Rewrite y as $(x^2 + 16)^{1/2}$, yielding

$$\dfrac{dy}{dx} = \dfrac{x}{(x^2 + 16)^{1/2}},$$

and apply the quotient rule, page 33.

7. A. Interchange x and y and solve for y.

8. A. Differentiate implicitly to get $4x - 4y^3\dfrac{dy}{dx} = 0$. Substitute $(-1, 1)$ to find $\dfrac{dy}{dx}$, the slope, at this point, and write the equation of the tangent: $y - 1 = -1(x + 1).$

9. C. Since $f(x) = x\ln x$,

$$f'(x) = 1 + \ln x, \qquad f''(x) = \dfrac{1}{x}, \qquad \text{and} \qquad f'''(x) = -\dfrac{1}{x^2}.$$

10. D. Since $\dfrac{dy^2}{d(\ln x)} = \dfrac{\dfrac{dy^2}{dx}}{\dfrac{d(\ln x)}{dx}}$ and $y^2 = \ln(x^2 + 1)$, we find

$$\dfrac{dy^2}{d(\ln x)} = \dfrac{\dfrac{2x}{x^2 + 1}}{\dfrac{1}{x}} = \dfrac{2x^2}{x^2 + 1} \qquad (x > 0).$$

11. A. If $f(t) = \dfrac{1}{t^2} - 4$ and $g(t) = \cos t$, then

$$f(g(t)) = \dfrac{1}{\cos^2 t} - 4 = \sec^2 t - 4,$$

and its derivative is $2\sec^2 t \tan t.$

12. **B.**   Since $e^{2 \ln x} = x^2$, we see that $f(x) = \frac{x^2}{x} = x$.

13. **B.**   Since $v = ks = \frac{ds}{dt}$, then $a = \frac{d^2s}{dt^2} = k\frac{ds}{dt} = kv = k^2s$.

14. **C.** $f'(x) = 4x^3 - 12x^2 + 8x = 4x(x - 1)(x - 2)$. To determine the signs of $f'(x)$, inspect the sign at any point in each of the intervals $x < 0$, $0 < x < 1$, $1 < x < 2$, and $x > 2$. The function increases whenever $f'(x) > 0$.

15. **D.**   Here $y' = \frac{1 - \ln x}{x^2}$, which is zero for $x = e$. Since the signs change from positive to negative as x increases through e, this critical value yields a relative maximum. Note that $f(e) = \frac{1}{e}$.

16. **E.**   Since $v = \frac{ds}{dt} = 5t^4 + 6t^2$ never changes signs, there are no reversals in motion along the line.

17. **E.**   Let $y = f(x)$ and solve for x. Then interchange x and y.

18. **C.**   Letting y be the length parallel to the wall and x the other dimension of the rectangle, we have

$$p = 2x + y \qquad \text{and} \qquad A = xy = x(p - 2x).$$

For $x = \frac{p}{4}$ we have $\frac{dA}{dx} = 0$, which yields $y = \frac{p}{2}$. Note that $\frac{d^2A}{dx^2} < 0$.

19. **B.**   We solve the differential equation $\frac{dy}{dx} = x^2$, getting $y = \frac{x^3}{3} + C$. Use $x = -1$, $y = 2$ to determine C.

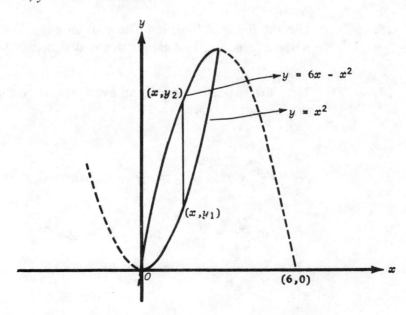

FIGURE AB4–1

20. D.   See Figure AB4–1. Since the area, A, of the ring equals $\pi(y_2^2 - y_1^2)$,

$$A = \pi[(6x - x^2)^2 - x^4] = \pi[36x^2 - 12x^3 + x^4 - x^4]$$

and

$$\frac{dA}{dx} = \pi(72x - 36x^2) = 36\pi x(2 - x),$$

where it can be verified that $x = 2$ produces the maximum area.

21. C.   The integral is equivalent to $\frac{1}{2}\int \frac{du}{u}$, where $u = 4 + 2\sin x$.

22. C.   Note that the given integral is of the type

$$\int \frac{dv}{a^2 + v^2} = \frac{1}{a}\tan^{-1}\frac{v}{a} + C.$$

23. E.   Draw a figure.

$$\int_{-2}^{3}|x - 1|\, dx = \int_{-2}^{1}(1 - x)\, dx + \int_{1}^{3}(x - 1)\, dx.$$

24. A.   This is of type $\int \frac{du}{u}$ with $u = \ln x$.

25. B.   If $u = \sqrt{x - 2}$, then $u^2 = x - 2$, $x = u^2 + 2$, $dx = 2u\, du$. When $x = 3$, $u = 1$; when $x = 6$, $u = 2$.

26. A.   Use the Rational Function Theorem on page 23 after expanding the denominator and noting that numerator and denominator are of the same degree.

27. C.   Note that x occurs only to an even power, so that $(-x, y)$ is on the graph whenever (x, y) is.

28. D. $\frac{dQ}{Q} = \frac{dt}{10}$ yields $\ln Q = \frac{t}{10} + C$, where $C = \ln Q_0$. So

$$\ln \frac{Q}{Q_0} = \frac{t}{10} \quad \text{or} \quad Q = Q_0 e^{t/10}.$$

29. A.   Here $\int_0^2 f(x)\, dx = \int_0^1 2x\, dx + \int_1^2 (3x^2 - 2)\, dx.$

**30. E.   See Figure AB4–2.

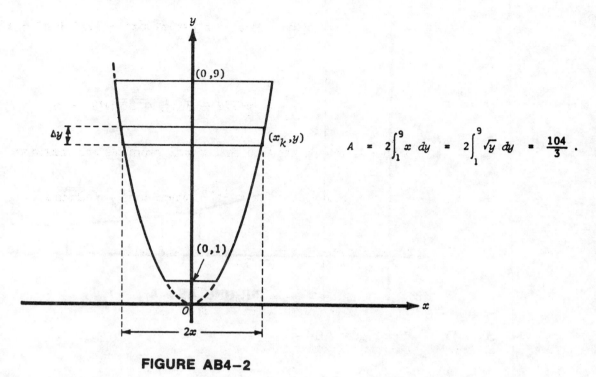

$$A = 2\int_1^9 x \, dy = 2\int_1^9 \sqrt{y} \, dy = \frac{104}{3}.$$

FIGURE AB4–2

**31. C.   See Figure AB4–3.

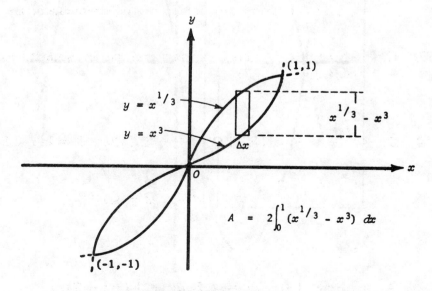

$$A = 2\int_0^1 (x^{1/3} - x^3) \, dx$$

FIGURE AB4–3

32. B. See Figure AB4-4.

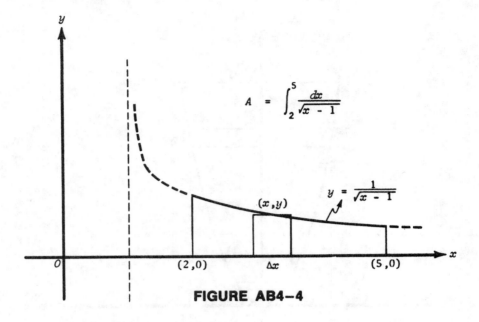

$$A = \int_2^5 \frac{dx}{\sqrt{x-1}}$$

$$y = \frac{1}{\sqrt{x-1}}$$

FIGURE AB4-4

33. D. See Figure AB4-5.

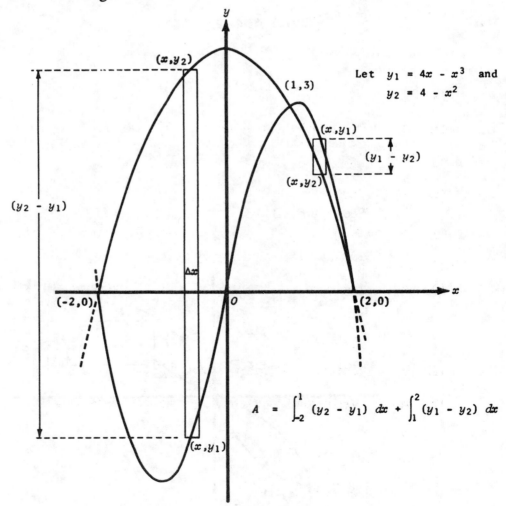

Let $y_1 = 4x - x^3$ and
$y_2 = 4 - x^2$

$$A = \int_{-2}^{1} (y_2 - y_1)\ dx + \int_{1}^{2} (y_1 - y_2)\ dx$$

FIGURE AB4-5

34. D.   The graph has a horizontal asymptote, $y = 3$, and two vertical asymptotes, $x = \pm 3$.

35. A.   See Figure AB4−6.

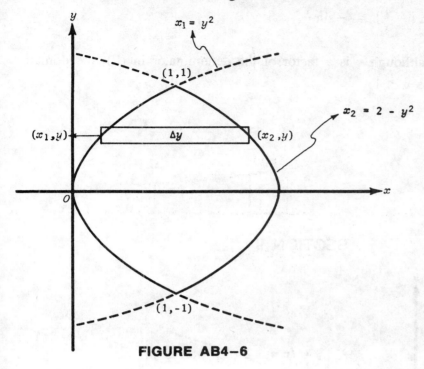

About the y-axis. Washer.

$\Delta V = \pi(x_2{}^2 - x_1{}^2)\,\Delta y$

$V = 2\pi\displaystyle\int_0^1 [(2 - y^2)^2 - y^4]\,dy$

$= 2\pi\displaystyle\int_0^1 (4 - 4y^2)\,dy$.

FIGURE AB4−6

36. E.   Note that $\lim\limits_{x \to 0} f(x) = f(0) = 1$.

37. B.   Note that $(0, 0)$ is on the graph, as are $(1, 2)$ and $(-1, -2)$. So only (B) and (E) are possible. Since $\lim\limits_{x \to \infty} y = \lim\limits_{x \to -\infty} y = 0$, only (B) is correct.

38. A. $f(g(x)) = 2 \ln e^{x/2} = 2 \cdot \dfrac{x}{2}$.

39. D. $(y_{av})_x = \dfrac{1}{e - 1}\displaystyle\int_1^e \ln x\, dx = \dfrac{1}{e - 1}\Big[x \ln x - x\Big]_1^e$, where $\int \ln x\, dx$ can be obtained by integrating by parts.

40. A.   Since $a = \dfrac{dv}{dt} = 6t$, $v = 3t^2 + C$, and since $v(0) = 1$, $C = 1$. Then $v = \dfrac{ds}{dt} = 3t^2 + 1$ yields $s = t^3 + t + C'$, and we can let $s(0) = 0$. Then we want $s(3)$.

41. B.   The area described equals $\displaystyle\int_1^e \dfrac{1}{x}\,dx = \ln e = 1$.

42. A.   Verify that, for each of the other equations,

$$\frac{d^2x}{dt^2} = -k^2x \qquad (k \text{ a constant})$$

while for $x = e^t + e^{-t}$,

$$\frac{d^2x}{dt^2} = x.$$

43. **E.** Since the equation can be rewritten as $\frac{y^2}{4} - (x + 1)^2 = 1$, it is of the type given in Equation 52 on page 486.

44. **D.** Note that $\sin\left(-\frac{1}{x}\right) = -\sin\frac{1}{x}$.

45. **C.** Note that, although x is a factor of the denominator in (E), the domain of $\frac{\sqrt{x - 1}}{x}$ is $x \geqq 1$.

SECTION II

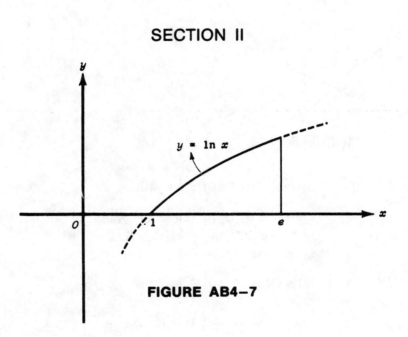

FIGURE AB4–7

1. See Figure AB4–7. The volume is equal to

$$2\pi \int_1^e x \ln x \, dx = 2\pi \left(\frac{x^2}{2} \ln x - \frac{x^2}{4}\right)\Big|_1^e$$

(where we integrated by parts); the volume is thus $\frac{\pi}{2}(e^2 + 1)$.

2. (a) $y = 0$ if $x = n\pi$, n an integer.

(b) The curve has maximum and minimum points respectively at $\left(2n\pi + \frac{\pi}{3}, \frac{3\sqrt{3}}{4}\right)$ and $\left(2n\pi - \frac{\pi}{3}, \frac{-3\sqrt{3}}{4}\right)$, where n is an integer.

(c) The coordinates of the inflection points are $(n\pi, 0)$, n an integer, and $\left(\cos^{-1}\left(-\frac{1}{4}\right), \pm\frac{3\sqrt{15}}{16}\right)$.

The curve is sketched in Figure AB4–8.

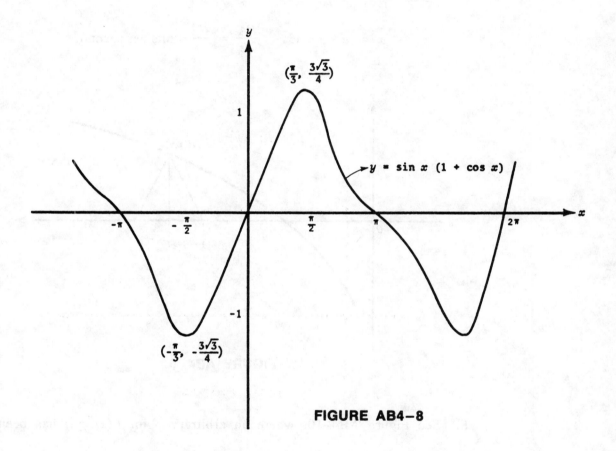

FIGURE AB4–8

3. Solve the differential equation $\frac{dV}{dt} = kV$. If V_0 is the initial volume, then $\ln \frac{V}{V_0} = kt$. When $t = 1$, $\frac{V}{V_0} = 0.9$. This yields $\ln \frac{V}{V_0} = t \ln 0.9$. Then

$$V = \frac{1}{2} V_0 \quad \text{when} \quad t = \frac{\ln 0.5}{\ln 0.9}.$$

4. See Figure AB4–9, where the length of a side of the equilateral triangle equals $2s$.

Since $(2s)^2 = s^2 + y^2$, $s = \dfrac{y}{\sqrt{3}}$. The area, A, of the triangle is ys or $\dfrac{y^2}{\sqrt{3}}$ or $\dfrac{1}{\sqrt{3}}(\ln x)^2$. Then

$$\frac{dA}{dt} = \frac{2}{\sqrt{3}} \cdot \frac{\ln x}{x} \cdot \frac{dx}{dt} = \frac{2}{\sqrt{3}} \cdot \frac{\ln x}{x} \cdot \sqrt{3} = \frac{2 \ln x}{x}$$

at any time t. When $y = 1$, that is, $\ln x = 1$,

$$x = e \qquad \text{and} \qquad \frac{dA}{dt} = \frac{2(1)}{e} \text{units per second.}$$

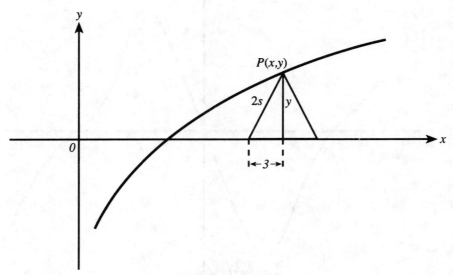

FIGURE AB4–9

5. See Figure AB4–10, where an arbitrary point $P(x_1, y_1)$ has been chosen, $x_1 > 0$.

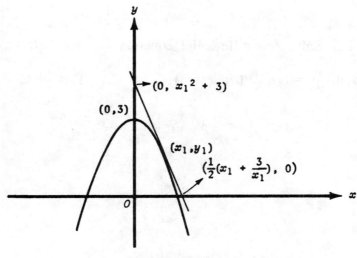

FIGURE AB4–10

Hint: The equation of the tangent at P is $y - y_1 = -2x_1(x - x_1)$. Replace y_1 by $3 - x_1^2$ to obtain the intercepts of this tangent:

$$x = \frac{x_1^2 + 3}{2x_1}, \qquad y = x_1^2 + 3.$$

The area A of the triangle is thus

$$A = \frac{1}{4}\left(x_1 + \frac{3}{x_1}\right)(x_1^2 + 3).$$

Verify that

$$\frac{dA}{dx_1} = \frac{1}{4}(x_1^2 + 3)\left(3 - \frac{3}{x_1^2}\right)$$

and that A is a minimum when $x_1 = 1$.

6. Here, $F(x) = \int_1^x \frac{\sin t}{t}\, dt.$

(a) $F'(x) = \frac{\sin x}{x}$ and $\lim\limits_{x \to 0} F'(x) = \lim\limits_{x \to 0} \frac{\sin x}{x} = 1.$

(b) From (a), $F'\left(\frac{1}{x}\right) = \frac{\sin \frac{1}{x}}{\frac{1}{x}}$ and thus $\lim\limits_{x \to 0} F'\left(\frac{1}{x}\right) = \lim\limits_{x \to 0} \frac{\sin \frac{1}{x}}{\frac{1}{x}} = \lim\limits_{x \to 0} x \sin \frac{1}{x}.$

Note that, since $-1 \leq \sin \frac{1}{x} \leq 1$ for all $x \neq 0$,

$$\text{if } x > 0, \qquad -x \leq x \sin \frac{1}{x} \leq x;$$

$$\text{if } x < 0, \qquad -x \geq x \sin \frac{1}{x} \geq x.$$

In either case, since $x \sin \frac{1}{x}$ is "squeezed" between two quantities each of which approaches zero as x does, so must $x \sin \frac{1}{x}$. Thus

$$\lim_{x \to 0} \frac{\sin \frac{1}{x}}{\frac{1}{x}} = 0.$$

Theorems used:
(1) If $a < b$ and $c > 0$, $ac < bc$, while if $c < 0$, $ac > bc$.
(2) The "Squeeze" Theorem: if $f(t) \leq g(t) \leq h(t)$ for all values of t near c and if $\lim\limits_{t \to c} f(t) = \lim\limits_{t \to c} h(t) = L$, then $\lim\limits_{t \to c} g(t) = L$.
(3) If $\lim\limits_{t \to c^-} g(t) = \lim\limits_{t \to c^+} g(t) = L$, then $\lim\limits_{t \to c} g(t) = L$.

BC Practice Examination 1

Section 1

1. $\lim\limits_{x \to \infty} \dfrac{3x^2 - 4}{2 - 7x - x^2}$ is

   (A) 3   **(B)** 1   **(C)** -3 **(D)** ∞ **(E)** 0

2. In using the method of partial fractions to decompose $\dfrac{4x^2 + 3x + 5}{(x - 1)(x^2 + 2)}$, one of the fractions obtained is

   (A) $\dfrac{3}{x - 1}$ **(B)** $\dfrac{3}{x^2 + 2}$ **(C)** $\dfrac{4}{x^2 + 2}$ **(D)** $\dfrac{3x}{x^2 + 2}$ **(E)** $\dfrac{x + 3}{x^2 + 2}$

3. $\lim\limits_{h \to 0} \dfrac{\cos\left(\dfrac{\pi}{2} + h\right)}{h}$ is

   (A) 1   **(B)** nonexistent   **(C)** 0   **(D)** -1 **(E)** none of these

4. The distance from the point $(6, -4)$ to the circle $x^2 + y^2 - 8y = 0$ is

   (A) $\sqrt{10}$ **(B)** 4   **(C)** 6   **(D)** 10   **(E)** none of these

5. The series

$$(x - 1) - \frac{(x - 1)^2}{2!} + \frac{(x - 1)^3}{3!} - \frac{(x - 1)^4}{4!} + \cdots$$

   converges

   (A) for all real x **(B)** if $0 \leqq x < 2$ **(C)** if $0 < x \leqq 2$
   (D) only if $x = 1$ **(E)** for all x except $0 < x < 2$

6. Suppose $\lim\limits_{x \to 2} f(x) = 5$. Then for any positive number ϵ there is a positive number δ such that, whenever $0 < |x - 2| < \delta$,

   (A) $|x - 5| < \epsilon$ (B) $|f(x) - 2| < \epsilon$ (C) $|f(2) - 5| < \epsilon$
   (D) $|f(x) \div 5| < \epsilon$ (E) $|f(x) - \epsilon| < 5$

7. The maximum value of the function $f(x) = x^4 - 4x^3 + 6$ on the closed interval $[1, 4]$ is

   (A) 1   (B) 0   (C) 3   (D) 6   (E) none of these

8. The equation of the cubic with a relative maximum at $(0, 0)$ and a point of inflection at $(1, -2)$ is

   (A) $y = x^3 - 3x^2$ (B) $y = -x^3 + 3x^2$ (C) $y = 4x^3 - 6x^2$
   (D) $y = x^3 - 3x$ (E) $y = 3x - x^3$

9. If $f(x)$ is continuous at the point where $x = a$, which of the following statements may be false?

   (A) $\lim\limits_{x \to a} f(x)$ exists.   (B) $\lim\limits_{x \to a} f(x) = f(a)$.   (C) $f'(a)$ exists.

   (D) $f(a)$ is defined.   (E) $\lim\limits_{x \to a^-} f(x) = \lim\limits_{x \to a^+} f(x)$.

10. Which of the following functions is *not* everywhere continuous?

    (A) $y = |x|$ (B) $y = \dfrac{x}{x^2 + 1}$ (C) $y = \sqrt{x^2 + 8}$ (D) $y = x^{2/3}$

    (E) $y = \dfrac{4}{(x + 1)^2}$

11. The equation of the tangent to the curve of $y = x^2 - 4x$ at the point where the curve crosses the y-axis is

    (A) $y = 8x - 4$ (B) $y = -4x$ (C) $y = -4$ (D) $y = 4x$
    (E) $y = 4x - 8$

12. If y is a differentiable function of x, then the slope of the curve of $xy^2 - 2y + 4y^3 = 6$ at the point where $y = 1$ is

    (A) $-\dfrac{1}{18}$ (B) $-\dfrac{1}{26}$ (C) $\dfrac{5}{18}$ (D) $-\dfrac{11}{18}$ (E) 2

13. If $x = 2t - 1$ and $y = 3 - 4t^2$, then $\dfrac{dy}{dx}$ is

    (A) $4t$ (B) $-4t$ (C) $-\dfrac{1}{4t}$ (D) $2(x + 1)$ (E) $-4(x + 1)$

14. A curve is given parametrically by $x = e^t$ and $y = 2e^{-t}$. The equation of the tangent to the curve at $t = 0$ is

    (A) $2x + y = 0$ (B) $x + 2y = 5$ (C) $y = -2x + 5$
    (D) $2x + y = 4$ (E) $y = 2x$

15. If, for all x, $f'(x) = (x - 2)^4(x - 1)^3$, it follows that the function f has

(A)   a relative minimum at $x = 1$
(B)   a relative maximum at $x = 1$
(C)   both a relative minimum at $x = 1$ and a relative maximum at $x = 2$
(D)   neither a relative maximum nor a relative minimum
(E)   relative minima at $x = 1$ and at $x = 2$

16. If a particle's motion along a straight line is given by

$$s = t^3 - 6t^2 + 9t + 2,$$

then s is increasing

(A)   when $1 < t < 3$ (B)   when $-1 < t < 3$ (C)   for all t
(D)   when $|t| > 3$ (E)   when $t < 1$ or $t > 3$

17. Which of the following statements is *not* true of the graph of $y^2 = x^3 - x$?

(A)   It is symmetric to the x-axis.
(B)   It intersects the x-axis at 0, 1, and -1.
(C)   It exists only for $|x| \leq 1$.
(D)   It has no horizontal asymptotes.
(E)   It has no vertical asymptotes.

18. The area in the first quadrant bounded by the curve $y = x^2$ and the line $y - x - 2 = 0$ is equal to

(A) $\dfrac{3}{2}$ (B) $\dfrac{2}{3}$ (C) $\dfrac{7}{6}$ (D) $\dfrac{10}{3}$ (E) $\dfrac{9}{2}$

19. If differentials are used for the evaluation, then $\sqrt[4]{15}$ is approximately equal to

(A)   2.97   (B)   1.97   (C)   2.03   (D)   1.99   (E)   1.94

20. The curve of the equation $(x^2 - 1)y = x^2 - 4$ has

(A)   one horizontal and one vertical asymptote
(B)   two vertical but no horizontal asymptotes
(C)   one horizontal and two vertical asymptotes
(D)   two horizontal and two vertical asymptotes
(E)   neither a horizontal nor a vertical asymptote

21. If $\dfrac{dy}{dx} = \cos x \cos^2 y$ and $y = \dfrac{\pi}{4}$ when $x = 0$, then

(A) $\tan y = \sin x + 1$ (B) $\tan y = -\sin x + 1$

(C) $\sec^2 y = \sin x + 2$ (D) $\tan y = \dfrac{1}{2}(\cos^2 x + 1)$

(E) $\tan y = \sin x - \dfrac{\sqrt{2}}{2}$

22. If $f(x) = \cos x \sin 3x$, then $f'\left(\dfrac{\pi}{6}\right)$ is equal to

(A) $\dfrac{1}{2}$ (B) $-\dfrac{\sqrt{3}}{2}$ (C) 0 (D) 1 (E) $-\dfrac{1}{2}$

23. The region in the first quadrant bounded by the x-axis, the y-axis, and the curve of $y = e^{-x}$ is rotated about the x-axis. The volume of the solid obtained is equal to

(A) π (B) 2π (C) $\dfrac{1}{2}$ (D) $\dfrac{\pi}{2}$ (E) none of these

24. If $y = x^2 \ln x$ $(x > 0)$, then y'' is equal to

(A) $3 + \ln x$ (B) $3 + 2 \ln x$ (C) $3 \ln x$ (D) $3 + 3 \ln x$
(E) $2 + x + \ln x$

25. $\displaystyle\int_0^1 \dfrac{x\,dx}{x^2 + 1}$ is equal to

(A) $\dfrac{\pi}{4}$ (B) $\ln \sqrt{2}$ (C) $\dfrac{1}{2}(\ln 2 - 1)$ (D) $\dfrac{3}{2}$ (E) $\ln 2$

26. $\displaystyle\int_0^{\pi/2} \sin^2 x\,dx$ is equal to

(A) $\dfrac{1}{3}$ (B) $\dfrac{\pi}{4} - \dfrac{1}{4}$ (C) $\dfrac{\pi}{2}$ (D) $\dfrac{\pi}{2} - \dfrac{1}{3}$ (E) $\dfrac{\pi}{4}$

27. The acceleration of a particle moving along a straight line is given by $a = 6t$. If, when $t = 0$, its velocity, v, is 1 and its distance, s, is 3, then at any time t

(A) $s = t^3 + 3$ (B) $s = t^3 + 3t + 1$ (C) $s = t^3 + t + 3$

(D) $s = \dfrac{t^3}{3} + t + 3$ (E) $s = \dfrac{t^3}{3} + \dfrac{t^2}{2} + 3$

28. If $y = f(x^2)$ and $f'(x) = \sqrt{5x - 1}$, then $\dfrac{dy}{dx}$ is equal to

(A) $2x\sqrt{5x^2 - 1}$ (B) $\sqrt{5x - 1}$ (C) $2x\sqrt{5x - 1}$

(D) $\dfrac{\sqrt{5x - 1}}{2x}$ (E) none of these

29. $\displaystyle\lim_{x \to 0^+} x^x$
(A) $= 0$ (B) $= 1$ (C) $= e$ (D) $= \infty$ (E) does not exist

30. $\displaystyle\int_0^1 xe^x\,dx$ equals

(A) 1 (B) -1 (C) $2 - e$ (D) $\dfrac{e^2}{2} - e$ (E) $e - 1$

31. A 26-ft ladder leans against a building so that its foot moves away from the building at the rate of 3 ft/sec. When the foot of the ladder is 10 ft from the building, the top is moving down at the rate of r ft/sec, where r is

(A) $\dfrac{46}{3}$ (B) $\dfrac{3}{4}$ (C) $\dfrac{5}{4}$ (D) $\dfrac{5}{2}$ (E) $\dfrac{4}{5}$

32. The length of the curve $y = 2x^{3/2}$ between $x = 0$ and $x = 1$ is equal to

(A) $\dfrac{2}{27}(10^{3/2})$ (B) $\dfrac{2}{27}(10^{3/2} - 1)$ (C) $\dfrac{2}{3}(10^{3/2})$

(D) $\dfrac{4}{5}$ (E) none of these

33. If $\dfrac{dx}{dt} = kx$, and if $x = 2$ when $t = 0$ and $x = 6$ when $t = 1$, then k equals

(A) $\ln 4$ (B) 8   (C) e^3 (D) 3   (E) none of these

34. The locus of the polar equation $r = 2 \sec \theta$ is

(A) a circle   (B) a vertical line   (C) a horizontal line
(D) a parabola   (E) an oblique line through the pole

35. If $f(x) = \sqrt{1 - \dfrac{2}{x}}$ and $g(x) = \dfrac{1}{x}$ $(x \neq 0)$, then the derivative of $f(g(x))$

(A) $= 0$ (B) $= \dfrac{1}{x^2 \sqrt{1 - \dfrac{2}{x}}}$ (C) does not exist

(D) $= -\dfrac{1}{\sqrt{1 - 2x}}$ (E) $= \dfrac{-2}{(x^2 - 2)^{3/2}}$

36. If $x > 0$, then the derivative of x^{2x} equals

(A) $4x$ (B) $(2x)^{2x}(1 + \ln x)$ (C) $2(1 + \ln x)$ (D) $(2x) \cdot x^{2x-1}$
(E) $2x^{2x}(1 + \ln x)$

37. Which of the following functions does not satisfy the hypotheses of Rolle's theorem on the interval $[0, 1]$?

(A) $f(x) = \dfrac{x^2 - x}{2x - 1}$ (B) $f(x) = x^2 - x$ (C) $f(x) = x^3 - x^2$

(D) $f(x) = (x - 1)(e^x - 1)$ (E) $f(x) = \dfrac{x^2 - x}{x + 1}$

38. $\lim\limits_{n \to \infty} \dfrac{1}{n}\left[\dfrac{1}{n^2} + \left(\dfrac{2}{n}\right)^2 + \left(\dfrac{3}{n}\right)^2 + \cdots + \left(\dfrac{n}{n}\right)^2\right]$ is

(A) $\ln 2$ (B) $\dfrac{1}{4}$ (C) $\dfrac{1}{3}$ (D) 0   (E) ∞

39. The first-quadrant area under the curve $y = \dfrac{1}{\sqrt{1 - x^2}}$ and bounded at the right by $x = 1$ is

(A) ∞ (B) $\dfrac{\pi}{2}$ (C) $\dfrac{\pi}{4}$ (D) 2   (E) none of these

40. The curve of $y = \dfrac{1-x}{x-3}$ is concave up when

(A) $x > 3$ (B) $1 < x < 3$ (C) $x > 1$ (D) $x < 1$ (E) $x < 3$

41. The area of the largest isosceles triangle that can be drawn with one vertex at the origin and with the others on a line parallel to and above the x-axis and on the curve $y = 27 - x^2$ is

(A) 108 (B) 27 (C) $12\sqrt{3}$ (D) 54 (E) $24\sqrt{3}$

42. The average (mean) value of $\tan x$ on the interval from $x = 0$ to $x = \dfrac{\pi}{3}$ is

(A) $\ln 2$ (B) $\dfrac{3}{\pi}\ln 2$ (C) $\ln\dfrac{1}{2}$ (D) $\dfrac{9}{\pi}$ (E) $\dfrac{\sqrt{3}}{2}$

43. A particle moves along the curve given parametrically by $x = \tan t$ and $y = \sec t$. At the instant when $t = \dfrac{\pi}{6}$, its speed equals

(A) $\sqrt{2}$ (B) $2\sqrt{7}$ (C) $\dfrac{2\sqrt{5}}{3}$ (D) $\dfrac{2\sqrt{13}}{3}$

(E) none of these

44. If $y = \sin(x^2 - 1)$ and $x = \sqrt{u^2 + 1}$, then $\dfrac{dy}{du}$ equals

(A) $\dfrac{u\cos u^2}{\sqrt{1+u^2}}$ (B) $\cos u^2$ (C) $\dfrac{\cos u^2}{2\sqrt{1+u^2}}$ (D) $2u\cos u^2$

(E) $\dfrac{u\cos(x^2-1)}{\sqrt{1+u^2}}$

45. The coefficient of x^3 in the Taylor series of $\ln(1-x)$ about $x = 0$ (the so-called Maclaurin series) is

(A) $-\dfrac{2}{3}$ (B) $-\dfrac{1}{2}$ (C) $-\dfrac{1}{3}$ (D) 0 (E) $\dfrac{1}{3}$

BC Practice Examination 1, Section II

1. Prove that, if $x > 0$, then $x > \ln(1 + x)$, stating any theorems you use.

2. If the parabola $2y = x^2 - 6$ is rotated about the y-axis, it generates a paraboloid of revolution. Suppose a vessel having the shape of this paraboloid contains water to a depth of 4 ft. Determine the amount of water which must be removed to lower the surface by 2 ft.

3. The base of a solid is the region in the first quadrant bounded by the axes and the line $2x + 3y = 10$, and each cross-section perpendicular to the x-axis is a semicircle. Find the volume of the solid.

4. Let u and v be functions defined for all real numbers and satisfying the conditions:
   (a) $u(a + b) = u(a) \cdot u(b)$ for all reals a and b;
   (b) $u(x) = 1 + xv(x)$;
   (c) $\lim_{x \to 0} v(x) = 1$.

Prove that u has a derivative for each x and that $u'(x) = u(x)$.

5. Let $f(x)$ be the power series $\sum_{n=1}^{\infty} \dfrac{x^n}{n(n + 1)}$.

   (a) Write the power series for $f'(x)$, using sigma notation.
   (b) Find the interval of convergence of the series in (a), showing your work.
   (c) Evaluate $f'(0)$.

6. The motion of a particle in a plane is given parametrically (in terms of time t) by $x = \dfrac{1}{t}$ and $y = \ln t$, where $t > 0$.
   (a) Find the speed of the particle when $t = 1$.
   (b) Find the acceleration vector, **a**, when $t = 1$.
   (c) Find an equation which expresses y in terms of x and sketch its graph.
   (d) Show **v** and **a** when $t = 1$.

Answers to BC Practice Examination 1

SECTION I

The explanation for any question not given below will be found in the answer section to AB Practice Examination 1 on pages 366 to 369. Identical questions in Section I of Practice Examinations AB1 and BC1 have the same number. For example, explanations of the answers for Questions 1, 3, and 4, not given below, are given in Section I of Examination AB1, answers 1, 3, and 4 (page 366).

1. C	**8.** A	**15.** A	**22.** E	**29.** B	**36.** E	**43.** C
2. B	**9.** C	**16.** E	**23.** D	**30.** A	**37.** A	**44.** D
3. D	**10.** E	**17.** C	**24.** B	**31.** C	**38.** C	**45.** C
4. C	**11.** B	**18.** D	**25.** B	**32.** B	**39.** B	
5. A	**12.** A	**19.** B	**26.** E	**33.** E	**40.** E	
6. D	**13.** B	**20.** C	**27.** C	**34.** B	**41.** D	
7. D	**14.** D	**21.** A	**28.** A	**35.** D	**42.** B	

2. B.   Find A, B, and C such that $\dfrac{4x^2 + 3x + 5}{(x - 1)(x^2 + 2)} = \dfrac{A}{x - 1} + \dfrac{Bx + C}{x^2 + 2}$.

$$4x^2 + 3x + 5 = A(x^2 + 2) + (Bx + C)(x - 1). \qquad (*)$$

Letting $x = 1$ yields $12 = A(3)$. Letting $x = 0$ yields $5 = 2A - C$. So $A = 4$ and $C = 3$. To find B we can equate coefficients of the second degree in (*): $4 = A + B$, so that $B = 0$.

5. A.   We use the Ratio Test, page 294:

$$\lim_{n \to \infty} \left| \frac{(x - 1)^{n+1}}{(n + 1)!} \cdot \frac{n!}{(x - 1)^n} \right| = \lim_{n \to \infty} \frac{1}{n + 1} |x - 1|,$$

which equals zero if $x \neq 1$. The series also converges if $x = 1$.

6. D.   This is the definition of $\lim\limits_{x \to c} f(x)$ (see page 21).

13. B. $\dfrac{dx}{dt} = 2$ and $\dfrac{dy}{dt} = -8t$. Use the fact that $\dfrac{dy}{dx} = \dfrac{dy/dt}{dx/dt}$.

14. D.   Since $\dfrac{dy}{dx} = -2e^{-2t}$, at $t = 0$ the slope equals -2. Also at $t = 0$, $x = 1$ and $y = 2$. So the equation of the tangent is $y - 2 = -2(x - 1)$.

19. B.   Let $y = \sqrt[4]{x}$, $x = 16$, and $dx = -1$. Then

$$dy = \frac{1}{4}x^{-3/4}\, dx = \frac{1}{4 \cdot 8}(-1) \approx -0.03.$$

So $\sqrt[4]{15} \approx \sqrt[4]{16} - 0.03$.

21. A.   Separate the variables to obtain $\dfrac{dy}{\cos^2 y} = \cos x\, dx$ and solve $\int \sec^2 y\, dy = \int \cos x\, dx$. Use the given condition to obtain the particular solution.

26. E.   Use formula (20) on page 483 to rewrite as

$$\frac{1}{2} \int_0^{\pi/2} (1 - \cos 2x)\, dx = \frac{1}{2}\left(x - \frac{\sin 2x}{2}\right)\Bigg|_0^{\pi/2}$$

$$= \frac{1}{2}\left(\frac{\pi}{2}\right).$$

29. B.   Let $y = x^x$ and take logarithms. $\ln y = x \ln x = \dfrac{\ln x}{1/x}$. As $x \to 0^+$, this function has the indeterminate form ∞/∞. Apply L'Hôpital's rule:

$$\lim_{x \to 0^+} \ln y = \lim_{x \to 0^+} \frac{1/x}{-1/x^2} = \lim_{x \to 0^+} (-x) = 0$$

So $y \to e^0$ or 1.

32. B. The arc length is given by the integral $\int_0^1 \sqrt{1 + 9x}$, which is equal to

$$\frac{1}{9} \cdot \frac{2}{3}(1 + 9x)^{3/2} \Big|_0^1 = \frac{2}{27}(10^{3/2} - 1).$$

34. B. The given equation is equivalent to $r \cos \theta = 2$ or to $x = 2$.

36. E. We let $y = x^{2x}$; then $\ln y = 2x \ln x$. Also,

$$\frac{1}{y} y' = \frac{2x}{x} + 2 \ln x = 2(1 + \ln x).$$

So $y' = 2x^{2x}(1 + \ln x)$.

38. C. The trick here is to recognize that the limit of the sum is equal to $\int_0^1 x^2 \, dx$. To see this, recall that

$$\lim_{n \to \infty} \sum_{k=1}^n f(x_k) \, \Delta x = \int_a^b f(x) \, dx$$

where the interval from a to b has been subdivided (or partitioned). Here, then, we see that Δx can be replaced by $\frac{1}{n}$, x_k by $\frac{k}{n}$, and $f(x_k)$ by x_k^2 to get the given limit. Observing that, as k varies from 1 to n, x_k takes on the values $\frac{1}{n}, \frac{2}{n}, \ldots, \frac{n}{n}$, we conclude that the interval in question here is from 0 to 1.

39. B. Evaluate

$$\int_0^1 \frac{1}{\sqrt{1 - x^2}} \, dx = \lim_{h \to 0^+} \int_0^{1-h} \frac{1}{\sqrt{1 - x^2}} \, dx = \lim_{h \to 0^+} \sin^{-1} x \Big|_0^{1-h}.$$

43. C. $|\mathbf{v}| = \sqrt{\left(\frac{dx}{dt}\right)^2 + \left(\frac{dy}{dt}\right)^2} = \sqrt{(\sec^2 t)^2 + (\sec t \tan t)^2}$. At $t = \frac{\pi}{6}$,

$$|\mathbf{v}| = \sqrt{\left(\frac{2}{\sqrt{3}}\right)^4 + \left(\frac{2}{\sqrt{3}} \cdot \frac{1}{\sqrt{3}}\right)^2}.$$

45. C. The power series for $\ln(1 - x)$, if $x < 1$, is $-x - \frac{x^2}{2} - \frac{x^3}{3} - \cdots$.

SECTION II

1. METHOD 1. To show that $x > \ln(1 + x)$, if $x > 0$, let $f(x) = x - \ln(1 + x)$ and observe that

$$f'(x) = 1 - \frac{1}{x + 1} = \frac{x}{x + 1}$$

and that $f'(x)$ exists at each positive x. Since f is continuous at $x = 0$, it follows that $f(x)$ is continuous for each $x \geqq 0$ and that the Mean Value Theorem holds. Then there is a number c, $0 < c < x$, such that $\dfrac{f(x) - f(0)}{x} = f'(c)$. So

$$\frac{x - \ln(1 + x) - 0}{x} = \frac{c}{c + 1}.$$

Since c is positive so is the right-hand member, and since $x > 0$ we have $x - \ln(1 + x) > 0$, or (for $x > 0$) $x > \ln(1 + x)$.

Theorems used:

(1) The Mean Value Theorem: If f is continuous on the closed interval $[a, b]$ and its derivative f' exists on the open interval (a, b), then there is a number c, $a < c < b$, such that $\dfrac{f(b) - f(a)}{b - a} = f'(c)$.

(2) If $p > q$ and $r > 0$, then $pr > qr$.

METHOD 2. Let $f(x) = x - \ln(1 + x)$ and note that

$$f'(x) = 1 - \frac{1}{1 + x} = \frac{x}{x + 1} \qquad \text{and} \qquad f''(x) = \frac{1}{(1 + x)^2}.$$

The only critical value here is $x = 0$, and $f''(0) > 0$ so that this critical value of x yields a minimum. (In fact, $f'' > 0$ for all x.) Since f and its derivatives are continuous for all $x \geqq 0$ and since the minimum value of f at 0 is 0, then $f(x) > 0$ for all $x > 0$, or $x > \ln(1 + x)$.

Theorems used:

(1) If a function f and its derivative f' are continuous on an interval $[a, b]$, if $f'(c) = 0$ ($a < c < b$), and if $f''(c) > 0$, then $f(c)$ is a relative minimum.

(2) If a function and its derivatives are continuous on an interval, if its curve is everywhere concave up, and if there is only one critical value on this interval, then f attains its absolute minimum at this critical value.

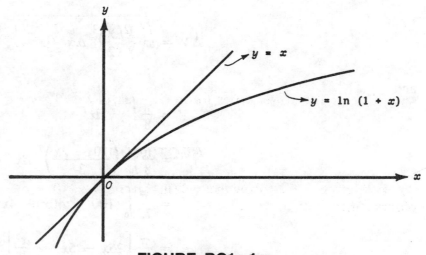

FIGURE BC1–1a

See Figure BC1–1a, where the curve of $y = \ln (1 + x)$ and the line $y = x$ are sketched. It is interesting to observe that $\ln (1 + x)$ may be regarded as the area under the curve $y = \dfrac{1}{1 + t}$ from $t = 0$ to $t = x$; that is,

$$\int_0^x \frac{1}{1 + t}\, dt = \ln (1 + x);$$

and that x may be interpreted as the area of a circumscribed rectangle of length x and height 1. The latter area is greater than the former for all $x > 0$. See Figure BC1–1b.

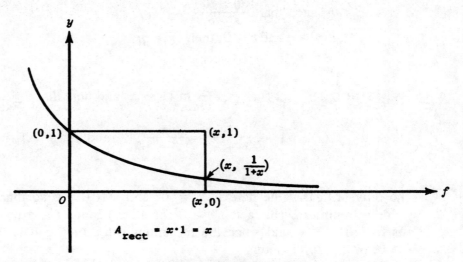

FIGURE BC1–1b

2. See the solution for Question 2 of AB Practice Examination 1, Section II, page 371.

3. The solid is sketched in Figure BC1–2. The volume of the slice is given by

$$\Delta V = \pi \cdot \frac{1}{2}\left(\frac{y}{2}\right)^2 \Delta x$$

and

$$V = \frac{\pi}{8}\int_0^5 y^2\, dx$$

$$= \frac{\pi}{8}\int_0^5 \left(\frac{10 - 2x}{3}\right)^2 dx$$

$$= \frac{\pi}{72}\int_0^5 (100 - 40x + 4x^2)\, dx$$

$$= \frac{\pi}{18}\left[25x - .5x^2 + \frac{x^3}{3}\right]_0^5 = \frac{125\pi}{54}.$$

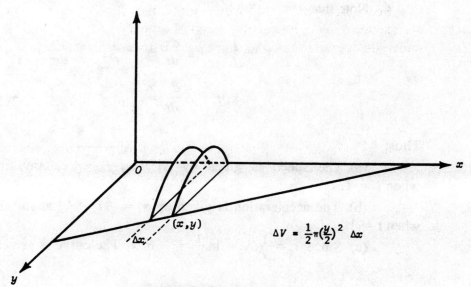

$$\Delta V = \frac{1}{2}\pi\left(\frac{y}{2}\right)^2 \Delta x$$

FIGURE BC1-2

4. See the solution for Question 6 of AB Practice Examination 1, Section II, page 373.

5. (a) Since

$$f(x) = \frac{x}{1\cdot 2} + \frac{x^2}{2\cdot 3} + \frac{x^3}{3\cdot 4} + \cdots + \frac{x^n}{n(n+1)} + \cdots,$$

and

$$f'(x) = \frac{1}{2} + \frac{x}{3} + \frac{x^2}{4} + \cdots + \frac{x^{n-1}}{n+1} + \cdots,$$

therefore

$$f'(x) = \sum_{n=1}^{\infty} \frac{x^{n-1}}{n+1}.$$

(b) We are the Ratio Test:

$$\lim_{n\to\infty}\left|\frac{x^n}{n+2}\cdot\frac{n+1}{x^{n-1}\cdot}\right| = \lim_{n\to\infty}\frac{n+1}{n+2}|x| = |x|.$$

The series therefore converges if $|x| < 1$. At $x = 1$, we get the series

$$\frac{1}{2} + \frac{1}{3} + \frac{1}{4} + \cdots,$$

which is the (divergent) harmonic series. At $x = -1$, we get

$$\frac{1}{2} - \frac{1}{3} + \frac{1}{4} - \frac{1}{5} + \cdots,$$

the (convergent) alternating harmonic series. So $f'(x)$ converges if $-1 \leqq x < 1$.

(c) $f'(0) = \frac{1}{2}$.

6. Note that

$$x = \frac{1}{t}; \qquad \frac{dx}{dt} = -\frac{1}{t^2}; \qquad \frac{d^2x}{dt^2} = \frac{2}{t^3};$$

$$y = \ln t; \qquad \frac{dy}{dt} = \frac{1}{t}; \qquad \frac{d^2y}{dt^2} = -\frac{1}{t^2}.$$

Thus:

(a) The speed is given by $|\mathbf{v}| = \sqrt{\frac{1}{t^4} + \frac{1}{t^2}}$ at any time t, and equals $\sqrt{2}$ when $t = 1$.

(b) The acceleration is given by $\mathbf{a} = \frac{2}{t^3}\mathbf{i} - \frac{1}{t^2}\mathbf{j}$ at any time, and equals $2\mathbf{i} - \mathbf{j}$ when $t = 1$.

(c) Since $t = \frac{1}{x}$, $y = \ln\frac{1}{x} = -\ln x$. The curve is sketched in Figure BC1–3.

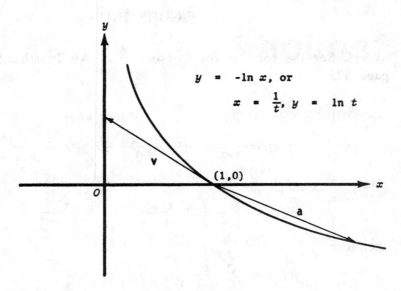

FIGURE BC1–3

(d) The particle is at $(1, 0)$ when $t = 1$ and travels in a clockwise direction along the curve as t increases. The vectors $\mathbf{v}$ and $\mathbf{a}$ when $t = 1$ are shown in the figure.

BC Practice Examination 2

Section I

1. $\lim\limits_{x \to \infty} \dfrac{20x^2 - 13x + 5}{5 - 4x^3}$ is

   (A) -5 **(B)** ∞ **(C)** 0 **(D)** 5 **(E)** 1

2. $\lim\limits_{x \to \pi/2} \dfrac{\cos x}{x - \dfrac{\pi}{2}}$ is

   (A) -1 **(B)** 1 **(C)** 0 **(D)** ∞ **(E)** none of these

3. $\lim\limits_{x \to 0} x \sin \dfrac{1}{x}$ is

   (A) 1 **(B)** 0 **(C)** ∞ **(D)** -1 **(E)** none of these

4. $\lim\limits_{h \to 0} \dfrac{\ln (2 + h) - \ln 2}{h}$ is

   (A) 0 **(B)** $\ln 2$ **(C)** $\dfrac{1}{2}$ **(D)** $\dfrac{1}{\ln 2}$ **(E)** ∞

5. If $y = \dfrac{x - 3}{2 - 5x}$, then $\dfrac{dy}{dx}$ equals

   (A) $\dfrac{17 - 10x}{(2 - 5x)^2}$ **(B)** $\dfrac{13}{(2 - 5x)^2}$ **(C)** $\dfrac{x - 3}{(2 - 5x)^2}$ **(D)** $\dfrac{17}{(2 - 5x)^2}$

   (E) $\dfrac{-13}{(2 - 5x)^2}$

6. For which function is $\displaystyle\sum_{n=0}^{\infty} \frac{(-1)^n x^{2n}}{(2n)!}$ the Taylor series about 0?

    (A) e^x (B) e^{-x} (C) $\sin x$ (D) $\cos x$ (E) $\ln(1 + x)$

7. If $f(x) = x \cos \dfrac{1}{x}$, then $f'\left(\dfrac{2}{\pi}\right)$ equals

    (A) $\dfrac{\pi}{2}$ (B) $-\dfrac{2}{\pi}$ (C) -1 (D) $-\dfrac{\pi}{2}$ (E) 1

8. If $xy^2 - 3x + 4y - 2 = 0$ and y is a differentiable function of x, then $\dfrac{dy}{dx}$ equals

    (A) $-\dfrac{1 + y^2}{2xy}$ (B) $\dfrac{3}{2y + 4}$ (C) $\dfrac{3}{2xy + 4}$ (D) $\dfrac{3 - y^2}{2xy + 4}$

    (E) $\dfrac{5 - y^2}{2xy + 4}$

9. If $x = \sqrt{1 - t^2}$ and $y = \sin^{-1} t$, then $\dfrac{dy}{dx}$ equals

    (A) $-\dfrac{\sqrt{1 - t^2}}{t}$ (B) $-t$ (C) $\dfrac{t}{1 - t^2}$ (D) 2 (E) $-\dfrac{1}{t}$

10. If y is a differentiable function of x, then the derivative of $\sin^2(x + y)$ with respect to x is

    (A) $2[\sin(x + y)]\dfrac{dy}{dx}$ (B) $[\cos^2(x + y)]\left(1 + \dfrac{dy}{dx}\right)$

    (C) $[\sin 2(x + y)]\left(1 + \dfrac{dy}{dx}\right)$ (D) $\cos^2(x + y)$ (E) $2\sin(x + y)$

11. The equation of the tangent to the curve $y = e^x \ln x$, where $x = 1$, is

    (A) $y = ex$ (B) $y = e^x + 1$ (C) $y = e(x - 1)$
    (D) $y = ex + 1$ (E) $y = x - 1$

12. If differentials are used for computation, then $\sqrt[3]{63}$ is approximately equal, to the nearest hundredth, to

    (A) 4.00 (B) 3.98 (C) 3.93 (D) 3.80 (E) 3.88

13. The hypotenuse AB of a right triangle ABC is 5 ft, and one leg, AC, is decreasing at the rate of 2 ft/sec. The rate, in square feet per second, at which the area is changing when $AC = 3$ is

    (A) $\dfrac{25}{4}$ (B) $\dfrac{7}{4}$ (C) $-\dfrac{3}{2}$ (D) $-\dfrac{7}{4}$ (E) $-\dfrac{7}{2}$

14. The derivative of a function f is given for all x by

$$f'(x) = x^2(x + 1)^3(x - 4)^2.$$

The set of x for which f is a relative maximum is

(A) $\{0, -1, 4\}$ (B) $\{-1\}$ (C) $\{0, 4\}$ (D) $\{1\}$
(E) none of these

15. Which one of the following series converges?

(A) $\sum_{n=1}^{\infty} \dfrac{1}{\sqrt{n}}$ (B) $\sum_{n=1}^{\infty} \dfrac{1}{n}$ (C) $\sum_{n=1}^{\infty} \dfrac{1}{2n + 1}$ (D) $\sum_{n=1}^{\infty} \dfrac{n}{n^2 + 1}$

(E) $\sum_{n=1}^{\infty} \dfrac{1}{n^2 + 1}$

16. If the displacement from the origin of a particle on a line is given by $s = 3 + (t - 2)^4$, then the number of times the particle reverses direction is

(A) 0   (B) 1   (C) 2   (D) 3   (E) none of these

17. If a particle moves in a plane so that $x = 2 \cos 3t$ and $y = \sin 3t$, then its speed on the interval $0 \leq t < \dfrac{\pi}{2}$ is a maximum when t equals

(A) 0   (B) $\dfrac{\pi}{6}$ (C) $\dfrac{\pi}{4}$ (D) $\dfrac{\pi}{3}$ (E) $\dfrac{\pi}{2}$

18. The maximum value of the function $f(x) = xe^{-x}$ is

(A) $\dfrac{1}{e}$ (B) e (C) 1   (D) -1 (E) none of these

19. A rectangle of perimeter 18 in. is rotated about one of its sides to generate a right circular cylinder. The rectangle which generates the cylinder of largest volume has area, in square inches, of

(A) 14   (B) 20   (C) $\dfrac{81}{4}$ (D) 18   (E) $\dfrac{77}{4}$

20. If $f'(x)$ exists on the closed interval $[a, b]$, then it follows that

(A) $f(x)$ is constant on $[a, b]$
(B) there exists a number c, $a < c < b$, such that $f'(c) = 0$
(C) the function has a maximum value on the open interval (a, b)
(D) the function has a minimum value on the open interval (a, b)
(E) the Mean Value Theorem applies

21. $\displaystyle\int_1^2 (3x - 2)^3 \, dx$ is equal to

(A) $\dfrac{16}{3}$ (B) $\dfrac{63}{4}$ (C) $\dfrac{13}{3}$ (D) $\dfrac{85}{4}$ (E) none of these

22. $\int_{\pi/4}^{\pi/2} \sin^3 \alpha \cos \alpha \, d\alpha$ is equal to

   (A) $\dfrac{3}{16}$ (B) $\dfrac{1}{8}$ (C) $-\dfrac{1}{8}$ (D) $-\dfrac{3}{16}$ (E) $\dfrac{3}{4}$

23. $\int x \cos x^2 \, dx$ equals

   (A) $\sin x^2 + C$ (B) $2 \sin x^2 + C$ (C) $-\dfrac{1}{2} \sin x^2 + C$

   (D) $\dfrac{1}{4} \cos^2 x^2 + C$ (E) $\dfrac{1}{2} \sin x^2 + C$

24. $\int_{0}^{1} \dfrac{e^x}{(3 - e^x)^2} \, dx$ equals

   (A) $3 \ln (e - 3)$ (B) 1 (C) $\dfrac{1}{3 - e}$ (D) $\dfrac{e - 2}{3 - e}$

   (E) none of these

25. If we let $x = 2 \sin \theta$, then $\int_{0}^{2} \dfrac{x^2 \, dx}{\sqrt{4 - x^2}}$ is equivalent to

   (A) $4 \int_{0}^{1} \sin^2 \theta \, d\theta$ (B) $\int_{0}^{\pi/2} 4 \sin^2 \theta \, d\theta$ (C) $\int_{0}^{\pi/2} 2 \sin \theta \tan \theta \, d\theta$

   (D) $\int_{0}^{2} \dfrac{2 \sin^2 \theta}{\cos^2 \theta} \, d\theta$ (E) $4 \int_{\pi/2}^{0} \sin^2 \theta \, d\theta$

26. $\int \dfrac{x^2 + x - 1}{x^2 - x} \, dx$ equals

   (A) $x + 2 \ln |x^2 - x| + C$ (B) $\ln |x| + \ln |x - 1| + C$
   (C) $1 + \ln |x^2 - x| + C$ (D) $x + \ln |x^2 - x| + C$
   (E) $x - \ln |x| - \ln |x - 1| + C$

27. $\int_{-1}^{1} (1 - |x|) \, dx$ equals

   (A) 0 (B) $\dfrac{1}{2}$ (C) 1 (D) 2 (E) none of these

28. $\lim\limits_{x \to \infty} x^{1/x}$

   (A) $= 0$ (B) $= 1$ (C) $= e$ (D) $= \infty$ (E) does not exist

29. The general solution of the differential equation $\dfrac{dy}{dx} = \dfrac{1 - 2x}{y}$ is a family of

   (A) straight lines   (B) circles   (C) hyperbolas   (D) parabolas
   (E) ellipses

30. If $F'(x) = G'(x)$ for all x and k is a constant, then it is necessary that

(A) $F(x) = G(x) + k$ (B) $F(x) = G(x)$ (C) $F(k) = G(k)$
(D) $F(x) = kG(x)$ (E) $F(x) = G(x + k)$

31. The area enclosed by the curve of $y^2 = x^3$ and the line segment joining $(1, 1)$ and $(4, -8)$ is given by

(A) $2\int_0^1 x^{3/2}\, dx + \int_1^4 (-3x + 4 - x^{3/2})\, dx$

(B) $2\int_0^1 x^{3/2}\, dx + \int_1^4 (4 - 3x + x^{3/2})\, dx$ (C) $\int_0^4 (4 - 3x - x^{3/2})\, dx$

(D) $2\int_0^1 x^{3/2}\, dx + \int_{-8}^1 \left(\dfrac{4 - y}{3} - y^{2/3}\right) dy$ (E) $\int_1^4 (x^{3/2} + 3x - 4)\, dx$

32. The first-quadrant area bounded below by the x-axis and laterally by the curves of $y = x^2$ and $y = 4 - x^2$ equals

(A) $6 - 2\sqrt{3}$ (B) $\dfrac{10}{3}$ (C) $\dfrac{8}{3}(2 - \sqrt{2})$ (D) $\dfrac{4\sqrt{2}}{3}$

(E) none of these

33. The equation of the curve shown is $y = \dfrac{4}{1 + x^2}$. What does the area of the shaded region of the rectangle equal?

(A) $4 - \dfrac{\pi}{4}$ (B) $8 - 2\pi$ (C) $8 - \pi$ (D) $8 - \dfrac{\pi}{2}$

(E) $2\pi - 4$

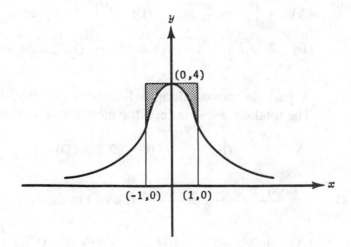

34. A curve is given parametrically by the equations $x = t$, $y = 1 - \cos t$. The area bounded by the curve and the x-axis on the interval $0 \le t \le 2\pi$ is equal to

(A) $2(\pi + 1)$ (B) π (C) 4π (D) $\pi + 1$ (E) 2π

35. The volume of an "inner tube" with inner diameter 4 ft and outer diameter 8 ft is, in cubic feet,

    (A) $4\pi^2$ (B) $12\pi^2$ (C) $8\pi^2$ (D) $24\pi^2$ (E) $6\pi^2$

36. The volume generated by rotating the region bounded by $y = \dfrac{1}{\sqrt{x}}$, $x = 1$, $x = 4$, and $y = 0$ about the y-axis is

    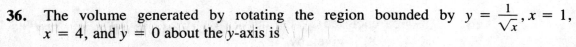

    (A) $\dfrac{28\pi}{3}$ (B) $\pi \ln 4$ (C) $\dfrac{64\pi}{3}$ (D) $\dfrac{14\pi}{3}$ (E) $\dfrac{16\pi}{3}$

37. If $f(x)$ and $g(x)$ are both continuous functions on the closed interval $[a, b]$, if $f(a) = g(a)$ and $f(b) = g(b)$, and if, further, $f(x) > g(x)$ for all x, $a < x < b$, then it follows that

    (A) $\displaystyle\int_a^b f(x)\, dx \geqq 0$ (B) $\displaystyle\int_a^b |g(x)|\, dx > \int_a^b |f(x)|\, dx$

    (C) $\displaystyle\int_a^b |f(x)|\, dx > \int_a^b |g(x)|\, dx$ (D) $\displaystyle\int_a^b f(x)\, dx > \int_a^b g(x)\, dx$

    (E) none of the preceding is necessarily true

38. The length of the arc of $y = \dfrac{1}{2}(e^x + e^{-x})$ from $x = 0$ to $x = 1$ equals

    (A) $\dfrac{1}{2}(e + e^{-1})$ (B) $\dfrac{1}{2}(e - e^{-1})$ (C) $e - \dfrac{1}{e}$ (D) $e + \dfrac{1}{e} - 2$

    (E) $\dfrac{1}{2}\left(e + \dfrac{1}{e}\right) - 1$

39. The length of one arch of the cycloid $x = \theta - \sin\theta$, $y = 1 - \cos\theta$ is given by the integral

    (A) $4\displaystyle\int_0^\pi \sin\dfrac{\theta}{2}\, d\theta$ (B) $2\displaystyle\int_0^\pi \sin\dfrac{\theta}{2}\, d\theta$ (C) $2\displaystyle\int_0^{2\pi} \sin\dfrac{\theta}{2}\, d\theta$

    (D) $2\sqrt{2}\displaystyle\int_0^\pi \sqrt{1 - \cos\theta}\, d\theta$ (E) none of these

40. A particle moves along a line with velocity, in feet per second, $v = t^2 - t$. The total distance, in feet, traveled from $t = 0$ to $t = 2$ equals

    (A) $\dfrac{1}{3}$ (B) $\dfrac{2}{3}$ (C) 2   (D) 1   (E) $\dfrac{4}{3}$

41. $\displaystyle\lim_{n\to\infty} \dfrac{\sqrt{1} + \sqrt{2} + \cdots + \sqrt{n}}{n^{3/2}}$ is equal to the definite integral

    (A) $\displaystyle\int_0^1 \sqrt{x}\, dx$ (B) $\displaystyle\int_0^1 \dfrac{1}{\sqrt{x}}\, dx$ (C) $\displaystyle\int_1^2 \sqrt{x}\, dx$ (D) $\displaystyle\int_0^1 \dfrac{1}{x}\, dx$

    (E) none of these

42. If $f'(x) = h(x)$ and $g(x) = x^3$, then $\dfrac{d}{dx} f(g(x)) =$

    (A) $h(x^3)$ (B) $3x^2 h(x)$ (C) $h'(x)$ (D) $3x^2 h(x^3)$ (E) $x^3 h(x^3)$

43. If $\frac{dy}{dx} = y \tan x$ and $y = 3$ when $x = 0$, then, when $x = \frac{\pi}{3}$, $y =$

(A) $\ln \sqrt{3}$ (B) $\ln 3$ (C) $\frac{3}{2}$ (D) $\frac{3\sqrt{3}}{2}$ (E) 6

44. The Cartesian equation of the polar curve $r = 2 \sin \theta + 2 \cos \theta$ is

(A) $(x - 1)^2 + (y - 1)^2 = 2$ (B) $x^2 + y^2 = 2$ (C) $x + y = 2$
(D) $x^2 + y^2 = 4$ (E) $y^2 - x^2 = 4$

45. $\int_0^\infty e^{-x/2} \, dx =$

(A) $-\infty$ (B) -2 (C) 1   (D) 2   (E) ∞

BC Practice Examination 2, Section II

1. (a) Sketch the region bounded by the curves of $\sqrt{x} + \sqrt{y} = 3$ and $x + y = 5$.

(b) Find the area of this region.

2. Sketch the graph of $y = x \ln x$, after finding
(a) the domain;
(b) intercepts;
(c) the coordinates of any relative maximum or minimum points;
(d) the coordinates of any points of inflection;
(e) the behavior of y for large x;
(f) $\lim y$ as $x \to 0^+$.

3. A particle moves along a line so that its displacement from the origin at any time t is given by $x = \frac{7}{2} e^{-4t} \sin 2t$.

(a) Prove: $a + 8v + 20x = 0$, where v is the velocity and a the acceleration of the particle.

(b) Show that relative maximum and minimum values of x occur when $\tan 2t = \frac{1}{2}$.

4. Let f be the function defined by $f(x) = \frac{1}{(1 + x)^2}$.

(a) Write the first four terms and the general term of the Maclaurin series expansion (about $x = 0$) of $f(x)$.

(b) Find the interval of convergence for the series in part (a), showing your work.

(c) Evaluate $f\left(\frac{1}{4}\right)$.

(d) How many terms of the series are needed to approximate $f\left(\frac{1}{4}\right)$ with an error not exceeding 0.05? Justify your answer.

5. A curve is given parametrically by

$$\begin{cases} x = 2a \cot \theta, \\ y = 2a \sin^2 \theta. \end{cases}$$

(a) Determine the Cartesian equation of the curve (i.e., find a single equation in x and y).

(b) Sketch the curve.

(c) Find the area bounded by the curve and the x-axis.

6. (a) Let the functions f, u, and v be differentiable for all x, and let $f'(x) = u(x)v(x)$. Suppose that, for a number c, $u(c) = 0$ but $v(c) \neq 0$. Show that if $u'(c)v(c) > 0$ then $f(c)$ is a relative minimum, while if $u'(c)v(c) < 0$ then $f(c)$ is a relative maximum. State any theorems you use.

(b) If $f'(x) = (3x - 3)(x^2 + 2)^3(x^2 - 2x + 3)^{3/2}$, find the x-coordinates of any relative maxima or minima of f by using the theorem in part (a).

Answers to BC Practice Examination 2

SECTION I

The explanation for any question not given below will be found in the answers to AB Practice Examination 2 on pages 382 to 389. Identical questions in Sections I of Practice Examinations AB2 and BC2 have the same number.

1. C	9. E	17. B	25. B	33. B	41. A
2. A	10. C	18. A	26. D	34. E	42. D
3. B	11. C	19. D	27. C	35. E	43. E
4. C	12. B	20. E	28. B	36. A	44. A
5. E	13. D	21. D	29. E	37. D	45. D
6. D	14. E	22. A	30. A	38. B	
7. A	15. E	23. E	31. B	39. C	
8. D	16. B	24. E	32. C	40. D	

6. D.   See Series (3) on page 311.

9. E.   **Here,**

$$\frac{dy}{dx} = \frac{\dfrac{dy}{dt}}{\dfrac{dx}{dt}} = \frac{\dfrac{1}{\sqrt{1 - t^2}}}{\dfrac{1}{2}\dfrac{(-2t)}{\sqrt{1 - t^2}}} = -\frac{1}{t}.$$

12. B.   If we let $f(x) = x^{1/3}$, then we want $f(64) + df$, where df is obtained when $x = 64$ and $dx = -1$.

15. E.   Each is essentially a p-series, $\sum \frac{1}{n^p}$. Such a series converges only if $p > 1$.

17. B.   The speed of the particle along the curve is given by

$$|\mathbf{v}| = \sqrt{\left(\frac{dx}{dt}\right)^2 + \left(\frac{dy}{dt}\right)^2}.$$

Here,

$$|\mathbf{v}| = \sqrt{36 \sin^2 3t + 9 \cos^2 3t}$$
$$= \sqrt{27 \sin^2 3t + 9 \sin^2 3t + 9 \cos^2 3t}$$
$$= \sqrt{27 \sin^2 3t + 9}.$$

The maximum value of $|\mathbf{v}|$ occurs when $\sin^2 3t = 1$, that is, when $\sin 3t = \pm 1$. From this we have $3t = \frac{\pi}{2}$ or $\frac{3\pi}{2}$, so that $t = \frac{\pi}{6}$ or $\frac{\pi}{2}$; but only $\frac{\pi}{6}$ is on the restricted interval.

25. B.   Note that, when $x = 2 \sin \theta$, $x^2 = 4 \sin^2 \theta$, $dx = 2 \cos \theta \, d\theta$, and $\sqrt{4 - x^2} = 2 \cos \theta$. Also,

$$\text{when } x = 0, \theta = 0;$$

$$\text{when } x = 2, \theta = \frac{\pi}{2}.$$

28. B.   Let $y = x^{1/x}$; then take logarithms. $\ln y = \frac{\ln x}{x}$. As $x \to \infty$, the fraction is of the form ∞/∞. $\lim\limits_{x\to\infty} \ln y = \lim\limits_{x\to\infty} \frac{1/x}{1} = 0$. So $y \to e^0$ or 1.

29. E.   Separating variables, we get $y \, dy = (1 - 2x) \, dx$. Integrating gives

$$\frac{1}{2}y^2 = x - x^2 + C$$

or

$$y^2 = 2x - 2x^2 + k$$

or

$$2x^2 + y^2 - 2x = k.$$

34. E.   The area, A, is represented by $\int_0^{2\pi} (1 - \cos t) \, dt = 2\pi$.

38. B. Since the length, s, is given by $\int_a^b \sqrt{1 + \left(\dfrac{dy}{dx}\right)^2}\, dx$, then

$$s = \int_0^1 \sqrt{1 + \left[\frac{1}{2}(e^x - e^{-x})\right]^2}\, dx$$

$$= \int_0^1 \sqrt{1 + \frac{1}{4}(e^{2x} - 2 + e^{-2x})}\, dx$$

$$= \int_0^1 \sqrt{\frac{1}{4}e^{2x} + \frac{1}{2} + \frac{1}{4}e^{-2x}}\, dx$$

$$= \int_0^1 \sqrt{\frac{1}{4}(e^x + e^{-x})^2}\, dx = \int_0^1 \frac{1}{2}(e^x + e^{-x})\, dx$$

$$= \frac{1}{2}(e^x - e^{-x})\Big|_0^1 = \frac{1}{2}\left[\left(e - \frac{1}{e}\right) - (1 - 1)\right] = \frac{1}{2}\left(e - \frac{1}{e}\right).$$

39. C. See Figure BC2–1.

$$\frac{ds}{d\theta} = \sqrt{\left(\frac{dx}{d\theta}\right)^2 + \left(\frac{dy}{d\theta}\right)^2} = \sqrt{(1 - \cos\theta)^2 + (\sin\theta)^2}$$

$$= \sqrt{2(1 - \cos\theta)} = \sqrt{2 \cdot 2\sin^2\frac{\theta}{2}} = 2\sin\frac{\theta}{2}.$$

So each arch has length $2\displaystyle\int_0^{2\pi} \sin\frac{\theta}{2}\, d\theta$.

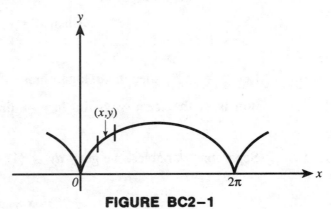

FIGURE BC2–1

41. A. Recall that $\displaystyle\lim_{n\to\infty} \sum_{k=1}^n f(x_k)\,\Delta x = \int_a^b f(x)\, dx$, where the subdivisions are on the interval $[a, b]$. Rewrite the given limit as follows:

$$\lim_{n\to\infty}\left[\left(\frac{1}{n}\right)^{1/2} + \left(\frac{2}{n}\right)^{1/2} + \left(\frac{3}{n}\right)^{1/2} + \cdots + \left(\frac{n}{n}\right)^{1/2}\right]\frac{1}{n}$$

and note that $\Delta x = \dfrac{1}{n}$, $x_k = \dfrac{k}{n}$, and $f(x_k) = (x_k)^{1/2}$. The interval in this case is from 0 to 1.

42. D. Here,

$$\frac{d}{dx} f(g(x)) = f'(g(x))g'(x) = h(g(x))g'(x) = h(x^3) \cdot 3x^2.$$

43. E. Separating variables yields $\frac{dy}{y} = \tan x$, so $\ln y = -\ln \cos x + C$. With $y = 3$ when $x = 0$, $C = \ln 3$. The general solution is therefore $(\cos x)\, y = 3$. When $x = \frac{\pi}{3}$,

$$\cos x = \frac{1}{2} \quad \text{and} \quad y = 6.$$

44. A. The given equation can be multiplied by r to yield

$$r^2 = 2r \sin \theta + 2r \cos \theta;$$

the Cartesian equation is obtained by using

$$x = r \cos \theta, \quad y = r \sin \theta, \quad \text{and} \quad x^2 + y^2 = r^2.$$

45. D. Evaluate

$$\lim_{b \to \infty} \int_0^b e^{-x/2}\, dx = -\lim_{b \to \infty} 2e^{-x/2}\Big|_0^b = -2(0 - 1).$$

SECTION II

   1. See the solution for Question 1 of AB Practice Examination 2, Section II, page 389.

   2. See the solution for Question 4 of AB Practice Examination 2, Section II, page 391.

   3. See the solution for Question 5 of AB Practice Examination 2, Section II, page 392.

   4. (a) $f(x) = \dfrac{1}{(1 + x)^2} = 1 - 2x + 3x^2 - 4x^3 + \cdots + (-1)^{n-1}nx^{n-1} \pm \cdots$.
   (b) We use the ratio test on the series of absolute values:

$$\lim_{n \to \infty} \left| \frac{(n + 1)x^n}{nx^{n-1}} \right| = \lim_{n \to \infty} \frac{n + 1}{n} |x| = |x|.$$

So the series converges if $|x| < 1$. At $x = 1$, we get $1 - 2 + 3 - 4 + \cdots$, which diverges; at $x = -1$, we get $1 + 2 + 3 + \cdots$, which also diverges. So the series converges only if $-1 < x < 1$.
   (c) $f\left(\frac{1}{4}\right) = \dfrac{1}{(1 + \frac{1}{4})^2} = 0.64$.

   (d) Since the series with $x = \frac{1}{4}$ is alternating, we want the first term omitted to be less than 0.05. For $n = 4$, the term is 0.0625; for $n = 5$, it is less than 0.02. So four terms are adequate.

5. (a) Since $x = 2a \cot \theta$,

$$x^2 = 4a^2 \cot^2 \theta = 4a^2(\csc^2 \theta - 1);$$

since $y = 2a \sin^2 \theta$,

$$\sin^2 \theta = \frac{y}{2a} \quad \text{and} \quad \csc^2 \theta = \frac{2a}{y}.$$

So

$$x^2 = 4a^2 \left(\frac{2a}{y} - 1\right) \quad \text{or} \quad x^2 + 4a^2 = \frac{8a^3}{y} \quad \text{and} \quad y = \frac{8a^3}{x^2 + 4a^2}.$$

(b) The curve is sketched in Figure BC2–2.

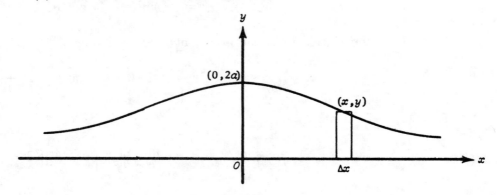

FIGURE BC2–2

(c) The area A bounded by the curve and the x-axis can be given either in rectangular coordinates or parametrically. In rectangular coordinates, we have

$$A = 2\int_0^\infty y \, dx = 2\int_0^\infty \frac{8a^3}{x^2 + 4a^2} \, dx$$

$$= \lim_{b \to \infty} 16a^3 \cdot \frac{1}{2a} \tan^{-1} \frac{x}{2a} \Big|_0^b$$

$$= 8a^2 \cdot \frac{\pi}{2} = 4\pi a^2.$$

6. (a) We are given that $f'(x) = u(x)v(x)$. Since $u(c) = 0$, $f'(c) = 0$. Now

$$f''(x) = u(x)v'(x) + u'(x)v(x) \quad \text{and} \quad f''(c) = u(c)v'(c) + u'(c)v(c).$$

Since $u(c) = 0$ while $v(c) \neq 0$, we see that

$$f''(c) = 0 + u'(c)v(c).$$

If $u'(c)v(c) > 0$, then the curve of f is concave up at $x = c$ and $f(c)$ is a relative min; if $u'(c)v(c) < 0$, then the curve is concave down and $f(c)$ is a relative max. (By the second-derivative test for relative extrema, if f' and f'' exist at every point in (a, b) and $f'(c) = 0$, where c is in (a, b), then, if $f''(c) < 0$, $f(c)$ is a relative max, while if $f''(c) > 0$, $f(c)$ is a relative min.)

   (b) If $f'(x) = (3x - 3)(x^2 + 2)^3(x^2 - 2x + 3)^{3/2}$, we can let $u(x) = (3x - 3)$ and $v(x)$ be the product of the other two factors in $f'(x)$. Note that $x = 1$ is the only critical value of f and that $u(1) = 0$, while $v(1) \neq 0$. We now examine $u'(x)v(x)$ when $x = 1$. Since $u'(x) = 3$,

$$u'(1)v(1) = 3(1 + 2)^3(1 - 2 + 3)^{3/2} = 3 \text{ times a positive number.}$$

By part (a), since $u'(1)v(1) > 0$, $f(1)$ is a relative min.

BC Practice Examination 3

Section I

1. $\lim\limits_{x \to 2} \dfrac{x^2 - 2}{4 - x^2}$ is

   (A) -2 (B) -1 (C) $-\dfrac{1}{2}$ (D) 0 (E) nonexistent

2. $\lim\limits_{x \to \infty} \dfrac{\sqrt{x} - 4}{4 - 3\sqrt{x}}$ is

   (A) $-\dfrac{1}{3}$ (B) -1 (C) ∞ (D) 0 (E) $\dfrac{1}{3}$

3. $\lim\limits_{x \to 0} \dfrac{\sin^2 \frac{x}{2}}{x^2}$ is

   (A) 4 (B) 0 (C) $\dfrac{1}{4}$ (D) 2 (E) nonexistent

4. $\lim\limits_{x \to 0} \dfrac{e^x - 1}{x}$ is

   (A) 1 (B) e (C) $\dfrac{1}{e}$ (D) 0 (E) nonexistent

5. The first four terms of the Taylor series about $x = 0$ of $\sqrt{1 + x}$ are

   (A) $1 - \dfrac{x}{2} + \dfrac{x^2}{4 \cdot 2} - \dfrac{3x^3}{8 \cdot 6}$ (B) $x + \dfrac{x^2}{2} + \dfrac{x^3}{8} + \dfrac{x^4}{48}$

   (C) $1 + \dfrac{x}{2} - \dfrac{x^2}{8} + \dfrac{x^3}{16}$ (D) $1 + \dfrac{x}{4} - \dfrac{x^2}{24} + \dfrac{x^3}{32}$

   (E) $-1 + \dfrac{x}{2} - \dfrac{x^2}{8} + \dfrac{x^3}{16}$

6. If $y = \dfrac{e^{\ln u}}{u}$, then $\dfrac{dy}{du}$ equals

   (A) $\dfrac{e^{\ln u}}{u^2}$ (B) $e^{\ln u}$ (C) $\dfrac{2e^{\ln u}}{u^2}$ (D) 1   (E) 0

7. If $y = \sin^3 (1 - 2x)$, then $\dfrac{dy}{dx}$ is

   (A) $3 \sin^2 (1 - 2x)$ (B) $-2 \cos^3 (1 - 2x)$ (C) $-6 \sin^2 (1 - 2x)$
   (D) $-6 \sin^2 (1 - 2x) \cos (1 - 2x)$ (E) $-6 \cos^2 (1 - 2x)$

8. If $f(u) = \tan^{-1} u^2$ and $g(u) = e^u$, then the derivative of $f(g(u))$ is

   (A) $\dfrac{2ue^u}{1 + u^4}$ (B) $\dfrac{2ue^{u^2}}{1 + u^4}$ (C) $\dfrac{2e^u}{1 + 4e^{2u}}$ (D) $\dfrac{2e^{2u}}{1 + e^{4u}}$

   (E) $\dfrac{2e^{2u}}{\sqrt{1 - e^{4u}}}$

9. The equation of the tangent to the curve of $xy - x + y = 2$ at the point where $x = 0$ is

   (A) $y = -x$ (B) $y = \dfrac{1}{2}x + 2$ (C) $y = x + 2$ (D) $y = 2$
   (E) $y = 2 - x$

10. Suppose $f(x) = 4x - 3$. To prove that $\lim\limits_{x \to 2} f(x) = 5$, we let ϵ be a positive number, then choose δ so that $|f(x) - 5| < \epsilon$ whenever $|x - 2| < \delta$. A suitable δ is

   (A) $\dfrac{\epsilon}{5}$ (B) $\dfrac{\epsilon}{3}$ (C) $\dfrac{\epsilon}{2}$ (D) ϵ (E) 4ϵ

11. If $y = x^2 e^{1/x}$ $(x \neq 0)$, then $\dfrac{dy}{dx}$ is

   (A) $xe^{1/x}(x + 2)$ (B) $e^{1/x}(2x - 1)$ (C) $\dfrac{-2e^{1/x}}{x}$
   (D) $e^{-x}(2x - x^2)$ (E) none of these

12. If differentials are used for computation, then $\sqrt[3]{127}$ is approximately equal to

   (A) 5.27   (B) 5.01   (C) 5.03   (D) 5.10   (E) 5.07

13. A point moves along the curve $y = x^2 + 1$ so that the x-coordinate is increasing at the constant rate of $\dfrac{3}{2}$ units per second. The rate, in units per second, at which the distance from the origin is changing when the point has coordinates $(1, 2)$ is equal to

   (A) $\dfrac{7\sqrt{5}}{10}$ (B) $\dfrac{3\sqrt{5}}{2}$ (C) $3\sqrt{5}$ (D) $\dfrac{15}{2}$ (E) $\sqrt{5}$

14. The set of x for which the curve of $y = 1 - 6x^2 - x^4$ has inflection points is

(A) $\{0\}$ (B) $\{\pm\sqrt{3}\}$ (C) $\{1\}$ (D) $\{\pm 1\}$
(E) none of these

15. If the position of a particle on a line at time t is given by

$$s = t^3 + 3t,$$

then the speed of the particle is decreasing when

(A) $-1 < t < 1$ (B) $-1 < t < 0$ (C) $t < 0$ (D) $t > 0$
(E) $|t| > 1$

16. A particle moves along the parabola $x = 3y - y^2$ so that $\dfrac{dy}{dt} = 3$ at all time t. The speed of the particle when it is at position $(2, 1)$ is equal to

(A) 0   (B) 3   (C) $\sqrt{13}$ (D) $3\sqrt{2}$ (E) none of these

17. The motion of a particle in a plane is given by the pair of equations $x = e^t \cos t$, $y = e^t \sin t$. The magnitude of its acceleration at any time t equals

(A) $\sqrt{x^2 + y^2}$ (B) $2e^t\sqrt{\cos 2t}$ (C) $2e^t$ (D) e^t (E) $2e^{2t}$

18. A rectangle with one side on the x-axis is inscribed in the triangle formed by the lines $y = x$, $y = 0$, and $2x + y = 12$. The area of the largest such rectangle is

(A) 6   (B) 3   (C) $\dfrac{5}{2}$ (D) 5   (E) 7

19. The abscissa of the first-quadrant point which is on the curve of $x^2 - y^2 = 1$ and closest to the point $(3, 0)$ is

(A) 1   (B) $\dfrac{3}{2}$ (C) 2   (D) 3   (E) none of these

20. By differentiating term-by-term the series

$$(x - 1) + \frac{(x - 1)^2}{4} + \frac{(x - 1)^3}{9} + \frac{(x - 1)^4}{16} + \cdots$$

the interval of convergence obtained is

(A) $0 \leq x \leq 2$ (B) $0 \leq x < 2$ (C) $0 < x \leq 2$ (D) $0 < x < 2$
(E) only $x = 2$

21. $\displaystyle\int \frac{x\,dx}{\sqrt{9 - x^2}}$ equals

(A) $-\dfrac{1}{2}\ln\sqrt{9 - x^2} + C$ (B) $\sin^{-1}\dfrac{x}{3} + C$ (C) $-\sqrt{9 - x^2} + C$

(D) $-\dfrac{1}{4}\sqrt{9 - x^2} + C$ (E) $2\sqrt{9 - x^2} + C$

22. $\int_{\pi/4}^{\pi/3} \sec^2 x \tan^2 x \, dx$ equals

(A) 5   (B) $\sqrt{3} - 1$ (C) $\dfrac{8}{3} - \dfrac{2\sqrt{2}}{3}$ (D) $\sqrt{3}$ (E) $\sqrt{3} - \dfrac{1}{3}$

23. $\int \dfrac{(y - 1)^2}{2y} \, dy$ equals

(A) $\dfrac{y^2}{4} - y + \dfrac{1}{2} \ln |y| + C$ (B) $y^2 - y + \ln |2y| + C$

(C) $y^2 - 4y + \dfrac{1}{2} \ln |2y| + C$ (D) $\dfrac{(y - 1)^3}{3y^2} + C$

(E) $\dfrac{1}{2} - \dfrac{1}{2y^2} + C$

24. $\int_{\pi/6}^{\pi/2} \cot x \, dx$ equals

(A) $\ln \dfrac{1}{2}$ (B) $\ln 2$ (C) $-\ln (2 - \sqrt{3})$ (D) $\ln (\sqrt{3} - 1)$

(E) none of these

25. $\int_{1}^{e} \ln x \, dx$ equals

(A) $\dfrac{1}{2}$ (B) $e - 1$ (C) $e + 1$ (D) 1   (E) -1

26. If $F(x) = \int_{1}^{x} \sqrt{t^2 + 3t} \, dt$, then $F'(x)$ equals

(A) $\dfrac{2}{3}[(x^2 + 3x)^{3/2} - 8]$ (B) $\sqrt{x^2 + 3x}$ (C) $\sqrt{x^2 + 3x} - 2$

(D) $\dfrac{2x + 3}{2\sqrt{x^2 + 3x}}$ (E) none of these

27. Which one of the following improper integrals converges?

(A) $\int_{-1}^{1} \dfrac{dx}{(x + 1)^2}$ (B) $\int_{1}^{\infty} \dfrac{dx}{\sqrt{x}}$ (C) $\int_{0}^{\infty} \dfrac{dx}{(x^2 + 1)}$

(D) $\int_{1}^{3} \dfrac{dx}{(2 - x)^3}$ (E) none of these

28. If $f(x)$ is continuous on the interval $-a \leqq x \leqq a$, then it follows that $\int_{-a}^{a} f(x) \, dx$

(A) equals $2\int_{0}^{a} f(x) \, dx$ (B) equals 0   (C) equals $f(a) - f(-a)$

(D) represents the total area bounded by the curve $y = f(x)$, the x-axis, and the vertical lines $x = -a$ and $x = a$

(E) is not necessarily any of these

29. If $\dfrac{dy}{dx} = \dfrac{y}{x}$ $(x > 0, y > 0)$ and $y = 3$ when $x = 1$, then

   (A) $x^2 + y^2 = 10$ (B) $y = x + \ln 3$ (C) $y^2 - x^2 = 8$
   (D) $y = 3x$ (E) $y^2 - 3x^2 = 6$

30. The area bounded by the cubic $y = x^3 - 3x^2$ and the line $y = -4$ is given by the integral

   (A) $\displaystyle\int_{-1}^{2} (x^3 - 3x^2 + 4)\, dx$ (B) $\displaystyle\int_{-1}^{2} (x^3 - 3x^2 - 4).dx$

   (C) $2\displaystyle\int_{0}^{2} (x^3 - 3x^2 - 4)\, dx$ (D) $\displaystyle\int_{-1}^{2} (4 - x^3 + 3x^2)\, dx$

   (E) $\displaystyle\int_{0}^{3} (3x^2 + x^3)\, dx$

31. The area bounded by the curve $y = \dfrac{1}{x + 1}$, the axes, and the line $x = e - 1$ equals

   (A) $1 - \dfrac{1}{e^2}$ (B) $\ln (e - 1)$ (C) 1 (D) 2

   (E) $2\sqrt{e + 1} - 2$

32. The area bounded by the curve $x = 3y - y^2$ and the line $x = -y$ is represented by

   (A) $\displaystyle\int_{0}^{4} (2y - y^2)\, dy$ (B) $\displaystyle\int_{0}^{4} (4y - y^2)\, dy$

   (C) $\displaystyle\int_{0}^{3} (3y - y^2)\, dy + \displaystyle\int_{0}^{4} y\, dy$ (D) $\displaystyle\int_{0}^{3} (y^2 - 4y)\, dy$

   (E) $\displaystyle\int_{0}^{3} (2y - y^2)\, dy$

33. The area in the first quadrant bounded by the curve with parametric equations $x = 2a \tan \theta$, $y = 2a \cos^2 \theta$, and the lines $x = 0$ and $x = 2a$ is equal to

   (A) πa^2 (B) $2\pi a^2$ (C) $\dfrac{\pi a}{4}$ (D) $\dfrac{\pi a}{2}$ (E) none of these

34. A solid is cut out of a sphere of radius 2 by two parallel planes each 1 unit from the center. The volume of this solid is

   (A) 8π (B) $\dfrac{32\pi}{3}$ (C) $\dfrac{25\pi}{3}$ (D) $\dfrac{22\pi}{3}$ (E) $\dfrac{20\pi}{3}$

35. The volume generated by rotating the region bounded by $y = \dfrac{1}{\sqrt{1 + x^2}}$, $y = 0$, $x = 0$, and $x = 1$ about the y-axis is

   (A) $\dfrac{\pi^2}{4}$ (B) $\pi\sqrt{2}$ (C) $2\pi(\sqrt{2} - 1)$ (D) $\dfrac{2\pi}{3}(2\sqrt{2} - 1)$

   (E) $1 + \dfrac{\pi^2}{4}$

36. If $G(x) = \int_0^{3x} \sqrt{\cos t}\ dt$, then the best approximation to $G'(0.3)$ among the following is

   (A) 0.788   (B) 0.885   (C) 2.365   (D) 2.932   (E) 3.000

37. A particle moves on a straight line so that its velocity at time t is given by $v = 4s$, where s is its distance from the origin. If $s = 3$ when $t = 0$, then, when $t = \frac{1}{2}$, s equals

   (A) $1 + e^2$ (B) $2e^3$ (C) e^2 (D) $2 + e^2$ (E) $3e^2$

38. The length of the arc of $y = \frac{1}{2}x^2 - \frac{1}{4}\ln x$ from $x = 1$ to $x = 4$ is

   (A) $\frac{15}{2} + \ln 2$ (B) $\frac{1}{2}(15 + \ln 2)$ (C) $\frac{1}{2}(17 + \ln 2)$ (D) $3\frac{15}{64}$

   (E) $\frac{45}{16}$

39. Suppose the function f is continuous on $1 \leqq x \leqq 2$, that $f'(x)$ exists on $1 < x < 2$, that $f(1) = 3$, and that $f(2) = 0$. Which of the following statements is *not* necessarily true?

   (A) The mean-value theorem applies to f on $1 \leqq x \leqq 2$.
   (B) $\int_1^2 f(x)\ dx$ exists.
   (C) There exists a number c in the closed interval $[1, 2]$ such that $f'(c) = 3$.
   (D) If k is any number between 0 and 3, there is a number c between 1 and 2 such that $f(c) = k$.
   (E) If c is any number such that $1 < c < 2$, then $\lim_{x \to c} f(x)$ exists.

40. $\lim_{n \to \infty} \left[\ln \left(1 + \frac{1}{n} \right) + \ln \left(1 + \frac{2}{n} \right) + \cdots + \ln \left(1 + \frac{n}{n} \right) \right] \cdot \frac{1}{n}$ is equal to the definite integral

   (A) $\int_0^1 \ln x\ dx$ (B) $\int_1^2 \ln \frac{1}{x}\ dx$ (C) $\int_1^2 \ln (1 + x)\ dx$

   (D) $\int_1^2 \ln x\ dx$ (E) none of these

41. A particle moves along a line so that its acceleration, a, at time t is $a = -t^2$. If the particle is at the origin when $t = 0$ and 3 units to the right of the origin when $t = 1$, then its velocity at $t = 0$ is

   (A) 0   (B) $\frac{1}{12}$ (C) $2\frac{11}{12}$ (D) $\frac{37}{12}$ (E) none of these

42. The curve with parametric equations $x = \sqrt{t - 2}$ and $y = \sqrt{6 - t}$ is

   (A) part of a circle   (B) a parabola   (C) a straight line
   (D) part of a hyperbola   (E) none of these

43. The distance from $(-4, 5)$ to the line $3y + 2x - 7 = 0$ is

(A) $\dfrac{9}{5}$ (B) $\dfrac{9}{\sqrt{13}}$ (C) 1   (D) $\sqrt{13}$ (E) none of these

44. The base of a solid is the region bounded by $x^2 = 4y$ and the line $y = 2$, and each plane section perpendicular to the y-axis is a square. The volume of the solid is

(A) 8   (B) 16   (C) 20   (D) 32   (E) 64

45. Which of the following functions could have the graph sketched below?

(A) $f(x) = xe^x$ (B) $f(x) = xe^{-x}$ (C) $f(x) = \dfrac{e^x}{x}$

(D) $f(x) = \dfrac{x}{x^2 + 1}$ (E) $f(x) = \dfrac{x^2}{x^3 + 1}$

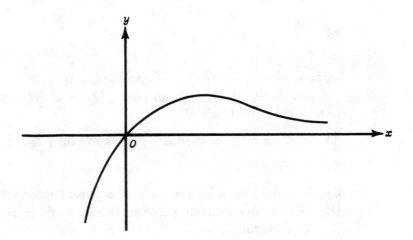

BC Practice Examination 3, Section II

1. An open box with a capacity of 4 ft³ and with a square base is to be made from sheet metal costing 50¢ per square foot. If welding costs 10¢ per foot and all the edges will need to be welded,

(a) find the dimensions for the box that will minimize the cost;

(b) show that the dimensions that yield minimum cost are independent of the costs of the sheet metal and of the welding.

2. The region bounded by the curve of the function $y = f(t)$, the t-axis, $t = 0$, and $t = x$ is rotated about the t-axis. The volume generated for all x is $2\pi(x^2 + 2x)$. Find $f(t)$.

3. (a) If $f(x) = |x|^3$, find $f'(0)$ by using the definition of derivative.

(b) Is $f'(x)$ continuous at $x = 0$? Why or why not?

(c) Is the function $x|x|$ continuous at $x = 0$? Why or why not?

(d) Evaluate $\displaystyle\int_{-1}^{1} (1 - |x|)\, dx$.

4. Sketch the curve of $y = e^{x/2} + e^{-x/2}$, and find its length from $x = 0$ to $x = 2$.

5. (a) Does the series

$$\frac{1}{2} + \frac{2^2}{2^2} + \frac{3^2}{2^3} + \frac{4^2}{2^4} + \cdots + \frac{n^2}{2^n} + \cdots$$

converge? Justify your conclusion.

(b) Find the interval of convergence of the series $\displaystyle\sum_{n=1}^{\infty} \frac{(x-1)^n}{n \cdot 2^n}$, showing your work.

6. (a) Find the function y for which $\dfrac{dy}{dx} = e^{x-y}$, given that when $x = 0$, $y = 1$.

(b) For what values of n does $\displaystyle\int_1^{\infty} \frac{dx}{x^{2-n}}$ converge? To what?

Answers to BC Practice Examination 3

SECTION I

The explanation for any question not given below will be found in the answers to AB Practice Examination 3 on pages 401 to 408. Identical questions in Sections I of Practice Examinations AB3 and BC3 have the same number.

1. E	**9.** E	**17.** C	**25.** D	**33.** A	**41.** D
2. A	**10.** A	**18.** A	**26.** B	**34.** D	**42.** A
3. C	**11.** B	**19.** B	**27.** C	**35.** C	**43.** E
4. A	**12.** C	**20.** B	**28.** E	**36.** C	**44.** D
5. C	**13.** B	**21.** C	**29.** D	**37.** E	**45.** B
6. E	**14.** E	**22.** E	**30.** A	**38.** B	
7. D	**15.** C	**23.** A	**31.** C	**39.** C	
8. D	**16.** D	**24.** B	**32.** B	**40.** D	

5. **C.** See Example 54 on pages 313 and 314.

10. **A.** $|4x - 3 - 5| < \epsilon \quad \rightarrow \quad 4|x - 2| < \epsilon \quad \rightarrow \quad |x - 2| < \dfrac{\epsilon}{4}$. Choose $\delta \leq \dfrac{\epsilon}{4}$.

12. **C.** If we let $y = \sqrt[3]{x}$, we want $y + dy$, where $x = 125$ and $\Delta x = dx = 2$. Then,

$$dy = \frac{dx}{3x^{2/3}}\bigg|_{x=125} \approx 0.03.$$

16. D. The speed, $|\mathbf{v}|$, equals $\sqrt{\left(\dfrac{dx}{dt}\right)^2 + \left(\dfrac{dy}{dt}\right)^2}$, and since $x = 3y - y^2$ we find

$$\frac{dx}{dt} = (3 - 2y)\frac{dy}{dt} = (3 - 2y)\cdot 3.$$

Then $|\mathbf{v}|$ is evaluated, using $y = 1$, and equals $\sqrt{(3)^2 + (3)^2}$.

17. C. Here,

$$\frac{dx}{dt} = e^t(\cos t - \sin t), \qquad \frac{dy}{dt} = e^t(\sin t + \cos t),$$

and

$$\frac{d^2x}{dt^2} = -2(\sin t)e^t, \qquad \frac{d^2y}{dt^2} = 2(\cos t)e^t;$$

and the magnitude of the acceleration, $|\mathbf{a}|$, is given by

$$|\mathbf{a}| = \sqrt{\left(\frac{d^2x}{dt^2}\right)^2 + \left(\frac{d^2y}{dt^2}\right)^2} = 2e^t.$$

Using the Ratio Test, page 294, we find that the series converges when $0 < x < 2$. Be sure to check the endpoints!

20. B. The new series is

$$1 + \frac{x - 1}{2} + \frac{(x - 1)^2}{3} + \frac{(x - 1)^2}{4} + \cdots.$$

22. E. The integral is equal to $\left.\dfrac{\tan^3 x}{3}\right|_{\pi/4}^{\pi/3} = \dfrac{1}{3}(3\sqrt{3} - 1)$.

27. C. $\displaystyle\int_0^\infty \frac{dx}{x^2 + 1} = \lim_{b\to\infty} \left.\tan^{-1} x\right|_0^b = \dfrac{\pi}{2}$. The integrals in (A), (B), and (D) all diverge to infinity.

29. D. Separate variables to get $\dfrac{dy}{y} = \dfrac{dx}{x}$, and integrate to get $\ln y = \ln x + C$. Since $y = 3$ when $x = 1$, $C = \ln 3$.

33. A. See Figure BC3–1.

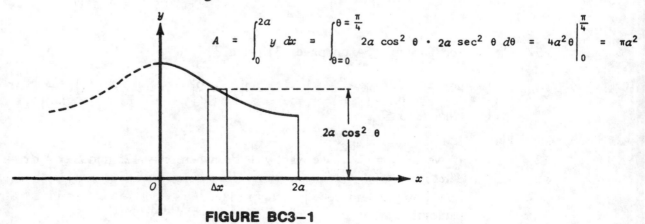

$$A = \int_0^{2a} y\ dx = \int_{\theta=0}^{\theta=\frac{\pi}{4}} 2a\cos^2\theta \cdot 2a\sec^2\theta\ d\theta = 4a^2\theta \Big|_0^{\frac{\pi}{4}} = \pi a^2$$

FIGURE BC3–1

36. C. $G'(x) = 3\sqrt{\cos 3x}$. Using a calculator, in the radian mode, we find that

$$G'(0.3) = 3\sqrt{\cos 0.9} = 2.365.$$

38. B.   The arc length is

$$\int_1^4 \sqrt{1 + \left(\frac{dy}{dx}\right)^2} \, dx = \int_1^4 \sqrt{1 + \left(x - \frac{1}{4x}\right)^2} \, dx$$

$$= \int_1^4 \sqrt{\left(x + \frac{1}{4x}\right)^2} \, dx = \int_1^4 \left(x + \frac{1}{4x}\right) \, dx$$

$$= \frac{15}{2} + \frac{1}{4} \ln 4 = \frac{15}{2} + \frac{1}{2} \ln 2.$$

39. C.   By the Mean Value Theorem, there is a number c in $[1, 2]$ such that

$$f'(c) = \frac{f(2) - f(1)}{2 - 1} = -3.$$

40. D.   We can rewrite the given limit as

$$\lim_{n \to \infty} \sum_{k=1}^n \ln \left(1 + \frac{k}{n}\right) \cdot \frac{1}{n}.$$

This can be viewed in two ways.

(1) $\lim_{n \to \infty} \sum_{k=1}^n \ln (1 + x_k) \, \Delta x$, where

$$\Delta x = \frac{1}{n}, \qquad x_k = \frac{k}{n}, \qquad f(x_k) = \ln (1 + x_k),$$

and the partition is for the interval from 0 to 1. This is equivalent to the integral $\int_0^1 \ln (1 + x) \, dx$. However, the latter does not appear in the question as a choice.

(2) $\lim_{n \to \infty} \sum_{k=1}^n (\ln x_k) \, \Delta x$, where Δx again equals $\frac{1}{n}$; but now

$$x_k = 1 + \frac{k}{n} \qquad \text{and} \qquad f(x_k) = \ln x_k.$$

The partitioned interval here is from 1 to 2. This infinite sum is equivalent to $\int_1^2 \ln x \, dx$, which is (D).

42. A.   Here $x^2 + y^2 = 4$, whose locus is a circle; but since the given equations imply x and y both nonnegative, the curve defined is in the first quadrant.

44. D. See Figure BC3–2.

$$\Delta V = (2x)^2 \, \Delta y = 4x^2 \, \Delta y$$
$$= 16y \, \Delta y$$

$$V = \int_0^2 16y \, dy$$

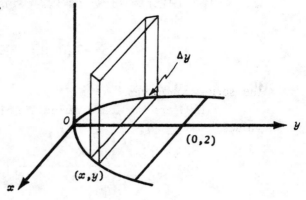

FIGURE BC3–2

SECTION II

1. See the solution for Question 1 of AB Practice Examination 3, Section II, pages 408–409.

2. See the solution for Question 3 of AB Practice Examination 3, Section II, page 410.

3. See the solution for Question 5 of AB Practice Examination 3, Section II, pages 410–411.

4. See Figure BC3–3. Recall that $ds^2 = dx^2 + dy^2$. So

$$s = \int_0^2 \sqrt{dx^2 + \frac{1}{4}(e^x - 2 + e^{-x}) \, dx^2}$$

$$= \int_0^2 \sqrt{\frac{1}{4}(e^{x/2} + e^{-x/2})^2 \, dx^2}$$

$$= \int_0^2 \frac{1}{2}(e^{x/2} + e^{-x/2}) \, dx$$

$$= e^{x/2} - e^{-x/2} \Big|_0^2 = e - \frac{1}{e}.$$

$y = e^{x/2} + e^{-x/2}$

$y = e^{x/2}$

$y = e^{-x/2}$

FIGURE BC3–3

5. (a) We apply the Ratio Test. Since

$$\lim_{n \to \infty} \frac{(n + 1)^2}{2^{n+1}} \cdot \frac{2^n}{n^2} = \frac{1}{2},$$

the series converges.

(b) Here we have, from the Ratio Test applied to the series of absolute values,

$$\lim_{n \to \infty} \left| \frac{(x - 1)^{n+1}}{(n + 1) \cdot 2^{n+1}} \cdot \frac{n \cdot 2^n}{(x - 1)^n} \right| = \lim_{n \to \infty} \frac{n}{2(n + 1)} |x - 1| = \frac{1}{2} |x - 1|.$$

So the series converges if $|x - 1| < 2$; that is, if $-1 < x < 3$.

At $x = -1$ we get the series $-1 + \frac{1}{2} - \frac{1}{3} + \frac{1}{4} - \cdots$, which converges. At $x = 3$ we get $1 + \frac{1}{2} + \frac{1}{3} + \frac{1}{4} + \cdots$, which diverges. So the series converges if $-1 \leqq x < 3$.

6. (a) Hint: Since $e^{x-y} = \frac{e^x}{e^y}$, we can separate variables, getting $e^y \, dy = e^x \, dx$. Then $y = \ln |e^x + e - 1|$.

(b) If $n < 1$, $\displaystyle\int_1^{\infty} \frac{dx}{x^{2-n}}$ converges to $\dfrac{1}{1 - n}$.

BC Practice Examination 4

Section I

1. $\lim\limits_{x \to 0} \dfrac{x^3 - 3x^2}{x}$ is

    (A) 0 **(B)** ∞ **(C)** −3 **(D)** 1 **(E)** 3

2. $\lim\limits_{h \to 0} \dfrac{\sin\left(\frac{\pi}{2} + h\right) - 1}{h}$ is

    (A) 1 **(B)** −1 **(C)** 0 **(D)** ∞ **(E)** none of these

3. What is $\lim\limits_{x \to 0} \dfrac{x - \tan x}{x - \sin x}$?

    (A) −2 **(B)** −1 **(C)** 0 **(D)** 2 **(E)** Limit does not exist.

4. $\lim\limits_{x \to \infty} x \tan \dfrac{\pi}{x}$ is

    (A) 0 **(B)** 1 **(C)** $\dfrac{1}{\pi}$ **(D)** π **(E)** ∞

5. If $y = \ln \dfrac{x}{\sqrt{x^2 + 1}}$, then $\dfrac{dy}{dx}$ is

    (A) $\dfrac{1}{x^2 + 1}$ **(B)** $\dfrac{1}{x(x^2 + 1)}$ **(C)** $\dfrac{2x^2 + 1}{x(x^2 + 1)}$ **(D)** $\dfrac{1}{x\sqrt{x^2 + 1}}$

    (E) $\dfrac{1 - x^2}{x(x^2 + 1)}$

6. If $y = \sqrt{x^2 + 16}$, then $\dfrac{d^2y}{dx^2}$ is

   (A) $-\dfrac{1}{4(x^2 + 16)^{3/2}}$ (B) $4(3x^2 + 16)$ (C) $\dfrac{16}{\sqrt{x^2 + 16}}$

   (D) $\dfrac{2x^2 + 16}{(x^2 + 16)^{3/2}}$ (E) $\dfrac{16}{(x^2 + 16)^{3/2}}$

7. If $x = a \cot \theta$ and $y = a \sin^2 \theta$, then $\dfrac{dy}{dx}$, when $\theta = \dfrac{\pi}{4}$, is equal to

   (A) $\dfrac{1}{2}$ (B) -1 (C) 2 (D) $-\dfrac{1}{2}$ (E) $-\dfrac{1}{4}$

8. The equation of the tangent to the curve $2x^2 - y^4 = 1$ at the point $(-1, 1)$ is

   (A) $y = -x$ (B) $y = 2 - x$ (C) $4y + 5x + 1 = 0$
   (D) $x - 2y + 3 = 0$ (E) $x - 4y + 5 = 0$

9. The set of all x for which the power series $\displaystyle\sum_{n=0}^{\infty} \dfrac{x^n}{(n + 1)\cdot 3^n}$ converges is

   (A) $\{-3, 3\}$ (B) $|x| < 3$ (C) $|x| > 3$ (D) $-3 \leqq x < 3$
   (E) $-3 < x \leqq 3$

10. If $y = \sqrt{\ln (x^2 + 1)}$ $(x > 0)$, then the derivative of y^2 with respect to $\ln x$ is equal to

   (A) $\dfrac{2x}{x^2 + 1}$ (B) $\dfrac{2}{x^2 + 1}$ (C) $\dfrac{2x}{\ln x(x^2 + 1)}$ (D) $\dfrac{2x^2}{x^2 + 1}$

   (E) none of these

11. If $f(t) = \dfrac{1}{t^2} - 4$ and $g(t) = \cos t$, then the derivative of $f(g(t))$ is

   (A) $2 \sec^2 t \tan t$ (B) $\tan t$ (C) $2 \sec t \tan t$ (D) $\dfrac{2}{t^3 \sin t}$

   (E) $-\dfrac{2}{\cos^3 t}$

12. When a series is used to approximate $\displaystyle\int_0^{0.2} e^{-x^2}\, dx$, the value of the integral, to four decimal places, is

   (A) -0.0392 (B) 0.1974 (C) 0.1920 (D) 0.2000
   (E) none of these

13. What is the polar equation of the graph?

   (A) $r = 2 \cos \theta + 2$
   (B) $r = 4 \cos \theta$
   (C) $r = 4 \cos \theta + 2$
   (D) $r = 4 \cos 2\theta$
   (E) $r = 4 \cos \theta + 4$

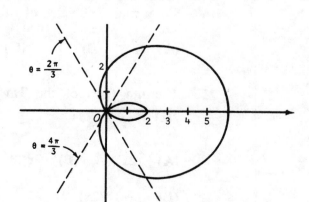

14. The radius of a sphere is 10 in.; if the radius is increased by 0.01 in., then the approximate change, in cubic inches, in the volume is

(A) 0.4π (B) $\dfrac{4}{9}\pi$ (C) 4π (D) 8π (E) 40π

15. A relative maximum value of the function $y = \dfrac{\ln x}{x}$ is

(A) 1   (B) e (C) $\dfrac{2}{e}$ (D) $\dfrac{1}{e}$ (E) none of these

16. If a particle moves on a line according to the law $s = t^5 + 2t^3$, then the number of times it reverses direction is

(A) 4   (B) 3   (C) 2   (D) 1   (E) 0

17. A particle moves counterclockwise on the circle $x^2 + y^2 = 25$ with constant speed of 2 ft/sec. Its velocity vector, **v**, when the particle is at (3, 4), equals

(A) $-\dfrac{1}{5}(8\mathbf{i} - 6\mathbf{j})$ (B) $\dfrac{1}{5}(8\mathbf{i} - 6\mathbf{j})$ (C) $-2\sqrt{3}\mathbf{i} + 2\mathbf{j}$

(D) $2\mathbf{i} - 2\sqrt{3}\mathbf{j}$ (E) $-2\sqrt{2}(\mathbf{i} - \mathbf{j})$

18. A rectangular pigpen is to be built against a wall so that only three sides will require fencing. If p feet of fencing are to be used, the area of the largest possible pen is

(A) $\dfrac{p^2}{2}$ (B) $\dfrac{p^2}{4}$ (C) $\dfrac{p^2}{8}$ (D) $\dfrac{p^2}{9}$ (E) $\dfrac{p^2}{16}$

19. Let $\mathbf{R} = a \cos kt\mathbf{i} + a \sin kt\mathbf{j}$ be the (position) vector $x\mathbf{i} + y\mathbf{j}$ from the origin to a moving point $P(x, y)$ at time t, where a and k are positive constants. The acceleration vector, **a**, equals

(A) $-k^2\mathbf{R}$ (B) $a^2k^2\mathbf{R}$ (C) $-a\mathbf{R}$ (D) $-ak^2(\cos t\mathbf{i} + \sin t\mathbf{j})$
(E) $-\mathbf{R}$

20. The region bounded by the parabolas $y = x^2$ and $y = 6x - x^2$ is rotated about the x-axis so that a vertical line segment cut off by the curves generates a ring. The value of x for which the ring of largest area is obtained is

(A) 4   (B) 3   (C) $\dfrac{5}{2}$ (D) 2   (E) $\dfrac{3}{2}$

21. The nth term of the Taylor series expansion about $x = 0$ of the function $f(x) = \dfrac{1}{1 + 2x}$ is

(A) $(2x)^n$ (B) $2x^{n-1}$ (C) $\left(\dfrac{x}{2}\right)^{n-1}$ (D) $(-1)^{n-1}(2x)^{n-1}$

(E) $(-1)^n(2x)^{n-1}$

22. $\int \dfrac{e^u}{4 + e^{2u}}\, du$ is equal to

(A) $\ln (4 + e^{2u}) + C$ (B) $\dfrac{1}{2} \ln |4 + e^{2u}| + C$ (C) $\dfrac{1}{2} \tan^{-1} \dfrac{e^u}{2} + C$

(D) $\tan^{-1} \dfrac{e^u}{2} + C$ (E) $\dfrac{1}{2} \tan^{-1} \dfrac{e^{2u}}{2} + C$

23. $\int_2^4 \dfrac{du}{\sqrt{16 - u^2}}$ equals

(A) $\dfrac{\pi}{12}$ (B) $\dfrac{\pi}{6}$ (C) $\dfrac{\pi}{4}$ (D) $\dfrac{\pi}{3}$ (E) $\dfrac{2\pi}{3}$

24. When the method of partial fractions is used to decompose $\dfrac{2x^2 - x + 4}{x^3 - 3x^2 + 2x}$, one of the fractions obtained is

(A) $-\dfrac{5}{x - 1}$ (B) $-\dfrac{2}{x - 1}$ (C) $\dfrac{1}{x - 1}$ (D) $\dfrac{2}{x - 1}$

(E) $\dfrac{5}{x - 1}$

25. If we replace $\sqrt{x - 2}$ by u, then $\int_3^6 \dfrac{\sqrt{x - 2}}{x}\, dx$ is equivalent to

(A) $\int_1^2 \dfrac{u\, du}{u^2 + 2}$ (B) $2 \int_1^2 \dfrac{u^2\, du}{u^2 + 2}$ (C) $\int_3^6 \dfrac{2\, u^2}{u^2 + 2}\, du$

(D) $\int_3^6 \dfrac{u\, du}{u^2 + 2}$ (E) $\dfrac{1}{2} \int_1^2 \dfrac{u^2}{u^2 + 2}\, du$

26. $\int_2^4 \dfrac{du}{(u - 3)^2}$ equals

(A) 2   (B) 1   (C) -1 (D) -2 (E) none of these

27. Which of the following improper integrals diverges?

(A) $\int_0^\infty e^{-x^2} dx$ (B) $\int_{-\infty}^0 e^x dx$ (C) $\int_0^1 \dfrac{dx}{x}$ (D) $\int_0^\infty e^{-x} dx$

(E) $\int_0^1 \dfrac{dx}{\sqrt{x}}$

28. If $\dfrac{dQ}{dt} = \dfrac{Q}{10}$ and $Q = Q_0$ when $t = 0$, then Q, at any time t, equals

(A) $Q_0 \cdot 10^t$ (B) $Q_0 \cdot e^{10t}$ (C) $Q_0 \cdot t^{1/10}$ (D) $Q_0 \cdot e^{t/10}$

(E) $\dfrac{10 Q_0}{10 - t}$

29. If a function $f(x)$ is defined by

$$f(x) = \begin{cases} 2x & \text{for } x \leq 1 \\ 3x^2 - 2 & \text{for } x > 1 \end{cases},$$

then $\int_0^2 f(x)\ dx$ equals

(A) 6   (B) 5   (C) 4   (D) 3   (E) none of these

30. The area bounded by the parabola $y = x^2$ and the lines $y = 1$ and $y = 9$ equals

(A) 8   (B) $\dfrac{84}{3}$ (C) $\dfrac{64}{3}\sqrt{2}$ (D) 32   (E) $\dfrac{104}{3}$

31. The total area bounded by the curves $y = x^3$ and $y = x^{1/3}$ is equal to

(A) $\dfrac{1}{2}$ (B) $\dfrac{5}{6}$ (C) 1   (D) 2   (E) none of these

32. The area under the curve $y = \dfrac{1}{\sqrt{x-1}}$, above the x-axis, and bounded vertically by $x = 2$ and $x = 5$ equals

(A) 1   (B) 2   (C) 4   (D) ln 2   (E) 2 ln 2

33. The total area enclosed by the curves $y = 4x - x^3$ and $y = 4 - x^2$ is represented by

(A) $\displaystyle\int_{-2}^{2}(4x - x^3 + x^2 - 4)\ dx$ (B) $\displaystyle\int_{-2}^{2}(4 - x^2 - 4x + x^3)\ dx$

(C) $\displaystyle\int_{-2}^{1}(4x - x^3 + x^2 - 4)\ dx + \int_{1}^{2}(4 - x^2 - 4x + x^3)\ dx$

(D) $\displaystyle\int_{-2}^{1}(4 - x^2 - 4x + x^3)\ dx + \int_{1}^{2}(4x - x^3 + x^2 - 4)\ dx$

(E) $\displaystyle\int_{-2}^{1}(4x - x^3 + x^2 - 4)\ dx$

34. The region bounded by $y = \tan x$, $y = 0$, and $x = \dfrac{\pi}{4}$ is rotated about the x-axis. The volume generated equals

(A) $\pi - \dfrac{\pi^2}{4}$ (B) $\pi(\sqrt{2} - 1)$ (C) $\dfrac{3\pi}{4}$ (D) $\pi\left(1 + \dfrac{\pi}{4}\right)$

(E) none of these

35. The volume obtained by rotating the region bounded by $x = y^2$ and $x = 2 - y^2$ about the y-axis is equal to

(A) $\dfrac{16\pi}{3}$ (B) $\dfrac{32\pi}{3}$ (C) $\dfrac{32\pi}{15}$ (D) $\dfrac{64\pi}{15}$ (E) $\dfrac{8\pi}{3}$

36. The first-quadrant region bounded by $y = \frac{1}{\sqrt{x}}$, $y = 0$, $x = q$ $(0 < q < 1)$, and $x = 1$ is rotated about the x-axis. The volume obtained as $q \to 0^+$ equals

(A) $\frac{2\pi}{3}$ (B) $\frac{4\pi}{3}$ (C) 2π (D) 4π (E) none of these

37. A curve is given parametrically by the equations

$$x = 3 - 2 \sin t \quad \text{and} \quad y = 2 \cos t - 1.$$

The length of the arc from $t = 0$ to $t = \pi$ is

(A) $\frac{\pi}{2}$ (B) π (C) $2 + \pi$ (D) 2π (E) 4π

38. If 40 grams of a radioactive substance decomposes to 20 grams in 2 years, then, to the nearest gram, the amount left after 3 years is

(A) 10   (B) 12   (C) 14   (D) 16   (E) 17

39. The average (mean) value of $y = \ln x$ over the interval from $x = 1$ to $x = e$ equals

(A) $\frac{1}{2}$ (B) $\frac{1}{2(e-1)}$ (C) $\frac{2}{e-1}$ (D) $\frac{1}{e-1}$

(E) none of these

40. A particle moves along a line with acceleration $a = 6t$. If, when $t = 0$, $v = 1$, then the total distance traveled between $t = 0$ and $t = 3$ equals

(A) 30   (B) 28   (C) 27   (D) 26   (E) none of these

41. $\lim\limits_{n \to \infty} \frac{1}{n}\left(\cos \frac{2}{n} + \cos \frac{4}{n} + \cos \frac{6}{n} + \cdots + \cos \frac{2n}{n}\right)$ equals

(A) $\sin 2$ (B) $\frac{1}{2} \sin 2$ (C) $-\sin 2$ (D) $\frac{1}{2}(\sin 2 - \sin 1)$

(E) none of these

42. If $x = f(t)$ is the law of motion for a particle on a line, the particle is said to be in *simple harmonic motion* if the acceleration $\frac{d^2x}{dt^2} = -k^2x$, k a constant. Which of the following equations does *not* define a simple harmonic motion?

(A) $x = e^t + e^{-t}$ (B) $x = 4 \cos 2t$ (C) $x = 2 \sin (\pi t + 2)$
(D) $x = \sin 2t + \cos 2t$ (E) $x = \cos (\omega t + \phi)$, ω and ϕ constants

43. $\int_0^1 x^2 e^x \, dx =$

(A) $-3e - 1$ (B) $-e$ (C) $e - 2$ (D) $3e$ (E) $4e - 1$

44. The graph of $y = \sin \frac{1}{x}$ is symmetric to

(A) the x-axis (B) the y-axis (C) the line $y = x$
(D) the origin (E) none of these

45. Which of the following functions has a graph which is asymptotic to the y-axis?

(A) $y = \dfrac{x}{x^2 - 1}$ (B) $y = \dfrac{x}{x^2 + 1}$ (C) $y = x - \dfrac{2}{x}$

(D) $y = \ln(x - 1)$ (E) $y = \dfrac{\sqrt{x - 1}}{x}$

BC Practice Examination 4, Section II

1. A point $P(x, y)$ moves along the curve $y = \ln x$ so that its x-coordinate increases at the rate of $\sqrt{3}$ units per second. If P is one vertex of an equilateral triangle whose other two vertices are on the x-axis, find how fast the area of the triangle is changing when the y-coordinate of P is 1.

2. If a tangent is drawn to the parabola $y = 3 - x^2$ at any point other than the vertex, then it forms a right triangle with the axes. At what point on the curve should the tangent be drawn to form the triangle of least area?

3. Suppose $F(x) = \displaystyle\int_1^x \frac{\sin t}{t}\, dt$.

(a) Determine $\displaystyle\lim_{x \to 0} F'(x)$, if it exists.

(b) Prove that $\displaystyle\lim_{x \to 0} F'\left(\frac{1}{x}\right)$ exists and equals zero. State any theorems you use.

4. A particle moves along the x-axis so that its acceleration at time t is given by $a = \dfrac{1}{v}$. If its initial velocity v_0 equals 2 and it starts at the origin,

(a) find its velocity at any time t if the particle moves in the positive direction;
(b) find the distance it travels in the first 6 secs.

5. (a) Sketch the curve of $y = e^{-x}$ and find, if it exists, the area in the first quadrant under the curve.

(b) Find, if it exists, the volume obtained if this area is revolved about the x-axis.

6. (a) Does the series $\displaystyle\sum_{j=1}^{\infty} je^{-j}$ converge? Justify your answer.

(b) Express $\displaystyle\lim_{n \to \infty} \sum_{k=1}^{\infty} \frac{1}{n} e^{-k/n}$ as a definite integral, and evaluate the integral.

Answers to BC Practice Examination 4

SECTION I

The explanation for any question not given below will be found in the answers to AB Practice Examination 4 on pages 420 to 426. Identical questions in Sections I of Practice Examinations AB4 and BC4 have the same number.

1.	A	**9.**	D	**17.**	A	**25.**	B	**33.**	D	**41.**	B
2.	C	**10.**	D	**18.**	C	**26.**	E	**34.**	A	**42.**	A
3.	A	**11.**	A	**19.**	A	**27.**	C	**35.**	A	**43.**	C
4.	D	**12.**	B	**20.**	D	**28.**	D	**36.**	E	**44.**	D
5.	B	**13.**	C	**21.**	D	**29.**	A	**37.**	D	**45.**	C
6.	E	**14.**	C	**22.**	C	**30.**	E	**38.**	C		
7.	D	**15.**	D	**23.**	D	**31.**	C	**39.**	D		
8.	A	**16.**	E	**24.**	A	**32.**	B	**40.**	A		

3. A. Use L'Hôpital's rule twice:

$$\lim_{x \to 0} \frac{x - \tan x}{x - \sin x} = \lim_{x \to 0} \frac{1 - \sec^2 x}{1 - \cos x} = \lim_{x \to 0} \frac{-2 \sec^2 x \tan x}{\sin x}$$

$$= \lim_{x \to 0} \frac{-2 \sec^2 x}{\sin x} \frac{\sin x}{\cos x} = \lim_{x \to 0} -2 \sec^3 x = -2.$$

7. D. $\dfrac{dy}{dx} = -2 \sin^3 \theta \cos \theta$.

9. D. Use the Ratio Test:

$$\lim_{n \to \infty} \left| \frac{x^{n+1}}{(n + 2) \cdot 3^{n+1}} \cdot \frac{(n + 1) \cdot 3^n}{x^n} \right| = \lim_{n \to \infty} \frac{n + 1}{n + 2} \cdot \frac{1}{3} |x| = \frac{|x|}{3},$$

which is less than 1 if $-3 < x < 3$. When $x = -3$, we get the convergent alternating harmonic series.

12. B.

$$\int_0^{0.2} e^{-x^2} \, dx = \int_0^{0.2} \left[1 - x^2 + \frac{x^4}{2!} - \frac{x^6}{3!} + \cdots \right] dx$$

$$= x - \frac{x^3}{3} + \frac{x^5}{5 \cdot 2!} - \frac{x^7}{7 \cdot 3!} + \cdots \Bigg|_0^{0.2} = 0.1974 + R,$$

where the first three terms were used to obtain 0.1974. Since the remainder, R, is for an alternating convergent series, R is less than (the absolute value of) the first term omitted. So

$$R < \frac{0.2^7}{42} < 0.0001.$$

13. **C.** Note that, when $\theta = 0$, $r = 6$; when $\theta = \frac{\pi}{2}$, $r = 2$; when $\theta = \frac{2\pi}{3}$, $r = 0$; when $\theta = \pi$, $r = -2$. These points satisfy only (C).

14. **C.** V, the volume, equals $\frac{4}{3}\pi r^3$, so that

$$\Delta V \simeq \frac{dV}{dr} dr = 4\pi r^2 \, dr = 4\pi.$$

17. **A.** Let $\dot{x} = \frac{dx}{dt}$ and $\dot{y} = \frac{dy}{dt}$. Then

$$2x\dot{x} + 2y\dot{y} = 0 \quad \text{and} \quad \dot{y} = -\frac{3}{4}\dot{x} \quad \text{at the point } (3, 4).$$

Using, also, the fact that the speed is given by $|\mathbf{v}| = \sqrt{\dot{x}^2 + \dot{y}^2} = 2$, we have $\dot{x}^2 + \dot{y}^2 = 4$, yielding $\dot{x} = \frac{+8}{-5}$ and $\dot{y} = \frac{-6}{+5}$ at the given point. Since the particle moves counterclockwise, the velocity vector, $\mathbf{v}$, at $(3, 4)$ must be $-\frac{8}{5}\mathbf{i} + \frac{6}{5}\mathbf{j}$.

19. **A.** $\mathbf{v} = -ak \sin kt\mathbf{i} + ak \cos kt\mathbf{j}$, and

$$\mathbf{a} = -ak^2 \cos kt\mathbf{i} - ak^2 \sin kt\mathbf{j} = -k^2\mathbf{R}.$$

21. **D.** The series is $1 - 2x + 4x^2 - 8x^3 + 16x^4 - \cdots$.

23. **D.** $\displaystyle\int_2^4 \frac{du}{\sqrt{16 - u^2}} = \sin^{-1}\frac{u}{4}\Big|_2^4 = \frac{\pi}{3}$. The given integral *is* improper. See Example 27, page 258.

24. **A.** Assume that

$$\frac{2x^2 - x + 4}{x(x - 1)(x - 2)} = \frac{A}{x} + \frac{B}{x - 1} + \frac{C}{x - 2}.$$

Then

$$2x^2 - x + 4 = A(x - 1)(x - 2) + Bx(x - 2) + Cx(x - 1).$$

Since we are looking for B, we let $x = 1$:

$$2(1) - 1 + 4 = 0 + B(-1) + 0; \ B = -5.$$

26. **E.** We must be careful to note that this is an a *improper* integral (for which the function becomes infinite at $x = 3$, on the interval of integration). This integral diverges.

27. **C.** Check to verify that each of the other improper integrals converges.

34. A.   See Figure BC4−1.

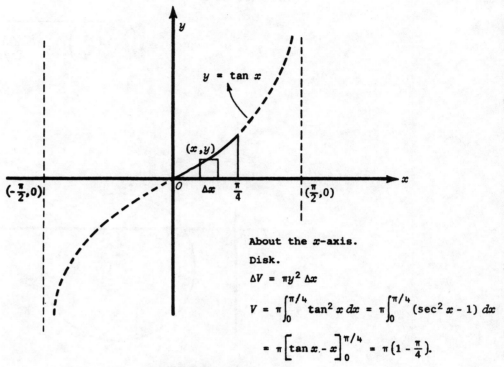

About the x-axis.

Disk.

$\Delta V = \pi y^2 \, \Delta x$

$V = \pi \displaystyle\int_0^{\pi/4} \tan^2 x \, dx = \pi \int_0^{\pi/4} (\sec^2 x - 1) \, dx$

$\qquad = \pi \left[\tan x - x \right]_0^{\pi/4} = \pi \left(1 - \dfrac{\pi}{4} \right).$

FIGURE BC4−1

36. E.   See Figure BC4−2.

About the x-axis.   Disk.

$\Delta V = \pi y^2 \, \Delta x = \pi \dfrac{1}{x} \, \Delta x$

$V = \displaystyle\lim_{q \to 0^+} \pi \int_q^1 \dfrac{1}{x} \, dx.$

Note that $V \to \infty$ as $q \to 0^+$.

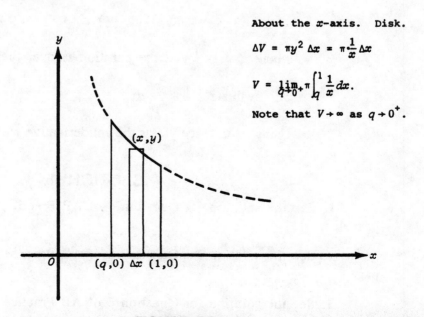

FIGURE BC4−2

37. D. See Figure BC4–3, which shows that we seek the length of a semicircle of radius 2 here. The answer can, of course, be found by using the formula for arc length:

$$s = \int_0^\pi \sqrt{\left(\frac{dx}{dt}\right)^2 + \left(\frac{dy}{dt}\right)^2}\, dt.$$

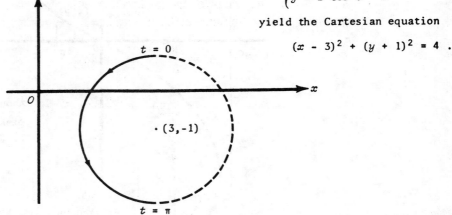

The parametric equations

$$\begin{cases} x = 3 - 2\sin t \\ y = 2\cos t - 1 \end{cases}$$

yield the Cartesian equation

$$(x - 3)^2 + (y + 1)^2 = 4 .$$

FIGURE BC4–3

38. C. If Q is the amount at time t, then $Q = 40e^{-kt}$. Since $Q = 20$ when $t = 2$, $k = -0.3466$. We now find Q when $t = 3$, from $Q = 40e^{-(0.3466)3}$, getting $Q = 14$ to the nearest gram.

41. B. Note that the given limit can be rewritten as

$$\lim_{n \to \infty} \sum_{k=1}^{n} \cos 2x_k\, \Delta x,$$

where $\Delta x = \frac{1}{n}$ and the partition is over the interval from 0 to 1. Then we evaluate $\int_0^1 \cos 2x\, dx$.

43. C. Using parts twice yields the antiderivative $x^2 e^x - 2xe^x + 2e^x$.

SECTION II

1. See the solution for Question 4 of AB Practice Examination 4, Section II, page 428.

2. See the solution for Question 5 of AB Practice Examination 4, Section II, page 428.

3. See the solution for Question 6 of AB Practice Examination 4, Section II, page 429.

4. (a) Since $\frac{dv}{dt} = \frac{1}{v}$, or $v\,dv = dt$, it follows that

$$\frac{v^2}{2} = t + C.$$

Thus $v = \sqrt{2t + 4}$.

   (b) The distance traveled in the first 6 sec is

$$\int_0^6 \sqrt{2t + 4}\,dt = \frac{56}{3}.$$

5. (a) The curve is sketched in Figure BC4–4. The area A in the first quadrant under the curve, if it exists, is given by the improper integral

$$A = \int_0^\infty e^{-x}\,dx = \lim_{b\to\infty}\int_0^b e^{-x}\,dx$$

$$= \lim_{b\to\infty} -e^{-x}\Big|_0^b = -\lim_{b\to\infty}\left(\frac{1}{e^b} - 1\right) = 1.$$

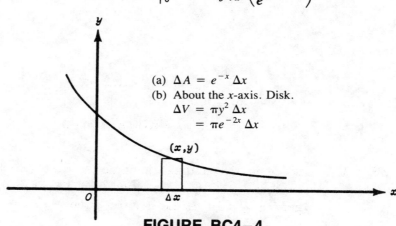

(a) $\Delta A = e^{-x}\,\Delta x$
(b) About the x-axis. Disk.
   $\Delta V = \pi y^2\,\Delta x$
   $= \pi e^{-2x}\,\Delta x$

(x, y)

FIGURE BC4–4

So the area does exist and equals 1.

   (b) The volume V obtained by rotating the first-quadrant area about the x-axis is given by

$$V = \pi\int_0^\infty e^{-2x}\,dx = \pi\lim_{b\to\infty}\int_0^b e^{-2x}\,dx$$

$$= -\frac{\pi}{2}\lim_{b\to\infty} e^{-2x}\Big|_0^b$$

$$= -\frac{\pi}{2}\lim_{b\to\infty}\left(\frac{1}{e^{2b}} - 1\right) = \frac{\pi}{2}.$$

So the volume also exists.

6. (a) We use the Integral Test:

$$\int_1^\infty xe^{-x}\,dx = \lim_{b\to\infty} -e^{-x}(1 + x)\Big|_1^b = -\lim_{b\to\infty}\left(\frac{1 + b}{e^b} - \frac{2}{e}\right),$$

and by applying L'Hôpital's rule to the first fraction within the parentheses the limit equals $\frac{2}{e}$. Since the improper integral converges, so does the given series.

(b) $\lim_{n\to\infty} \sum_{k=1}^{n} \frac{1}{n} e^{-k/n} = \lim_{n\to\infty} \sum_{k=1}^{n} e^{-k/n} \cdot \frac{1}{n}$. This is just the limit of the Riemann sum $\sum_{k=1}^{n} f(x_k) \, \Delta x$ with

$$\Delta x = \frac{1}{n}, \qquad x_k = \frac{k}{n}, \qquad f(x_k) = e^{-x_k}, \qquad \text{and} \qquad a = 0, b = 1,$$

with the interval $0 \le x \le 1$ partitioned into equal subintervals. Therefore it equals the integral $\int_0^1 e^{-x} \, dx$, whose value is $1 - \frac{1}{e}$.

Appendix: Formulas and Theorems for Reference

Algebra

1. QUADRATIC FORMULA. The roots of the quadratic equation

$$ax^2 + bx + c = 0 \quad (a \neq 0)$$

are given by

$$x = \frac{-b \pm \sqrt{b^2 - 4ac}}{2a}.$$

2. BINOMIAL THEOREM. If n is a positive integer, then

$$(a + b)^n = a^n + na^{n-1}b + \frac{n(n-1)}{1 \cdot 2} a^{n-2}b^2 + \frac{n(n-1)(n-2)}{1 \cdot 2 \cdot 3} a^{n-3}b^3$$
$$+ \cdots + nab^{n-1} + b^n.$$

3. REMAINDER THEOREM. If the polynomial $Q(x)$ is divided by $(x - a)$ until a constant remainder R is obtained, then $R = Q(a)$. In particular, if a is a root of $Q(x) = 0$, then $Q(a) = 0$.

Geometry

The sum of the angles of a triangle is equal to a straight angle (180°).

PYTHAGOREAN THEOREM

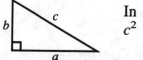

In a right triangle,
$c^2 = a^2 + b^2$.

In the following formulas,

A	is	area	B	is	area of base
S		surface area	r		radius
V		volume	C		circumference
b		base	l		arc length
h		height or altitude	θ		central angle (in radians)
s		slant height			

4. Triangle: $A = \frac{1}{2}bh.$

5. Trapezoid: $A = \frac{1}{2}(b_1 + b_2)h.$

6. Parallelogram: $A = bh.$

7. Circle: $C = 2\pi r;\ A = \pi r^2.$

8. Circular sector: $A = \frac{1}{2}r^2\theta.$

9. Circular arc: $l = r\theta.$

10. Cylinder:

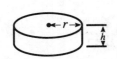

$$V = \pi r^2 h = Bh.$$
$$S\ (\text{lateral}) = 2\pi rh.$$
$$\text{Total surface area} = 2\pi r^2 + 2\pi rh.$$

11. Cone:

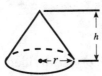

$$V = \frac{1}{3}\pi r^2 h = \frac{1}{3}Bh.$$
$$S\ (\text{lateral}) = \pi r\sqrt{r^2 + h^2}.$$
$$\text{Total surface area} = \pi r^2 + \pi r\sqrt{r^2 + h^2}.$$

12. Sphere:
$$V = \frac{4}{3}\pi r^3.$$
$$S = 4\pi r^2.$$

Trigonometry

Basic Identities
13. $\sin^2\theta + \cos^2\theta = 1.$
14. $1 + \tan^2\theta = \sec^2\theta.$
15. $1 + \cot^2\theta = \csc^2\theta.$

Sum and Difference Formulas
16. $\sin(\alpha \pm \beta) = \sin\alpha\cos\beta \pm \cos\alpha\sin\beta.$
17. $\cos(\alpha \pm \beta) = \cos\alpha\cos\beta \mp \sin\alpha\sin\beta.$
18. $\tan(\alpha \pm \beta) = \dfrac{\tan\alpha \pm \tan\beta}{1 \mp \tan\alpha\tan\beta}.$

DOUBLE-ANGLE FORMULAS

19. $\sin 2\alpha = 2 \sin \alpha \cos \alpha$.

20. $\cos 2\alpha = \cos^2 \alpha - \sin^2 \alpha = 2 \cos^2 \alpha - 1 = 1 - 2 \sin^2 \alpha$.

21. $\tan 2\alpha = \dfrac{2 \tan \alpha}{1 - \tan^2 \alpha}$.

HALF-ANGLE FORMULAS

22. $\sin \dfrac{\alpha}{2} = \pm \sqrt{\dfrac{1 - \cos \alpha}{2}}$; $\sin^2 \alpha = \dfrac{1}{2} - \dfrac{1}{2} \cos 2\alpha$.

23. $\cos \dfrac{\alpha}{2} = \pm \sqrt{\dfrac{1 + \cos \alpha}{2}}$; $\cos^2 \alpha = \dfrac{1}{2} + \dfrac{1}{2} \cos 2\alpha$.

REDUCTION FORMULAS

24. $\sin (-\alpha) = -\sin \alpha$; $\cos (-\alpha) = \cos \alpha$.

25. $\sin \left(\dfrac{\pi}{2} - \alpha\right) = \cos \alpha$; $\cos \left(\dfrac{\pi}{2} - \alpha\right) = \sin \alpha$.

26. $\sin \left(\dfrac{\pi}{2} + \alpha\right) = \cos \alpha$; $\cos \left(\dfrac{\pi}{2} + \alpha\right) = -\sin \alpha$.

27. $\sin (\pi - \alpha) = \sin \alpha$; $\cos (\pi - \alpha) = -\cos \alpha$.

28. $\sin (\pi + \alpha) = -\sin \alpha$; $\cos (\pi + \alpha) = -\cos \alpha$.

If a, b, c are the sides of triangle ABC, and A, B, C are respectively the opposite interior angles, then:

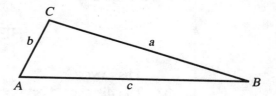

29. LAW OF COSINES. $c^2 = a^2 + b^2 - 2ab \cos C$.

30. LAW OF SINES. $\dfrac{a}{\sin A} = \dfrac{b}{\sin B} = \dfrac{c}{\sin C}$.

31. The area $A = \dfrac{1}{2} ab \sin C$.

GRAPHS OF TRIGONOMETRIC FUNCTIONS

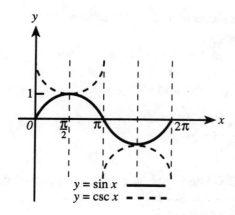

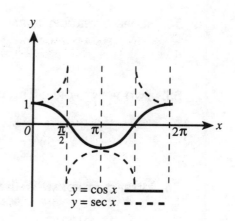

The four functions sketched above, sin, cos, csc, and sec, all have period 2π.

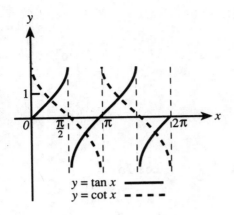

$y = \tan x$ ———
$y = \cot x$ - - - -

The two functions tan and cot have period π.

INVERSE TRIGONOMETRIC FUNCTIONS

$y = \sin^{-1} x = \text{Arc sin } x$ implies $x = \sin y$, where $-\dfrac{\pi}{2} \leqq y \leqq \dfrac{\pi}{2}$.

$y = \cos^{-1} x = \text{Arc cos } x$ implies $x = \cos y$, where $0 \leqq y \leqq \pi$.

$y = \tan^{-1} x = \text{Arc tan } x$ implies $x = \tan y$, where $-\dfrac{\pi}{2} < y < \dfrac{\pi}{2}$.

Analytic Geometry

RECTANGULAR COORDINATES

DISTANCE

32. The distance d between two points, $P_1(x_1, y_1)$ and $P_2(x_2, y_2)$, is given by

$$d = \sqrt{(x_2 - x_1)^2 + (y_2 - y_1)^2}.$$

EQUATIONS OF THE STRAIGHT LINE

33. POINT-SLOPE FORM. Through $P_1(x_1, y_1)$ and with slope m:

$$y - y_1 = m(x - x_1).$$

34. SLOPE-INTERCEPT FORM. With slope m and y-intercept b:

$$y = mx + b.$$

35. TWO-POINT FORM. Through $P_1(x_1, y_1)$ and $P_2(x_2, y_2)$:

$$y - y_1 = \frac{y_2 - y_1}{x_2 - x_1}(x - x_1).$$

36. INTERCEPT FORM. With x- and y-intercepts of a and b respectively:

$$\frac{x}{a} + \frac{y}{b} = 1.$$

37. GENERAL FORM. $Ax + By + C = 0$, where A and B are not both zero. If $B \neq 0$, the slope is $-\frac{A}{B}$; the y-intercept; $-\frac{C}{B}$, the x-intercept, $-\frac{C}{A}$.

DISTANCE FROM POINT TO LINE

38. Distance d between a point $P(x_1, y_1)$ and the line $Ax + By + C = 0$ is

$$d = \left| \frac{Ax_1 + By_1 + C}{\sqrt{A^2 + B^2}} \right|.$$

EQUATIONS OF THE CONICS

CIRCLE

39. With center at $(0, 0)$ and radius r: $x^2 + y^2 = r^2$.

40. With center at (h, k) and radius r: $(x - h)^2 + (y - k)^2 = r^2$.

PARABOLA

41. With vertex at $(0, 0)$ and focus at $(p, 0)$: $y^2 = 4px$.

42. With vertex at $(0, 0)$ and focus at $(0, p)$: $x^2 = 4py$.

With vertex at (h, k) and axis

43. parallel to x-axis, focus at $(h + p, k)$: $(y - k)^2 = 4p(x - h)$.

44. parallel to y-axis, focus at $(h, k + p)$: $(x - h)^2 = 4p(y - k)$.

ELLIPSE

With major axis of length $2a$, minor axis of length $2b$, and distance between foci of $2c$:

45. Center at $(0, 0)$, foci at $(\pm c, 0)$, and vertices at $(\pm a, 0)$:

$$\frac{x^2}{a^2} + \frac{y^2}{b^2} = 1.$$

46. Center at $(0, 0)$, foci at $(0, \pm c)$, and vertices at $(0, \pm a)$:

$$\frac{y^2}{a^2} + \frac{x^2}{b^2} = 1.$$

47. Center at (h, k), major axis horizontal, and vertices at $(h \pm a, k)$:

$$\frac{(x - h)^2}{a^2} + \frac{(y - k)^2}{b^2} = 1.$$

48. Center at (h, k), major axis vertical, and vertices at $(h, k \pm a)$:

$$\frac{(y - k)^2}{a^2} + \frac{(x - h)^2}{b^2} = 1.$$

For the ellipse, $a^2 = b^2 + c^2$, and the eccentricity $e = \frac{c}{a}$, which is *less* than 1.

HYPERBOLA

With real (transverse) axis of length $2a$, imaginary (conjugate) axis of length $2b$, and distance between foci of $2c$:

49. Center at $(0, 0)$, foci at $(\pm c, 0)$, and vertices at $(\pm a, 0)$:

$$\frac{x^2}{a^2} - \frac{y^2}{b^2} = 1.$$

50. Center at $(0, 0)$, foci at $(0, \pm c)$, and vertices at $(0, \pm a)$:

$$\frac{y^2}{a^2} - \frac{x^2}{b^2} = 1.$$

51. Center at (h, k), real axis horizontal, vertices at $(h \pm a, k)$:

$$\frac{(x - h)^2}{a^2} - \frac{(y - k)^2}{b^2} = 1.$$

52. Center at (h, k), real axis vertical, vertices at $(h, k \pm a)$:

$$\frac{(y - k)^2}{a^2} - \frac{(x - h)^2}{b^2} = 1.$$

For the hyperbola, $c^2 = a^2 + b^2$, and the eccentricity $e = \frac{c}{a}$, which is *greater* than 1.

POLAR COORDINATES

RELATIONS WITH RECTANGULAR COORDINATES

53. $x = r \cos \theta$;
$y = r \sin \theta$;
$r^2 = x^2 + y^2$;
$\tan \theta = \frac{y}{x}$.

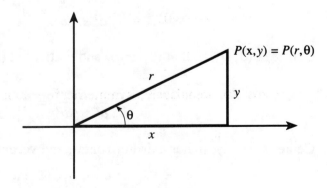

$P(x,y) = P(r,\theta)$

SOME POLAR EQUATIONS

54. $r = a$ circle, center at pole, radius a.
55. $r = 2a \cos \theta$ circle, center at $(a, 0)$, radius a.
56. $r = 2a \sin \theta$ circle, center at $(0, a)$, radius a.
57. $\left.\begin{array}{l} r = a \sec \theta \\ \text{or } r \cos \theta = a \end{array}\right\}$ line, $x = a$.
58. $\left.\begin{array}{l} r = b \csc \theta \\ \text{or } r \sin \theta = b \end{array}\right\}$ line, $y = b$.

roses (four leaves)

59. $r = \cos 2\theta$.

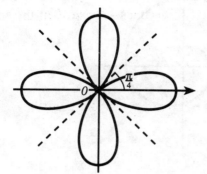

60. $r = \sin 2\theta$.

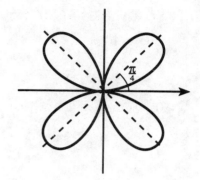

cardioids (specific examples below)

61. $r = a(1 \pm \cos \theta)$.

62. $r = a(1 \pm \sin \theta)$.

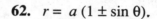

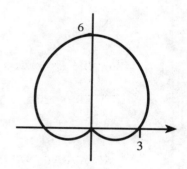

$r = 4(1 - \cos \theta)$

$r = 3(1 + \sin \theta)$

63. $r^2 = \cos 2\theta$, lemniscate, symmetric to x-axis.

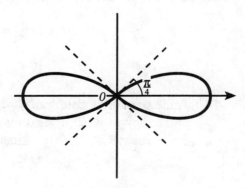

$(r^2 = \sin 2\theta$ is a lemniscate symmetric to $y = x$.)

64. $r = \theta$, (double) spiral of Archimedes

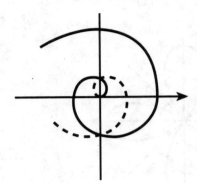

For $\theta > 0$, the curve consists only of the solid spiral.

65. $r\theta = a \, (\theta > 0)$, hyperbolic (or reciprocal) spiral

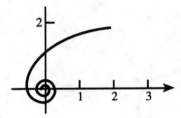

This curve is for $r\theta = 2$.
Note that $y = 2$ is an asymptote.

Exponential and Logarithmic Functions

PROPERTIES

e^x	$\ln x \; (x > 0)$
$e^0 = 1;$	$\ln 1 = 0;$
$e^1 = e;$	$\ln e = 1;$
$e^{x_1} \cdot e^{x_2} = e^{x_1 + x_2};$	$\ln (x_1 \cdot x_2) = \ln x_1 + \ln x_2;$
$\dfrac{e^{x_1}}{e^{x_2}} = e^{x_1 - x_2};$	$\ln \dfrac{x_1}{x_2} = \ln x_1 - \ln x_2;$
$e^{-x} = \dfrac{1}{e^x}.$	$\ln x^r = r \ln x \quad (r \text{ real}).$

INVERSE PROPERTIES

$f(x) = e^x$ and $f^{-1}(x) = \ln x$ are inverses of each other:

$$f^{-1}(f(x)) = f(f^{-1}(x)) = x;$$

$$\ln e^x = e^{\ln x} = x \quad (x > 0).$$

GRAPHS

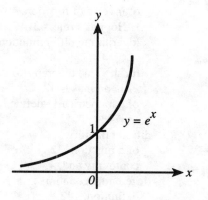

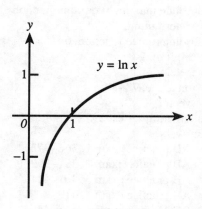

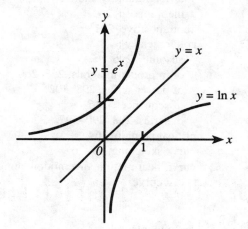

Index